MODELING OF PHOTONIC DEVICES

MODELING OF PHOTONIC DEVICES

VITTORIO M. N. PASSARO
EDITOR

Nova Science Publishers, Inc.
New York

For permission to use material from this book please contact us:
Telephone 631-231-7269; Fax 631-231-8175
Web Site: http://www.novapublishers.com

LIBRARY OF CONGRESS CATALOGING-IN-PUBLICATION DATA

Passaro, Vittorio M. N.
Modeling of photonic devices / Vittorio M.N. Passaro.
p. cm.
ISBN 978-1-60456-980-3 (hardcover)
1. Photonics. 2. Optoelectronic devices. I. Title.

TA1520. P37 2009
621.36--dc22

2008037499

Published by Nova Science Publishers, Inc. ✦ New York

CONTENTS

Preface		**vii**
Chapter 1	Photonic Signal Processors and Sensors *Vittorio M. N. Passaro*	**1**
Chapter 2	Slot and Hollow Waveguides *Michele Lagioia, Renzo Loiacono and Vittorio M. N. Passaro*	**79**
Chapter 3	Directional Couplers *Vittorio M. N. Passaro*	**117**
Chapter 4	Ring Resonators *Francesco De Leonardis and Vittorio M. N. Passaro*	**143**
Chapter 5	Silicon Modulators *Vittorio M. N. Passaro and Francesco De Leonardis*	**165**
Chapter 6	Diffraction Gratings *Giuseppe Tandoi and Vittorio M. N. Passaro*	**187**
Chapter 7	Nonlinear and Dispersive Effects in SOI Optical Waveguides *Francesco De Leonardis and Vittorio M. N. Passaro*	**231**
Chapter 8	Raman Effect in SOI Optical Structures *Francesco De Leonardis and Vittorio M. N. Passaro*	**251**
Chapter 9	Thermo-Optic Effect in Silicon Structures *Francesca Magno and Vittorio M. N. Passaro*	**293**
Chapter 10	Gain in Optically or Electrically Pumped Erbium-Doped Silicon Nanoclusters *Roberto Pagano and Vittorio M. N. Passaro*	**311**
Chapter 11	1D Photonic Bandgap as Delay Line *Francesco De Leonardis and Vittorio M. N. Passaro*	**341**
Chapter 12	Grating Devices for Optical Sensing *Renzo Loiacono and Vittorio M. N. Passaro*	**359**

Chapter 13	Multi-Quantum-Well Solar Cells *Francesco De Leonardis and Vittorio M. N. Passaro*	**383**
Index		**401**

PREFACE

The purpose of this new book is to show the main techniques and strategies needed to design and model simple and complex photonic devices, which can be used for telecommunications, signar processing or sensing applications. In fact, the modeling approaches and techniques of photonic devices are not yet well standardized. In many fields, in particular Silicon Photonics or Polymer Photonics, a few experimental devices have been presented in literature, but their fundamental modeling and design aspects are still completely or partially ignored. The high-quality research and review chapters are written by the members of the Photonics Research Group.

Chapter 1 - In this chapter, an overview of guided-wave optical signal processors for optical signal processing, space applications and telecommunications systems is presented. In particular, some aspects of mathematical modeling of these devices are reviewed. Moreover, many significant examples of photonic sensors for detection of electric fields, angular velocities, biochemical substances are briefly presented, by putting into evidence the relevant modeling and design techniques. Finally, an approach to evaluate the sensitivity of rib waveguides is presented.

Chapter 2 - In this chapter, an overview on low index core silicon waveguides is presented. In particular, the state-of-the-art for silicon slot waveguides is discussed. Numerical analysis for the optimization of SOI slot waveguides is described for both linear and nonlinear applications using a finite element method. The linear behaviour has been investigated at both 1.55 and 3.39μm, considering different values for geometrical parameters. For non linear applications, the waveguide dimensions and the nonlinear medium have been chosen in order to maximize the optical Kerr effect at 1.55μm using a novel iterative method. Possible fabrication tolerances have been included in the study. Afterwards, a brief overview on hollow-core waveguides is presented, and the theoretical analysis of a micron-sized device is described. An ideal and a real structure, inherently to the verticality of lateral sidewalls of the core, are considered. The dimensions chosen for the study are smaller than those recently presented in literature.

Chapter 3 - In this chapter, an overview on the modeling of optical directional couplers for many applications of Photonics is presented. At first, a detailed investigation of silicon-on-insulator optical rib waveguides is carried out to find the best design parameters. Then, the modeling of the coupling mechanism both in coupled straight waveguides and in optical resonators is developed. Many numerical examples are presented and discussed.

Chapter 4 - In this chapter, nonlinear optical propagation of optical pulses in silicon-on-insulator passive ring resonator is theoretically investigated. Two photon absorption and free carrier absorption are taken into account to investigate the performance limits of passive resonators under pulsed excitation. Moreover, a model including the fundamental physical effects of active ring resonators (ring lasers) for understanding the optical mode behavior and control of the oscillating mode in semiconductor injection ring lasers is briefly reviewed. The fully physical model is derived from a quantum-mechanical approach and do not depend on any semi-empirical approximation. The ring laser behavior is related to its physical and technological constraints, as the backscattering

Chapter 5 - In this chapter, an overview on silicon optical modulators is presented. In particular, some aspects of mathematical modeling of these devices are reviewed. In the first part of the chapter, the authors have focused on a fully integrated multiphysics approach for the rigorous design of metal-oxide-semiconductor modulators. Then, modeling techniques for silicon modulators based on p-n junction are shown by optimizing structures already demonstrated in literature. Finally, a novel approach to all-optical modulation in silicon waveguides is briefly presented as based on Raman effect.

Chapter 6 - In this chapter, an overview of diffraction gratings is presented. In particular, some aspects of mathematical modelling of these photonic components are reviewed. In the first part the authors have focused their attention on the main grating structures, configurations, technologies and applications. Then, a description of the grating behaviour is presented by using the Coupled Mode Theory. Finally, the main numerical methods to solve Maxwell's equations in periodic media are presented, i.e., Finite-Difference Time-Domain (FDTD), Finite Element Method (FEM), Transfer Matrix Method (TMM), and Floquet-Bloch Theory (FBT), with some design examples of grating-based devices.

Chapter 7 - In this chapter, nonlinear optical propagation of ultrafast pulses in silicon-on-insulator rib waveguides is theoretically investigated. Two photon absorption, free carrier dispersion, self- and cross-phase modulation induced by Kerr effect, walk-off, group velocity dispersion, third-order dispersion, self-steeping and polarization coupling are taken into account by a very general modeling under sub-picosecond regime. Pulse compression as induced by the soliton generation are presented and discussed.

Chapter 8 - In this chapter, stimulated Raman scattering, two photon absorption, free carrier dispersion, self- and cross-phase modulation induced by Kerr effect, walk-off, pump depletion and polarization coupling are investigated theoretically in silicon-on-insulator (SOI) optical waveguides. The propagation of both pump and Stokes pulses in picosecond and sub-picosecond regimes is presented. Comparisons between Raman amplification in straight waveguides and in optical resonators iscarried out. Finally, the guidelines for the design of Raman ring lasers based on SOI technology are proposed.

Chapter 9 - In this chapter, an extensive analysis of silicon guided-wave structures modulated by thermo-optic effect is carried out by a multiphysics integrated approach. Results are obtained and presented for silicon-on-insulator rib waveguides, as well as for silicon waveguide arrays and silicon slot waveguides. Conditions to achieve best performance in terms of high switching frequency and low power consumption are derived.

Chapter 10 - In this chapter, the rate equation system describing the interaction between silicon nanoclusters and erbium ions is theoretically investigated. The system solution is found both for optical and electrical pumping. Then, a possible DBR laser cavity design is proposed. Moreover, a simplified model of rate equation is presented, that permits one to obtain a very fast gain calculation, and the evaluation of the saturation power of the system. The main purpose of this chapter is to investigate the theoretical conditions to reach a net gain inside this material system, i.e. to find the critical physical parameters for both pumping conditions and to make a comparison between them.

Chapter 11 - In this chapter, some aspects of the modeling of mono-dimensional (1D) photonic bandgap structures are considered. Floquet-Bloch theory is developed and applied to guided-wave multi quantum well structures. Particular applications for delay lines are presented, assuming an optimized multi quantum well waveguide.

Chapter 12 - In this chapter, applications for a Bragg grating based photonic sensor are discussed. The main differences between fiber Bragg gratings and waveguide Bragg gratings are discussed and an overview on the current state of the technology is given for the latter category. Modeling of a refractometric sensor based on shallow Bragg gratings in silicon on insulator sub micrometric rib waveguides is introduced along with numerical results obtained from the model. The possibility of using third order instead of first order grating is discussed and performance compared, thus overcoming fabrication problems due to sub-micrometer scale features. A detection limit of approximately 10-4 RIU has been calculated for a 173 μm long grating. Strategies to further improve this value have been discussed, too. Finally, influence of fabrication tolerances on optimized gratings has been investigated.

Chapter 13 - In this chapter, a generalized and self-consistent modeling for analyzing multi-quantum-well solar cells is presented. Continuity equations, Poisson, and Schrödinger equations are introduced in the model with the aim to consider the carriers dynamics inside the device. Theoretical investigations on behaviour and performance of multi-quantum-well solar cells are presented in terms of carrier concentrations and electric fields produced inside the device.

In: Modeling of Photonic Devices
Editor: Vittorio M. N. Passaro

ISBN 978-1-60456-980-3

Chapter 1

PHOTONIC SIGNAL PROCESSORS AND SENSORS

Vittorio M. N. Passaro*

Photonics Research Group, Dipartimento di Elettrotecnica ed Elettronica,
Politecnico di Bari, via Edoardo Orabona n. 4, 70125 Bari, Italy

ABSTRACT

In this chapter, an overview of guided-wave optical signal processors for optical signal processing, space applications and telecommunications systems is presented. In particular, some aspects of mathematical modeling of these devices are reviewed. Moreover, many significant examples of photonic sensors for detection of electric fields, angular velocities, biochemical substances are briefly presented, by putting into evidence the relevant modeling and design techniques. Finally, an approach to evaluate the sensitivity of rib waveguides is presented.

Keywords: Optical Signal Processors, Photonic Sensors, Guided-Wave Optical Devices, Modeling.

1. INTRODUCTION

In recent years, the interest in *Integrated Optical* (IO) *Technologies* [1-2] has grown rapidly. In fact, considerable advantages are allowed by *Photonics* over its discrete, micro-optical and fiber-based counterparts, in terms of compactness, reliability, performance, resistance to mechanical vibrations, and significant cost reduction.

Wireless networks have a central role in the modern information society. The development of wireless networks surged with the introduction of cellular telephony. Now that cellular telephony has matured, mobile users are increasingly demanding interactive

* E-mail address: passaro@deemail.poliba.it

multimedia services, such as teleconferencing, video transmission, and high bit rate data streams that require large bandwidths. To increase the bandwidth of future wireless networks, it is necessary to develop transmission systems more effective than regular omni-directional antennas. Optical processing of radiofrequency (RF) signals can significantly increase the performance of these systems. It will make them able to provide broadband wireless services to mobile users. Moreover, radar systems performance in terms of reliability, jamming robustness, and beam forming flexibility can be significantly enhanced by the use of optical guided-wave devices to process RF signals.

The invention of laser sources and optical fibers allows the transfer of a huge mass of data by modulating optical signals propagating in optical fibers. Until 1995, the bit rate of an optical communication system was limited to 10 Gb/s due to two general reasons. One reason is the dispersive and non linear effects occurring in optical fibers. The other is the limits set by the regeneration of transmitted signals in the electrical domain. A very significant increase in transmission capacity of optical telecommunication systems has been achieved with the introduction of optical amplifiers with which optical signals are amplified in the optical domain. The use of multiplexing techniques such as Wavelength Division Multiplexing (WDM) and Optical Time Division Multiplexing (OTDM) permit a very significant increase in the transmission capacity as well. WDM systems with a capacity of 1.6 Tb/s (by multiplexing 160 channels at 10 Gb/s) are commercially available and systems having a capacity of 6.4 Tb/s (160 channels at 40 Gb/s) are under development. Utilization of optical signal processing in high-capacity backbone optical networks is vital. It is important to achieve very high modulation bit rates, effective signal routing and switching in the optical domain, as well as application of sophisticated multiplexing techniques to transmitted signals.

Great research efforts have been recently dedicated to investigations of new IO technologies applicable to space exploration. These applications can especially benefit from the unique features offered by these new technologies. IO technologies have definitely matured. They positively impact the ground-based equipment of the space segment. This is due to their intrinsic characteristics, such as higher efficiency, lower losses, lower power consumption and smaller sizes with respect to those of standard electronic technology. Moreover, some of these benefits appear to be even more attractive in space-based subsystems, where some features can be highlighted including hardware mass reduction, reduced losses, higher integration level, better electromagnetic compatibility, down-link reduction and on-board data processing.

Moreover, the actual interest for a large number of topics in the Silicon Photonics research field is very high and of increasing importance [3-4]. A very important technology company such as Intel Corp. is developing many research programs in Photonics, with the aim to use silicon as a fundamental technological platform for all photonic functions in any telecommunications system link (including generation, modulation, filtering, wavelength conversion, switching, amplification and detection of light). In fact, silicons low cost, vast existing microelectronics infrastructure and its technology allowing large integration of electronic and photonic devices, make a Silicon-on-Insulator (SOI) technological platform a very attractive prospective for Photonics. Silicon photonics' most promising applications are in the fields of optical interconnects, low-cost telecommunications and optical sensors. Optical interconnects have been demonstrated as very important because they permit one to achieve a significant bandwidth increase in a variety of application areas including rack-to-rack, board-to-board, chip-to-chip and intrachip applications. SOI optical interconnects may permit one to overcome the bandwidth limitations due to frequency-dependent losses imposed by copper interconnects.

Actually, employment of electronic and optical micro- and nano-sensors in a great number of application fields such as medicine, microbiology, particle physics, automotive, environmental safety and defence is also receiving a very strong interest. Sensor and actuator monolithic integration in micro-electro-mechanical systems (MEMSs), nano-electro-mechanical systems (NEMSs) and micro-opto-electro-mechanical systems (MOEMSs) has been achieved in a great number of sensing devices. Photonic sensors have attracted a great attention because of their immunity to electromagnetic interference, good compactness and robustness and high compatibility with fibre networks, but also because they usually exhibit shorter response time and higher sensitivities with respect to MEMS/MOEMS devices [5]. Different integrated optical chemical and biochemical sensors have been proposed adopting prevalently CMOS-compatible technological platforms.

Thus, the tasks of modeling and simulation of photonic devices and systems is becoming more and more significant [6]. However, a large number of modeling techniques for Photonics is not yet well standardized and many aspects of simulation tasks are still open. In general, different well-known modeling techniques include beam propagation methods (BPM), finite difference time domain (FDTD), as well as coupled mode theory (CMT), transfer matrix methods (TMM), Floquet-Bloch theory (FBT) for periodic structures [7] and method of lines (MoL) [8]. Powerful commercial CAD tools are often used in literature, as those produced by Silvaco Corp., Apollo Photonics Inc., Optiwave Systems Inc., RSoft Design Group. In this frame, the Photonics Research Group of Politecnico di Bari (URL: http://dee.poliba.it/photonicsgroup) is very active and internationally recognized for modeling and simulation of photonic devices and systems. For many years, a number of home-made modeling techniques and numerical approaches have been developed to design and simulate various photonic devices, including grating-assisted directional couplers, optical modulators, thermo-optic devices, guided-wave sensors, nonlinear devices and so on.

In this chapter, integrated optical circuits and devices for optical signal processing and sensing are reviewed, by particularly focusing on applications in microwave photonics, optical telecommunications, space engineering, biochemistry, and medicine.

2. Optical Signal Processing for Microwave Photonics

Optical signal processing by guided-wave devices permits the realization of highly performing systems capable of generating and detecting elaborate RF signals. This highly attractive prospective has generated a specific research area commonly called Microwave Photonics [9-12]. The common approach adopted in RF signal processing is performed either at RF signal frequency or at intermediate frequency. The intermediate frequency signal is produced through down-conversion operation, as in Figure 1(a). This approach has limited flexibility because a change in the band of signals to be processed requires a design change of the RF circuit and/or use of different hardware technology. By converting the RF signal to an optical one by direct (or external) laser modulation, it is possible first to perform signal processing in the optical domain. Subsequently, the processed optical signal is converted to an RF one by an optical receiver, shown in Figure 1(b). The advantages of RF signal optical processing include immunity to electromagnetic interference (EMI), high speed, parallel signal processing capability and the possibility to realize low-loss delay lines. Main applications of optical signal processors in microwave photonics include signal filtering, multi-gigabit per second analog-to-digital (A/D) conversion and beam forming for phased array antennas. Moreover, photonic signal processing can be effectively adopted in high performing frequency converters and mixers [13], signal correlators [14], and arbitrary waveform generators [15].

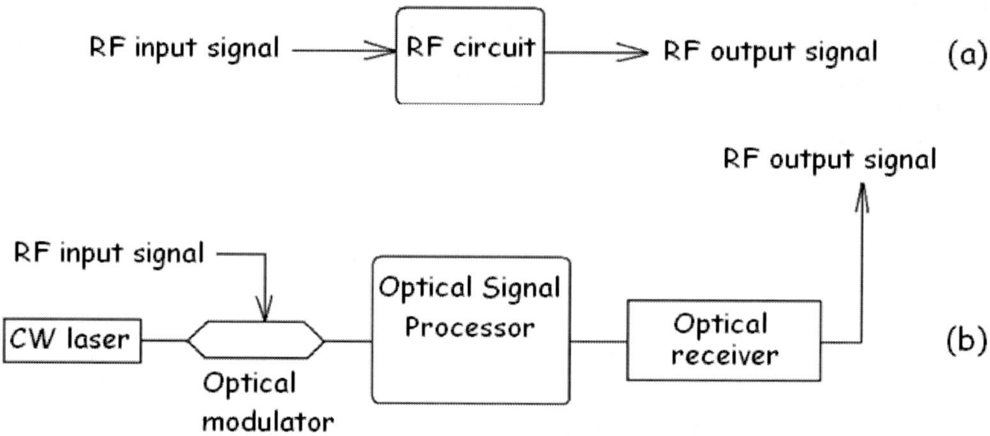

Figure 1. (a) Common approach for RF signal processing; (b) RF signal processing in optical domain.

In this section, main advances in the field of integrated optical technologies for RF signal filtering, conversion of large bandwidth analog electronic signals to digital format and generation of electronic signals controlling the direction of a signal transmitted by a phased-array antenna are reviewed.

2.1. Integrated Optical Circuits for RF Signal Filtering

Microwave filtering in optical domain has been revealed to be a very powerful technique for the processing of high-frequency signals. The transfer function as a ratio between output sequence *Z*-transform and input sequence *Z*-transform of any RF, discrete-time, filter can be written in the form:

$$H(z)=\frac{\sum_{r=0}^{n} w_r z^{-r}}{1-\sum_{k=1}^{m} b_k z^{-k}} \tag{1}$$

where z^{-1} represents the basic delay between the samples. If $b_k=0$ for all k values, the filter is usually called *transversal*. RF transversal filters require three essential functions: delay, multiplication and summation. All these functions can be implemented in the optical domain by guided-wave devices.

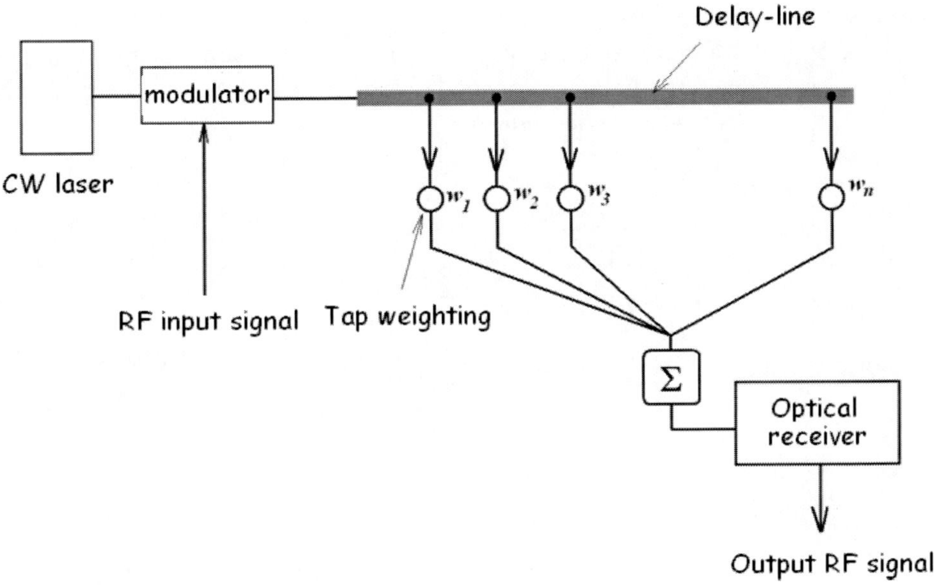

Figure 2. Architecture of transversal filters for RF optical processing using one laser source.

Optical delay lines are realized by using either optical waveguides or optical fibers. They exhibit propagation losses significantly lower than those exhibited by other delay media such as surface acoustic wave (SAW), micro-strip or superconducting delay lines. Moreover, optical delay lines losses are independent from the modulating frequency in both microwave and millimeter-wave frequency ranges. None of the other delay media exhibits this property [16]. A common architecture for designing of a RF optical transversal filter is shown in Figure 2. The tapped delay line structure consists of an optical delay line with taps distributed at constant intervals along its length. An optical signal modulated by a RF input signal is tapped, weighted, and then combined. The tap weighting coefficients are, in general, complex quantities. Then, the optical signal exiting from a combining element is sent to an optical receiver where it is converted to a RF output signal. Tapping, weighting, and combining correspond to sampling, multiplying, and summing operations in conventional digital filters. The transfer function of a RF filter is a function of both delays and weights.

An alternative block diagram for the implementation of a RF optical filter is shown in Figure 3. It implies the use of an array of laser sources producing optical signals with different amplitudes. These amplitudes are used to implement the coefficients, w_r, given at the numerator of Eq. (1). These generated optical signals are multiplexed and the resulting signal is modulated by an input RF signal. The modulated optical signal is sent to a dispersive device which introduces a wavelength-dependent delay. Next, the output from the dispersive device is converted to the RF domain by an optical receiver [17].

RF photonic filters can be grouped into two classes: optically coherent and optically incoherent. In the latter, the intensity of the optical signal is used, and hence only positive tap weighting coefficients are possible. In the former, the optical phase can be used to perform negative weights, leading to a higher flexibility in the transfer function design. Negative weights are very important for RF filter design because only low-pass filters are possible with positive taps. To obtain a pass-band transfer function, negative taps are essential. An optically coherent RF photonic filter requires very accurate phase stability. Hence better performance is possible with this IO technology.

A coherent optical transversal filter using a single laser source architecture is proposed in Figure 2 [18]. It is a silica-on-silicon, single-mode waveguide with height 6 μm and width 6 μm. The relevant IO circuit is shown in Figure 4.

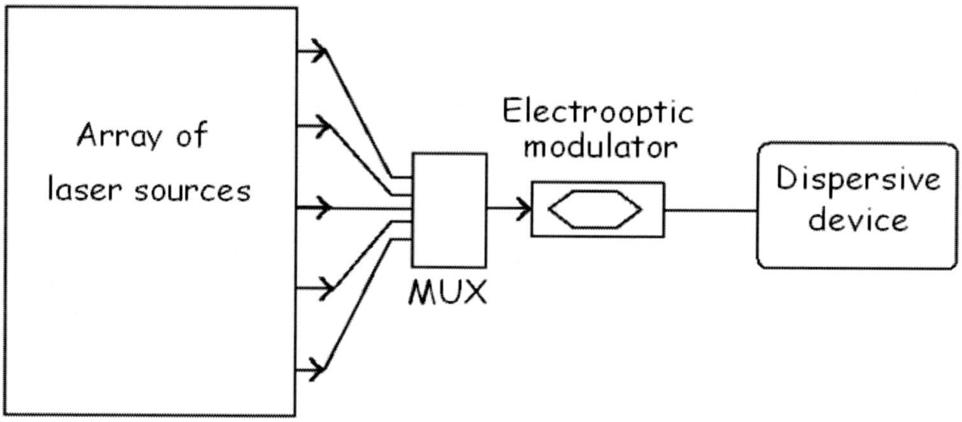

Figure 3. Architecture of transversal filters for RF optical processing using an array of laser sources.

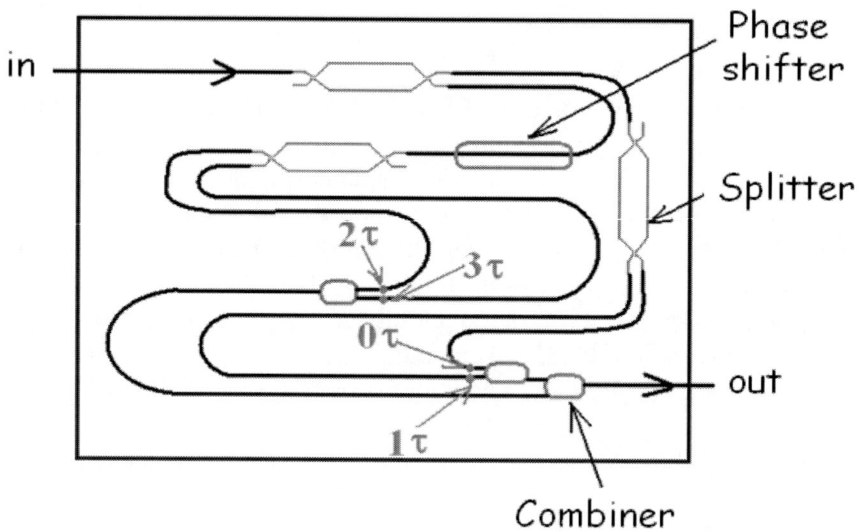

Figure 4. Integrated optical circuit for RF signal filtering realized in silica-on-silicon technology.

By tunable optical power splitters and phase shifters, arbitrary complex tap coefficients can be obtained and so the filter can have arbitrary frequency characteristics. In a 4-tap optically coherent RF filter, the input signal is converted to four signals using three tunable splitters which are configured in a tree structure. These four signals exhibit reciprocal delays equal to 0τ, 1τ, 2τ, 3τ (τ ~ 200ps), and are combined using three directional couplers configured in a tree structure as well. The tunable splitters are implemented by symmetrical Mach-Zehnder interferometers; they are tuned using thermo-optic effect. Each phase shifter is a waveguide with a thin-film heater deposited on it. The experimental RF optical filter can be used to process broadband signals over 4 THz [18]. It exhibits an attenuation ratio of more than 20 dB between pass-band and stop-band. The use of coupled ring resonators realized by a silica-on-silicon guiding structure for pass-band filtering of RF signals has been also demonstrated [19]. High stop-band rejection is obtained due to the use of coupled ring resonators. This usage provides additional filtering features such as sharp roll off, flat pass-band, and low insertion losses. The RF optical filter consists of three evanescently coupled ring resonators, as shown in Figure 5 [19]. The radius of each ring is 4.25 mm. This filter exhibits a typical bandwidth of 1.5 GHz and a free spectral range of 7.5 GHz. A thermo-optic phase shifter has been integrated on each ring to control the optical phase. The phase shifters can be used to compensate the length variation of each resonator due to fabrication tolerances or to shape the filter transfer function. A very high-order microring resonator applicable to RF optical filtering has been proposed [20].

As previously explained, the impossibility of implementing negative tap weighting coefficients is a severe limitation of incoherent RF optical filters. To overcome this limitation, a differential detection scheme has been presented [21]. According to this approach, positive and negative parts of the filter are implemented by all positive structures and filter subtraction is performed by step detection. Using this design strategy, either high-pass or low-pass filter has been implemented. The possibility to realize negative taps in the incoherent regime using the property of gain saturation in a semiconductor optical amplifier (SOA) has been also demonstrated [22].

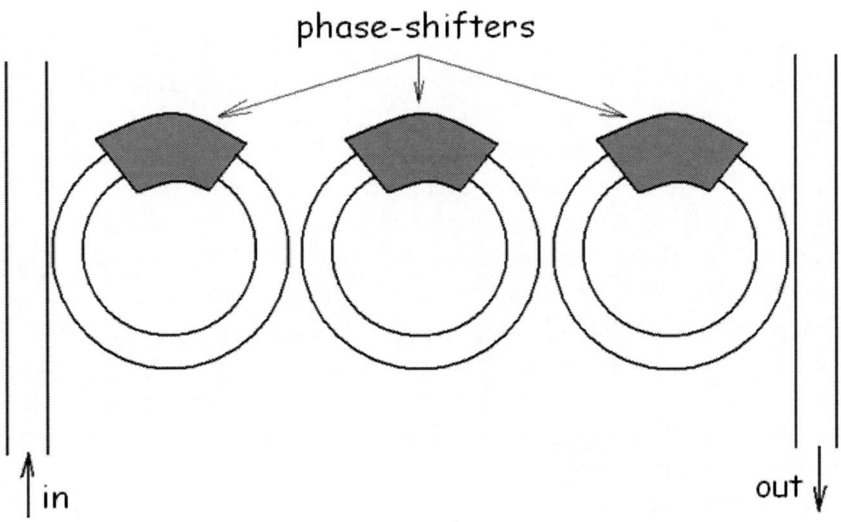

Figure 5. Coupled resonator in silica-on-silicon technology for pass-band filtering of RF signals.

Architectures based on an array of lasers for optical signals generation (see Figure 3) have been employed for RF optical filter implementation [23-24]. To realize negative taps, an approach in which a RF modulating signal in a Mach-Zehnder electro-optic modulator undergoes a π-phase inversion, has been proposed. Employing the linear part of the modulator transfer function with positive slope (device biased at V_{+}) or negative slope (device biased at

V_-), incoherent RF photonic filters with negative coefficients have been implemented, as sketched in Figure 6. By this approach, a six-tap RF photonic filter with three positive and three negative weights has been demonstrated. This RF optical filter exhibits a free spectral range, around 1.5 GHz.

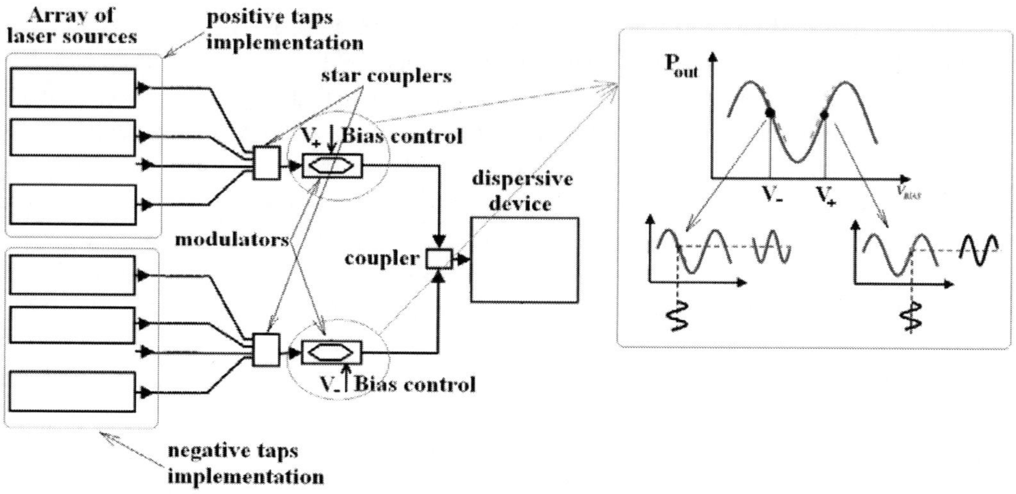

Figure 6. Incoherent RF photonic filters with negative coefficients realized by biasing Mach-Zehnder electro-optic modulators.

2.2. Integrated Optical Analog-to-Digital Converters

High-resolution ($\geq$ 12 bits) analog-to-digital converters (ADCs) having bandwidths around several GHz represent a very attractive prospective in a great number of application fields. Examples of such fields are ranging from spread spectrum telecommunications systems to advanced radar systems, as well as from radio astronomy to real-time measurements on modern communication systems.

Until now, a fully electronic implementation of ADCs simultaneously having high resolution and large bandwidth has encountered a great number of problems. These problems essentially relate to sampling aperture jitter. For example, electronic ADCs with bandwidth around 1 GHz exhibit a resolution lower than nine effective bits, whereas ADCs having 12 effective bits have a maximum bandwidth around 100 MHz [25].

The use of optical techniques for sampling of electronic analog signals can overcome some limitations imposed on ADCs performance by aperture jitter [26]. In fact, sampling with aperture jitter of less than 50 fs is very difficult to achieve by employing conventional electronic techniques. However, it can be easily obtained using optical sampling systems based on short-pulse laser sources [27]. Precisely timed optical clock pulses provided by a pulsed laser can be used to sample an electrical signal via an optical integrated device (usually a Mach-Zehnder modulator).

Consequently, electrical signal variations are converted into optical pulse intensity changes. The resulting modulated optical pulses are sent to a high speed photodiode that convert them to electric domain pulses. Subsequently, they can be quantized and processed (see Figure 7). To assure a high ADC bandwidth, it is necessary that either the modulator or the photodiode exhibits a bandwidth of the order of some GHz. Moreover, the photodiode has to be linear over the whole range of operating power, it cannot exhibit spill-over from one pulse to the next and it should have low noise.

Linearization of modulator transfer function can be achieved by using a Mach-Zehnder modulator having a 2×2 directional coupler at the output. Both outputs of the modulator are converted to the electric domain and changed to the digital form. At this point, it is possible to invert the modulator nonlinear transfer function by a digital post-processing [28-30]. Optical sampling does not reduce the rate at which the electronic circuits have to quantize the signal provided by photodiode.

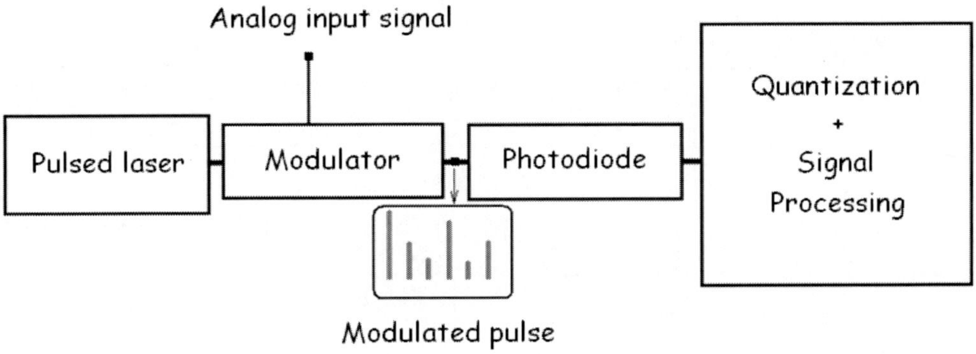

Figure 7. Basic architecture of an optically sampled ADC.

A reduction of the operating frequency of ADC electronic section can be achieved by first demultiplexing the data stream after the modulation, and then by sending adjacent modulated pulses to an array of photodiodes and quantization circuits, as in Figure 8. The demultiplexing operation can be done in time domain [27], or wavelength domain [31-32], using a laser source capable to generate pulses at different wavelengths [33]. Recently, progress towards a silicon integrated version of the wavelength demultiplexed optical sampling ADC has been reported [34].

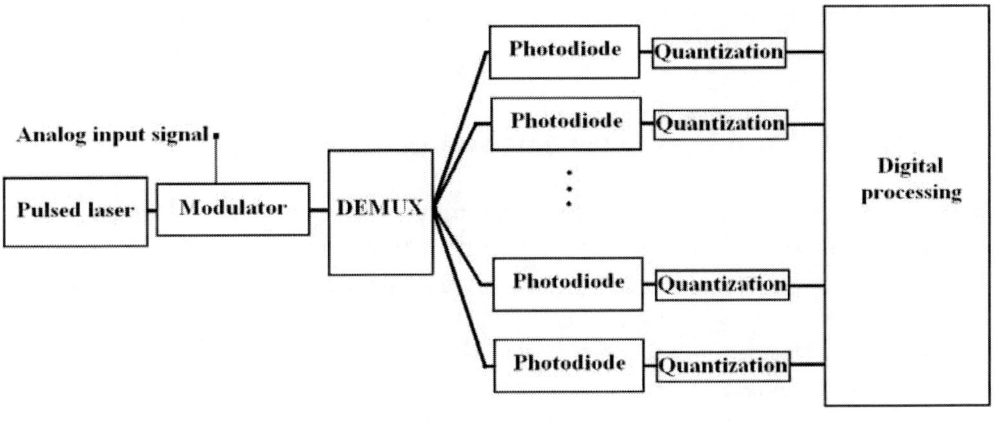

Figure 8. Demultiplexed optical sampling ADC architecture.

Not only sampling operation, but also quantization can be performed through ADCs in the optical domain as well. To this aim, essential optical components for optical quantization are Mach-Zehnder modulators, whose electrodes must differ in length by a factor of 2. Therefore, the maximum number of exhibited transfer functions within the ADC operating range is equal to 0, 2, 4, 8, and so on (see Figure 9). As shown in Figure 10, pulses from an optical clock are modulated with appropriately designed modulators. The outputs from these modulators are converted to the electrical domain using an array of photodiodes. The electronic signals output from the photodiodes are compared with a preset threshold voltage (V_T). As a result, both sampling and quantization of an input electronic signal ranging from -V_m to +V_m can be achieved [35-36].

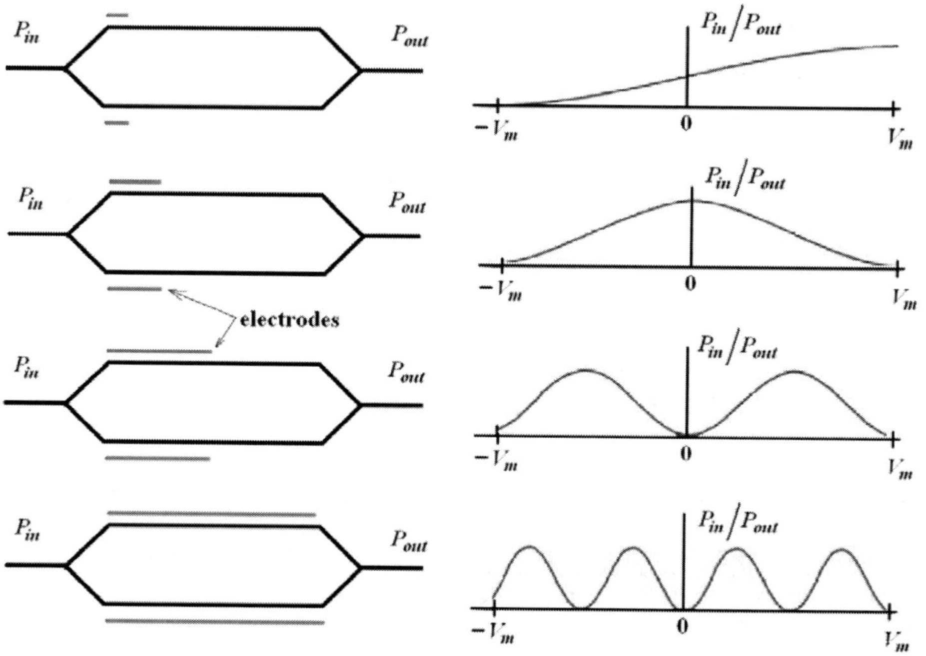

Figure 9. Mach-Zehnder optical modulators having different electrodes (by a factor of 2 in length) and transfer functions.

2.3. Integrated Optical Beam Forming Networks

Beam forming networks are systems that control the direction of a signal transmitted by a phased-array antenna. They consist of an array of multiple closely-spaced regular antennas, called phased-array antenna elements. The pitch distance *d* between adjacent antenna elements determines the width of the beam. This distance is generally of the order of the transmitted signal wavelength, whereas the differential phase delay between the signals applied to adjacent elements controls the beam angle. Unlike regular antennas, phased-array antennas can transmit information in a narrow dedicated beam towards users in one particular location, and then rapidly switch the beam towards users at a different location. Phased-array antennas have the potential to increase the system capacity of modern wireless networks. They also can be used in the field of radar systems, where the advantages of phased-array antennas over regular antennas are very apparent. Examples of such advantages are improved target-tracking capability, increased detection range, and reduced interference levels. In the past decade, the interest in phased-array antennas has spread through many other areas, such as satellite communications, radio astronomy and automotive applications, where phased-array antennas can be used to avoid collision between cars and to regulate cruise-control systems.

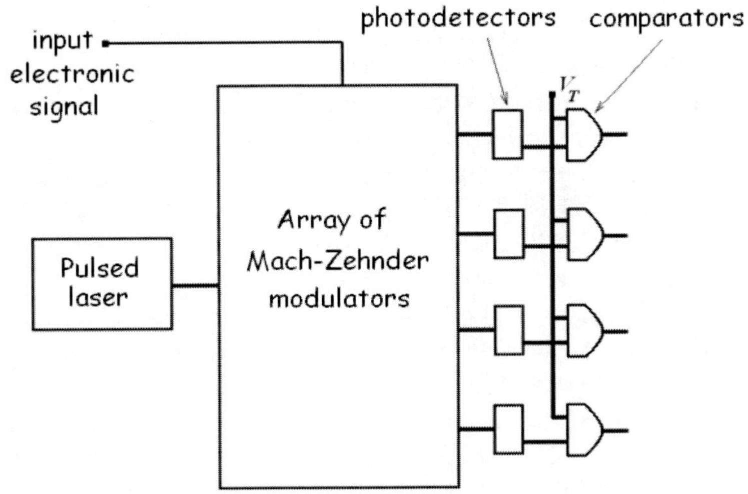

Figure 10. ADC with sampling or quantization operations performed in optical domain.

Integrated optical beam formers have been reported in almost all Photonics integrated technology platforms, such as polymers [37-39], silica [40-43], $LiNbO_3$ [44], GaAs and InP [45-46]. Figure 11 shows the schematic of an acousto-optic guided-wave integrated circuit for beam forming and shaping networks [47].

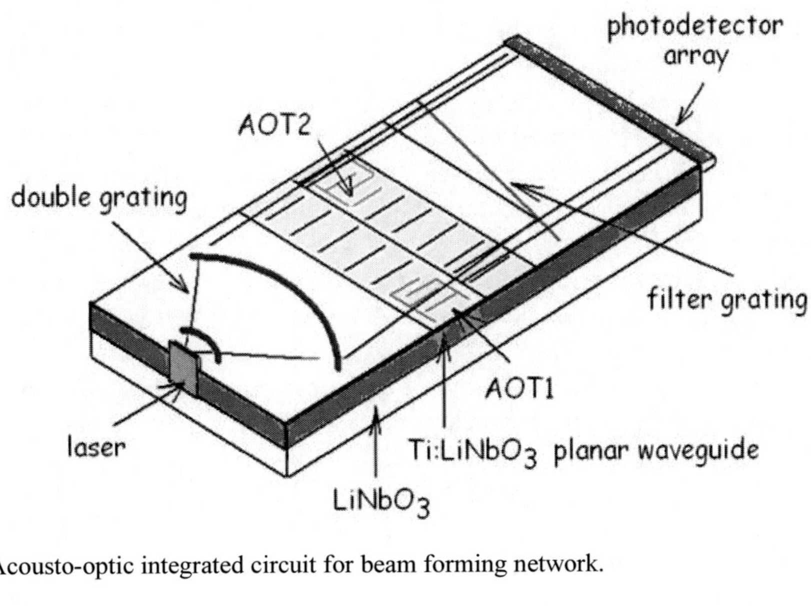

Figure 11. Acousto-optic integrated circuit for beam forming network.

The circuit is based on a Ti:$LiNbO_3$ planar waveguide with sizes 40×1×15 mm^3. A butt-coupled diode laser, at λ = 0.85 μm, generates a TM-polarized laser beam which, after the expansion and collimation by a double nonlinear grating, impinges on two acousto-optic transducers, together with the TE-polarized beam generated through polarization conversion. The two transducers excite contra-directional surface acoustic waves (SAW) [48-49] and are designed at an optimal center frequency f_c = 400 MHz. The beams generated by the double grating are incident at the Bragg angle and eight output beams are present at the output of the acousto-optic interaction region, four having TM and four having TE polarization. A filter grating with a reflection efficiency of 97% eliminates some of these beams, while the others are detected on the opposite side of the waveguide by a photodiode array.

The radiofrequency beam formed by the signals detected in the array can be linearly frequency-controlled. This is achieved by applying frequency changes to both electrical signals of acousto-optic transducers with respect to the center frequency. The changes are within 8.74 MHz, which result in covering the space from -90° to 90° with respect to the direction normal to array. The performance of this integrated optical circuit can be so summarized: instantaneous bandwidth of 174 kHz, modulation bandwidth of 20 MHz, signal to noise ratio of 20 dB, linear dynamic range of 10 dB. The beam is driven at the doubled frequency of the carrier, 800 MHz (L band).

The advantages of this architecture are remarkable in terms of power consumption (< 120 mW for one chip), dimensions and weight (< 20 g), with respect to electronic architectures, even involving Monolithic Microwave Integrated Circuits (MMIC) and optical fibers. Moreover, this optical circuit should not induce any quantization phase error and has a small error on the beam position, about 0.5°. In contrast, in the case of electronic architectures beam position error is greater than 1°. A number of chips, each one containing the control system of 100 antenna elements, could be also considered for increasing the potential of array antenna. In case of 10 chips, the power consumption is about 1.2 W at a carrier frequency of about 1 GHz, sizes of 6 cm^3, and weight of 0.2 kg. By using gallium arsenide technology, the carrier frequency rises up to about 2 GHz but at the expenses of much higher power consumption (≈ 54 W). The design of this device involves the use of home-made CAD tools based on both matrix approach and coupled mode theory.

A significant improvement in terms of carrier frequency can be obtained by using electro-optic technologies (particularly based on lithium niobate), where the frequency and phase control of each antenna element depends on a guided-wave interferometer Mach-Zehnder modulator. In this fashion, it is possible to operate at higher frequencies close to 40 GHz with power consumption of 50 W, sizes less than 60 cm^3, and weight of about 2 kg.

A different approach to very compact integrated optical beam forming networks is based on true time-delay (TTD) switched-delay-line (SDL) and fabricated in III/V semiconductor compounds, i.e. InP substrate [50]. A SDL consists of two optical switches cascaded by a bypass line on one branch, and a delay line on the other branch. The delay line delays the optical signal by $\Delta\tau$. The first optical switch decides whether the input signal is guided to the bypass line or the delay line, as in Figure 12(a). Cascading n SDLs, a delay variable from 0 to $(2^n\text{-}1)\Delta\tau$ can be applied to the optical signal, as in Figure 12(b).

The architecture of the optical beam forming network is shown in Figure 13 [50]. Four optical carriers at four equally spaced wavelengths generated by a multi-wavelength laser source are multiplexed and then modulated. The modulated optical signal is de-multiplexed and the four obtained optical signals are satisfactorily delayed. The delayed signals are sent to four photo-detectors which convert the delayed optical signal to the RF domain. A multi-wavelength laser source and a Mach-Zehnder modulator are integrated in a single InP photonic integrated circuit (PIC) [51]. SDL circuits (constituted by delay lines and Mach-Zehnder interferometer switches) and a multiplexer/de-multiplexer realized by an arrayed waveguide grating (AWG), are integrated together in the single InP PIC. Error-free 42-GHz transmission of M-order QAM 155 Mb/s data has been achieved by the proposed optical beam-forming network.

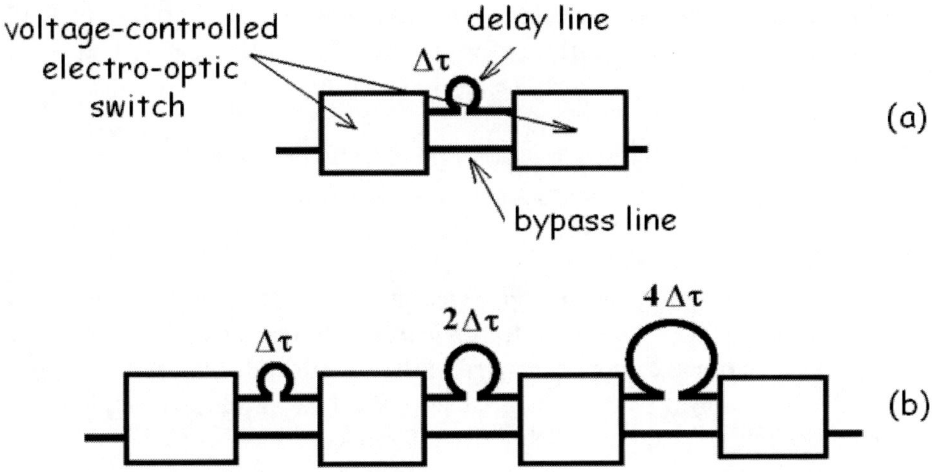

Figure 12. Optical switched-delay line structure.

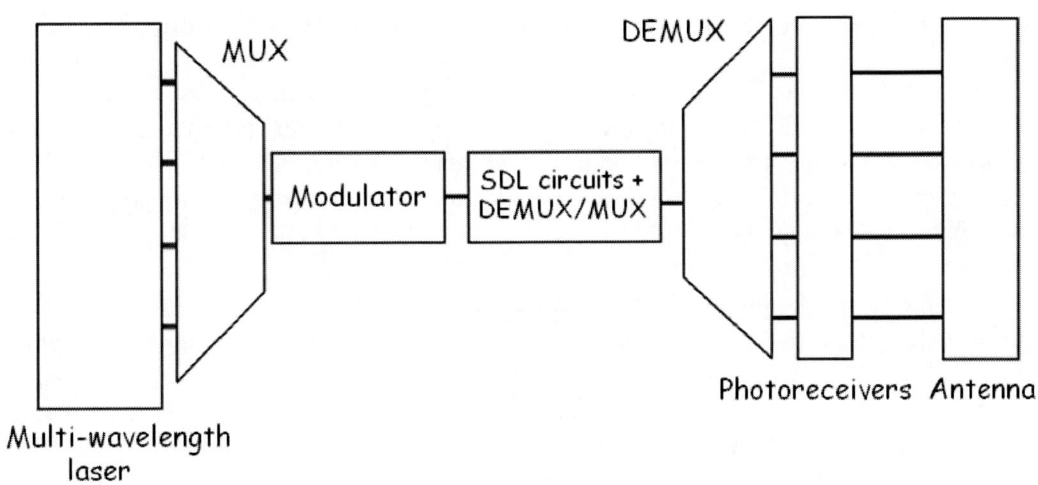

Figure 13. Optical beam-forming network based on a multi-wavelength laser source.

3. Optical Signal Processing for Optical Communications

IO devices have a central role in high-speed optical telecommunication systems. Generation, amplification, modulation, detection, multiplexing and de-multiplexing of optical signals are usually performed by guided-wave devices. Such devices are fabricated using a wide range of technological platforms, including ferroelectric materials ($LiNbO_3$), silicon and compounds (silica, SiON, silicon nitride, SOI), polymers, III/V semiconductor compounds (GaAs, InP). Cost reduction of these devices and enhancement of their performance are fundamental conditions for the wide usage of optical technologies in telecommunication networks. In particular, silicon employment for photonic devices (Silicon Photonics) has proven once again to be a very attractive prospective for cost reduction and performance enhancement of important optical components such as, modulators, switches and add-drop multiplexers [4], [52-53].

Due to the considerable increase in data traffic in backbone optical networks, the study of integrated devices for both switching and routing of all-optical Wavelength Division Multiplexing (WDM) signals has received great attention in the last decade. This interest is due to the fact that, in optical backbone networks, the cost of every network node including a cross connection, a switch, a router and a repeater, is kept high by necessities. These necessities arise from the need to convert the optical signal to the electrical domain, process it and then reconvert the processed optical signal to the electrical domain for retransmission. If the optical signal entering in the network node should be processed only in the optical domain, the cost of this node should be significantly reduced.

In this section, we review some main results reached in the field of modulation, filtering and switching by silicon photonic devices. Then, optical devices for switching and routing of all-optical WDM signals are briefly described.

3.1. Silicon Optical Devices for Modulation, Filtering and Switching

One of the most important requirements of any integrated optical technology is the ability to perform optical modulation and allowing encoding a bit stream onto the optical carrier provided by a laser source. Light can be modulated using different physical effects, depending on the material used to fabricate modulators. For high-speed modulation, electro-optic effect is commonly employed. The low cost of silicon, existence of vast microelectronics infrastructure and the possibility to integrate electronic and photonic devices, make this technological platform a very attractive prospective for integrated optics. Unfortunately, unstrained silicon does not exhibit electro-optic effect, thus to solve the problem of optical modulation, the free-carrier plasma dispersion effect is commonly adopted. This effect is adopted to electrically change the silicon refractive index by varying the electron and hole distribution within the material. In fact, refractive index shift is related to free carrier distribution change by the well-known semi-empirical Soref's relationship valid at 1550 nm [54]:

$$\Delta n_{Si} = -8.8\times10^{-22}\Delta N_e - 8.5\times10^{-18}\left[\Delta N_h\right]^{0.8} \qquad (2)$$

where ΔN_e and ΔN_h are the changes of electron and hole concentrations (in cm^{-3}), respectively.

The free-carrier concentration in silicon devices can be varied in principle by carrier injection, accumulation or depletion. In recent years, reverse biased p-n junctions, forward biased p-i-n junctions and metal-oxide-semiconductor (MOS) structures operating in accumulation regime have been employed for modulators using either interferometer or resonant configurations [55-69] (see Figure 14). Many of these devices are capable to elaborate bit streams with rates greater than 1 Gb/s and bandwidths above a few GHz. By integrating a forward biased p-i-n junction in a SOI waveguide, a high modulation efficiency can be achieved [64], but at the expenses of a significant DC electrical power consumption. On the other hand, reverse biased p-n junctions and MOS structures have quasi-zero DC power consumption [62-63]. Among SOI optical modulators having a sub-micrometer guiding structure, those employing p-n junctions offer the lowest modulation efficiency. Sub-micrometer MOS modulators permits to simultaneously achieve very low DC power consumption, high modulation efficiency and high speed [69]. Very recently, a silicon modulator adopting a sub-micrometer guiding structure and based on reverse biased p-n junction has been designed and fabricated [70]. This device exhibits a bandwidth of about 20 and can be used to process digital signals up to 30 Gb/s.

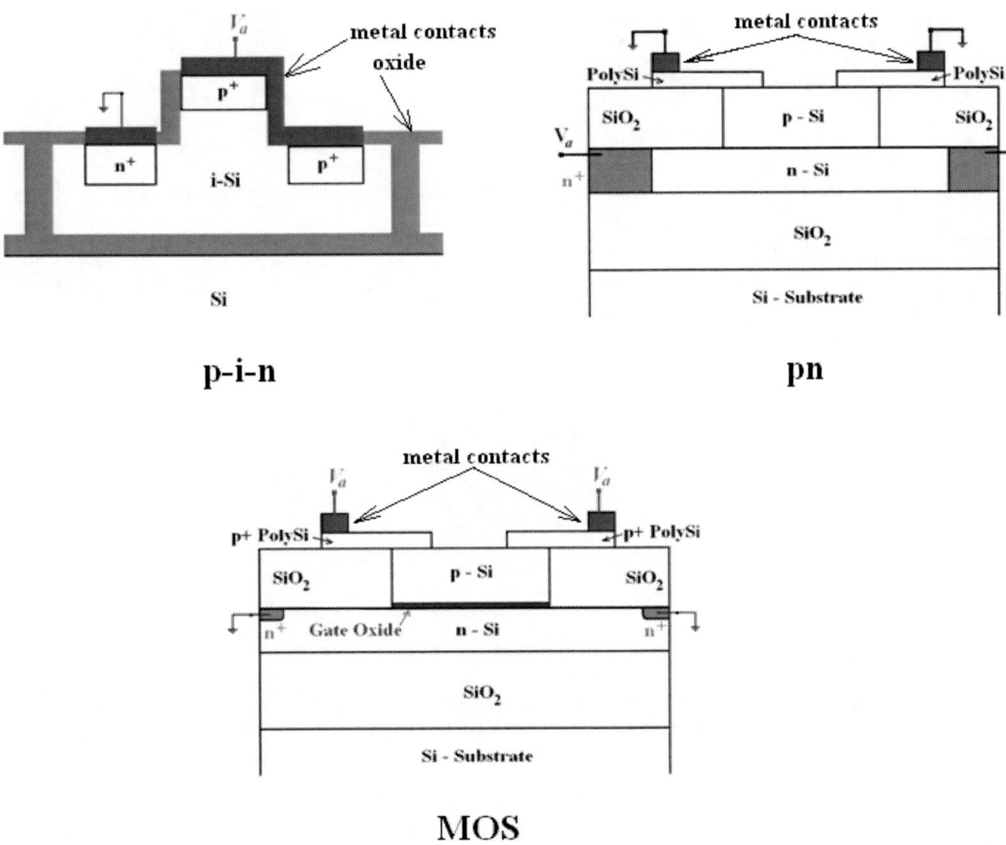

Figure 14. Main configurations for silicon modulators.

Modeling techniques based on commercial CAD tools, as that by Silvaco Corp and BPM approaches, are usually used in literature for design of these devices, where a different calculation environment is used for either electrical or optical domain. An integrated approach to multiphysics modelling has been also successfully applied, as it is explained in chapter 5 of this book.

Moreover, Silicon technology permits the fabrication of highly compact optical microring, micro-racetrack and microdisk structures that have been extensively explored for filtering applications. Optical add-drop filters are of particular interest to modern WDM optical networks. Add-drop filters are four ports optical components. They permit the addition or drop a channel to or from a WDM signal (a dropped signal is not present in the WDM signal exiting from a through port). Their main requirements include low input-to-drop-port loss (< 3 dB), high in-band extinction in the through port (> 30 dB), and wide free spectral range (FSR) (> 30 nm). If an add-drop filter is implemented by a microring, micro-racetrack or microdisk structure, a channel corresponding to one micro-cavity resonance wavelength can be added or dropped, as depicted in Figure 15.

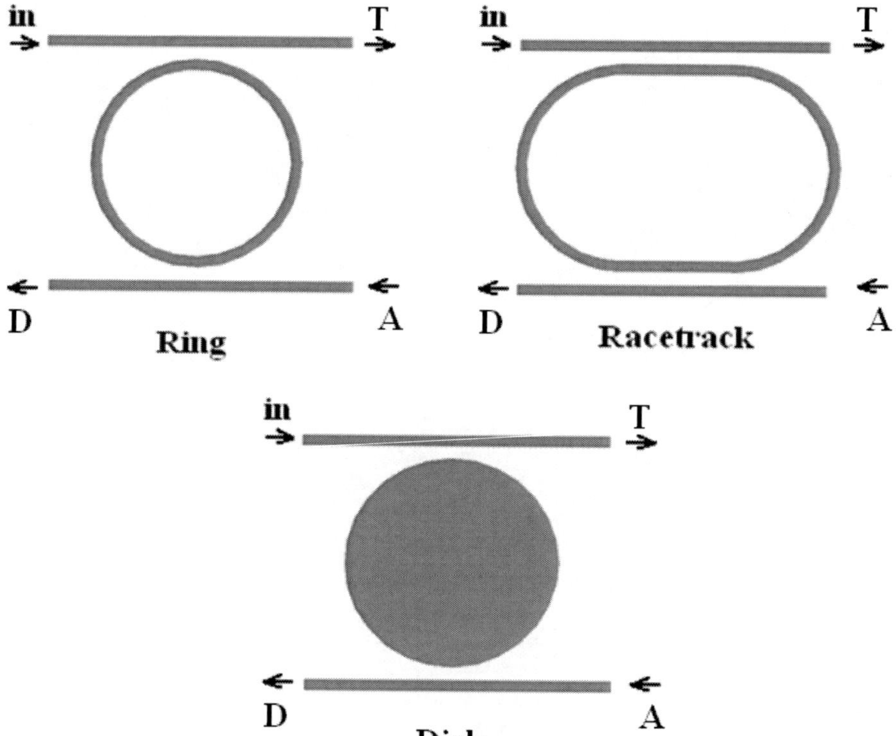

Figure 15. Microring, micro racetrack and microdisk resonators for optical add-drop multiplexers (A: add port; D: drop port; T: through port).

At wavelength of 1550 nm, silicon has a refractive index around 3.48, while it is around 1.44 for silicon oxide. Thus, large refractive index contrast assured by SOI technological platform allows fabricating microring or micro racetrack resonator with a radius of a few of microns. For instance, microring resonators having a small radius of 5 μm have been fabricated both at Interuniversity Microelectronics Centre (IMEC) in Belgium and at NTT

Microsystem Integration Laboratories in Japan. In both cases, a Si-wire waveguide, whose cross section is sketched in Figure 16, is fabricated starting from a SOI chip [71-72]. A very important parameter to quantify the performance of these micro-cavities is the quality factor, *Q*, defined as:

$$Q = \frac{\lambda_0}{\Delta\lambda} \tag{3}$$

where λ_0 is the resonance wavelength and $\Delta\lambda$ is the 3-dB bandwidth of relevant resonance peak. Examples include a microring resonator having a radius of 10 μm and *Q* of 20000 [72] and a micro-racetrack resonator with a radius of 5 μm and *Q* of 8000 [73]. A three stage microring filter as fabricated in silicon-rich silicon nitride (SiN, with refractive index 2.2) with optical resonator frequencies matched to better than 1 GHz, has been also reported [74]. This filter exhibits 17 dB through rejection, 2.5 dB drop loss and 40 GHz bandwidth. Details on modelling of optical resonators are presented in other chapters of this book.

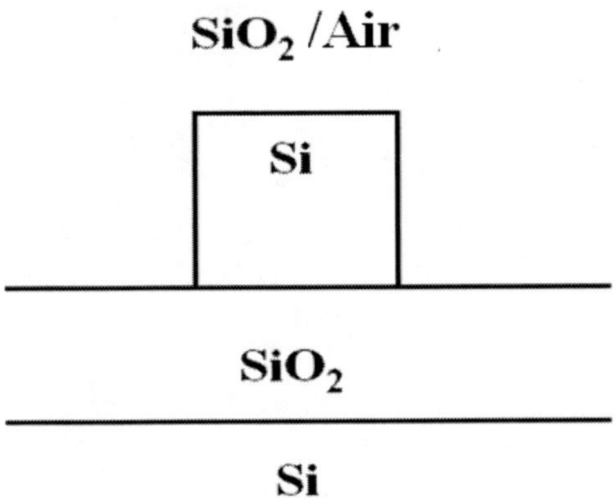

Figure 16. Si-wire guiding structure cross section.

Furthermore, the thermo-optic coefficient is defined as the derivative of the refractive index with respect to temperature. The value of thermo-optic coefficient of Silicon is very high. This value is 1.86×10^{-4} K^{-1} at 1550 nm. Consequently, the effective index of an optical mode propagating in a SOI waveguide can be significantly changed by a temperature shift induced by the heater. A significant effort of research has been devoted to design and fabricate thermo-optically controlled optical switches in SOI waveguides. A Si-wire waveguide with height 300 nm and width 300 nm has been used in 1×1, 1×2 and 1×4 thermo-optical switches [75]. These devices have an architecture based on Mach-Zehnder interferometer in which the phase difference between optical signals propagating in both interferometer arms are produced by a heater placed on one of the two arms (see Figure 17). The switching power and extinction ratio of this device are around 90 mW and 20 dB, respectively. An in depth static and dynamic investigation of the thermo-optic effect in rib SOI waveguides has been reported in literature [76-77] by using a multiphysics integrated approach, as it is reported in chapter 9.

3.2. Switching and Routing of All-Optical WDM Signals

Routing and transmission are the basic functions required to move packets through a telecommunications network. In actual optical communication networks, packet routing is performed using digital electronics, while high-capacity WDM transmission and optical circuit-switched systems are used for information transport between routers. The motivation to extend the use of optical signal processing to packet switching is due to keep data in the optical domain as much as possible until bits have to be manipulated at the endpoints. Optical packet switches attract research interest because they remedy the actual bandwidth mismatch between WDM optical transmission and electronic routers that are presently used to switch, route and forward optical data packets.

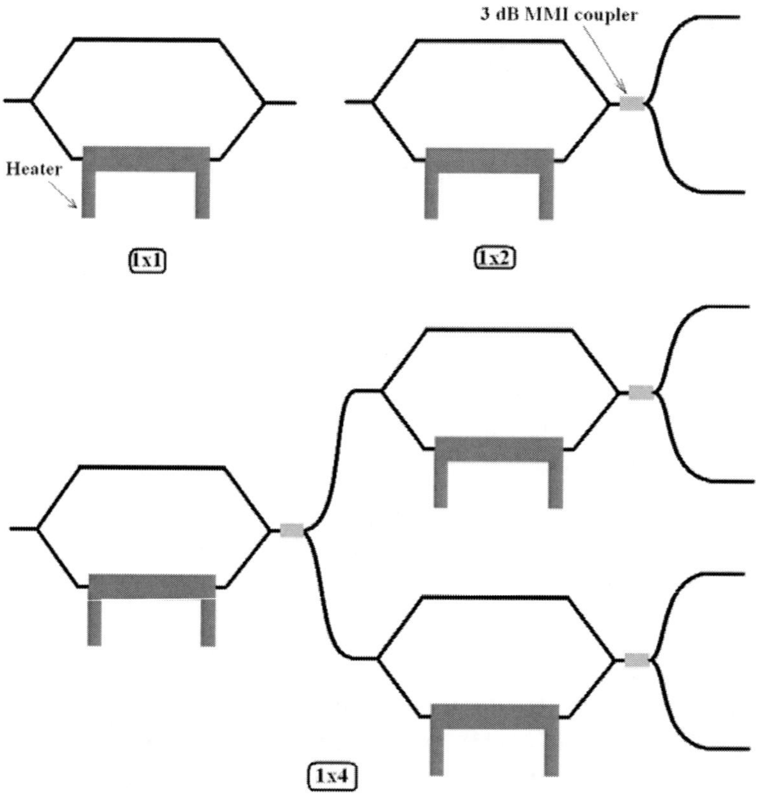

Figure 17. 1 x 1, 1 x 2 and 1 x 4 optical switches based on silicon photonic wire waveguides.

The use of all-optical packet switching allows the maintenance of high data rates, transparency and efficiency of the optical network [78]. Fundamental building blocks in all-optical packet switching are all-optical wavelength converters and optical flip-flop memory elements [79].

All-optical wavelength conversion is usually performed by exploiting nonlinear gain and index dynamics in semiconductor optical amplifiers (SOAs). In SOA-based wavelength converters, the relatively slow recovery of the carrier density (typically several hundreds of picoseconds) is dominated by the electron–hole recombination time, limiting the maximum operating speed. To overcome this drawback, the use of fiber Bragg gratings (FBG) or waveguide filters has been demonstrated [80-81]. Wavelength conversion at 100 Gb/s has been achieved by using a long SOA (2 mm) in combination with a FBG [82]. Differential Mach-Zehnder interferometers with SOAs in both arms have been used to realise fast wavelength conversion [83]. An interferometer wavelength converter using only one SOA has been also proved [84] as well as a wavelength converter capable of processing optical signals at a rate of 80 Gb/s [85]. The realization is made possible by using only one SOA and an optical pass-band filter and selecting the blue-shifted sideband converted light.

A very attractive configuration to implement an optical flip-flop is based on two pulsed micro-lasers (radius of 8 μm) coupled by a passive waveguide [86], realized by InP/InGaAsP technology as sketched in Figure 18. Each laser has two lasing modes: clockwise (CW) and counter-clockwise (CCW). The system has two stable states. In the first stable state (state A), CW propagating light from laser 1 is injected via the waveguide into laser 2, where it is amplified. Injected light competes with the laser 2 self-oscillations for available power, forcing light to circulate only in the CW direction in laser 2. Conversely, in the second stable state (state B), CCW propagating light from laser 2 is injected via the waveguide into laser 1, forcing light to circulate only in the CCW direction in laser 1. Thus, a fully integrated architecture on InP substrate employing two SOAs and an AWG capable to implement an optical flip-flop has been demonstrated [87].

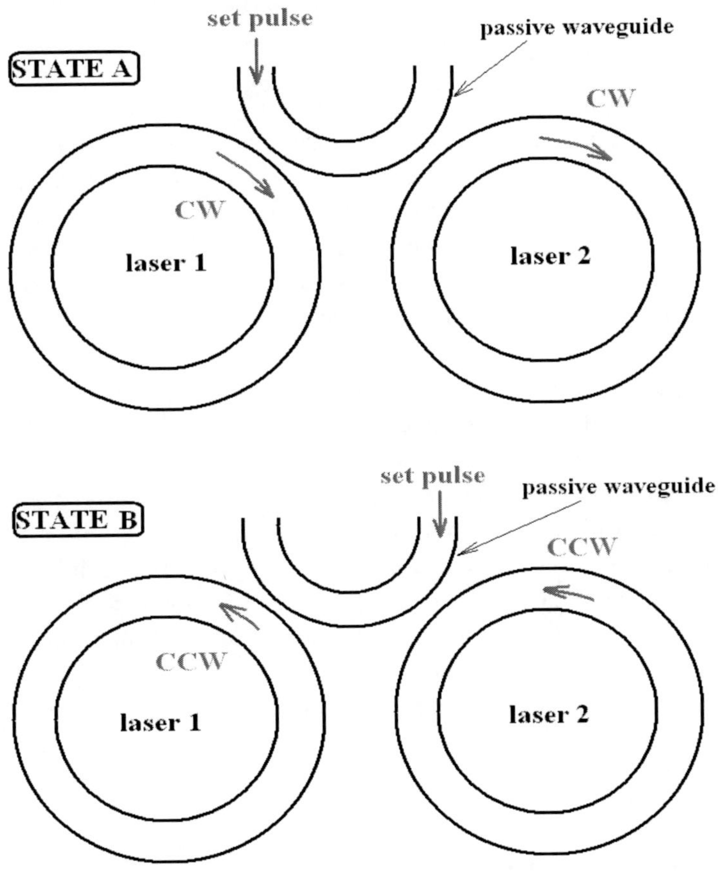

Figure 18. Stable states in an optical flip-flop implemented by two coupled ring lasers.

4. Optical Signal Processing for Space Applications

It is clear that Photonics has definitely reached the technological maturity with a number of applications on industrial scale. Moreover, IO devices are also expected to positively impact the space segment technology due to their intrinsic characteristics which make them attractive with respect to standard electronic technologies. In particular, specific advantages that could be achieved in the space segment by using guided-wave devices are:

- smaller sizes due to lower wavelengths involved, giving overall payload mass reduction and, then, both launch cost and satellite cost reduction;
- reduced losses, related to higher efficiency, giving less power consumption with a consequent reduction of both solar panel area and battery mass, and, then, satellite mass and cost reduction;
- better electromagnetic compatibility, allowing operation of various equipments in small volumes;
- down link reduction, due to either data decimation or data pre-processing in earth observation applications, allowing both space segment and ground segment cost reduction;
- on board data processing, allowing real-time direct-to-user data delivery services to a large number of users.

The most mature technology for guided-wave signal processing is that based on lithium niobate ($LiNbO_3$), which can be advantageously used in space-borne applications because of the low sensitivity of optical attenuation, refractive index, photorefractive effect, and polarization to ionizing irradiation and to thermal variations. However, $LiNbO_3$ technology shows a number of performance limitations in terms of maximum chip size to be processed, power consumption, resistance to optical damage, and influence of technological conditions, such as annealing time and temperature and cooling procedure on the electro-optic and photorefractive effects.

Examples of fiber optics space applications are reported in literature [88]. They include launch rockets, space shuttle vehicles, space platforms and satellites for telecommunications systems, sensors and signal processing. Among the various advantages to be mentioned, wide frequency bandwidth of optical fibers, very high number of information channels, low weight of optical fibers and their high electromagnetic compatibility. In the future, a further decrease of the fiber cost in large systems is expected. The fiber losses are no more the major concern as they were because their attenuation has been remarkably reduced. Both coupler and connector losses should continue to decrease. Smart sensors will be more and more used in aerospace applications. In this section some attractive configurations of $LiNbO_3$ and GaAs-based guided-wave devices for applications in the field of earth observation and remote sensing are briefly reviewed.

4.1. Earth Observation

Recent studies show that evolving systems for earth observation applications will progress from their present stage of decision support to monitor systems providing both management information and commercial benefits. In this frame, IO technologies can significantly contribute to the improvement of on-board capabilities required in these new systems.

In particular, the unique features offered by integrated optics may be deemed to be attractive especially in Digital Signal Processing architectures. The main advantages of integrated optics-based solutions are:

- reduction of data flow to be transmitted to ground due to on-board performance of amplitude extraction, azimuth compression and multi-look summation. This reduction can be estimated in a reduction of one order of magnitude , enabling a wider class of users to operate their own ground receiving stations with reduced costs;
- reduction of time required from data acquisition by the instrument and data delivery to user, achieved by skipping the bottlenecks of data transmission to ground station, image processing and data re-launch through terrestrial network or data delivery in magnetic tapes. Data could be delivered directly by thc satellite in real time for users;
- much simpler elaboration to be performed at ground, since the most expensive operations are already performed on-board, reducing both user cost and complexity.

In order to evaluate the potential impact of Integrated Optics in the field of earth observation, a reference scenario consisting of small satellites each carrying at least a payload relying on a Synthetic Aperture Radar (SAR) instrument can be considered. SAR is a powerful technique for reconstructing ground images whose main advantage is the possibility of obtaining high-resolution images at relatively long transmission wavelengths and with reduced antenna sizes. SAR applications can be easily implemented by using optical devices. This is because the synthesis of an object image is obtained by correlating the received echoes of a radar signal with a reference signal [89]. This operation is equivalent to the optical reconstruction of the Fresnel diffraction pattern of the same object when illuminated with coherent light. A number of studies have been proposed in literature to find out optical solutions, either guided-wave or bulk, for the reconstruction of SAR images from moving airborne or space-borne platforms. Acousto-optic transducers and coupled-charge devices (CCD) can increase the advantages of the optical approach in terms of compactness. In these devices the image reconstruction is obtained by spatial and/or temporal integration.

Figure 19 shows a compact semiconductor integrated optical solution for the reconstruction of two-dimensional SAR images on air-borne or space-borne platforms [90]. This optical processor makes use of the spatial and temporal integration in order to execute the fundamental range and azimuth data compression, coming from the echoes collected by an air-borne SAR antenna. A laser diode is butt-coupled to the planar waveguide and emits TM-polarized light at the free-space optical wavelength of 0.85 μm. The laser beam is expanded and collimated by a double diffraction grating having a nonlinear groove profile, which induces also the TE-polarized wave. An acousto-optic Bragg cell, obtained on a ZnO additional layer on the planar waveguide to improve its piezoelectric effect, deflects the incident beams generating two diffracted and two un-diffracted beams. The un-diffracted beams are filtered by a reflection grating, the TM diffracted beam is coupled by a lens array to a channel waveguide array, and the TE diffracted beam is suppressed by the same lens array. Therefore, the TM beam is out-coupled by another grating and focused along one line in order to illuminate a CCD matrix. This illumination is modulated by a transmission mask which reproduces the reference chirp function of the transmitted SAR signal, in close proximity to the CCD.

The laser beam is modulated by the reference chirp function, while the base-band amplitude modulated signal of the echoes received from the object, illuminated by the radar carrier, is applied to the electrodes of the Bragg cell. Then, the Bragg cell executes a temporal correlation, i.e. a data compression between the reference and the received signal. This processor has been designed at the carrier frequency of 10 GHz, having bandwidth of 50 MHz, pulse repetition frequency of 1110 pps, average distance of the object of 10 km, Bragg cell length of 1 cm, range resolution of 126 cm for a 800x800 CCD matrix [90]. The transducer bandwidth is 210 MHz, with a center frequency of 650 MHz. The azimuth resolution, depending on the number of pixels in the CCD matrix, is equal to 45 cm. This optical architecture could be fabricated by using either gallium arsenide or lithium niobate substrate, the latter being more promising for reduction of power consumption.

However, each acousto-optic (AO) architecture for SAR data reconstruction is intrinsically limited by the length of acousto-optic cell, along which the fundamental range data compression occurs. This length depends on the size of the substrate crystal. This means that SAR images from platforms very far from the illuminated area (space-borne) could not be reconstructed with useful resolution by using only one processor.

However, they could be reconstructed by using a large number of equal processors (about 80) working in parallel. For example, in the GaAs-based architecture of Figure 19, a number of parallel processors can be employed in order to obtain an equivalent very long Bragg cell, for the processing of strongly delayed data streams coming from long-distance point targets (typically 800 km). In this case, the overall processing data time should be 230 μs, with a pulse repetition frequency of 2000 Hz, signal-to-noise ratio of about 10 dB, range and azimuth resolution of about 30 m, radiometric resolution and accuracy of about 3 dB, sizes less than 500 cm^3, weight less than 4 kg, power consumption of the order of 10 W (using lithium niobate) or 70 W (using gallium arsenide technology). The radiometric resolution and accuracy, which in a number of applications represent significant parameters, can be improved up to about 1 dB by using pseudo-random binary antipodal sequences as reference radar signal instead of conventional chirp function. In these optical approaches, the higher-level functions of control of Doppler centroid and SCANSAR system in advanced SAR (ASAR) can be obtained by using an adaptive transmission mask at the input of the CCD matrix, which stores the previous and actual range and azimuth information. Further details can be found in literature [91].

An alternative processing architecture, as presented in Figure 20, is able to overcome the intrinsic limitations of the AO-based SAR processors [92]. It is based on the electro-optic effect which occurs in ferroelectric materials when an electric signal is applied to electrodes. This basic interferometer configuration includes four Mach-Zehnder modulators, each one characterized by an aperiodic inverse traveling-wave electrode. The electrode shape is designed to reproduce the signal that is obtained by multiplying the base-band received echoes by two different reference signals, implementing the appropriate range and azimuth data compression. Then, these two signals are independently time integrated by two fast photo-detectors. The sampled signals produced by the photo-detectors are electronically filtered to obtain the compressed signal without any amplitude distortion. Finally, the signals can be multiplexed and recorded on a two-dimensional buffer to reproduce the sampled correlation functions for range and azimuth data compression.

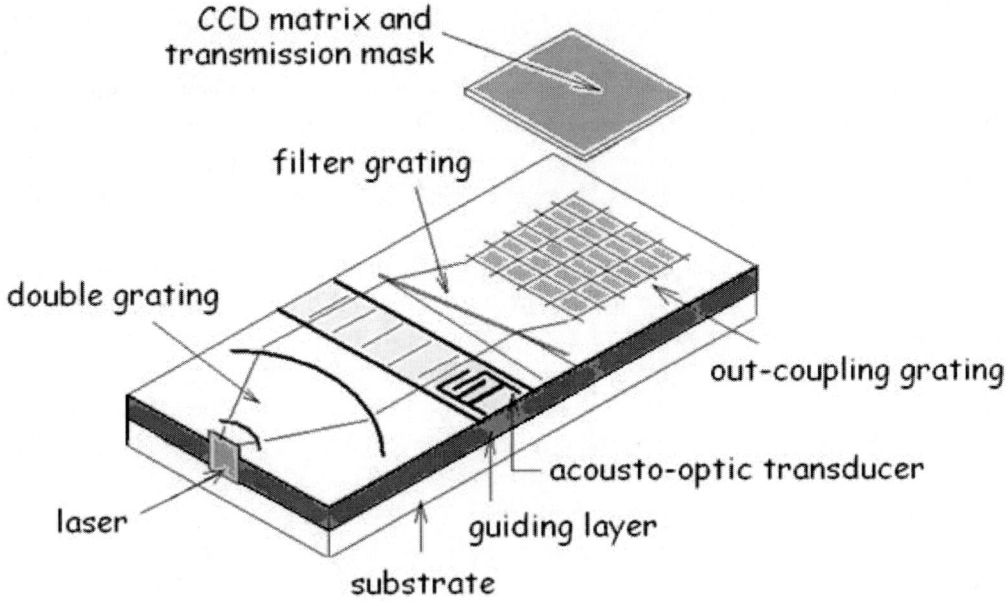

Figure 19. Acousto-optic processor for SAR applications.

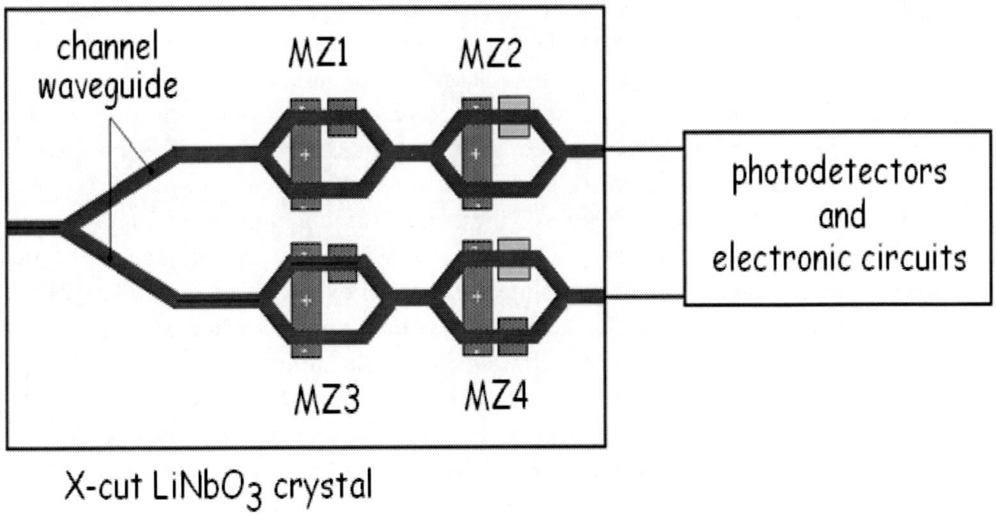

Figure 20. Architecture of SAR electro-optic integrated processor.

Since the data serial multiplication between the received echoes and reference signals can occur with very long data stream, the processor can in principle operate either in airborne or space-borne applications. The processor reproduces the images coming from ground surface by using a conventional chirp pulse as reference signal, with an antenna beam at 90° from the platform moving direction. At the optical wavelength of 1.3 μm, the mid-band frequency is 410 MHz, with a response time/bandwidth ratio of 10 ns/GHz, a linear dynamic range of about 50 dB and a signal-to-noise ratio of 8 dB. The circuit on lithium niobate planar waveguide has been designed by considering both titanium in-diffusion and proton exchange fabrication techniques in lithium niobate substrate. The former has shown the best performance in terms of dynamic range and optical losses. The circuit was simulated in a number of different working conditions, corresponding to some NASA/ESA missions (AIR-C, ERS1, SAR-X, SEASAT). The best range resolution is about 10 m in airborne missions, and 60 m in space-borne ones, with an azimuth resolution of 1.5 m and 6 m, respectively. Further improvements could be obtained in terms of signal-to-noise ratio (90%) and SAR image resolution (50%) if binary pseudo-random sequences with narrow bandwidth are used as reference signals. However, this optical architecture, if used in space-borne applications, presents strong limitations in terms of radiometric resolution ($\approx$ 4 dB), spatial range and azimuth resolution ($\approx$ 50 m), intrinsic limits (with no control of Doppler centroid or use of SCANSAR), chip integration (no more than 64 elements on the same substrate), synchronization sensitivity, limited optical processing. To simulate the behavior of traveling-wave modulator aperiodic electrodes, Laplace transform was used.

4.2. Remote Sensing

Another example of potential and unique advantages offered by photonic devices to earth observation is given through the use of an integrated optic phase comparator. This comparator is conceived to implement a pattern recognition scheme in remote sensing applications.

The actual digital systems for data acquisition by satellite remote sensing can involve very large and powerful computers for the classification and identification of the large amount of information coming from the earth surface. In order to reduce this amount of data, it is crucial to adopt on-board data pre-processing. A guided-wave optical pre-processor, which is particularly advantageous in space applications, requires less power consumption, weight and overall dimensions as compared with its electronic counterparts. Figure 21 shows the schematic of a guided-wave optical pre-processor based on a lithium niobate substrate and titanium indiffusion (Ti:$LiNbO_3$) and post-annealing proton exchange technologies [93].

In the functional scheme of this device, a laser beam at the free-space optical wavelength of λ=0.6328 μm is coupled into a channel waveguide and divided in two waves by a Y-branch. The two waves undergo phase changes induced by a set of electro-optic modulators. Then, the waves are coupled into a planar region and made to interfere properly in a holographic zone. The processor operation is based on the holographic subtraction effect, which arises between two waves interfering with different phases.

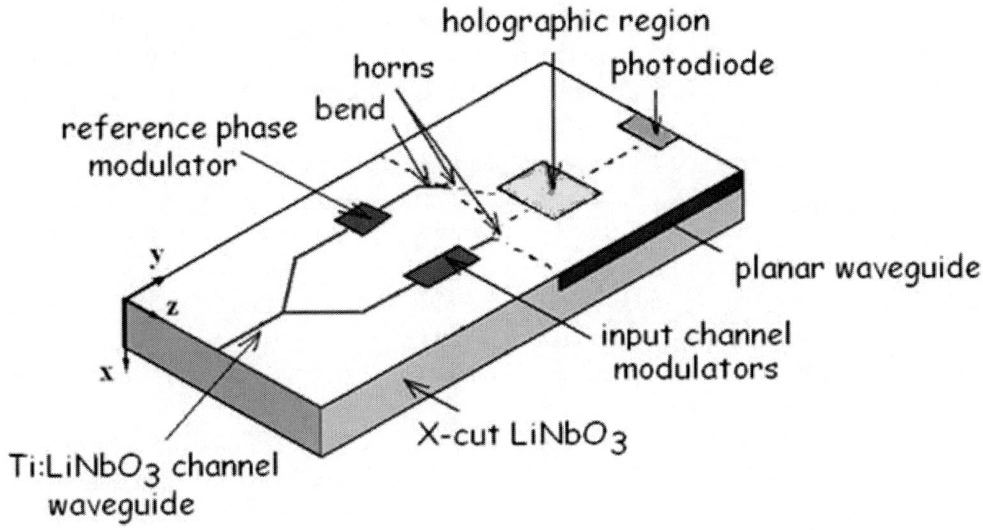

Figure 21. Guided-wave optical preprocessor.

Continuous writing and reading of holographic gratings induced by the change of voltages applied on the electro-optic transducers are so possible. High efficiency and low optical loss can be obtained when the photorefractive susceptibility of the planar waveguide is enhanced in the holographic region by iron doping. The overall sizes of the device are $15\times1\times2.5$ mm^3. The circuit performance was evaluated in the frequency range within 100 kHz with an optical power of 1 mW at the input of the channel waveguide. This evaluation was conducted while considering the radiation, attenuation and scattering losses. The results include a time constant of 1.1 μs, a holographic writing time of 1.22 μs and a change of the equivalent phase noise of 0.17°.

Thus, the equivalent noise does not allow detection of signals lower than the threshold of 11 mV at the output. The circuit can work also with continuous changes of the reference signal, within a bandwidth of 10 kHz. The intrinsic weight of the optical chip is low, about 20 g, and the power consumption is about 100 mW. The dynamic performance of this device is fundamentally limited by the holographic writing time, of the order of 1 μs. The best results for improving this characteristic time have been obtained by using a Ti-indiffused proton exchanged (TIPE) planar waveguide with Fe-doping. This technological solution is convenient because it offers a significant number of degrees of freedom in the design and fabrication of guiding structure. This is a significant improvement in the holographic efficiency and dynamics [94].

5. Guided-Wave Optical Sensors

The field of applications of physical, chemical and biochemical sensing is extremely wide. In recent years, guided-wave optical devices have been investigated and experimented in a number of specific applications due to their significant advantages in terms of high sensitivity, low power consumption, high electromagnetic compatibility and immunity to environmental conditions. During the last three decades, electric and magnetic field photonic sensors have been widely investigated [95]. In these sensors, a microwave signal provided by an antenna modulates an optical signal generated by an optical source. Recently, a new approach for E-field measurements based on two coupled cavities, which could measure the resonance wavelength shift due to the applied electric field, has been investigated, with encouraging results in terms of sensitivity and bandwidth [96].

Angular velocity sensing by optical signals was proposed in 1963 for the first time [97]. The development of electrically pumped semiconductor ring lasers during the 1980s and 1990s has enabled the fabrication of very compact, active optical angular velocity sensors for gyroscope applications [98]. In these sensors, the optical propagation takes place in guiding structures realized using semiconductor materials [99-100].

The first optical biochemical sensor was based on the measurement of changes in absorption spectrum and was developed for the measurement of CO_2 and O_2 concentrations. Since this first example, optical technologies have been widely adopted to overcome some problems connected with the use of electrochemical techniques for bio-sensing analysis, such as low sensitivity, large dimensions and influence of electromagnetic interference. In recent years, biochemical sensing by guided-wave optical devices has received a considerable attention. Different chemical and biochemical photonic sensors have been proposed. Examples of these sensors are those based on directional couplers [101], Mach-Zehnder interferometers [102], Bragg gratings [103], and microring resonators [104-105]. Furthermore, IO devices have been proposed for sensing other physical quantities, such as temperature, pressure, acceleration or micro displacements (see for example [106-108]), as well as for chemical and biochemical substances, such as ammonia, glucose, DNA, water salinity, organic solvents, and so on. In this section, a brief review of some important examples of photonic sensors is presented.

5.1. E-Field Photonic Sensors

In general, the development of photonic technologies for optical sensors offers an efficient and advantageous way to measure electromagnetic fields. For this purpose, crystal properties, like the Pockels electro-optic effect, can be often used. In general, these sensors do not suffer from electromagnetic pollution, because they are based on dielectric materials, and consequently they can be used for RF field measurements in high voltage settings, monitoring of remote fields, Specific Absorption Rate (SAR) and low frequency measurements, giving information about both phase and frequency of sensing field without disturbing it. Optical sensors give the possibility of effectuate measurements characterized by good spatial resolution, large bandwidth and, especially, low invasiveness. Towards traditional electronic technologies, these advantages are discriminating in several applications.

Optical sensors for electric fields are principally of two types: sensors that convert an electric field in an optical signal through a laser or a LED, and sensors that operate the conversion by means of optical modulators using electro-optic crystals. The former have the advantage of a greater sensitivity and the disadvantage of a limited battery life, while the latter does not suffer from electromagnetic interference, presents small perturbation of measured signal, flat frequency response, high bandwidth and operative time not limited by batteries [109]. For applications that require particularly high fields, the probe head can be fabricated externally to the dielectric material. Photonic probes are very long-lasting and impulsive or great amplitude fields do not damage them, even if design parameters are exceeded. The head of passive probes do not enclose electronic circuits and supply [110-111].

For their dielectric nature, fibre optics (used to carry the signal out-coming from the laser and from the crystal) cover a great area of interest in the field of electromagnetic measurements. Advantages include galvanic isolation of the sensor from earth potential, low electromagnetic interference sensitivity and very small dimensions. The fibre optic keeps in contact with the sensitive region, sensor head, which measures the changes produced by a field applied to the material. Designed photonic probes give large bandwidth and low dispersion to guarantee high fidelity to Electromagnetic Pulse effect (EMP), keeping unchanged both phase and amplitude information. Photonic sensors can be based on a variety of physical principles and designs depending on the particular optical material. The main technologies proposed in literature are based on lithium niobate ($LiNbO_3$), III/V semiconductors (e.g. GaAs), silica and other silicon compounds (SiO_2, SiON, Si, SOI) and Potassium Titanyl Phosphate (KTP). The schematic architecture of the most photonic E-field sensors is sketched in Figure 22.

The architecture includes a laser source, an optical fiber to link the semiconductor laser to the sensor section and a sensing device composed of an optical modulator.

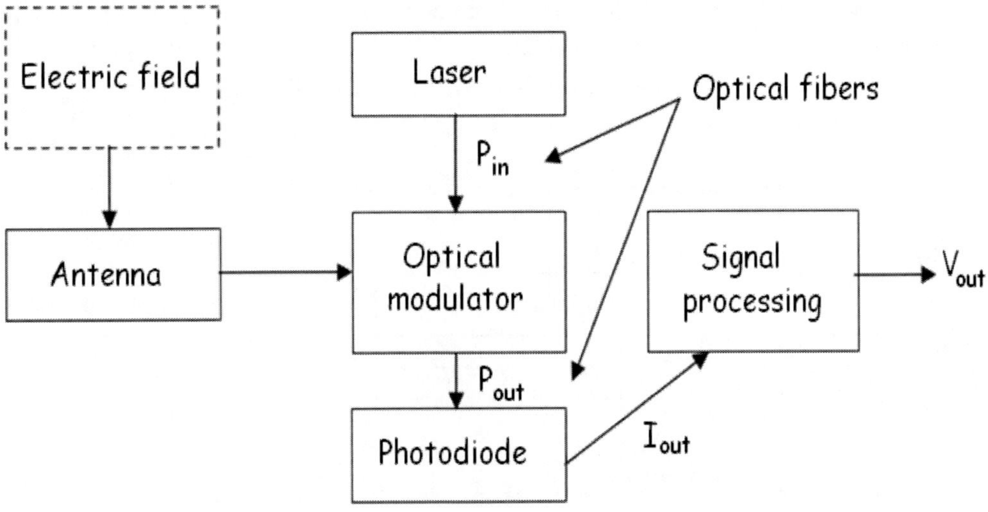

Figure 22. Schematic general architecture of an E-field sensor.

The intensity of light passed through the optical modulator is linearly modulated by the applied voltage, proportional to the external electric field. The light transmitted at the end of the modulator is collected by a photo-detector. The photo-generated output current is processed by the signal processing circuit to obtain an output voltage proportional to the external electric field under measurement. In Figure 23 four typical schemes of modulators used as electric field probes are shown. Their transfer functions are reported in literature [111].

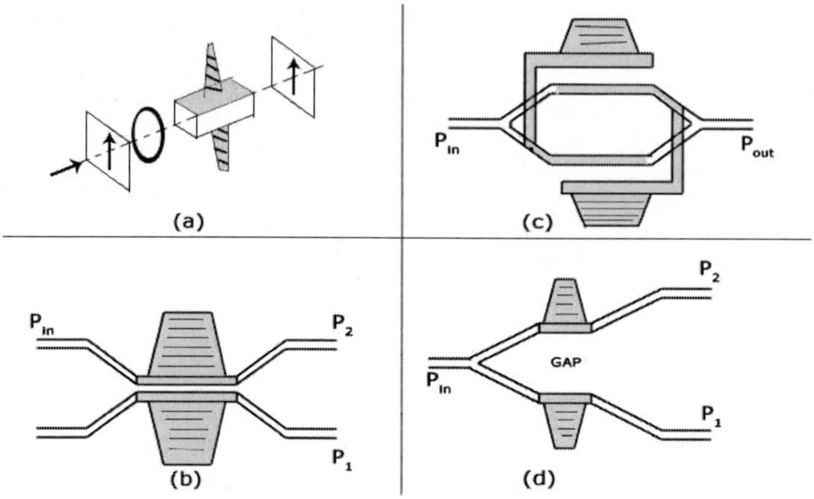

Figure 23. Passive photonic modulators used as electromagnetic field probes. (a) Pockels cell; (b) Four gate coupler; (c) Mach-Zehnder interferometer; (d) Three gate coupler.

In order to define the performance of photonic sensors, the evaluation of the maximum noise level at the photo-detector is critical, defined as noise equivalent field (NEF) [111]. The National Institute for Standards and Technology (NIST) has fabricated and tested some photonic systems for electromagnetic field measurements. The main features of these systems are reported in Table 1.

Table 1. Performance of photonic modulators in electromagnetic field sensors

System	Type 1	Type 2	Type 3
Source wavelength	GaAlAs SRD 807nm	GaAlAs laser 830nm	YAG diode 1320nm
Pd (dBm)	-22	-19	0
Fibre (1)	PM e MM	MM	PM e SM
Modulator	Three gates coupler LiNbO3	LiNbO3 bulk Pockels cell	Bi4Ge3O12 bulk Pockels cell
Vπ (V)	10	265	2100
Sm (dB)	-15	-44.6	-62.6
Antenna	2 cm with RTD (2)	15 cm with RTD	15 cm with RTD
ST (dB)	-23	-105	-104
Bandwidth (MHz)	0,02 ÷ 1000 500 ÷ 3000	0,01 ÷ 200	0,01 ÷ 1000
NEF (V·m-1·H·m-1/2)	0.1	0.2	1
Dynamic range (dB)	65	68	72

([1]) MM = multimode, SM = single-mode, PM = polarization maintaining

([2]) RTD = resistive tapered dipole

Type 1 is a 2 cm tapered dipole directly deposited on a modified three gates directional coupler. The dipole is fabricated with a Nickel-Chrome thin film.

Type 2 employs three 15 cm dipoles orthogonally arranged to have isotropic detection. Dipoles are fabricated on a planar fused silicon substrate on which a resistive material thin film is tapered.

Type 3 is a 15 cm cylinder dipole fabricated by tapering the resistive film depth during deposition. The system is designed for monitoring high field levels in EMP environments.

(a) *Pockels cell*: when the phase shift is $-\pi/2$, the modulator is in its operative point, where maximum sensitivity and linearity are ensured. Modulator voltage characteristic is limited by the electro-optic coefficient value, the minimum crystal dimensions to consider it as lumped, and the maximum crystal length allowed for obtaining high operative frequency. The minimum V_π value for 5 GHz operations is about 100V. For a modulator that employs a $LiNbO_3$ crystal with z propagation axis in order to improving thermal stability, $V_\pi = 250$V. Increasing V_π and decreasing the modulator sensitivity in order to measure high electric fields is relatively easy, using for example a $Bi_4Ge_3O_{12}$ (BGO) crystal with V_π of the order of 2100V.

(b) *Four gates directional coupler*: in the interaction region the two waveguides are close enough to allow coupling between their evanescent fields. By changing the relative propagation constants by means of electro-optic effect, the field can bounce back and forth between the two waveguides. The great difficulty in using this directional coupler for electromagnetic measurements is the need to find an active voltage operative point.

(c) *Mach-Zehnder interferometer*: typically, in order to delineate waveguide channels and operate the titanium diffusion into $LiNbO_3$ crystal surface within the fabrication of guided-wave optical (GWO) modulator, photolithography is employed. The phase shift is introduced by the different optical path lengths in two modulator arms. This configuration presents a greater sensitivity, but achieving the designed phase shift during fabrication and monitoring temperature dependence is rather complicated.

(d) *Three gates directional coupler*: this configuration has more advantages with respect to the previous one. It has high tolerance towards the use of laser with small wavelength shift and a larger linear range compared to other modulators.

In lithium niobate technology, the electro-optic modulator is generally a Mach-Zehnder (MZ) type. Examples of miniaturized voltage sensors based on electro-optic effect in $LiNbO_3$ have been proposed [112-113]. In some devices the minimum electric field detectable is of the order of 0.22 mV/m at 50 MHz and 0.079 mV/m at 750 MHz. The 3dB bandwidth measured is about 300 MHz. The sensor proposed [114] leads to detect the isotropic electric field, with a bandwidth of 10 GHz, sensitivity of 22 mV/m and sensing element length of 8 mm. The sensing part of the sensor is constituted by three MZ interferometers oriented along orthogonal x,y,z axes. The interferometers are obtained from Ti diffusion in a z-cut $LiNbO_3$ substrate with sizes $55\times1\times0.5$ mm^3. The laser source is a YAG diode at 1.3 μm with a optical power equal to 25 mW. To increase the modulation efficiency and obtain a good trade-off between sensitivity and bandwidth, a MZ interferometer with segmented electrodes has been proposed and patented [115], using a polarization maintaining fiber (PMF) at the input, as in Figure 24. The device is fabricated in X-cut $LiNbO_3$. The length of the dipole is equal to the substrate length, i.e. 30 mm. The bandwidth and measured sensitivity are 3 GHz and 1mV/m, respectively. Electric field sensors with this configuration have been used to monitor electromagnetic field generated by annular phased-array applicator SIGMA-60 of the system BSD-2000, adopted for cancer treatment by hyperthermia [116].

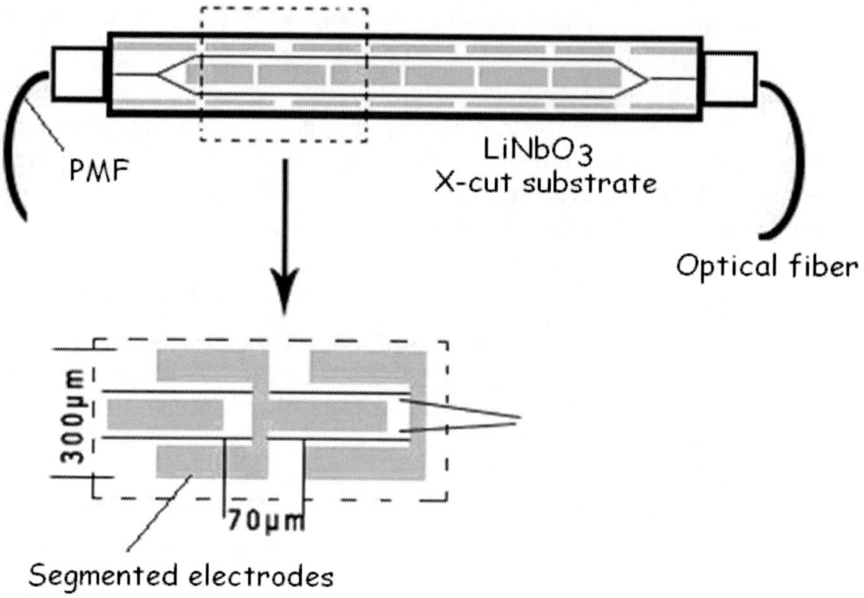

Figure 24. MZ modulator with segmented electrodes for E-field sensing application.

Another approach to optically sensing E-fields is based on a MZ interferometer without electrodes. This is possible by inducing a reverse poling of z-cut $LiNbO_3$ crystal in the region in which an interferometer arm is present [117]. The potential of this solution is to avoid the perturbations induced by the metal and reduce the capacitive effect influencing the processing speed. The typical performance of this device is a sensitivity of 0.22 V/m/√Hz and a bandwidth of 1 GHz. Another possible realization of devices without electrodes is based on Periodically Poled Lithium Niobate (PPLN). PPLN is realized with a Bragg grating where the inverse polarization layers have a width of 20 µm, a grating period of 40 µm and a total length of the Bragg devices around 20 mm. This sensor leads to obtain both larger bandwidth and sensitivity. The experimental results have shown a bandwidth of 400 MHz and a sensitivity of about 0.25 V/m/√Hz. Coherence modulation of light provided by a low-coherence optical source can be exploited to design a multi-channel E-field optical sensor [118] capable to measure different applied electric fields, as in Figure 25.

In this case, electrical signals can change the optical delays provided by modulators which are based on velocity mismatch introduced by an electro-optic $LiNbO_3$ crystal between TE–like and TM–like modes. Coherence modulation based E-field optical sensors are used for low frequency measurement within 20 kHz and exhibit good linearity. In general, these sensors based on Ti-diffused $LiNbO_3$ exhibit only weak electro-optic effect, thus requiring long electrodes (usually 20 mm) and limiting their bandwidth to about 1 GHz.

To overcome this limitation, an alternative approach to optically sensing E-fields has been proposed, based on traveling-wave semiconductor modulators. In fact, the use of semiconductor modulators instead of $LiNbO_3$ modulators is an advantage, since significantly shorter interaction lengths are required for similar modulation efficiencies for broadband field measurements up to frequencies of about 1 GHz. Thus, GaAs-based MZ modulators operating at 1.3 µm are becoming available. The operation of these modulators is similar to that of lithium niobate modulators discussed above. Better phase matching between optical and electrical signals is expected using GaAs travelling-wave modulators instead of their lithium niobate counterpart. This should allow for higher bandwidth devices without sacrificing the modulation efficiency. However, obtaining low optical insertion loss is a remaining challenge for GaAs MZ modulators.

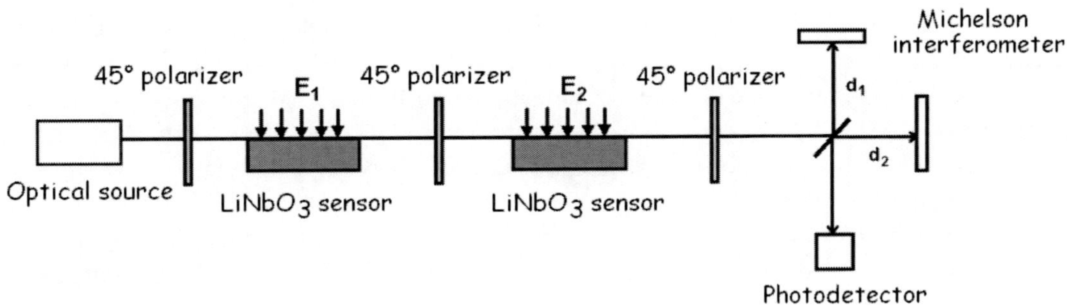

Figure 25. E-field sensor based on coherence modulation.

Another important III/V semiconductor optical modulator is the electro-absorption (EA) waveguide modulator [119]. Semiconductor EA waveguide modulators fabricated in either a p-i-n or p-n junction structure, as shown in Figure 26, are also being considered for analogue fibre optic applications. Here, a reverse bias across the junction modulates the electric field in the waveguide and changes the absorption coefficient of the material.

The semiconductor EA waveguide modulator has an exponential transmission function for the optical power given by:

$$P_{out}(V) = P_{in} t_m \exp\left[-\Gamma \Delta\alpha(V) L\right] \qquad (4)$$

where P_{in} is the input power, t_m is the modulator transmission loss factor, L is the waveguide length, Γ is the overlap integral between the optical mode and active absorbing layer, and $\Delta\alpha$ is the absorption coefficient change. In quantum well structures, $\Delta\alpha$ is related to Quantum Confined Stark Effect (QCSE). In thick (>500 Å) layers or bulk semiconductors, $\Delta\alpha$ is due to Franz-Keldysh effect (FKE). QCSE modulators typically have a large $\Delta\alpha$ at low applied voltages, but also a relatively small Γ, while FKE modulators typically have small $\Delta\alpha$ and large Γ.

It can be seen that these two parameters equally affect the modulator performance and, therefore, both devices can obtain comparable modulation performance. Pre-biasing the EA modulator in the quasi linear region enables analog operation, similarly to MZ modulators. The modulator linearity can be assessed by measuring the harmonic and inter-modulation distortion around a given bias.

Several QCSE and FKE modulators have been tested for use in electromagnetic environment (EME) monitoring systems. Liquid phase epitaxy (LPE) has been used to fabricate 1.32 μm FKE modulators based on an InGaAsP active absorbing waveguide layer. Typically, the ridge-waveguide modulator has a device length of 300 μm, waveguide thickness of 0.4 μm, device capacitance of 0.2 pF, and $\Gamma = 0.7$. For digital applications, an extinction ratio larger than 30 dB at less than 10V has been obtained. The 1.52 μm (MOCVD) QCSE modulators use an InGaAs/InP quantum-well ridge-waveguide structure.

Typically, these devices have a length of 650 μm, $\Gamma = 0.1$, and extinction ratio larger than 30 dB at applied bias less than 10V. The performance of FKE and QCSE modulators show important aspects in terms of relative transmission, linearity, absorption change, and inter-modulation products versus applied voltage.

These aspects indicate that the QCSE modulators are preferable in cases when an E-field sensor requires high sensitivity and large bandwidth. Typically, photonic E-field sensors based on QCSE modulators can have a frequency range from 10 MHz to 6 GHz, minimum detectable field below 100 mV/m within the whole frequency range, and maximum sensitivity at 0.9 GHz with a minimum detectable field strength of 16 mV/m.

The silica sensor represents another alternative to realise a photonic E-field sensor. The sensing element is constituted by a region of optical fiber in which the cladding layer is removed and substituted with a layer of Polymer Dispersed Liquid Crystal (PDLC), deposited directly on the fiber core, as shown in Figure 27 [120].

The PDLC layer influences the characteristics of the evanescent optical field. When the external electric field is applied to the fiber, the dipoles of the PDLC change their orientation and become transparent to the field. These changes in the properties of the PDLC induce a change of optical power reaching the fiber end. Therefore, the external electric field is evaluated by measuring the optical intensity at the output.

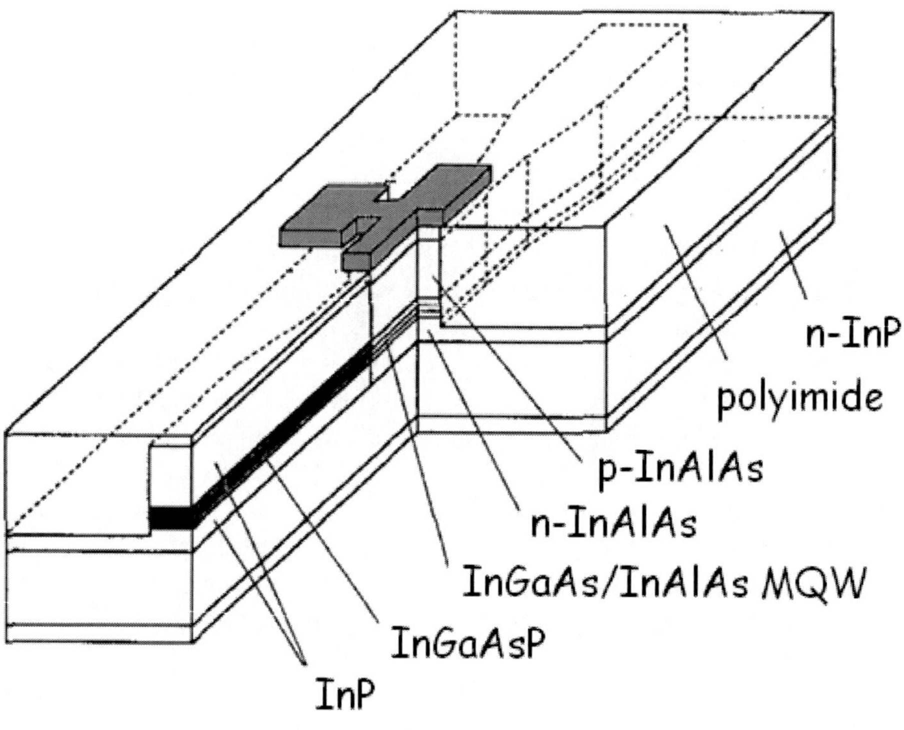

Figure 26. Electro-absorption modulator structure.

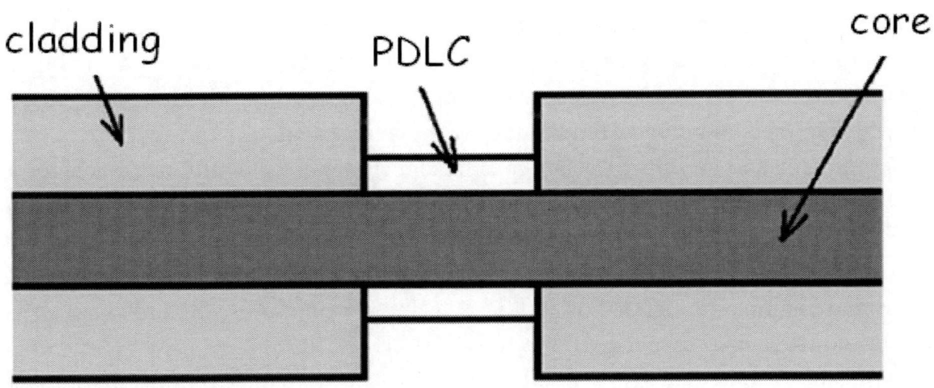

Figure 27. E-field sensor structure based on PDLC.

Generally the PDLC used is PMMA/E7, but other materials, exhibiting optical property modification produced by an electric field exposure, are still possible, as electro-chromic ones. Surrounding a tapered optical fiber by a solution containing an electro-chromic solute, it is possible to realize an electric field sensor capable to operate in the low frequency range [121]. Another sensor architecture involving optical fibres is based on evanescent coupling between the optical fiber core and a slab multimode waveguide. The refractive index of the waveguide is larger than the fiber core. Thus, changing the refractive index of waveguide by means of electro-optic effect, it is possible to change the coupling efficiency and, then, the optical power. Heating induced in metallic materials or strain induced in electrostrictive and piezoelectric ceramics can be exploited in extrinsic electric field optical fiber sensors [122-124].

Since KTP material has a very large electro-optic coefficient, high threshold to optical damage and low sensitivity to thermal effects, it can be considered an efficient candidate for photonic sensors. The architecture of a E-field sensor based on KTP material is sketched in Figure 28.

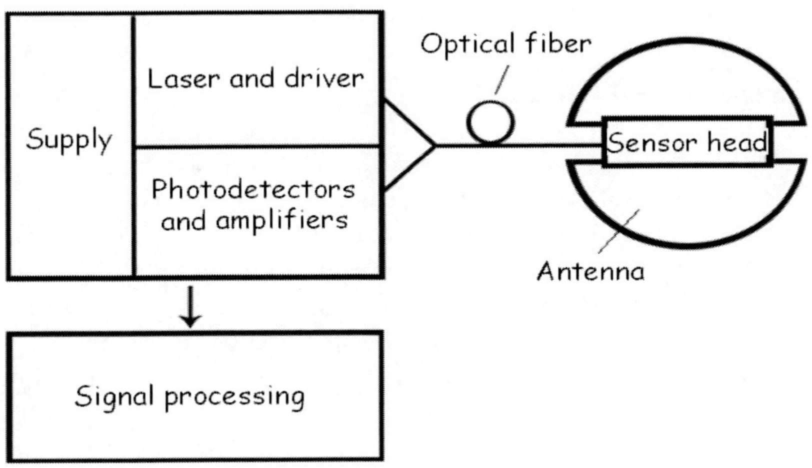

Figure 28. Schematic architecture of E-field sensor based on KTP crystal.

The optical E-field sensor head constituted by a KTP layer placed between two metal semi-sphere antennas is optically connected to a Transceiver Unit by means of a single-mode optical fiber (in transmission) and by means of a graded index multimode optical fiber to the receiver [125]. The optical beam is modulated by the antenna signal while it travels twice into the crystals (a mirror and an optical prism are present in the architecture to couple the optical beams to the multimode fiber). In the Transceiver Unit, the modulated optical signal is converted to an electric signal by means of the photo-detector. The experimental measurements made on a sensor with a KTP crystal thickness of 370 μm, length 7.2 mm, width 4 mm and antenna diameter 9.6 cm, show a linear behaviour for fields larger than 100 kV/m, with a flat frequency response in the range 70 kHz ÷ 40 MHz.

In general, these sensors show a good sensitivity, but they also require a complex large architecture, not suitable for high performance optoelectronic integration.Usually, the trend in the design of E-field sensors is to individuate the most appropriate material to achieve the best trade-off between conflicting requirements, i.e. high sensitivity and optoelectronic integration. In this sense, reduced architecture sizes, as allowed by standard SOI technology, is believed to bring a great improvement with respect to the state-of-the-art, significantly improving the integration scale as well as performance and cost [3-4]. The ubiquitous silicon Microelectronics chip is taken for granted in modern society. There has been much research involved in producing these high technology devices.

Continued developments in silicon and, more recently, in $Si_{1-x}Ge_x$ semiconductor alloy and strained silicon technology continue to advance the frontiers of device integration, complexity, and speed. These advances have been driven by application requirements in switching technology (i.e. computers) and high-speed electronics (i.e. wireless telecommunications).

Therefore, E-field sensors based on SOI technology should move towards the future direction to integrate on the same silicon substrate the optical sensor with the front-end electronics, such as photo-detector and trans-impedance amplifier. In the following section, architecture and performance of an innovative concept of SOI E-field sensor are described in some detail.

5.2. Dual-Cavity E-Field Sensor

Optical microcavity structures are of great current interest for photonic applications. These structures can exhibit high quality factor, and small modal volumes. Intriguing potential applications for such cavity structures are in optical modulations, switching and sensors. Microdisk and microring resonators have been for example proposed as sensitive chemical sensors and biosensors due to their large Q factors [126].

It is well known that the resonance lineshape of a traditional microring is symmetrical with respect to its resonant wavelength, typically having a Lorentzian shape. However, it can be modified by using a ring resonator coupled to an input/output waveguide incorporating two partially reflecting elements (gratings) [127]. This dual-cavity structure can produce a sharp asymmetric Fano-resonance line shape, in which the slope between the zero and the unity transmission is greatly enhanced compared with that of a conventional ring resonator. This sharply asymmetric line shape provides greatly improved slope sensitivity. SOI technology is proposed to realise a MOS capacitor highly sensitive to the external electric field to be measured. The basic architecture of this approach is shown in Figure 29.

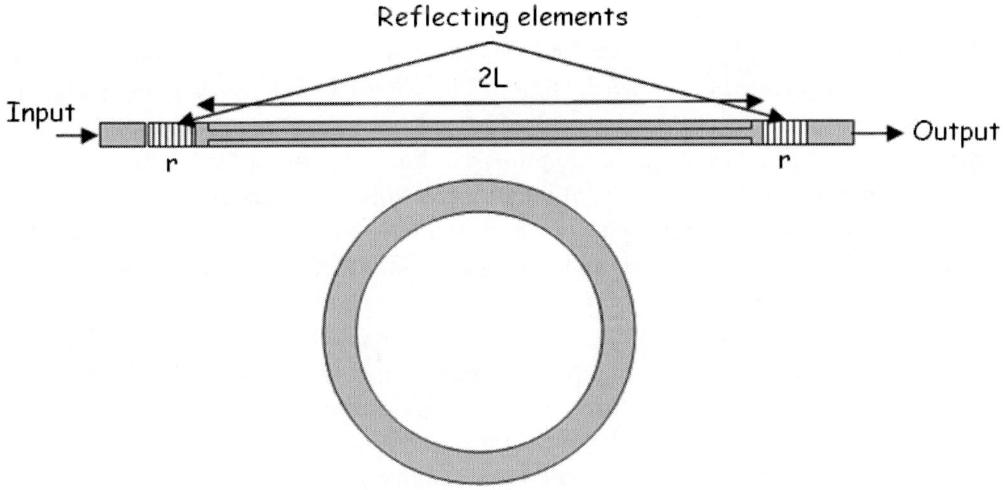

Figure 29. Schematic architecture of dual-cavity field sensor.

The device is constituted by a ring resonator, an output waveguide, coupled to the ring through a coupling coefficient k, and two reflecting elements (gratings) placed at the straight waveguide edges, at a distance 2L from each other. The sensing element is the Fabry-Perot cavity constituted by the output waveguide between the two reflectors and by the antenna dipole, whose two electrodes are on the straight waveguide, as in Figure 29. To analyse the spectral behaviour of this configuration, the scattering matrix method can be used [127]. The transmission coefficient T_r relating the incoming and outgoing optical field components at the input port (a_1 and b_1) to those at the output port (a_2 and b_2) can be obtained from:

$$\begin{bmatrix} b_2 \\ a_2 \end{bmatrix} = \frac{1}{j\sqrt{1-r^2}} \begin{bmatrix} -1 & -r \\ r & 1 \end{bmatrix} \times \begin{bmatrix} e^{j\delta} & 0 \\ 0 & e^{-j\delta} \end{bmatrix} \times \begin{bmatrix} 1-\dfrac{j\gamma}{\omega-\omega_0} & -\dfrac{j\gamma}{\omega-\omega_0} \\ \dfrac{j\gamma}{\omega-\omega_0} & 1+\dfrac{j\gamma}{\omega-\omega_0} \end{bmatrix} \times$$
$$\times \begin{bmatrix} e^{j\delta} & 0 \\ 0 & e^{-j\delta} \end{bmatrix} \times \frac{1}{j\sqrt{1-r^2}} \begin{bmatrix} -1 & -r \\ r & 1 \end{bmatrix} \times \begin{bmatrix} a_1 \\ b_1 \end{bmatrix} = M_T \times \begin{bmatrix} a_1 \\ b_1 \end{bmatrix} \qquad (5)$$

where ω_0 is the center angular frequency of the ring, ω represent the angular frequency of the input beam, γ is the half width of the ring resonance, r is the amplitude reflection coefficient (which can be different for each grating) and δ is the phase shift that the waveguide mode acquires as it propagates along L (see Figure 29). The transmittivity coefficient is given by:

$$T_r = \left| \frac{1}{M_{T,22}} \right|^2 \tag{6}$$

The advantage of this structure with respect to a simple ring resonator without the reflecting elements, is clearly shown in Figure 30, where $k = 0.1$ is assumed. The figure shows the transmittivity spectrum for $r = 0$ and $r = 0.5$ and different values of L. This plot evidences the potentials of the architecture proposed. For $r = 0$, the shape of transmittivity has a Lorentzian shape (typical of any conventional ring resonator). In case of $r = 0.5$ and $L = 4.7$ mm, the profile of the transmittivity shows a asymmetric Fano resonant line shape with an increasing slope between zero and unity transmission, if compared with the Lorentzian resonance. This characteristic can be used for highly sensitive detection of the resonance shift. In fact, the Fano resonance is due to complex interference occurring in the structure, as formed by Fabry-Perot cavity between the reflecting elements and ring resonator. Compared with a conventional ring resonator where waves propagate in only one direction, the added partially reflecting elements introduce backward propagating waves that perturb the transmitted wave phase and, hence, lead to a complex interference. Anyway, the best behaviour of the structure in terms of slope critically depends on the relative positions of the resonant frequencies of both ring and Fabry-Perot cavity. For each ring radius, the relative position of the resonant frequencies can be controlled by means of the distance between the two reflecting elements. In fact, Figure 30 shows also the cases with $r = 0.5$, $L = 2$ mm, or $r = 0.5$, $L = 10$ mm, where the slope of the Fano lineshape is less that Lorentzian profile.

It is also shown that the wavelength shift scale can be modified by using different ring radii (smaller changes with larger ring radii, e.g. from 150 to 1500 μm). It is possible to demonstrate that the optimum distance between two reflectors to increase the slope with respect to the conventional ring resonator must verify the following relationship:

$$2\delta = (2n + 0.5)\pi \tag{7}$$

where n is an integer number. The transmittivity spectrum sketched for different values of r coefficient assuming a length L satisfying the relationship (7) and $k = 0.1$, shows that the slope of the resonant line shape at the resonant wavelength can be strongly enhanced by increasing the reflection coefficient r of both gratings.

Therefore, ring radius, semi-distance between the gratings L, reflection coefficient r and coupling coefficient k are parameters to be appropriately selected to satisfy sensor requirements over sensitivity and intrinsic bandwidth. Thus, it is clear as the resonant spectrum of the structure depends on the effective refractive index of the waveguide and ring resonator modes. In fact, in our sensor configuration a shift of the resonant spectrum is produced by changing the refractive index of the waveguide between the two reflectors. In this situation, the most efficient choice is to include the antenna dipole on the output waveguide, not on the ring as in previous GaAs-based architectures.

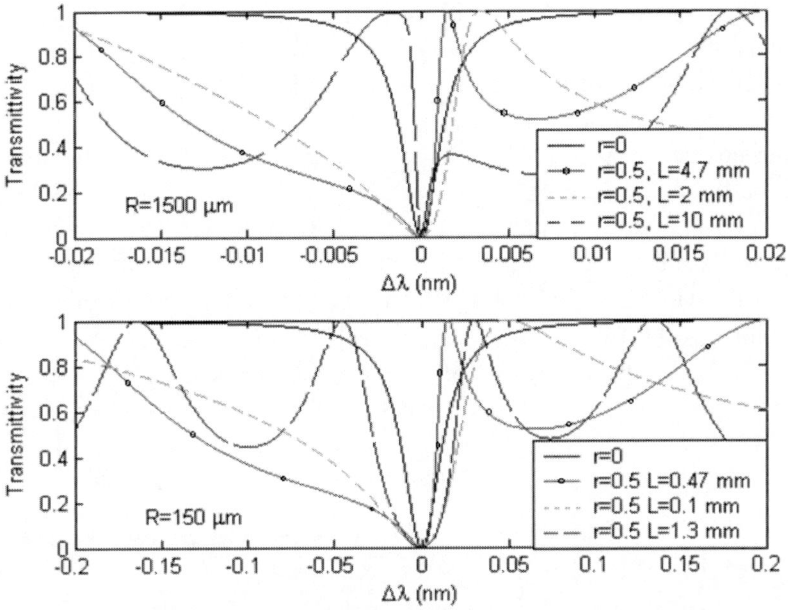

Figure 30. Transmittivity spectrum for various L and r.

The waveguide structure of the sensor is based on free-carrier plasma dispersion as in a conventional p-i-n diode, enhanced by a MOS structure utilizing clouds of charges accumulated at the surfaces of the gate oxide to achieve the required index change. It has been demonstrated [128] that when the capacitor is biased in accumulation regime, the majority carriers of the silicon electrodes are transported into and out the optical path and accumulated on the capacitor. Thus, the relatively slow carrier recombination processes, that limit the speed of p-i-n diode devices, does not significantly influence the dynamics of the sensor. The cross-section of the output waveguide (first variant) is similar to that presented in [128] and is shown in Figure 31(a), while the ring resonator cross section is sketched in Figure 31(b).

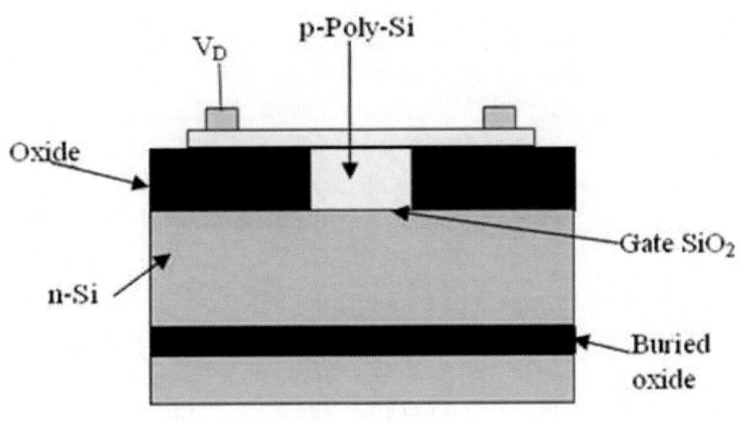

Figure 31(a). Cross-section view of output waveguide.

The waveguide can include a n-type crystalline silicon slab and a p-type doped poly-silicon rib with a thin gate oxide sandwiched between them. The structure includes a wide poly-silicon layer on the top of oxide layers on both sides of the poly-silicon rib, as in Figure 31(a).

Aluminium contacts are assumed as deposited on the top of this poly-silicon layer. The oxide regions on the either side of the rib maintain optical confinement and prevent optical field from penetrating into the contact area. The field distributions for both polarizations have been obtained by 2D full-vectorial finite element method (FV-FEM). In this case, poly-silicon rib and gate oxide widths are assumed both as 2.5 μm, total poly-silicon thickness at the waveguide middle as 0.9 μm, poly-silicon layer 10.5 μm wide, 1.4 μm n-type crystalline silicon slab and p-type doped poly-silicon rib with 12 nm gate oxide sandwiched between them. In this example, the effective indices are 3.246023 and 3.242125, respectively.

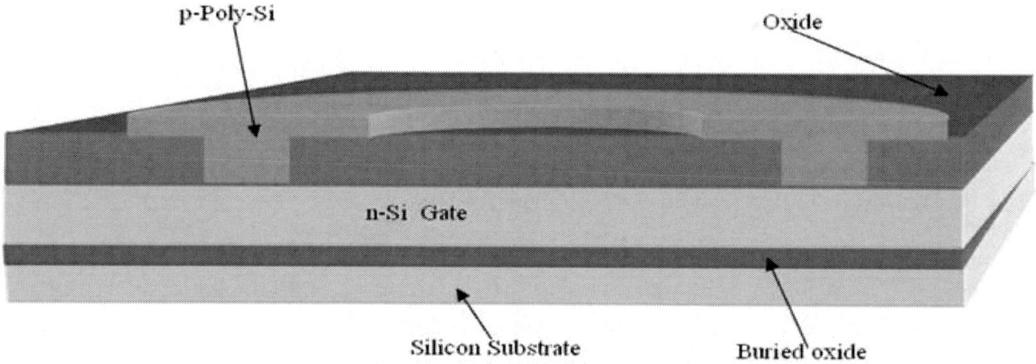

Figure 31(b). Cross-section view of ring resonator.

Both optical modes are highly confined, i.e. they do not penetrate into the contact area reducing the metal contact loss. Since quasi-TE mode lead to an best optical confinement in the rib structure, then it is more convenient to use this optical mode to increase the sensitivity of the structure to the clouds of charge and, thus, to the external electric field. Another investigated possibility is that to design a polarization-insensitive device.

The MOS-type straight waveguide has been also alternatively designed (second variant) in order to support only the quasi-TE fundamental mode with slab height h=1.4 μm, poly-Si rib height 1.6 μm, total rib height 1.7 μm, and rib width w =1.8 μm, as in Figure 32.

Propagation properties have been evaluated by using FEM and a mixed numerical procedure based on conformal transformation. The cross-section includes a n-type crystalline silicon slab with $N_D=1.7\times10^{16}$ cm^{-3}, a p-type doped poly-silicon rib with $N_A=3.5\times10^{16}$ cm^{-3} and a thin oxide sandwiched between them, 10 nm thick. A buried oxide buffer layer, t_{SiO2}=1 μm, has been considered. Moreover, the structure includes two strongly doped poly-silicon thin layers (100 nm) to assure ohmic contacts with the metallized regions, one on the rib top and the other for ground contact. Aluminium contacts 0.5μm thick are assumed. The waveguide parameters have been selected in order to reduce as much as possible the influence of the metal on the propagating field and the loss due to p-type highly doped region. The p- and n- doping levels have been chosen in order to have an efficient MOS behavior without inducing high optical losses.

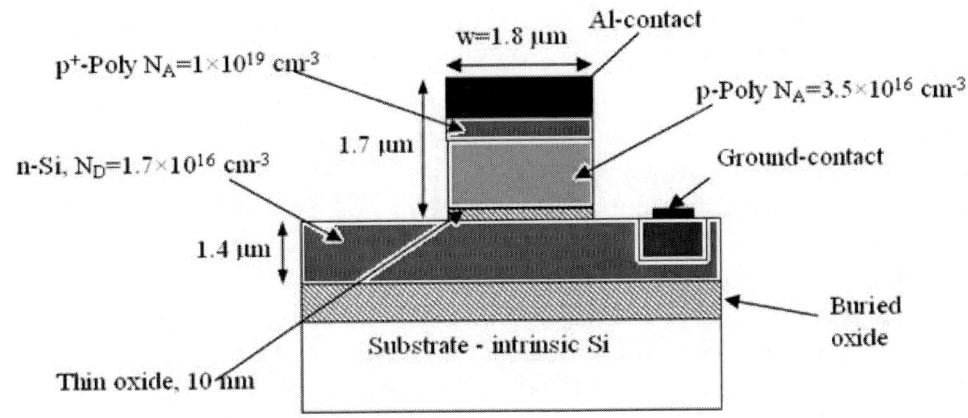

Figure 32. Cross-section of the MOS-type SOI straight waveguide.

In general, the guidelines in design of the straight waveguide are related to both choice of type and concentration of doping of the bulk silicon and poly-silicon and optimisation of the optical mode spot size to increase its overlap with the charge layers close to the thin oxide.The optimum region of operation of the MOS capacitor is the accumulation one. In fact, if the applied voltage quickly changes (high frequency), then the minority carriers in case of inversion regime cannot be generated at the oxide semiconductor interface to obtain the carrier equilibrium density. As a result, inversion layer is not formed. In this case, the applied voltage will cause a deep depletion layer of the semiconductor, then the MOS capacitance (and device sensitivity) will be very small. In the accumulation regime, it is important to prevent the depletion effect of the poly-silicon since it induces a negative slope in the C-V characteristic. Therefore the poly depletion effect should induce a decreasing of the slope of $\Delta n_{eff}\left(V_a\right)$ and, again, a reduction of the sensor sensitivity.

To avoid this depletion effect, we could in principle choice a waveguide having either <n-bulk, p-poly> or <p-bulk, n-poly> combination. However, the two possibilities are not equivalent because of their different behaviour in the frequency domain. In fact, the former solution <n-bulk, p-poly> leads to optimise the sensor bandwidth due to the higher mobility of the accumulated electrons versus holes. The optical simulations performed by FEM have also taken into account the charge layers on top and bottom of the thin oxide. The thickness and the free carrier concentration of the charge layers have been evaluated by means of our electric model as a function of the applied voltage and the induced refractive index change calculated. By this way, effective index $n_{eff,w}$ and $dn_{eff,w}/dQ$ can be evaluated for each value of applied voltage V_a. To validate the numerical procedure and have confidence with results, results given by our optical and electrical simulations have been compared with theoretical and experimental values presented by Intel Company [128] by assuming the same structure with 1.4 μm n-type crystalline silicon slab and p-type doped poly-silicon rib with a 12 nm gate oxide sandwiched between them. The poly-silicon rib and the gate oxide widths are both 2.5 μm, and the total poly-silicon thickness at the waveguide centre is 0.9 μm. The structure presents a wide (10.5 μm) poly-silicon layer on the top of oxide layers on both sides of the poly-silicon rib. The comparison for both electrical and optical parameters has shown a very good agreement [96].

Directional coupling between the straight waveguide and the ring resonator has been investigated by BPM, by considering the straight section 100 μm long and ring resonator radius of 150 μm. Highly efficient coupling between these two high-confinement waveguides was found, the effective coupling length being much less than 100μm, so making negligible the influence of the electrodes far from this coupling region. A coupling efficiency ranging from 2 to 1.3% has been achieved for a gap g in the range 2 ÷ 2.5μm. The results have been also compared with those obtained by coupled mode theory approach for curvilinear waveguides, showing a good agreement.

The propagation loss has been accurately calculated by using FEM and verifying results by BPM. The absorption as induced by doping has been considered by using the Soref's equations [54], then the imaginary part of refractive index of each layer has been evaluated by using the relationship $\mathrm{Im}(n) = \Delta\alpha \cdot \lambda / 4\pi$, and this index used to calculate again the mode effective index and its propagation loss factor. Loss coefficients for our structure were about 0.16 cm^{-1} (ring) and 0.176 cm^{-1} (straight waveguide). An additional loss of 0.5 dB/cm (0.115 cm^{-1}) due to propagation into bulk silicon has been also considered. Moreover, we have evaluated a scattering loss due to sidewall roughness of 0.43 dB/cm for the ring resonator. Finally, total optical loss of the order of 1.56 dB/cm for the ring resonator has been evaluated.

As previously discussed, the resonant wavelengths of the ring resonator depends on the effective index of waveguide modes. This index can vary due to application of an electric field on the antenna electrodes according to the following relationship:

$$\Delta n_{eff} = \frac{dn_{eff}}{dQ}\frac{dQ}{dV}\Delta V = \frac{dn_{eff}}{dQ}C(V)\cdot(V_D - V_{FB}) \tag{8}$$

where dn_{eff}/dQ_{acc} is the device charge efficiency, strongly affected by the gate position within the optical mode and by waveguide sizes, while V_{FB} represents the flat-band voltage of the MOS capacitor. $C(V) = dQ/dV$ is the capacitance of the MOS structure and depends on the applied voltage. In particular, if we assume $C(V) = C_{acc}$ = cost, the device works in accumulation regime, being Q_{acc} the accumulated gate charge as induced by the external electric field (corresponding voltage V_D) and C_{acc} the accumulation capacitance. We have developed a very generalized model for the calculation of MOS capacitance in any operation regime (depletion, accumulation) and for any structure parameter (doping, oxide thickness, oxide charge, temperature,…). This allows us to design the sensor in each regime where $C = C(V_D, V_{FB})$, i.e. for any voltage and dynamics to be measured. Thus, the device simulation shows that, increasing the applied voltage starting from the V_{FB} value, the spectrum shifts to larger wavelengths around the resonance, being the transmittivity peaks strongly influenced even by very small changes of the voltage. Therefore, highly sensitive detection of the external electric field can be made by either monitoring the shift of the resonant wavelength using diffraction grating or CCD matrix (wavelength interrogation), or by measuring the output light intensity change at a fixed wavelength (amplitude interrogation). To further increase the sensitivity of the sensor, it is appropriate to use a laser source having an operational wavelength matched to the point of maximum gradient of the resonance line shape.

The design of the whole sensor requires in general a number of trade-offs and feedback links among all modeled parameters of the architecture components regards to their optical and electrical behavior. After defining the design guidelines for the waveguides and the sensor conceptual geometry, a number of choices have been made in order to improve the device performance as much as possible. The main characteristics considered as required in environmental measurements are: very high stability and sensitivity, linearity, and frequency broadband response. The first choice has been the definition of the cross-section of the output waveguide (FP cavity). At this step, the target was the design of single-mode waveguide with optical field peak close to the thin oxide, to achieve better sensitivity of MOS structure. The next step was to determine the best operation region of the sensor in order to maximize again the sensitivity. In this situation, the minimum value of the external field detectable is $\Delta E_{\min} = \Delta T_r / S_E$, where ΔT_r is the minimum change of optical power (normalized to the input power) that can be measured by the photo-detector, depending on its signal-noise ratio. The term S_E is defined as:

$$S_E = \frac{\partial(T_r)}{\partial E} = \frac{\partial(T_r)}{\partial V_a} \cdot \frac{\partial V_a}{\partial E} = \frac{\partial(T_r)}{\partial \delta_1} \cdot \frac{\partial(\delta)}{\partial V_a} \cdot G = S_V \cdot G = \frac{\partial(T_r)}{\partial \delta_1} \cdot \left(k_0 \cdot 2L_1 \cdot C(V_a) \cdot \frac{dn_{eff}}{dQ} \right) \cdot G$$

where G is the effective length of the dipole antenna connected to the sensor electrode. Therefore, to reduce $\Delta E_{\min}$, it is needed to design the structure to maximize the term S_E. In order to increase S_E a large G is also required. In the voltage range around -1÷1 V, the MOS capacitor works in depletion region and significant changes of transmittivity cannot be detected. This is due to the minimum value of the MOS capacitance, giving a low variation of effective index. For voltages less than about 1 V, the MOS capacitor is in inversion region and the sensitivity is still low.

Therefore, the best region is the accumulation one ($V_a > V_{FB}$). The transmittivity is improved by decreasing the oxide layer thickness from 15 to 5 nm. In these calculations the quantum mechanics effect (tunneling through the oxide layer), which occurs for very thin layers (< 10 nm) is also considered. To incorporate this quantum effect, a compact formula based on a detailed quantum mechanics analysis has been used. It gives the oxide effective thickness which is used to correct the physical thickness in the classical model.

To obtain directly the value of the applied voltage from the optical detected power, a linear characteristic of the transmittivity versus applied voltage (T_r,V_a) is needed. To this aim, we can bias the MOS capacitor with a constant voltage and evaluate the range of external voltages where the (T_r,V_a) characteristic is approximated by a linear function. By biasing the device with 5 V, the characteristic is linear in the range -1÷1 V around 5V, assuming an absolute error less than 0.05 (amplitude interrogation).

Finally, an estimation of the device bandwidth has been obtained. In this device, the relatively slow carrier recombination process, which limits the speed of p-i-n diode structures in SOI technology, does not significantly influence the MOS structure dynamics. Then, the sensor bandwidth is governed by antenna output impedance, MOS capacitance and parasitics effects in the charge layer. Assuming an antenna resistance of 50 Ω, the device 3-dB bandwidth has been evaluated as about 500 MHz.

Some considerations have been also made about the thermal effects over the sensor behavior. These effects are significant along the FP cavity, where the SOI waveguide heating induced by Al electrodes occurs. Because of the temperature increase in the waveguide, the effective index change is modified as:

$$\Delta n_{eff} = \frac{dn_{eff,w}}{dT}\Delta T + \frac{dn_{eff,w}}{dQ}\frac{\partial C}{\partial T}\Delta T \cdot V_a + \frac{dn_{eff,w}}{dQ}C(V_a)\cdot V_a + \frac{\partial}{\partial T}\left(\frac{dn_{eff,w}}{dQ}\right)\Delta T \cdot C(V_a)\cdot V_a$$

By applying again a classical model of MOS capacitor, we have verified that the changes of MOS capacitance and $dn_{eff,w}/dQ$ with temperature are negligible in a large range of values. The main contribution is the first one, due to thermo-optic effect in SOI waveguides. It has been found by calculating the temperature and thermal gradient distributions in the waveguide cross section by a FEM thermal model. Then, the refractive index of each layer has been changed by considering the thermo-optic coefficient of the relevant material (i.e. 1.86 10^{-4} K^{-1} for silicon and 10^{-5} K^{-1} for silicon oxide) and, finally, the mode effective index has been calculated again. The effective index change with temperature is rigorously linear and the estimated value for our structure has been $dn_{eff,w}/dT$ = 1.645 10^{-4} K^{-1}. Therefore, a temperature change of 10 K will produce in our structure a very small wavelength shift, around 11.3 10^{-3} nm. This could be compensated by a Peltier cell, or taken into account during the calibration step of the device [96].

Another significant variant of dual-cavity sensor architecture consists of substituting the MOS structure with a PIN diode, similarly to [129]. The PIN cross section is sketched in Figure 33 and its doping concentration profile was obtained by 2D FEM [130].

To evaluate the change of free carrier concentration versus applied voltage, a numerical code based on Maxwell's equations and Boltzmann transport theory has been used, by assuming the absence of magnetic fields and a constant density of states. The problem is formulated in the model using three variables, i.e. ψ, n, ρ. These variables are related to each other by means of the following three Poisson semiconductor equations:

$$-\nabla\cdot(\varepsilon\nabla\psi) = q(p-n+N)$$

$$-\nabla\cdot\mathbf{J}_n = -qR_{SRH} \qquad (9)$$

$$-\nabla\cdot\mathbf{J}_p = qR_{SRH}$$

where ψ indicates the electrostatic potential, q is the elementary charge, p and n are the hole and electron concentrations, respectively, and N is the fixed charge. The term R_{SRH} represents the Shockley-Read-Hall recombination, depending on hole and electron concentrations and carrier lifetimes.

It is possible to write the electron and hole current densities, J_n and J_p, in terms of ψ, p and n by means of $\mathbf{J}_n = -qn\mu_n\nabla\psi + qD_n\nabla n$ and $\mathbf{J}_p = -qp\mu_p\nabla\psi - qD_p\nabla p$, where μ_n and μ_p are the carrier mobilities, and D_n and D_p are the carrier diffusivities for electrons and holes, respectively. This nonlinear problem has been solved using 2D FEM, assuming symmetric boundary conditions for boundaries in contact with the insulator, Neumann boundary conditions for boundaries far away from the active device area and fixed electrostatic potential for boundaries in contact with the metal. It is worth noting that, increasing the applied voltage, an enhancement of free hole concentration in the intrinsic regions of PIN structure (see regions I, II, III in Figure 33) can be observed. This increment of free carrier concentration induces a refractive index change according to well-known Soref's equation (2). The average refractive index change in regions I, II, III, has been evaluated versus applied voltage ranging from 0 to 0.5V. The maximum change of refractive index is seen as supported by the rib layer in contact with the p^+ region, where the voltage is directly applied. Complementary behavior could be observed for free electrons. However, the refractive index change evaluated for $V_a = 0.5$ V in the intrinsic regions induces a very small effective refractive index change.

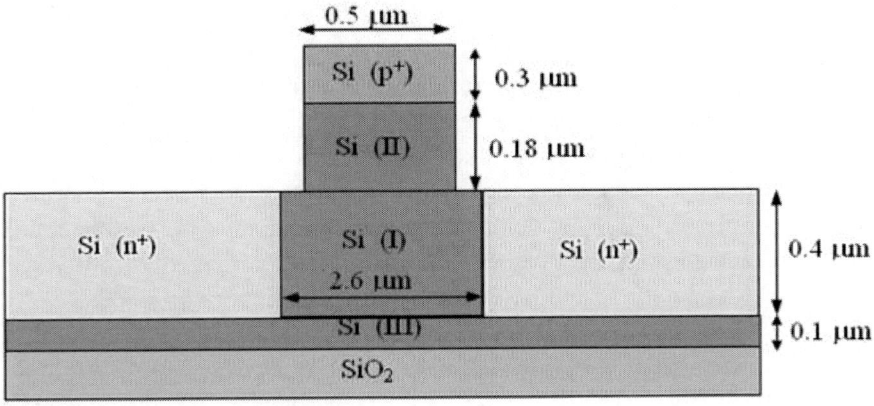

Figure 33. Cross section of PIN waveguide.

In fact, optical simulations performed by means of 2D FEM demonstrate a small change as $\Delta n_{eff} = n_{eff}(0.5V) - n_{eff}(0V) = -10^{-6}$. Then, larger voltage values are needed to induce significant refractive index changes and resulting resonance wavelength shifts to be measured.

6. Photonic Sensors for Biochemical Applications

Photonic and electronic micro- and nano-sensors are emerging as very attractive devices to be employed in a great number of application fields such as medicine, microbiology, particle physics, automotive, environmental safety and defence.

In the last few years, a great effort has been devoted to sensor and actuator monolithic integration and micro-electro-mechanical system (MEMS) and micro-opto-electro-mechanical system (MOEMS) development. In the same period, optical sensors have attracted considerable attention because of their immunity to electromagnetic interference, good compactness, robustness and high compatibility with fiber networks, but also because they usually exhibit shorter response time and higher sensitivities with respect to MEMS/MOEMS devices. Sensors used for biochemical or biomedical applications (e.g. bio-molecules concentration measurement, DNA molecules investigation and pH estimation) must reach some specific performance requirements which are significantly different from those typical for other sensors. In particular, a biosensor has to be contextually highly sensitive and selective to the analyte being detected, to be biocompatible as well as immune to environmental influences, such as pressure or temperature changes. Photonic technologies enable to satisfy these requirements allowing to fabricate very compact, highly performing and low cost biosensors.

A biochemical sensor is a device consisting of a biologically or biophysically-derived sensing element (usually named bio-receptor) integrated with a physical transducer that transforms a measurand into the output signal, as in Figure 34.

As bioreceptor, molecular species (antiboby, enzyme, protein or nucleic acid) or a biological system (cells, tissue) are commonly employed to recognize biochemical substances. Transduction mechanisms are usually electrochemical, mass-based or optical. Electrochemical detection is commonly based on the chemical potential of particular species in solution (analytes) measured by comparison with a reference electrode. Mass-based detection exploits the change in oscillation frequency of a piezoelectric crystal which depends on applied electric signal frequency and crystal mass.

A bioreceptor is a biological molecular specie or a biological system that adopts a biochemical mechanism for analyte recognition. The most used bioreceptors are based on antibody-antigen, enzymatic, nucleic acid, or cellular interactions [131]. The antibodies are complex biomolecules formed by hundreds of aminoacids arranged in ordered sequence. Exposing immune system cells to antigen substances, antibodies are produced.

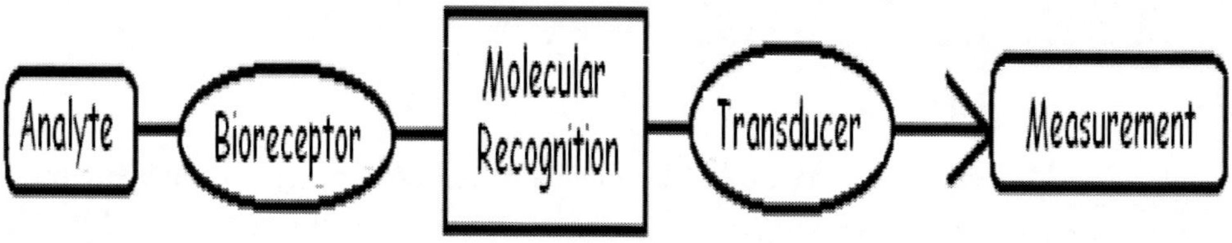

Figure 34. Biochemical /biomedical sensors operating principle.

The interaction between an antibody and the relevant antigen is highly specific because their molecular structures are complementary and antigen-antibody bond is very stable. Therefore antigen-antibody interaction is very rapid and antigen-antibody complex is characterized by a reasonable lifetime. Antigen-antibody reaction specificity enables to use antibodies as specific detectors capable to sense the presence of interesting analyte, also when its amount is very small and if a great number of other chemical substances are present in the sample.

6.1. Optical Transduction Mechanisms

Fluorescence phenomena are related to an energetic transition from an excited state to the ground state producing photon emission. At equilibrium, organic molecules have the lowest possible energy. When a molecule absorbs electromagnetic energy, it moves to an excited energy level (usually a singlet state). The energy level reached by the molecule absorbing electromagnetic energy depends on the incident radiation wavelength. A molecule in an excited state tends to reach a lower energy level. Relaxation of the molecule from an excited state may occur in different manner. If relaxation takes place by photon emission without any change of spin multiplicity (from a singlet state to another singlet state), the transition is known as fluorescence. Emitted photon wavelength depends on loss energy by the molecule in the relaxation process.

The principle scheme of typical fluorescence-based biosensor is shown in Figure 35 [132]. In this sensor, the optical signal provided by a laser propagates in an optical fiber until fiber section on which antibodies are immobilized. The optical radiation produced by antigen fluorescence is sent back to the detection system. This sensor can be used to detect the carcinogen benzo[a]pyrene (BaP) or to distinguish BaP and benzopyrene tetrol (BPT) [133].

Fluorescence-based optical sensors are used in biochemical analyses of other chemical compounds and in *in vivo* and *in vitro* analyses of tissues and cells (for example, neoplastic cells). Moreover, fluorescence is widely used in DNA sequencing (individuation of nucleotides sequence constituting DNA molecules), because it enables to reach a high readout speed [134].

Surface plasmon resonance is a charge-density oscillation that may exist at the interface of two media with dielectric constants of opposite signs, for instance a metal and a dielectric. The charge density wave is associated with an electromagnetic wave, the field vectors of which reach its maximum at the interface and decay evanescently into both media. This surface plasma wave (SPW) is a TM-polarized wave (magnetic vector is perpendicular to SPW propagation direction and parallel to interface plane). When this wave interacts with a sample, SPW phase changes. Measuring this phase shift it is possible, for example, to investigate the interaction kinetics between a protein and an antibody and detect DNA hybridization [135-136].

Raman spectroscopy is based on Raman effect that results from energy exchange between incident photon and scattering molecules. In Rayleigh (or elastic) scattering, incident photon energy does not change after the collision and the scattered photon preserves the identical frequency as the incident photon. In Raman (or inelastic) scattering, energy is transferred from the molecule to the photon or vice versa. When the scattered photon has more energy than the incident photon, the process is referred to as Raman to Stokes scattering. When the scattered photon has less energy than the incident photon, the process is referred to as Raman to anti-Stokes scattering. Raman signals are usually weak and require powerful sources and sensitive detectors. Only a limited number of biological molecules contributes to tissues fluorescence whereas most biological molecules are Raman active, so Raman spectroscopy can overcome some limitations of fluorescence, for example, in cancer diagnosis.

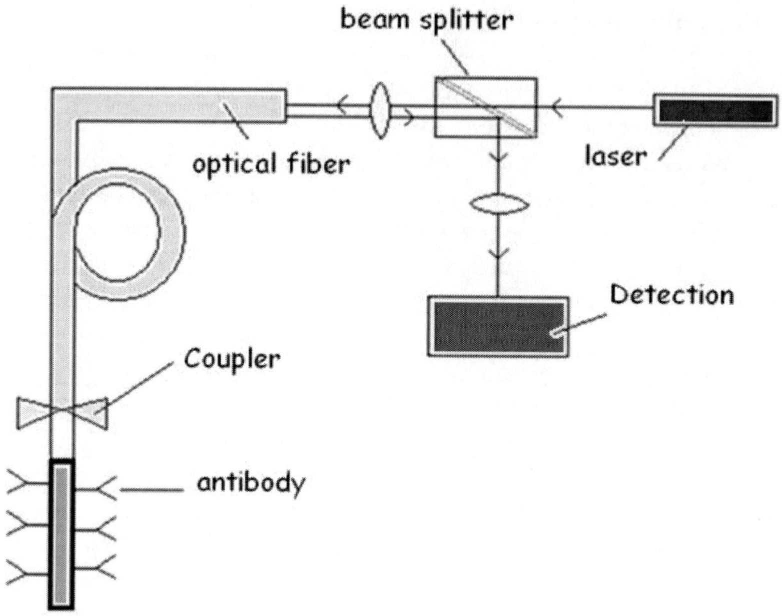

Figure 35. Typical fluorescence-based photonic sensor.

Figure 36 shows a typical apparatus used in Raman spectroscopy [137]. It consists of a laser (wavelength 785 nm, power 300 mW) providing the optical signal used to induce Raman effect, a fiber that collects optical signals exiting from the sample, and a spectrograph connected to a processing unit.

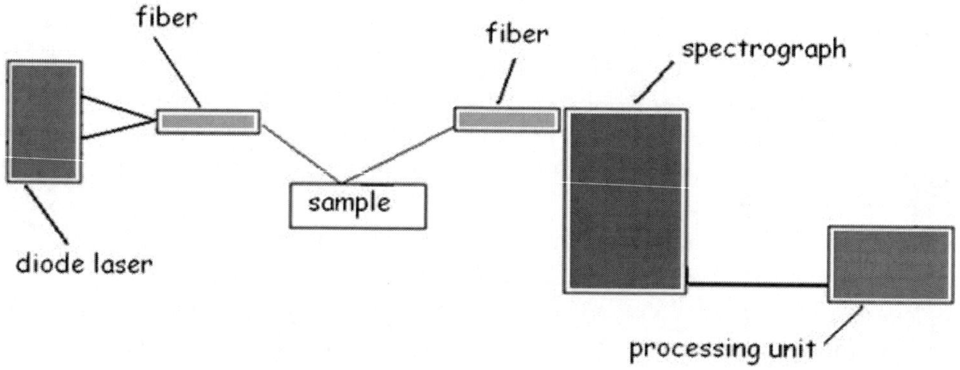

Figure 36. Typical apparatus for Raman spectroscopy.

For instance, Raman spectroscopy has been used for human immunodeficiency virus (HIV) detection [138] and DNA fragments sequencing [139]. The Lambert-Beer law relates solution absorbance A to analyte dispersed in this solution concentration C, by the following relation:

$$A = \varepsilon LC \tag{10}$$

where L is the optical path length and ε is the wavelength-dependent molar absorption coefficient. Measuring solution absorbance, it is possible to estimate some bioanalyte concentration in aqueous solutions. This transduction mechanism is used in a number of integrated optical and fiber biosensors to sense biological pathogens or gases (as oxygen or carbon dioxide) concentrations and measure pH in solutions [140].

Concentration of some biochemical substances (for example, glucose) in aqueous solutions can be measured by monitoring the solution refractive index in wavelength infrared region. This transduction mechanism is widely used in guided-wave optical sensing because it permits to induce a phase shift in optical signal propagating in the guiding structure.

6.2. Integrated Optical Biosensors

Integrated optical sensors usually adopt a waveguide to confine optical power. These devices are based on evanescent field sensing. Although light is confined within the guiding film of the waveguide, there is a part of the guided light that travels through a region extending outside into the medium surrounding the waveguide, and can interact with external environment. This kind of optical sensors exhibits very interesting advantages with respect to other optical sensors, such as compactness, very high sensitivity and possibility of mass production. In a large number of integrated optical biosensors proposed in the last few years, the analyte presence either produces a shift in waveguide cover medium refractive index (*homogeneous sensing*) or determines the thickness increase of a molecular layer deposited on guiding film/cover medium interface (*surface sensing*). Both phenomena affect the propagating optical mode effective index which is measured in different ways, according with biosensor architecture.

In general, interaction between sample and optical signal propagating in the sensor produces an optical mode effective index change and, then, a change of its phase. To convert this phase shift in an amplitude change, interferometer architectures are commonly used. Among these architectures, widely used Mach-Zehnder assures very high sensitivity. In this kind of integrated optical biosensors (see Figure 37), input optical signal is split by a Y-junction in two signals having the same power.

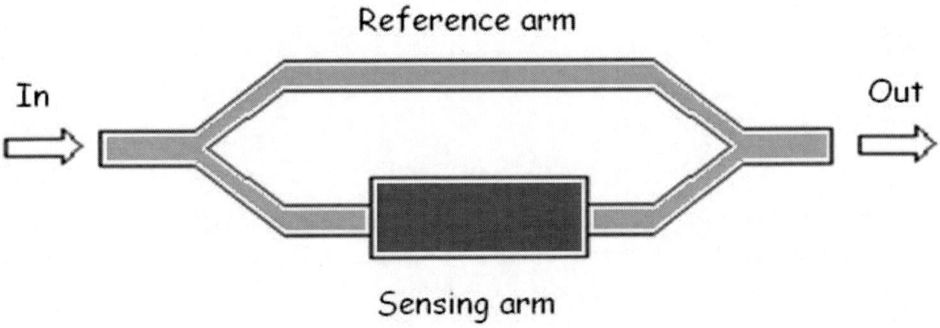

Figure 37. Mach-Zehnder interferometer architecture.

These two signals propagate in the reference and sensing arms (in this last arm the interaction between sample and optical signal takes place). After the propagation in both arms, the two optical signals accumulate a phase shift $\Delta\varphi$. Phase shifted optical signals interfere by an output Y-branch and normalized optical power at the output of this Y-branch is given by:

$$\frac{P_{out}}{P_{in}} = 1 + cos(\Delta\varphi) \tag{11}$$

where

$$\Delta\varphi = 2\pi\frac{L}{\lambda}(N_s - N_r) \tag{12}$$

L is the reference and sensing arm length, λ is the optical signal wavelength, N_s is the effective index of mode propagating in sensing arm and N_r is the effective index of mode propagating in reference arm. In the last two decades, a great number of integrated optical sensors based on Mach-Zehnder interferometers (MZI) have been realized to detect different biochemical species, such as organic compounds [141-142] and proteins [143-145].

CMOS-compatible technologies are often employed for fabrication of these sensors (guiding film in silicon, silicon nitride or silicon oxynitride), but glasses [146-147] and III-V semiconductor compounds [148] technologies have been also proposed. The sensitivity S of MZI sensors can be expressed, assuming that $\Delta\varphi$ varies in a narrow range around $\pi/2$, as:

$$S = -\frac{2\pi}{\lambda}LS_w \tag{13}$$

where S_w is the waveguide sensitivity, defined in case of homogeneous sensing, as:

$$S_w = \frac{\partial N}{\partial n_C} \tag{14}$$

and, in case of surface sensing, as:

$$S_w = \frac{\partial N}{\partial d} \tag{15}$$

where N is the effective index of optical mode propagating in guiding structure, n_C the cover medium refractive index and d the thickness of molecular layer deposited on guiding film/cover medium interface (this will be clarified in the last section on waveguide sensitivity). Moreover, sensitivity of integrated optical MZI sensors depends on sensing arm length L, i.e. it is limited only by the undesired shift in propagating mode effective index due to, for example, any temperature change. Thus, a trade-off between device length and sensitivity has to be usually found. A MZI optical biosensor has been fabricated, capable to monitor antigen-antibody immunoreactions with a minimum detectable change in cover medium refractive index equal to 7×10^{-6} [102]. A detection limit equal to 1.5×10^{-6} has been obtained by a MZI optical biosensor fabricated in a BGG36 glass substrate [149].

Other configurations adopted for interferometer integrated optical biosensors are those based on Young interferometers. Using a four-channel Young interferometer, the simultaneous and independent monitoring of tree antigen-antibody reactions has been achieved [150]. In this device the generated interference pattern is recorded by a CCD camera and analyzed by Fast Fourier Transform (FFT) algorithm, obtaining a detection limit of 8.5×10^{-8}.

Anti-Resonant Reflecting Optical Waveguide (ARROW) is a five-layer guiding structure where light is confined within the core layer by total internal reflection at the cover medium/core interface and by anti-resonant reflection (reflectivity of around 99.96%) due to the presence of the two cladding layers underneath the core, as it is shown in Figure 38. ARROW is a leaky guiding structure with virtual mono-mode behaviour, where higher order modes are filtered out by loss discrimination due to the low reflectivity assured to these modes by both cladding layers. The refractive indices and thicknesses of the cladding layers have to be properly designed for the working wavelength to assure high reflectivity and, therefore, good guiding characteristics. ARROWs are usually fabricated in a CMOS-compatible technology and the materials commonly used include silicon for substrate, silicon oxide for core and second cladding layer and silicon nitride for first cladding.

ARROWs exhibit low losses, permit larger dimensions than conventional total internal reflection-based waveguides (micrometers instead of nanometers), enable to confine optical power in low index material (as silicon oxide) and exhibit good sensitivity. These features are quite attractive for sensing applications, so ARROWs employment in highly sensitive and low cost integrated optical sensors has been discussed [151-153]. A Mach-Zehnder interferometer adopting an ARROW as guiding structure has been used to fabricate a 15 mm long optical biosensor, with detection limit 2×10^{-5} [154].

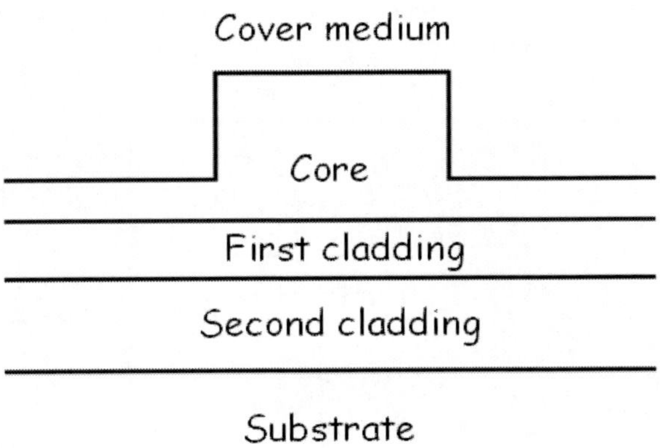

Figure 38. ARROW waveguide structure.

Another possible approach is based on hollow waveguides. They permit to simultaneously confine the optical power and the solution to be analyzed in the waveguide core constituted by a low-index liquid substance. In this kind of waveguides, the optical signal is confined inside the core (whose refractive index is around 1.33) by two cladding layers designed to produce an anti-resonant reflection (see Figure 39). A liquid core hollow waveguide has been adopted to realize an integrated sensor capable to detect refractive index changes in liquid substance constituting the core [155]. This sensor exhibits a minimum detectable change in core medium refractive index of 9×10^{-4}.

Moreover, integrated optical biosensors exploiting surface plasmon resonance (SPR) are constituted by a single mode waveguide produced, for example, by thermal ion-exchange in a glass substrate. A small region of the waveguide is coated with a thin metal film. This metal-coated part of the sensor can support a SPW and form the interaction region with the solution containing the analyte.

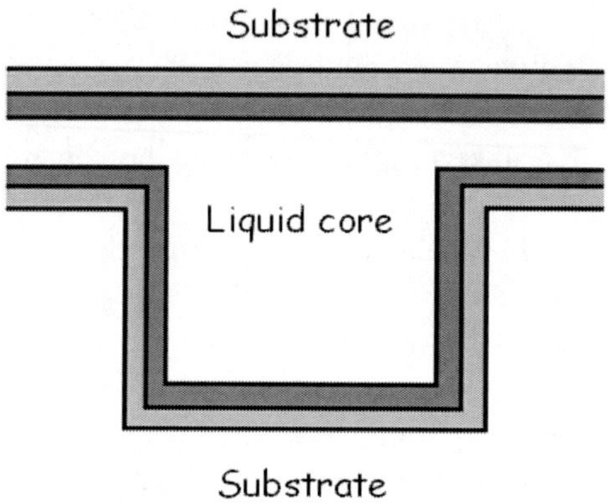

Figure 39. Hollow waveguide cross section.

The structure incorporates a Y-junction used as 3 dB power splitter and a quasi-TM mode is excited in the input waveguide. After passing through the 3 dB power divider, this mode couples across the step discontinuity between the metal coated and uncoated waveguide regions exciting the SPW. After propagating through the metal-coated region of the sensor, SPW excites the optical mode in the uncoated output waveguide (see Figure 40).

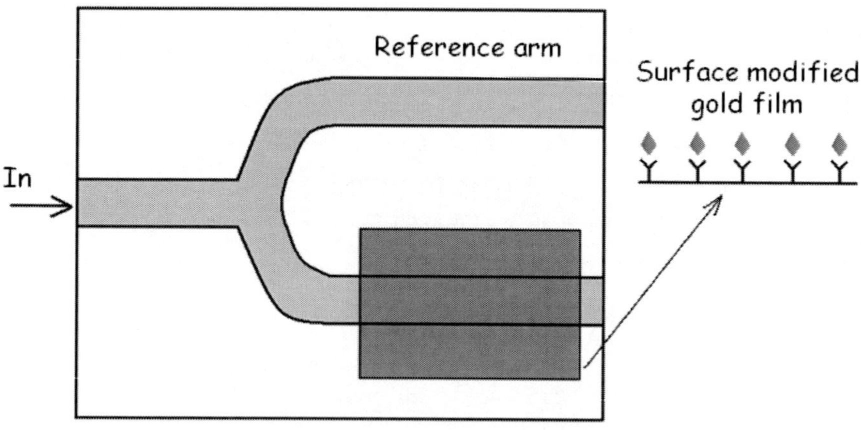

Figure 40. Integrated optical biosensor exploiting SPR scheme.

An integrated optical biosensor for simazine detection has been designed and fabricated [156]. This sensor exhibits a minimum detection of 1×10^{-4} and a detection limit of 0.1 μg/l for simazine concentration measurements. A minimum detectable change in cover medium refractive index of 1.2×10^{-6} is achievable by a SPR based integrated biosensor [157]. This sensor is able to detect 2 nm of human chorigonadotropin in 1 ml of 1% bovine serum albumin solution.

It is well known that a Bragg grating is a 3D periodic structure (Figure 41) whose refractive index profile can be expressed as:

$$n(x,y,z) = n_0(x,y) + \Delta n(x,y,z) \tag{16}$$

where $n_0(x,y)$ is the waveguide index distribution and $\Delta n(x,y,z)$ is the periodic refractive index perturbation, whose period Λ along z has to satisfy the following relation:

$$\Lambda = \frac{m}{2N}\lambda_c \tag{17}$$

where m is the grating order, λ_c is the centre wavelength of the spectral response and N is the effective index of the propagating mode. From Eq. (17) it is evident that a change in refractive index of guiding structure cover medium (and consequently in the propagating mode effective index) produces a centre wavelength shift of grating reflection and transmission spectra. This shift permits to measure biochemical quantities affecting the cover medium refractive index. Bragg grating employment as building block for fabrication of highly sensitive and very compact integrated optical biosensors has been proposed [103, 158]. For example, a 76 μm long first order Bragg grating has been fabricated using a Si_3N_4/SiO_2 waveguide and minimum detectable change in refractive index cover medium has been calculated as 4×10^{-5} [103].

Bragg grating-based device sensitivity is given by:

$$S \equiv \frac{\partial \lambda_c}{\partial n_C} = S_w \frac{\lambda_c}{2N} \tag{18}$$

where n_C is the cover medium refractive index and S_w is the already defined waveguide sensitivity, depending on sensing adopted mechanism (homogeneous or surface sensing). Using SOI sub-micrometer silicon-wire waveguides (see Figure 16), a very high S_w is obtainable, with a value of 0.31 experimentally proved [159]. For optical modes propagating in Si-wire waveguides whose effective indices are around 2 and λ_c = 1550 nm, a device sensitivity $S \sim 120$ nm is obtainable.

Moreover, microring, microdisk and micro-racetrack resonators are emerging as very attractive building-blocks for micrometer-scale Photonics, essentially because of their compactness and versatility. In fact, a great number of optical functions related to telecommunications systems even different from filtering, such as modulation, amplification and switching, can be performed using guided-wave micro-resonators. The use of optical micro-resonators in sensing applications is a relatively new research field. Their advantages are related to the device size reduction by some orders of magnitude, which does not compromise the device sensitivity, because the large photon lifetime within the resonator at the resonance provides an equivalently long interaction length to achieve a detectable phase shift. Light coupled to micro-resonators is confined within the structure by total internal reflection or as whispering gallery modes (WGMs), forming high quality factor resonant modes.

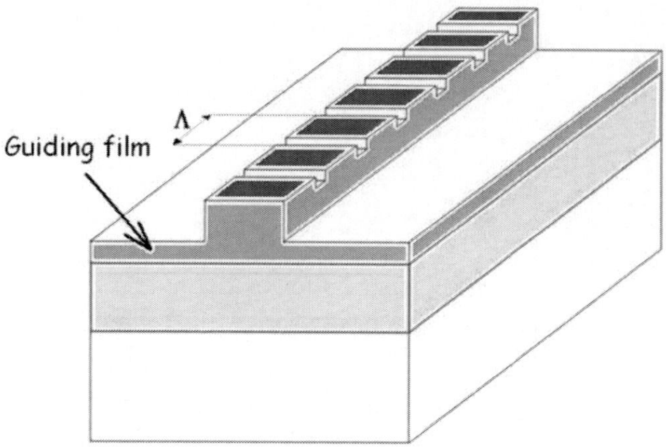

Figure 41. Bragg grating guided-wave optical structure.

Any interaction with the evanescent tail of the optical field (usually the change in waveguide cover medium refractive index) affects the guided mode, and thus changes the cavity resonance, as sketched in Figure 42. This change in resonance can be detected with very high sensitivity by optimizing both microcavity design and observation method. The capability of a microring resonator with SOI wire waveguide and radius of 3 μm to sense ammonia concentration changes has been theoretically demonstrated [160], whereas a design strategy to optimize Si_3N_4/SiO_2 microring-based integrated optical sensors has been proposed [161].

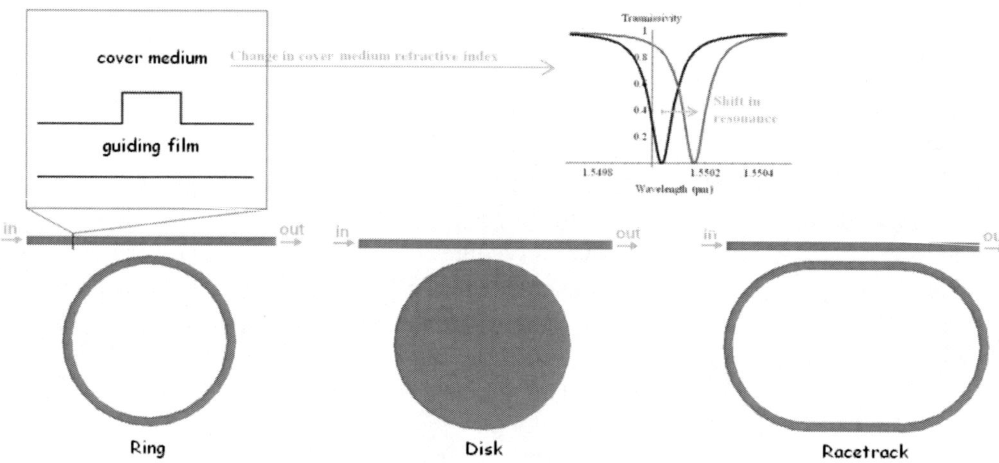

Figure 42. Microcavity-based integrated optical architectures for sensing.

A micro-racetrack resonator fabricated using polymeric materials, in which the resonance slope has been enhanced by introducing two partially reflecting elements (Figure 43), has been adopted for biochemical sensing of glucose [126]. This configuration produces a Fano-resonant lineshape that greatly enhances the device sensitivity, with a minimum detectable concentration change of glucose solution around 0.024%, or 24 mg/dl.

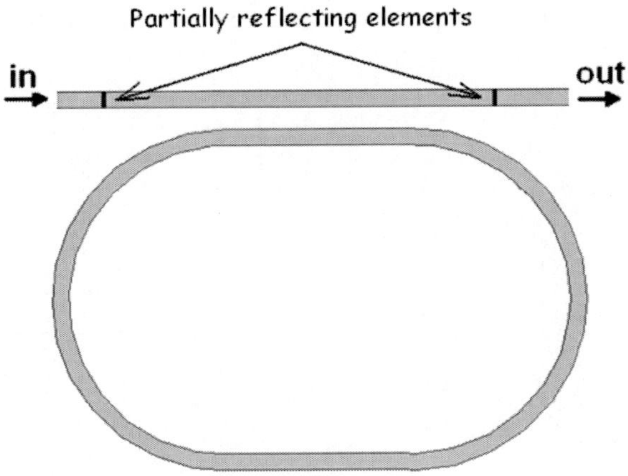

Figure 43. Micro-racetrack resonator with enhanced sensitivity by two partially reflecting elements.

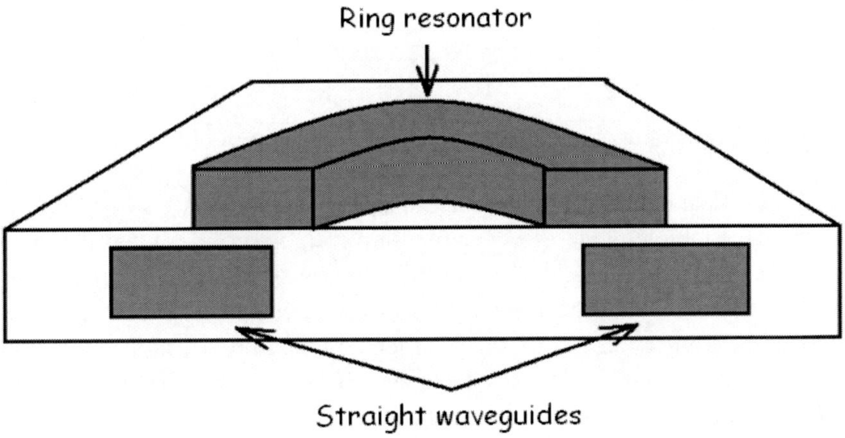

Figure 44. Scheme of vertical coupling between microring and straight waveguides.

A lowering of detection limit of glucose concentration measurement is obtainable by increasing the microcavity quality factor Q around 20,000, as estimated for the microring resonator fabricated using polymeric materials [104]. Thus, a detection limit of 0.915 mg/dl in glucose concentration measurement (homogeneous sensing) and 250 pg/mm^2 in spavidin molecule detection (surface sensing) can be achieved. Finally, the capability of this sensor to detect shifts of propagating mode effective index around 10^{-7} has been also proved.

Vertical coupling between straight waveguide and microring resonator is employed in other biosensors [162], as in Figure 44. Microring quality factor around 12,000 and detection limit of 1.8×10^{-5} in terms of minimum detectable change in effective index, can be obtained.

Adopting a Si_xN_y/SiO_2 waveguide, a racetrack resonator having a radius of 2 mm has been fabricated for protein detection [163]. This sensor is capable to detect very low concentrations of avidin in phosphate-buffered saline solution (detection limit of 0.1 nM). Adopting a microdisk resonator fabricated by CMOS-compatible technology (Si_3N_4/SiO_2 waveguide) and with radius of 15 μm, measurements of cover medium refractive index change as low as 10^{-4} were experimentally demonstrated [164].

Table 2. Comparison among architectures of integrated optical biosensors

Year	Author	Architecture	Technology	Detection Limit	Total Length
1997	B. Drapp et al. [149]	MZI	Glass	1.5×10^{-6}	15 000
1999	R.D. Harris et al. [156]	SPR	Glass	1×10^{-4}	2 000
2002	E. Krioukov et al. [164]	Disk resonator	CMOS (*)	1×10^{-4}	30
2003	F. Prieto et al. [102]	MZI	CMOS (*)	7×10^{-6}	30 000
2003	A. Ymeti et al. [150]	Young interferometer	CMOS (*)	8.5×10^{-8}	≈ 50 000
2003	F. Prieto et al. [153]	MZI (ARROW)	CMOS (*)	2×10^{-5}	15 000
2004	S. Campopiano et al. [155]	Hollow waveguide	CMOS (*)	9×10^{-4}	200 000
2005	W.C.L. Hopman et al. [103]	Bragg grating	CMOS (*)	4×10^{-5}	76
2006	F. Dell'Olio et al. [160]	Ring resonator	SOI	1×10^{-4}	6
2006	A. Yalçinc et al. [162]	Ring resonator	Glass	1.8×10^{-5}	120

(*) CMOS-compatible technology.

In Table 2 significant integrated optical biosensors are compared in terms of detection limit (minimum detectable change of cover medium refractive index) and device length. We note that interferometer configurations are usually the most sensitive but requiring very long structures (tens of mm). Photonic sensors based on micro-cavities are significantly more compact (with reduction in length of three orders of magnitude) and exhibit a sensitivity comparable to that of Mach-Zehnder based sensors. Bragg grating based biosensors are quite compact and very sensitive. Further, their sensitivity can be significantly enhanced using sub-micrometer SOI waveguides as guiding structures.

6.3. Coupled SOI Slot Waveguides

Nowadays, there is a current trend in SOI photonic integrated circuits and devices to move toward smaller dimensions for cost efficiency achievement and device performance improvement. In this scenario, a silicon nanometer guiding structure usually referred as slot waveguide [165] is attracting a considerable attention. In optical slot waveguides, fabricated in SOI technology by either e-beam or deep-UV lithography [166], the electric field discontinuity at the interface between high index contrast materials enables high optical confinement inside a nanometer-size area of low index material (e.g. air, silicon oxide, aqueous solutions, silicon nanocrystals, thermo-optic or electro-optic polymers).

This guiding structure can be realized nearing two silicon wires with nanometer sizes, as in Figure 45(a), supporting quasi-TE and quasi-TM modes, whose typical profiles are shown in Figure 45(b).

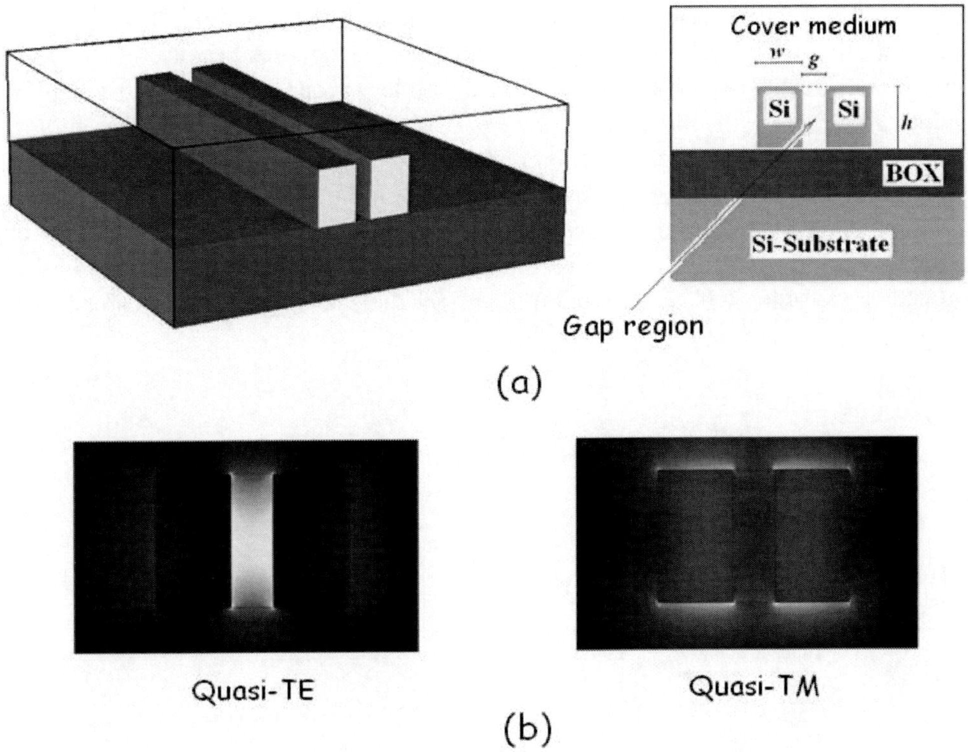

Figure 45. (a) Slot waveguide structure; (b) quasi-TE and quasi-TM modes confined in slot waveguide.

Thus, the field confinement in low index central region is very high for quasi-TE polarization. In the last three years, a great variety of optical devices has been proposed and realized using SOI slot waveguides, as it will be shown in chapter 2 of this book. Slot guiding structure effective index is very sensitive to cover medium refractive index changes. Adopting an aqueous solution as cover medium whose refractive index depends on analyte concentration (glucose or ethanol, for example), a waveguide effective index shift due to analyte concentration change can be observed. Monitoring slot guiding structure effective index in an appropriate integrated architecture (as Mach-Zehnder interferometer, ring resonator, Bragg grating, directional coupler, and so on), it is possible to fabricate highly sensitive and miniaturized integrated optical chemical sensors. Chemical and biochemical optical sensing have been proved by slot waveguide structures [167-168]. In this section, the use of slot waveguides to realize a SOI asymmetric directional coupler for optical chemical sensing is proposed. A mixed FEM+CMT procedure is employed for device modelling and optimization.

The proposed integrated sensor has an architecture as sketched in Figure 46. It includes two coupled parallel slot waveguides separated by a distance d between slot waveguide centres.This coupler is asymmetric because in one guiding structure (W1) a cover medium region (including also the slot gap) Teflon fluoropolymer (refractive index 1.31) is assumed as deposited, whereas in the other waveguide (sensing waveguide, W2) cover medium is constituted by an aqueous solution whose refractive index (around 1.33) depends on the chemical agent concentration (e. g. glucose) dispersed in the solution. When this concentration changes, a refractive index change of sensing waveguide cover medium occurs.

This variation range was assumed between 1.333 and 1.334, i.e. 10^{-3}. In turn, this shift induces a variation of optical power exiting from both waveguides because of the change of coupling conditions between the two optical modes propagating in the coupled slot waveguides. By measuring the optical powers exiting from both waveguides and/or their difference by two photodetectors, it is possible to estimate with good sensitivity the concentration of the chemical agent dispersed in the solution.

In a generic slot waveguide, the sensitivity S_w to cover medium refractive index change can be defined by Eq. (14). It depends on the electric field squared module fraction confined in the slot and, according with variational theorem for dielectric waveguides, can be written:

$$S_w \cong \frac{2n_c^0}{\eta_0 P} \iint_C |\mathbf{E}(x,y)|^2 = \frac{2n_c^0}{\eta_0} \Gamma_C \cdot \Pi \qquad (19)$$

where

$$\Gamma_C = \frac{\iint_C |\mathbf{E}(x,y)|^2}{\iint_\infty |\mathbf{E}(x,y)|^2} \text{ and } \Pi = \frac{\iint_\infty |\mathbf{E}(x,y)|^2}{P}$$

n_c^0 is the cover medium refractive index unperturbed value, η_0 is the free space impedance, P is the optical power carried by the propagating mode in the slot, C indicates the cover medium region (including also the slot gap region), and E is the electric field vector related to propagating mode (being x and y the transverse coordinates).

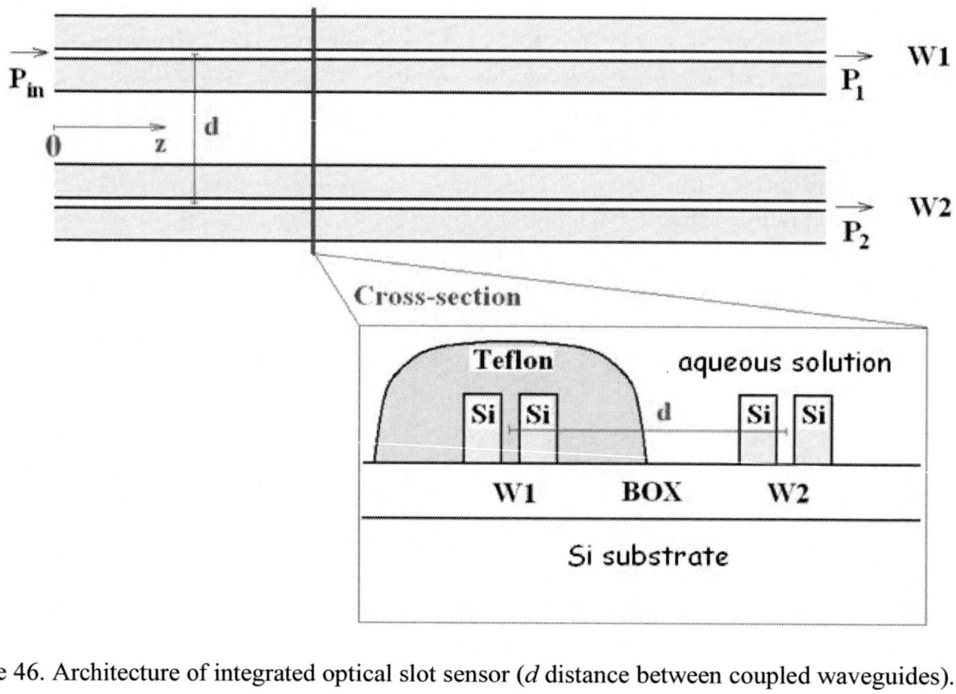

Figure 46. Architecture of integrated optical slot sensor (d distance between coupled waveguides).

Quasi-TE mode supported by a slot waveguide is highly confined in the gap region, as in Figure 45(b), and its confinement factor Γ_C in the cover medium (including gap region) is typically around 60-70 %, then its sensitivity to cover medium index change is usually very large. Quasi-TM mode is significantly less sensitive to cover index change and its confinement factor Γ_C is only around 40-50 %. SOI slot waveguide sensitivity was optimized by a very accurate FEM-based approach [167]. A guiding structure having a sensitivity $S_w > 1$ ($S_w = 1.0076$) for quasi-TE mode has been demonstrated, assuming silicon and silicon oxide refractive indices of 3.476 and 1.444, respectively. Sensitivity for quasi-TE mode of SOI slot waveguide adopted in our design is more than two times larger than that calculated for a Si_3N_4/SiO_2 slot waveguide adopted in another sensor [168]. Structure data are summarized in Table 3.

The directional coupler-based integrated sensor has been modelled by CMT developed for generic monomodal parallel waveguides [169] using symmetric and asymmetric supermodes. FEM has been adopted in calculation of mode effective indices because it assures a very high accuracy also for nanometer high index contrast guiding structures. In FEM mesh generation, triangular vector-elements are used, with about 100,000 elements. Changing the boundary condition from a perfect electric conductor to a perfect magnetic conductor, a negligible influence on the simulation results has been observed.

To validate the proposed modelling technique, the asymmetrical directional coupler behaviour has been 3D simulated by Eigenmode Expansion method (EME) [170] in case of quasi-TE polarization (more sensitive). Geometrical parameters reported in Table 3 have been assumed for coupled slot waveguides with $d = 1$ μm and aqueous solution as cover medium. In this simulation, cover medium refractive index is 1.333 for sensing waveguide (W2 in Figure 46), whereas it is 1.31 for the other one (W1).

Table 3. Parameters of coupled slot waveguides

Parameter	Value
Si-wire height (h)	324 nm
Si-wire width (w)	180 nm
Slot gap region width (g)	100 nm
Effective index (quasi-TE)	1.578638
Effective index (quasi-TM)	1.999899
Cover confinement factor (quasi-TE)	0.7644
Cover confinement factor (quasi-TM)	0.4117
Waveguide sensitivity (quasi-TE)	1.0076
Waveguide sensitivity (quasi-TM)	0.4040

Simulation of optical propagation within the coupler performed by 3D EME method is shown in Figure 47. For $z = 23.5$ μm we observe the maximum power transfer from *W1* to *W2* in terms of optical intensity distribution in the coupler cross-section, as shown in Figure 47(a)), whereas for $z = 47$ μm we observe the optical power as fully confined in W1, as in Figure 47(b). For $z = 23.5$ μm, normalized optical powers confined in W1 and W2 have been estimated as 0.3302 and 0.6698, respectively. For $z = 47$ μm, normalized powers confined in

W1 and W2 are equal to 0.9724 and 0.0276, respectively. Normalized optical powers confined in *W1* (P_1) and in *W2* (P_2) have been also calculated versus propagation direction and results are in Figure 48. The agreement between simulation results obtained by EME method and calculations performed by our modelling procedure (FEM+CMT) is very good. Then, coupling length between *W1* and *W2* is predicted to be $z = 23.33$ μm.

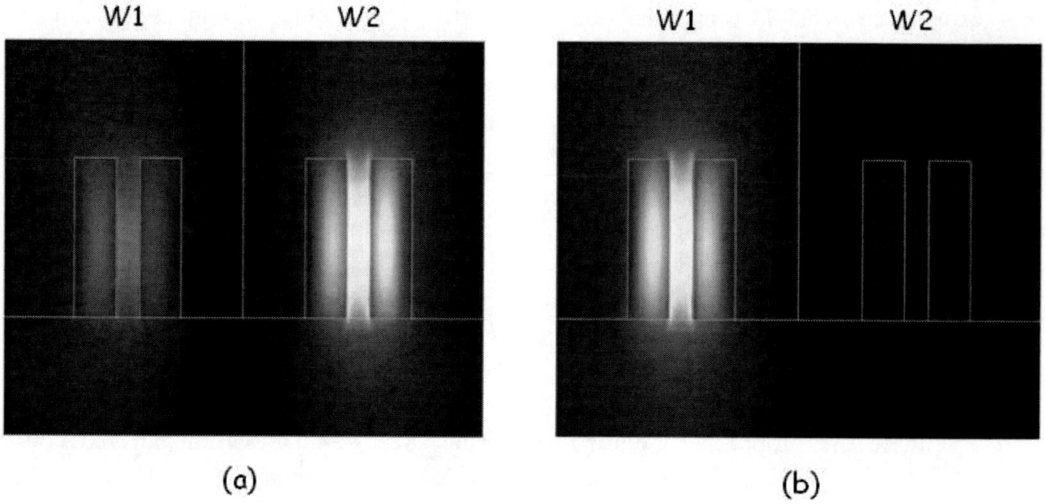

Figure 47. Optical intensity (quasi-TE mode by EME method) distribution in cross-section of asymmetrical directional coupler ($d = 1$ μm) at: (a) $z = 23.5$ μm; (b) $z = 47$ μm.

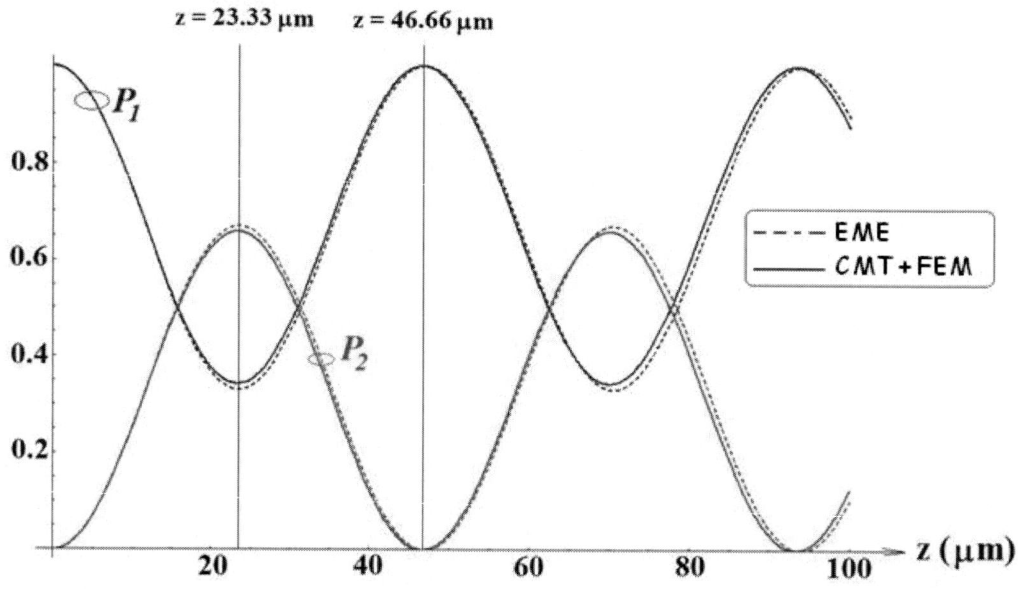

Figure 48. Normalized optical power confined in both coupled waveguides (with $d = 1$ μm) versus propagation length for quasi-TE mode (solid curves: CMT+FEM; dashed curves: 3D EME method).

For a given directional coupler length L, the normalized optical power exiting from the two waveguides depends on the sensing waveguide (*W2*) cover index n_{cs} ($1.333 \le n_{cs} \le 1.334$)

and distance d between the guiding structures. For $d = 0.8$, 1, 1.2 μm and quasi-TE and quasi-TM modes, dependence of propagation constant on n_{cs} has been found as highly linear, power P_1 varying more rapidly for quasi-TE than for quasi-TM as L changes. Thus, normalized power exiting from *W1* has a linear dependence on n_{cs} for each coupler length and polarization.

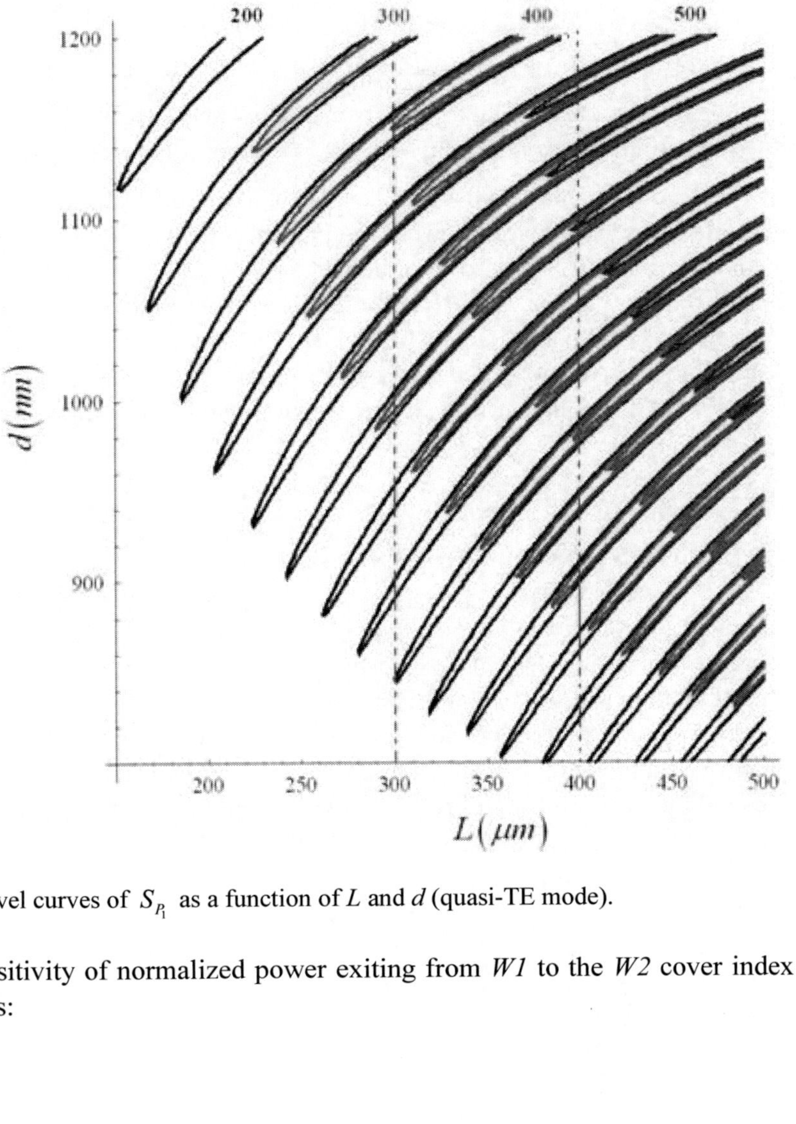

Figure 49. Level curves of S_{P_1} as a function of L and d (quasi-TE mode).

The sensitivity of normalized power exiting from *W1* to the *W2* cover index change can be defined as:

$$S_{P_1} = \left. \frac{\partial P_1}{\partial n_{cs}} \right|_{n_{cs} = n_{cs}^*} \tag{20}$$

where n_{cs}^* is a n_{cs} reference value (we assume $n_{cs}^* = 1.333$). It results S_{P_1} as significantly larger for quasi-TE than for quasi-TM mode and oscillating as L increases (oscillations are slower for quasi-TM mode than for quasi-TE one). Maxima values of S_{P_1} increase by increasing L, thus an arbitrary S_{P_1} value could be obtained by simply increasing L. However, an appropriate trade-off between coupler sensitivity and length has to be usually achieved. For d = 1 μm, larger sensitivity values can be obtained for either quasi-TE or quasi-TM modes. For example, assuming d = 1 μm and L = 313.4 μm, a sensitivity S_{P_1} = 215.29 is obtained for quasi-TE mode. S_{P_1} dependence on d and L has been investigated for quasi-TE mode by calculating S_{P_1} level curves, and results are sketched in Figure 49.

Sensitivity exhibits an oscillatory dependence on d and L and increasing both geometrical parameters (i.e. sensor total area) a sensitivity increase can be revealed. Assuming for S_{P_1} a target value of 500, we can observe that this sensitivity becomes achievable for lengths not smaller than 400 μm. Minimum d value to achieve S_{P_1} = 500 is equal to 1094.4 nm, minimum detectable refractive index change is 10^{-5}. and sensor total area is 1,200 μm^2. This value of minimum detectable refractive index is well five times lower than that reported for an integrated optical chemical sensor based on a conventional 10 mm long directional coupler [101], and about two times lower than that for a ring resonator-based optical biosensor, having a total area of 11,300 μm^2 [162]. Moreover, the theoretical minimum detectable refractive index is more than one order of magnitude lower than the experimental value reported elsewhere [168], having an area of 19,600 μm^2 and adopting a Si_3N_4/SiO_2 slot waveguide.

7. PHOTONIC SENSORS FOR GYRO APPLICATIONS

The measurements of rotation is of considerable interest in a number of application areas. For example, inertial navigation systems as used in aircraft and spacecraft depend critically on accurate inertial rotation sensors. A number of other applications of rotations sensors exist such as surveying, where the accurate determination of azimuth and geodetic latitude is important, geophysics application, earthquakes monitoring, robotics, medical applications, and so on. In general, the mains attributes required to gyroscopes in many applications are:

1. High sensitivity
2. Differential detection
3. Low cost batch fabrication
4. Vibration immunity from lack of moving parts
5. Small fabricated micro components
6. Most mature technology

Related to the above mentioned attributes, it is possible to find a variety of applications where some requirements are dominant with respect to others, as shown in Table 4. As it is evident from this table, in a number of applications such as automotive, robotics, medical, ecc, the requirements of small sizes, low cost and low power consumption are of fundamental importance.

The fiber gyroscopes constituted an initial response to the demands of miniaturizing, especially in the field of space technology which was ever more oriented to the use of small satellites of reduced sizes and cost. However, fiber optic gyroscopes have the drawbacks of not being fully integrable on a single chip and thus suffering the limitation of no compact package. Thus, the development of integrated photonic gyroscope has tried to move in the direction towards gyro miniaturization.

Table 4. Gyroscope applications

Requirements	Applications
High measurement sensitivity	Flight testing Antenna stabilization Aircraft and missile control
Shock and Vibration immunity	Ballistic control Platform stabilization Missile launch
Compact package	Robotics Autonomous Vehicle Control Missile guidance
Lower power consumption	Aerospace Missile
High reliability	Satellite Medical Rocket boosters
Low cost	Automotive Industrial GPS augmentation Rollover detection Medical

Nowadays, the possibility for fabricating integrated photonic gyroscopes may be said to emerge naturally from the confluence of several technologies that today can be considered mature. Thus, it is natural to consider the integrated photonic gyros as an alternative versus other approaches, as interferometer fiber optic gyros (IFOG), ring resonator fiber optic gyros (RFOG), Micro-Electro-Mechanical Systems (MEMS) or Micro-Opto-Electro-Mechanical Systems (MOEMS), at least in some applications. However, a literature review shows that only a few solutions have been proposed in the field of gyro photonic micro-sensors [171].

7.1. Integrated Optical Active Micro-Sensor

A fully integrated active optical gyroscope based on use of semiconductor ring laser (SRL) operating at 845 nm has been proposed, designed, accurately modelled [98] and also patented [172]. The gyro sensor in Figure 50 includes a AlGaAs-GaAs double quantum well (DQW) circular SRL with radius of 1.5 mm, which generates the two counter-propagating optical signals (laser threshold current is 7 mA, laser injection current is 17 mA and emitted optical power is 10 mW). Moreover, a circular directional coupler (bending radius 150 μm) having a reduced coupling efficiency (~ 0.2 %) which extracts the optical beams from SRL, an electro-optic phase modulator (electrodes length 2 mm) which imposes a $\pi/2$ phase shift between the two optical signals, a Y-junction (full aperture angle 2° and total length 8.7 mm) which acts as beam combiner and a photo-detector that generates the photocurrent for measuring the gyro angular rate, are also included. The whole optical circuit is integrated on a 15×3 mm^2 GaAs substrate. Active cavity optical gain is larger for quasi-TE than for quasi-TM modes, i.e. the ring laser emits quasi-TE modes only. This intrinsic effect of polarization selection produces the absence of any birefringence effect and coupling between the two polarizations. The guiding structure adopted in the sensor includes an $Al_{0.2}Ga_{0.8}As$ barrier to separate the GaAs quantum wells, two cladding layer to be realized in $Al_{0.2}Ga_{0.8}As$ and a $Al_{0.73}Ga_{0.27}As$ buffer. Waveguide structure is sketched in Figure 51.

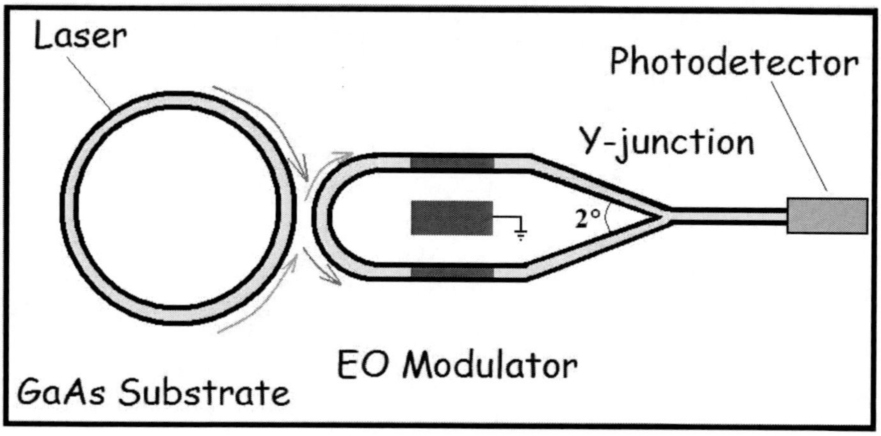

Figure 50. Miniaturized active optical gyro sensor architecture.

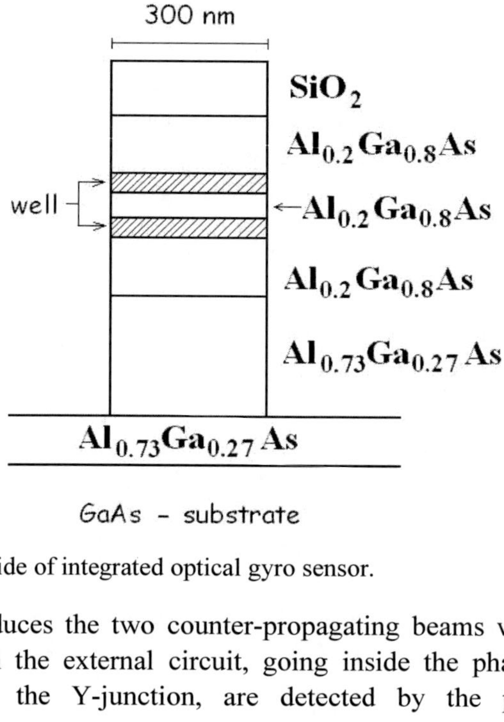

Figure 51. Optical waveguide of integrated optical gyro sensor.

The ring laser produces the two counter-propagating beams which, by the directional coupler, are coupled in the external circuit, going inside the phase modulator and, after recombination through the Y-junction, are detected by the photodetector. The two counter-propagating optical beams have different frequencies according to well-known Sagnac effect [173], when an angular rotation velocity directed around the axis normal to the plane of the ring is applied. The proposed configuration of the output coupler assures a low coupling efficiency, due to the strong asynchronism between output U-shaped waveguide and ring resonator. The advantage of having low coupling efficiency and low loss coefficients, consists of an increase of ring quality factor, improving the minimum angular velocity detectable (quantum limit). In fact, the quantum limit is the main source of noise which limits the sensitivity of the gyroscope sensor.

The geometry of U-shaped coupling device presents a much shorter coupling length than the curvature radius of the same output waveguide, which allows to avoid a periodic power exchange between the guide and the ring laser. The minimum separation between the external guide and the ring must be relatively narrow to get the desired coupling, since the fields are strongly confined within the ridge structures. The electro-optic modulator is a "push-pull' type and works by using two bias voltages. The first bias voltage is used to calibrate the device. In fact, it has a fixed value to produce a zero output corresponding to zero rotational velocity. Its determination is only affected by the detector noise components. The second bias voltage is used to distinguish the output signals of the detector, corresponding to two rotational velocities of equal module but having opposite directions. This integrated optical architecture is probably one of the most compact presented in literature up to date. However, some problems related to its fabrication technology (absorption and defects in GaAs, fabrication tolerances, sidewall roughness), optical losses, possible unidirectional operation of ring laser (mode competition effect) and quantum limit could represent serious drawbacks or limitations to its practical use.

In a SRL operating in bidirectional continuous-wave regime, the waveguide sidewall roughness redistributes the optical power between the counter-propagating beams. Thus, in SRL for gyro sensor, a mode locking between counter-propagating waves occurs. Dependence of total phase shift ψ between clockwise (CW) and counter-clockwise (CCW) optical signals on time can be modeled by:

$$\frac{d\psi}{dt} = 2\pi S\Omega + b\left[sin(\psi)\right] \quad (21)$$

where Ω is the rotation rate and b is the backscattering coefficient. For large Ω values ($b \ll 2\pi S\Omega$), ψ depends linearly on time and no stationary solution exists. In this case the frequency difference between CW and CCW signals linearly depends on rotation rate. If b is comparable to $2\pi S\Omega$ (reduced Ω values), the differential equation (21) exhibits a stationary solution and so the frequency difference between CW and CCW signals becomes equal to zero after a transitory regime. This produces a dead band within RLG static characteristic.

In the gyroscope sensor [98], the backscattering coefficient b has been estimated as 22.69 and lock-in limit (dead band delimiting angular rate) Ω_L as around 210 °/h. When angular rate to be measured is larger than Ω_L, the two optical signals combined in the Y-junction exhibit a frequency difference Δv. Power of the optical signal exiting from Y-junction periodically varies with a frequency equal to Δv. Applying Fast Fourier Transform (FFT) algorithm to photocurrent generated by the photodiode, it is possible calculate Δv and consequently the angular rate Ω. If $\Omega < \Omega_L$, the two optical signals entering in the Y-junction have the same frequency but they exhibit a constant phase shift ψ_r, i.e. the stationary solution of Eq. (21) and depends on Ω. To this phase shift is summed the fixed phase shift (equal to $\pi/2$) introduced by the optical modulator. The optical signal exiting from Y-junction has a constant optical power which depends on ψ_r according to the following expression:

$$\frac{P_{out}}{P_{in}} = \frac{1}{2}\left[1 + cos\left(\psi_r + \frac{\pi}{2}\right)\right] \tag{22}$$

where P_{out} is the optical power exiting from Y-junction and P_{in} is the optical power entering in Y-junction. Staring from photocurrent generated by the photodiode, it is possible to calculate ψ_r and consequently Ω. Maximum accuracy of this estimation is given by backscattering coefficient b estimation. Minimum angular rate detectable by this miniaturized gyro architecture is equal to 0.01 °/h (depending on optical power and linewidth of optical signals generated by SRL), whereas maximum detectable angular rate is limited by photodetector bandwidth, i.e. 10 GHz.

7.2. Photonic Nonlinear Gyro Sensor

Many nonlinear photonic devices in silicon are actually been demonstrated and developed. In particular, Stimulated Raman Scattering (SRS) is a significant effect in silicon to achieve a laser source. In addition, SRS used as the physical effect to realise an active cavity should lead to realise the same advantages obtained in other gyro sensors, such as small sizes and few components required to measure the value of angular velocity. Recently, a number of experimental and theoretical studies have been performed to demonstrate Raman amplification in SOI waveguides [174-175] and Raman laser operation [176-177].

In this section the novel concept of a gyro angular velocity sensor based on SRS is described [178]. The proposed modeling includes all physical nonlinear effects involved in the integrated structure, i.e. SRS, two photon absorption (TPA), free carrier absorption (FCA) where the free carriers are generated mainly by TPA of pump pulse, plasma dispersion effect, Self-Phase-Modulation (SPM) and Cross-Phase-Modulation (XPM) effects as induced by Kerr nonlinearity, backscattering and Sagnac effects. This theoretical approach has been obtained without any a-priori assumption. Details can be found in chapters 7 and 8 of this book.

The architecture of the optical angular velocity sensor based on SRS effect includes two pump lasers, a race-track resonator, two bus waveguides, two directional couplers, two phase modulators, two isolators and two photodetectors, as shown in Figure 52. The beams produced by two pump lasers P1 and P2 are injected into the waveguides and then coupled into the race-track resonator by means of directional couplers C1 and C2, respectively.

If each pump wave into the resonator has a power level larger than the threshold condition for the SRS effect, then two counter-propagating Stokes waves will be enhanced inside the cavity. Therefore, these Stokes waves are coupled outside to the bus waveguides, and then detected by photo-detectors PD1 and PD2. Finally, their photocurrents are processed by the electronic module. The isolators I1 and I2 avoid the detrimental injection of optical signals into the pump lasers produced by any back coupling. Finally, two intracavity thermo-optic phase modulators, M1 and M2, realise the optical dithering to overcome the lock-in effect. Thus, when a rotation around an axis normal to the device plane occurs, the two counter-propagating Stokes waves will experience a frequency difference proportional to the angular velocity (Sagnac effect) which can be extracted at the detection step by the electronic module.

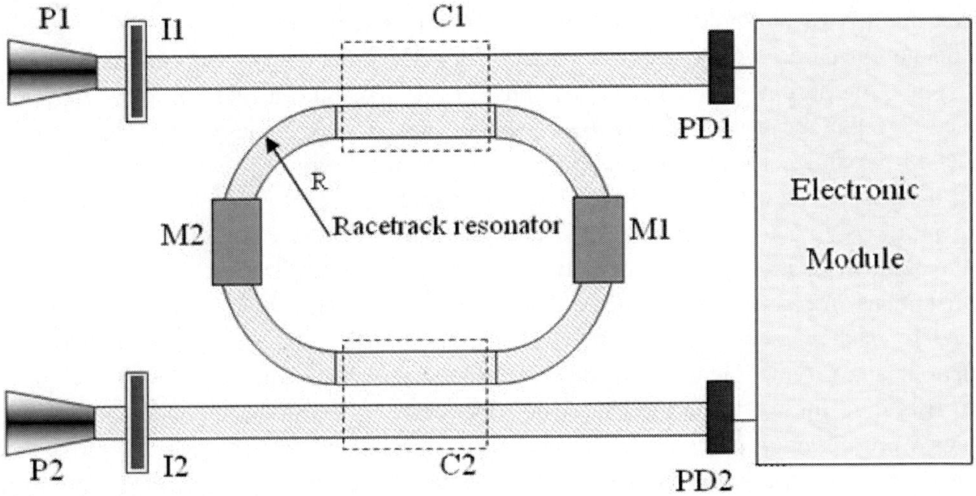

Figure 52. Architecture of optical angular velocity sensor based on Raman effect.

This nonlinear approach shows a number of significant advantages with respect to other integrated photonic configurations proposed in literature. In fact, since it can be classified as a active gyroscope due to the excitation of Stokes waves, it requires a cavity length shorter than that of a passive ring resonator sensor for a given value of sensitivity. In addition, our sensor presents the advantage to require a fewer components inside the architecture with respect to passive approach. In fact, it is possible to identify the value of angular velocity directly from photocurrents generated by two Stokes beams, without any synchronization circuit in feedback configuration, and noise reduction related to polarization conversion is not required. This last aspect essentially depends on the fact that the Raman gain is larger for TM than for TE polarization, inducing the Stokes waves to be aligned as quasi-TM. Note that the sensor proposed in this chapter presents a remarkable advantage also with respect to integrated sensors based on SRL. In fact, since two optical pumps are needed to excite the counter-propagating Stokes waves through SRS effect, mode competition is not a problem. It is interesting to put in evidence that, in general, the sensors based on a single ring laser present the same advantages in terms of compactness and sensitivity as this nonlinear device, but their behaviour can be definitely compromised by relatively high injection currents, where the self and cross gain saturations can induce one of the two counter-propagating laser beams to switch-off. Note that experimental or theoretical work has been not yet demonstrated in literature to exploit the SRS effect for a fiber optic gyro. This essentially depends on the large Brillouin gain coefficient, nearly three orders of magnitude better than the Raman-gain coefficient. Thus, in case of optical fiber technology, the Brillouin effect allows an optical gyro with reduced threshold and sizes to be achieved, compared with the Raman effect. On the contrary, the SRS effect is convenient for an integrated optical gyro sensor since, actually, a number of experimental investigations have demonstrated the Raman amplification in micrometer waveguides based on SOI technology.

8. SENSITIVITY ANALYSIS OF RIB WAVEGUIDES

In a number of photonic sensors, the chemical/physical quantity to be measured affects the propagating mode effective index, usually through a change of cover medium refractive index. Thus, measurement sensitivity depends on the distribution of evanescent optical field in the cover medium, so one of the most important design task is the waveguide optimization in order to maximize its sensitivity. In most of integrated optical sensors, a rib waveguide having a width of w, a total height of h_1 and an etch depth of h_1-h_2, is used, as shown in Figure 53(a).

Waveguide sensitivity can be again expressed by Eq. (14). A normalized analytic formalism was introduced to optimize the sensitivity of slab waveguides [179]. Formulas obtained for slab waveguides have been also used for rib and ridge waveguide sensitivity calculation, simplifying these kinds of waveguides into slab equivalent ones by the well-known Effective Index Method (EIM) [180]. However, it is important to observe that these formulas were obtained under specific assumption, i.e. only one of three refractive indices changes in the slab structure. If sensitivity formulas obtained for slab waveguides are used for rib or ridge ones, it is necessary to assume cover medium index change affects only one of the two refractive indices present in the equivalent slab structure. Under this assumption, a formula to derive ridge waveguide sensitivity has been obtained too [181].In this section an improved analytic procedure to calculate the sensitivity of a rib waveguide without the above discussed assumption is presented. Our simple approach is based on the application of EIM to the guiding structure, as in Figure 53(b). Results are compared with those derived by a numerical rigorous estimation carried out by FV-FEM [130].

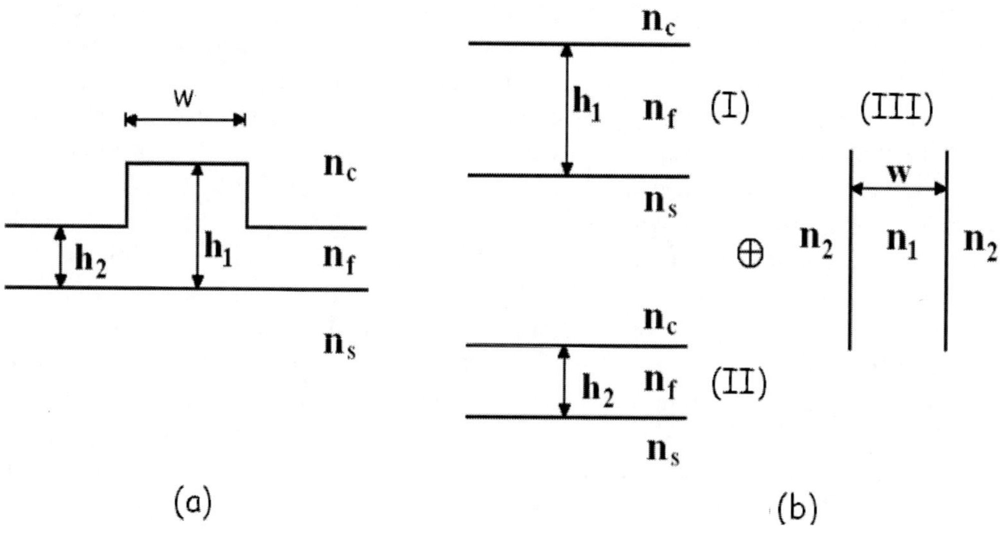

Figure 53.(a) Rib waveguide with a step-index profile; (b) Effective-index method for rib waveguides.

Effective indices of modes propagating in the rib waveguide are obtainable by EIM solving the characteristic equation of the slab waveguide with the same width w of the original rib waveguide (waveguide III in Figure 53(b), having effective index N). Either for quasi-TE or quasi-TM modes, this equation can be expressed as:

$$U\left[N, n_1(n_c), n_2(n_c)\right] = 0 \tag{23}$$

where n_1 is the effective index of a mode propagating in the slab waveguide with height h_1 (waveguide I), and n_2 is the effective index of a mode propagating in the slab waveguide with height h_2 (waveguide II). The homogeneous function U for rib waveguide fundamental quasi-TE and quasi-TM modes can be written as:

$$U\left[N, n_1(n_c), n_2(n_c)\right] = k_0 w \sqrt{n_1^2 - N^2} - 2arctan\left[\left(\frac{n_1^2}{n_2^2}\right)^{\rho} \sqrt{\frac{N^2 - n_2^2}{n_1^2 - N^2}}\right] \tag{24}$$

where k_0 is the vacuum wave-number, ρ=1 for rib waveguide fundamental quasi-TE mode and ρ=0 for rib waveguide fundamental quasi-TM mode. Waveguide sensitivity is calculated by deriving U with respect to N and n_c by:

$$S_w = \frac{\partial N}{\partial n_c} = -\frac{\partial U}{\partial n_c}\left(\frac{\partial U}{\partial N}\right)^{-1} \tag{25}$$

which is a consequence of implicit function theorem [182]. Thus, two analytic expressions for rib waveguide sensitivity have been obtained (for quasi-TE or quasi-TM) as:

$$S_w^{TE} = \frac{1}{N} \frac{Q + k_0 w n_2^3 A_1 \left(n_1 a_2 \sqrt{a_2} + n_2 a_1 \sqrt{a_2}\right)}{k_0 w n_2^3 \sqrt{a_2}\left(n_2 a_1 + n_1 a_2\right) + 2 a_1^2 a_2^2 \left(a_1 + a_2\right)} \tag{26}$$

$$S_w^{TM} = \frac{1}{N}\left[A_1 + \frac{2a_1\left(A_2 - A_1\right)}{\left(a_1 + a_2\right)\left(2 + k_0 w \sqrt{a_2}\right)}\right] \tag{27}$$

where

$$Q = 4a_1 a_2 A_2 n_1^2 - 4a_1 a_2 A_1 n_2^2 + 2a_2 A_1 a_1^2 a_2^2 + 2a_1 A_2 a_1^2 a_2^2 \tag{28}$$

$$a_1 = n_1^2 - N^2 \tag{29}$$

$$a_2 = N^2 - n_2^2 \tag{30}$$

$$A_i = n_i \frac{\partial n_i}{\partial n_c} \qquad \mathrm{i}=1,2 \tag{31}$$

A_i (i=1,2) coefficients, divided by n_i, represent the slab waveguide sensitivities and so can be calculated by a procedure similar as given in literature [179]. Effective indices n_i (i=1,2) are solutions of characteristic equation in the form:

$$V\left(n_i, n_c\right) = 0 \tag{32}$$

being V a function given by:

$$V\left(n_i, n_c\right) = h_i k_0 \sqrt{n_f^2 - n_1^2} - arctan\left[\left(\frac{n_f^2}{n_s^2}\right)^{\sigma} \sqrt{\frac{n_1^2 - n_s^2}{n_f^2 - n_1^2}}\right] - arctan\left[\left(\frac{n_f^2}{n_c^2}\right)^{\sigma} \sqrt{\frac{n_1^2 - n_c^2}{n_f^2 - n_1^2}}\right] \tag{33}$$

where n_c, n_f and n_s are rib waveguide cover, core and substrate refractive indices, respectively, Moreover, it occurs σ=1 for rib waveguide fundamental quasi-TM mode and σ =0 for rib waveguide fundamental quasi-TE mode. Then, the derivative of n_i (i=1,2) with respect to n_c is obtained as:

$$\frac{\partial n_i}{\partial n_c} = -\frac{\partial V}{\partial n_c}\left(\frac{\partial V}{\partial n_i}\right)^{-1} \tag{34}$$

similarly to formula (25). Two formulas for A_i (i=1,2) have been obtained for quasi-TE or quasi-TM polarization, respectively:

$$A_i^{TE} = \frac{n_c}{\sqrt{c_i}\left(1 + \frac{c_i}{f_i}\right)\left(h_i k_0 + \frac{1}{\sqrt{c_i}} + \frac{1}{\sqrt{s_i}}\right)} \tag{35}$$

$$A_i^{TM} = \frac{\left(2c_i + n_c^2\right)\left(\frac{c_i n_f^4 + f_i n_c^4}{f_i n_c^4}\right)^{-1} \frac{1}{n_c^3 \sqrt{c_i}}}{\left[\frac{h_i k_0}{n_f^2} + \frac{\left(c_i + f_i\right) n_c^2}{\left(c_i n_f^4 + f_i n_c^4\right)\sqrt{c_i}} + \frac{\left(s_i + f_i\right) n_s^2}{\left(s_i n_f^4 + f_i n_s^4\right)\sqrt{s_i}}\right]} \tag{36}$$

where

$$c_i = n_i^2 - n_c^2 \qquad \mathrm{i}=1,2 \tag{37}$$

$$s_i = n_i^2 - n_s^2 \qquad \mathrm{i}=1,2 \tag{38}$$

$$f_i = n_f^2 - n_i^2 \qquad i = 1,2. \tag{39}$$

Rib waveguide sensitivity, as defined by Eq. (14), can be also numerically estimated by varying the cover medium refractive index n_c in a narrow range and finding the relevant change of effective index N. To determine this effective index N, the rigorous approach based on full-vectorial FEM has been applied [130].

Then, different guiding structures have been considered to investigate the accuracy of proposed analytical method. In particular, sensitivity of waveguides using silicon oxide as substrate and silicon oxynitride (SiON), silicon nitride (Si_3N_4) or silicon as guiding film has been investigated. As cover medium, water (whose refractive index can be changed by mixing with ethanol [144]) and Poly-methyl-metha-crylate (PMMA) doped with Bromocresol Purple (whose refractive index is for example sensitive to ammonia concentration changes [105]) have been assumed. Considered rib waveguide parameters are summarized in Table 5.

Table 5. Optical and geometrical properties of considered rib waveguides

	Wavelength (λ)	n_s	n_f	n_c	h_1	w	h_1-h_2 variation range
SiON waveguide (cover: water)	633 nm	1.46	1.55	1.33	0.5 μm	2 μm	40-100 nm
Si_3N_4 waveguide (cover: water)	633 nm	1.46	2.00	1.33	0.6 μm	1 μm	20-80 nm
SOI waveguide (cover: PMMA)	1550 nm	1.44	3.48	1.476	1 μm	1.2 μm	100-300 nm

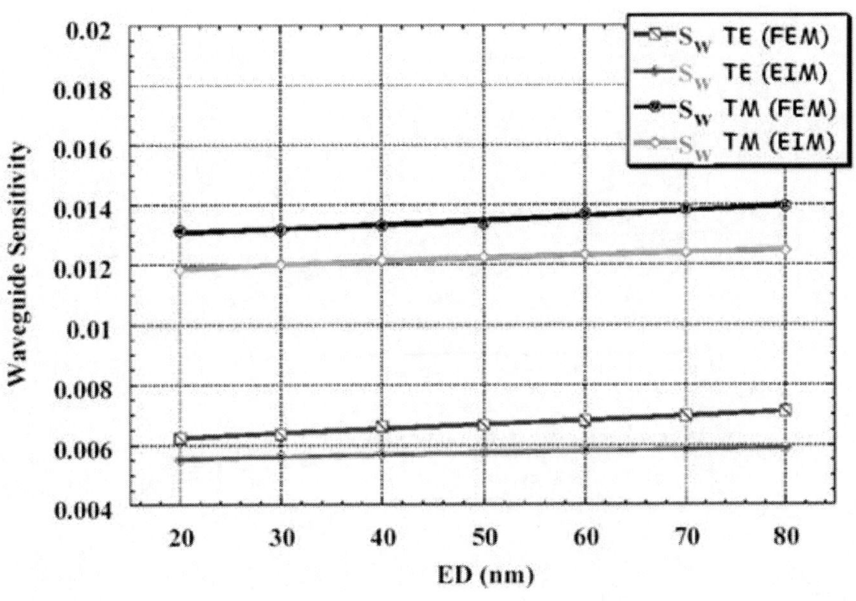

Figure 54. Silicon nitride (Si_3N_4) waveguide sensitivity dependence on *ED*, estimated by analytic approach and FEM analysis, for quasi-TE and quasi-TM modes.

For all considered guiding structures and for quasi-TE and quasi-TM modes, waveguide sensitivity dependence on etch depth ($ED = h_1$-h_2) has been investigated either by both proposed analytic formulas and FEM-based numerical approach, as sketched in Figures 54 and 55. Waveguides sensitivities calculated by analytical and numerical approaches exhibit similar dependence on etch depth in all considered cases. When *ED* increases, the difference between results obtained by both methods remains quite constant for low and medium index contrast waveguides (SiON and Si_3N_4 film), whereas it increases for high index contrast SOI waveguides.

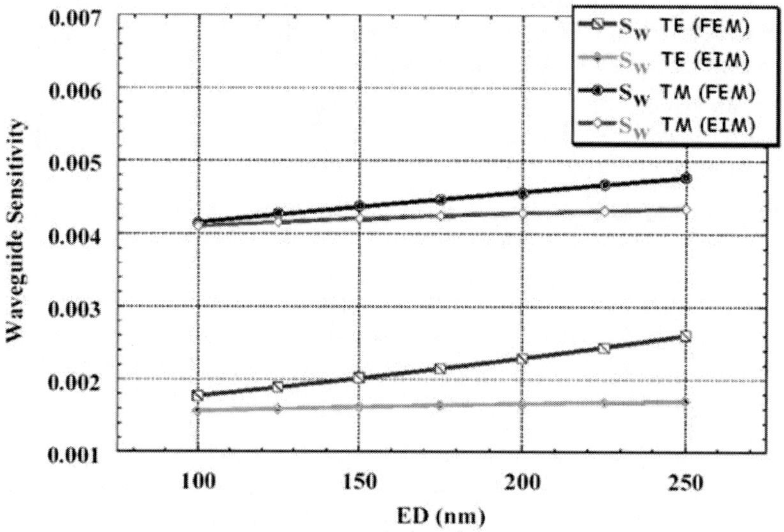

Figure 55. Silicon-on-Insulator (SOI) waveguide sensitivity dependence on *ED*, estimated by analytic approach and FEM analysis, for quasi-TE and quasi-TM modes.

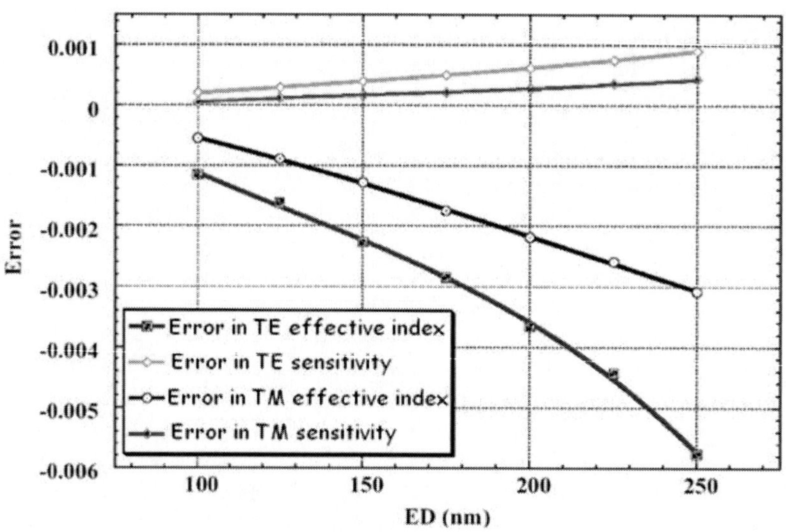

Figure 56. Absolute error in calculation of effective index and sensitivity by EIM approach in SOI waveguide, for both polarizations.

Then, both waveguide sensitivity estimation (analytic method) and waveguide fundamental mode effective indices calculation by EIM exhibit accuracies always at least of the same order of magnitude with respect to rigorous FEM evaluation. Moreover, for SOI waveguides, the sensitivity calculation carried out by EIM shows an accuracy one order of magnitude larger than that obtained for effective index estimation (see Figure 56). Thus, an accuracy around 10-4 has been proved for high index contrast guiding structures. This approach for waveguide sensitivity estimation is completely general and can be used to support the optimization process in a number of integrated optical sensors as realized by using different technologies, avoiding the use of much more memory and time consuming calculation methods, like FEM.

CONCLUSION

This chapter demonstrates a selection of a variety of optical signal processors, guided-wave devices and sensors to be used in a wide range of applications. Guided-wave devices have great potential and advantages over conventional electronics. These advantages are seen in high capacity telecommunications networks, millimeter wave signals filtering, microwave signals sampling, earth observation and sensing. Integrated optical circuits and systems are small in size and weight, immune to electromagnetic disturbances and offer high sensitivity and stability. They will substantially provide further advantages over equivalent electronic circuits. In this chapter some advances of electric field photonic sensors are also presented. In particular the description is focused on a novel photonic sensor approach for detecting weak and broadband electric fields, based on SOI technology. Improvements of sensitivity of the order of 3-4 times with respect to the conventional Fabry-Perot interferometers and at least one order of magnitude with respect to Mach-Zehnder interferometers should be so possible.

Moreover, this chapter shows the wide variety of photonic sensors for biochemical and biomedical applications proposed in the literature, since the fabrication of the first optical biochemical sensor based on the measurement of changes in absorption spectrum and developed for the measurement of CO_2 and O_2 concentration. Optical technologies permit to overcome some problems connected to the use of electrochemical techniques for biochemical analysis as low sensitivity, large dimensions and electromagnetic interference. Biosensor integrated optical architectures enable one to reach high sensitivity, selectivity and compactness. Among these architectures, those realized by CMOS-compatible technologies permit one to achieve low cost, full technological compatibility between optical sensor and read-out electronics and the possibility of mass fabrication.

Finally, integrated configurations based on microring and microdisk show very important advantages related to dimension reduction, without significantly affecting the device sensitivity. Finally, very recent advances in investigations of novel photonic sensors of angular velocity for gyro applications are briefly reviewed. A technique to estimate the rib waveguide sensitivity is also presented, as well as some aspects of modeling techniques.

REFERENCES

[1] Liu, J.-M., *Photonic Devices*, Cambridge University Press, 2005.

[2] Agrawal, G. P., *Lightwave Technology: components and devices*, John Wiley and Sons, 2004.

[3] Reed, G. T., *Nature*, 2004, 427, 595-596.

[4] Reed, G. T. and Knights, A.P., *Silicon Photonics: an introduction*, John Wiley and Sons, 2004.

[5] Motamedi, M. E. (Ed.), *MOEMS: Micro-Opto-Electro-Mechanical Systems*, SPIE Press, 2005.

[6] Piprek, J., *Optoelectronic Devices: Advanced Simulation and Analysis*, Springer, 2005.

[7] Passaro, V. M. N., *J. Lightwave Technology*, 2000, 18, 973-984.

[8] Scarmozzino, R., *et al.*, *IEEE J. Selected Topics in Quantum Electronics*, 2000, 6, 150-162.

[9] Vilcot, A., Cabon, B., Chazelas, J., *Microwave Photonics*, Kluwer Academic Publ., 2003.

[10] Seeds, A. J., Williams, K. J, *J. Lightwave Technology*, 2006, 24, 4628-4641.

[11] Seeds, A. J., *IEEE Trans. on Microwave Theory and Techniques*, 2002, 50, 877-887.

[12] Minasian, R. A., *IEEE Trans. on Microwave Theory and Techniques*, 2006, 54, 832-846.

[13] Roussell, H., Helkey, R., *IEEE Microwave and Guided-Wave Letters*, 1998, 8, 408-410.

[14] Hunter, D., Minasian, R., *Electronics Letters,* 1999, 35, 412-414.

[15] Chou, J., Han, Y., Jalali, B., *IEEE Photon. Technology Letters*, 2003, 15, 581-583.

[16] Jackson, K. P., Newton, S. A., *et al.*, *IEEE Trans. Microwave Theory and Techniques*, 1985, MTT-33, 193-209.

[17] Capmany, J., Pastor, D., Ortega, B., *IEEE Trans. Microwave Theory and Techniques*, 1999, 47, 1321-1326.

[18] Sasayama, K., Okuno, M., Habara, K., *J. Lightwave Technology*, 1991, 9, 1225-1230.

[19] Coppinger, F., Madsen, C. K., Jalali, B., *Microwave and Optical Technology Letters*, 1999, 21, 90-93.

[20] Little, B. E., Chu, S. T., *et al.*, *IEEE Photon. Technology Letters*, 2004, 16, 2263-2265.

[21] Sales, S., Capmany, J., Marti J., Pastor, D., *Electronics Letters*, 1995, 31, 1095-1096.

[22] Coppinger, F., Yegnanarayanan, S., Trinh, P. D., Jalali, B., *Electronics Letters*, 1997, 33, 973-975.

[23] Pastor, D., Capmany, J., Ortega, B., Martinez, A., Pierno, L., Varasi, M., *IEEE Photon. Technology Letters*, 2004, 16, 2126-2128.

[24] Capmany, J., Pastor, D., Martinez, A., *et al.*, *Optics Letters*, 2003, 28, 1415-1417.

[25] Valley, G. C., *Opt. Express*, 2007, 15, 1955-1982.

[26] Taylor, H. F., Taylor, M. J., Bauer, P. W., *Applied Physics Letters*, 1978, 32, 559-561.

[27] Juodawlkis, P. W., Twichell, J. C., Betts, G. E., *et al.*, *IEEE Trans. on Microwave Theory and Techn.*, 2001, 49, 1840-1853.

[28] Helkey, R., *IEEE Photon. Technology Letters*, 1999, 11, 599-601.

[29] Twichell, J. C., Helkey, R., *IEEE Photon. Technology Letters*, 2000, 12, 1237-1239.

[30] Clark, T. R., Currie, M., Matthews, P. J., *J. Lightwave Technology*, 2001, 19, 172-179.

[31] Clark, T. R., Kang, J. U., Esman, R. D., *IEEE Photon. Technology Letters*, 1999, 11, 1168-1170.

[32] Fok, M. P., Lee, K. L., Shu, C., *IEEE Photon. Technology Letters*, 2004, 16, 876-878.

[33] Kang, J. U., Esman, R. D., *Electronics Letters*, 1999, 35, 60-61.

[34] Kartner, F. X., Amataya, R., Barbastathis, G., *et al.*, *3rd IEEE Int. Conf. on Group IV Photonics Proc.*, 2006, 203.

[35] Taylor, H. F., *Proceedings of IEEE*, 1975, 63, 1524-1525.

[36] Walker, R. G., Bennion, I., Carter, A. C., *Electronics Letters*, 1989, 25, 1443-1444.

[37] Chen, Y., Wu, K., Zhao, F., Chen, R. T., *IEEE Antennas and Prop. Society Int. Symp. Proc.*, 2004, 4, 4324.

[38] Wang, X., Howley, B., Chen, M. Y., Basile, P., Chen, R. T., *Int. Photon. Research and Appl./Nanophoton. 2006 Technical Digest*, 2006, IMF4.

[39] Chen, Y., Wu, K., Zhao, F., Kim, G., Chen, R. T., *Proc. SPIE*, 2004, 5363, 125.

[40] LeGrange, J., Kasper, A., Madsen, C., Cappuzzo, M., Chen, E., Griffin, A., Laskowski, E. J., Rasras, M., *17th Ann. Meet. of IEEE LEOS Society Proc.*, 2004, 790.

[41] Horikawa, K., Ogawa, I., Ogawa, H., Kitoh, T., *IEEE MTT-S Int. Microwave Symposium Digest*, 1995, 65.

[42] Horikawa, K., Ogawa, I., Kitoh, T., Ogawa, H., *Opt. Fiber Commun. (OFC'96)*, 1996, 100.

[43] Rasras, M. S., Madsen, C. K., Cappuzzo, et al., *IEEE Photon. Technology Letters*, 2005, 17, 834-836.

[44] Horikawa, K., Nakasuga, Y., Origawa, H., *IEEE Trans. on Microwave Theory and Techniques*, 1995, 43, 2395-2401.

[45] Flamand, G., De Mesel, K., Moerman, I., Dhoedt, B., Hunziker, W., Kalmar, A., Baets, R., van Daele, P., Leeb, W., *IEEE Photon. Technology Letters*, 2000, 12, 876-878.

[46] Stulemeijer, J., van Vliet, F. E., Benoist, K. W., Maat, D. H. P., Smit, M. K., *IEEE Photon. Technology Letters*, 1999, 11, 122-124.

[47] Armenise, M. N., Passaro V. M. N., Noviello, G., *Applied Optics*, 1994, 33, 6194-6209.

[48] Armenise, M.N., Passaro, V.M.N., Impagnatiello, F., *J. Optical Society of America B*, 1991, 8, 443-448.

[49] Passaro, V.M.N., Matteo, A.M., *et al.*, *J. Lightwave Technology*, 1997, 15, 2114-2123.

[50] [50] Piqueras, M. A. *et al.*, *IEEE Trans. on Microwave Theory and Techn.*, 2006, 54, 887-899.

[51] den Besten, J. H., Caprioli, D., van Dijk, R., *et al.*, *IEEE/LEOS Symp. Proc.*, 2004, 95.

[52] Passaro, V. M. N. (Ed.), *Silicon Photonics*, Research Signpost, Kerala (India), 2006.

[53] Pavesi, L., Lockwood, D. J. (Eds.), *Silicon Photonics*, Springer, 2004.

[54] Soref, R. A., Bennett, B. R., *IEEE J. Quantum Electronics*, 1987, QE-23, 123-129.

[55] Liu, A., Jones, R., Liao, L., Samara-Rubio, D., Rubin, D., Cohen, O., Nicolaescu, R., Paniccia, M., *Nature*, 2004, 427, 615-618.

[56] Liu, A., Samara-Rubio, D., Liao, L., Paniccia, M., *IEEE J. Selected Topics in Quantum Electronics*, 2005, 11, 367-372.

[57] Liao, L., Liu, A., Jones, R., Rubin, D., Samara-Rubio, D., Cohen, O., Salib, M., Paniccia, M., *IEEE J. Quantum Electronics*, 2005, QE-41, 250-257.

[58] Samara-Rubio, D., Keil, U. D., Liao, L., Franck, T., Liu, A., Hodge, D. W., Rubin, D., Cohen, O., *J. Lightwave Technology*, 2005, 23, 4305-4314.

[59] Barrios, C. A., Lipson, M., *J. Applied Physics*, 2004, 96, 6008-6015.

[60] Png, C. E., Chan, S. P., Lim, S. T., Reed, G., *J. Lightwave Technology*, 2004, 22, 1573-1582.

[61] Barrios, C. A., Almeida, V. R., Panepucci, R. R., Schmidt, B. S., Lipson, M., *IEEE Photon. Technology Letters*, 2004, 16, 506-508.

[62] Liao, L., Samara-Rubio, D., Morse, M., Liu, A., Hodge, D., *Optics Express*, 2005, 13, 3129-3135.

[63] Gardes, F. Y., Reed, G. T., Emerson, N. G., Png, C. E., *Optics Express*, 2005, 13, 8845-8854.

[64] Xu, Q., Schmidt, B., Pradhan, S., Lipson, M., *Nature*, 2005, 435, 325-327.

[65] Gan, F., Kärtner, F. X., *IEEE Photon. Technology Letters*, 2005, 17, 1007-1009.

[66] Tu, X., Chen, S., Zhao, L., Sun, F., Yu, J., Wang, Q., *J. Lightwave Technology*, 2006, 24, 1000-1007.

[67] Xu, Q., Schmidt, B., Shakya, J., Lipson, M., *Optics Express*, 2006, 14, 9430-9435.

[68] Zhou, L., Poon, A. W., *Optics Express*, 2006, 14, 6851-6857.

[69] Barrios, C. A., *J. Lightwave Technology*, 2006, 24, 2146-2155.

[70] Liu, A., Liao, L., Rubin, D., Nguyen1, H., Ciftcioglul, B., Chetrit, Y., Izhaky, N., Paniccia, M., *Optics Express*, 2007, 15, 660-668.

[71] Bogaerts, W., *et al.*, *IEEE J. Selected Topics in Quantum Electronics*, 2005, 23, 401-412.

[72] Tsuchizawa, T. *et al.*, *IEEE J. Selected Topics in Quantum Electronics*, 2005, 11, 232-240.

[73] Dumon, P., Bogaerts, W., *et al., IEEE Photon. Technology Letters*, 2004, 16, 1328-1330.

[74] Barwicz, T., Popovic, M. A., Watts, M. R., Rakich, P. T., Ippen, E. P., Smith, H. I., *J. Lightwave Technology*, 2006, 24, 2207-2218.

[75] Chu, T., Yamada, H., Ispida, S., Arakawa, Y., *Optics Express*, 2005, 13, 5216-5226.

[76] Passaro, V. M. N., Magno, F., Tsarev, A. V., *Optics Express*, 2005, 13, 3429-3437.

[77] Magno F., Dell'Olio, F., Passaro, V. M. N., *Comsol Users Conf. Proc.*, Milano, 2006.

[78] Blumenthal, J., Bowers, D. J., Rau, J. E., Chou, L., Rangarajan, H.-F., Wang, S., Poulsen, W., *IEEE Communications Mag.*, 2003, 41, S23-S29.

[79] Liu, Y., Tangdiongga, E., *et al.*, *Proc. SPIE*, 2006, 6353, 635312-1.

[80] Yu, H. Y., *et al.*, *J. Lightwave Technology,* 1999, 17, 308-315.

[81] Dong, Y., *et al.*, *Proc. OFC'2000*, TuF2, 2000, 69.

[82] Ellis, A. D., *et al.*, *Electronics Letters*, 1998, 34, 1958-1959.

[83] Taijima, K., *Jpn. J. Applied Physics*, 1993, 32, L1726-L1749.

[84] Leuthold, J., *et al.*, *Optical and Quantum Electronics*, 2001, 33, 939-952.

[85] Li, Z., Liu, Y., *et al.*, *Electronics Letters*, 2005, 41, 1397-1399.

[86] Hill, M. T., Dorren, H. J. S., *et al.*, *Nature*, 2004, 432, 206-209.

[87] Hill, M. T., Dorren, *et al.*, *IEEE Photon. Technology Letters*, 2005, 17, 956-958.

[88] Soares, O., Proc. *Int. Work. on Innovation for Competitiveness*, 1997, 19-21.

[89] Horner, J. L., *Optical Signal Processing*, Academic Press Inc., 1987.

[90] Armenise, M. N., Passaro, V. M. N. *et al.*, *Optical Computing and Proc.*, 1992, 2, 79-93.

[91] Passaro, V.M.N., *Optical and Quantum Electronics*, 2002, 34, 893-913,

[92] Armenise, M. N., Passaro, V. M. N., *Optical Engineering*, 1994, 33, 1854-1862.

[93] Armenise, M. N., Passaro, V. M. N., *IEE Proceedings Part J*, 1990, 137, 347-356.

[94] Passaro, V.M.N., Magno, F., *Laser Physics*, 2007, 17, 231-243.

[95] Passaro, V. M. N., Dell'Olio, F., De Leonardis, F., *Progress in Quantum Electronics*, 2006, 30, 45-73.

[96] Passaro, V. M. N., De Leonardis, F., *IEEE J. Selected Topics in Quantum Electronics*, 2006, 12, 124-133.

[97] Macek, W. M., Davis, D. T. M., *Applied Physics Letters*, 1963, 2, 67-68.

[98] Sorel, M. *et al.*, *IEEE J. Quantum Electronics*, 2003, 39, 1187-1195.

[99] Armenise, M. N., Passaro, V. M. N. *et al.*, *J. Lightwave Technology,* 2001, 19, 1476-1494.

[100] Cao, H. *et al., Applied Physics Letters*, 2005, 86, 041101.

[101] Luff, B. J., Harris, R. D., Wilkinson, J. S., Wilson, R., Schiffrin, D. J., *Optics Letters*, 1996, 21, 618- 620.

[102] Prieto, F., Sepulveda, b., Calle, A., *et al.*, *Nanotechnology*, 2003, 14, 907-912.

[103] Hopman, W. C. L., Pottier, P., Yudistira, D., van Lith, J., Lambeck, P. V., De La Rue, R. M., Driessen, A., *IEEE J. Selected Topics in Quantum Electronics*, 2005, 11, 11-16.

[104] Chao, C. Y., Fung, W., Guo, L. J., *IEEE J. Selected Topics in Quantum Electronics*, 2006, 12, 134-142.

[105] Dell'Olio, F., Passaro, V. M. N., De Leonardis, F., *11th Int. Meeting on Chemical Sensors Proc.*, 2006, 27.

[106] Irace, A,. Breglio, G., *Optics Express*, 2003, 11, 2807-2812.

[107] Laine, J. P., Tapalian, H. C., Little, B. E., Haus, H. A., *Sensors and Actuators A*, 2001, 93, 1-7.

[108] Kiyat, I., Kocabas, C., Aydinli, A., *J. Micromech. and Microeng.*, 2004. 14, 374-381.

[109] Huang, S. J., Erickson, D. C., *IEEE Trans. on Power Delivery*, 1989, 4, 1579-1585.

[110] Kalinowski, D., Redlich, S., Jager, D., *12th LEOS Annual Meeting*, 1999, 1, 385.

[111] Kanda, M., Masterson, K.D., *Proceedings of IEEE*, 1992, 80, 209-215.

[112] Ito, H., *RandD Review of Toyota*, 2000, 35, art. 9.

[113] Kyosemi Corp, Optical electric field sensor, http://www.kyosemi.co.jp.

[114] Ichijo, A., patent n. JP2004219088, 2004.

[115] Tanabe, T., patent n. JP9089957, 1997.

[116] Wust, P., Berger, J., Fahling, H., Nadobny, J., Gellermann, J., Tilly, W., Rau, B., Petermann, K., Felix, R., *Int. J. Rad. Onc. Biol. Phys.*, 1999, 43, 927-937.

[117] Naghski, D.H., Boyd, J. T., Jackson, H. E., Sriram, S., Kngsley, S. A., Latess, J., *J. Lightwave Technology*, 2004, 12, 1092-1098.

[118] Gutiérrez–Martínez, C., Trinidad–García, G., Rodríguez–Asomoza, J., 2002, *IEEE Trans. on Instrumentation and Measurement*, 2002, 51, 985-989.

[119] Ido, T., Tanaka, S., Suzuki, M., Inoue, H., *Electronics Letters*, 1995, 31, 2124-2125.

[120] Tabib-Azar, M., Sutapun, B., Srikhirin, T., Lando, J., Adamovsky, G., *Sensors and Actuators A,* 2000, 84, 134-139.

[121] Fernandez-Valdivielso, C., Matias, I. R., Gorraiz, M., Arreguiand, F. J., Bariain C., Lopez-Amo, M., *Optical Fiber Sensors Conf. Tech. Dig.*, 2002, 499.

[122] Randa, J., Kanda, M., Orr, R. D., *IEEE Trans. on Electromagnetic Compatibility*, 1991, 33, 205-214.

[123] Koo, K. P., Sigel Jr., G. H., *IEEE J. Quantum Electronics*, 1982, QE-18, 670-675.

[124] Vohra, S. T., Bucholtz, F., Kersey, A. D., *8th Optical Fiber Sensors Conf.*, 1992, 418.

[125] Kopola, H., Thansandote, A., Chrostowski, J., Stuchly, S.S., *Conf. Prec. Electrom. Meas.*, 1990, 196.

[126] Chao, C., Guao, L. J., *Applied Physics Letters*, 2003, 83, 1527-1529.

[127] Fan, S., *Applied Physics Letters*, 2002, 80, 908-910.

[128] Liu, A., Jones, R., Samara-Rubio D., Rubin, D., Cohen, O., Nicolaescu, R., Paniccia, M., *Nature*, 2004, 427, 615-618.

[129] Reed, G. T., Png, C. E., *Materials Today,* 2005, 8, 40-50.

[130] Comsol Multiphysics by COMSOL AB © 2005, ver. 3.2, single license.

[131] Vo-Dinh, T., Allain, L., *Biomedical Photonics Handbook*, T. Vo-Dinh (Ed.), Boca Raton: CRC Press, 2003.

[132] Vo-Dinh, T., Tromberg, B. J., Griffin, G. D., Ambrose, K. R., Sepaniak, M. J., Gardenhire, E. M., *Applied Spectroscopy*, 1987, 41, 735-738.

[133] Vo-Dinh, T., Nolan, T., Cheng, Y. E., Sepaniak, M. J., Alarie, J. P., *Applied Spectroscopy*, 1990, 44, 128-134.

[134] Smith, L. M., Saunders, J. Z., Kaiser, R. J., Hughes, P., Dodd, C. R., Cornell, C. R., Heiner, C., Kent, S. B. H., Hood, L. E., *Nature*, 1986, 321, 674-679.

[135] Homola, J., Yee, S. S., Gaugliz, G., *Sensors and Actuators B: Chemical*, 1999, 54, 3-15.

[136] [136] Flanagan, M. T., Pantell, R. H., *Electronics Letters*, 1984, 20, 968-970.

[137] Mahadevan-Jansen, A., *Biomedical Photonics Handbook*, T. Vo-Dinh, Ed. Boca Raton: CRC Press, 2003.

[138] Isola, N. R., Stokes, D. L., Vo-Dinh, T., *Analitical Chemistry*, 1998, 70, 1352-1356.

[139] Vo-Dinh, T., Stokes, D. L., Griffin, G. D., Volkan, M., Kim, U. J., Simon, M. I., *J. Raman Spectroscopy*, 1999, 30, 785-793.

[140] Yotter, R. A., Lee, L. A., Wilson, D. M., *IEEE Sensors Journal*, 2004, 4, 395-411.

[141] Lillie, J. J., Thomas, M. A., Jokerst, N. M., Ralph, S. E., Dennis, K. A., Handerson, C. L., *J. Optical Society of America B*, 2006, 23, 642-651.

[142] Luo, D. H., Levy, R. A., Hor, Y. F., Federici, J. F., Pafchek, R. M., *Sensors and Actuators B: Chemical*, 2003, 92, 121-126.

[143] Luff, B. J., Wilkinson, J. S., Piehler, J., Hollenbach, U., Ingenhoff, J., Fabricius, N., *J. Lightwave Technology*, 1998, 16, 583-592.

[144] Brosinger, F., Freimuth, H., Lacher, M., Ehrfeld, W., Gedig, E., Katerkanp, A., Spener, F., Cammann, K., *Sensors and Actuators B: Chemical*, 1997, 44, 350-355.

[145] Weisser, M., Trovar, G., Mittler-Neher, S., Knoll, W., Brosinger, F., Freimuth, H., Lacher, M., Ehrfeld, W., *Biosensors and Bioelectronics*, 1999, 14, 405-411.

[146] Qui, Z., Matsuda, N., Itoh, K., Murabayashi, M., Lavers, C. R., *Sensors and Actuators B: Chemical*, 2002, 81, 254-258.

[147] Kerrat, R., Jaffrezic-Renault, N., Greco, P., Helmers, H., Benech, P., Rimet, R., *Sensors and Actuators B: Chemical*, 1996, 37, 7-11.

[148] Maisenholder, B., Zappe, H. P., Kunz, R. E., Riel, P., Moser, M., Edlinger, J., *Sensors and Actuators B: Chemical*, 1997, 38, 324-329.

[149] Drapp, B., Piehler, J., Brecht, A., Gauglitz, G., Luff, B. J., Wilkinson, J.S., Ingehoff, J., *Sensors and Actuators B: Chemical*, 1997, 38, 277-282.

[150] Ymeti, A., Kanger, J. S., Greve, J., Lambeck, P. V., Wijn, R., Heideman, R. G., *Applied Optics*, 2003, 42, 5649-5660.

[151] Prieto, F., Llobera, A., Jiménez, D., Doménguez, C., Calle, A., Lechuga, L. M., *J. Lightwave Technology*, 2000, 18, 966-972.

[152] Prieto, F., Lechuga, L. M., Calle, A., Llobera, A., Domínguez, C., *J. Lightwave Technology*, 2001, 19, 75-83.

[153] [Benaissa, K., Nathan, A., *Sensors and Actuators B: Chemical*, 1998, 65, 33-44.

[154] Prieto, F., Sepulveda, B., Calle, A., Llobera, A., Dominguez, C., Lechuga, L. M., *Sensors and Actuators B: Chemical*, 2003, 92, 151-158.

[155] Campopiano, S., Bernini, R., Zeni, L., Sarro, P. M., 2004, *Optics Letters*, 29, 1894-1896.

[156] [Harris, R. D., Luff, B. J., Wilkinson, J. S., Piehler, J., Brecht, A., Gauglitz, G., Abuknesha, R. A., *Biosensors and Bioelectronics*, 1999, 14, 377-386.

[157] Dostalek, J., Ctyroky, J., Homola, J., Brynda, E., Skalsky, M., Nekvindova, P., Spirkova, J., Skvor, J., Schrofel, J., *Sensors and Actuators B: Chemical*, 2001, 76, 8-12.

[158] Veldhuis, G. J., Berends, J. H., Heideman, R. G., Lambeck, P. V., *Pure and Applied Optics*, 1998, 7, L23-L26.

[159] Densmore, A., Xu, D.-X., Waldron, P., Janz, S., Cheben, P., *et al.*, *IEEE Photon. Technology Letters*, 2006, 18, 2218-2220.

[160] Passaro, V. M. N., Dell'Olio, F., De Leonardis, F., *Sensors*, 2007, 7, 2741-2749.

[161] Dell'Olio, F., Passaro, V. M. N., De Leonardis, F. 2006, *Proc. Int. Conf. on Transparent Optical Networks*, Nottingham, 4, 128.

[162] Yalçin, A., Popat, K. C., Aldridge, J. C., Desai, T. A., *et al., IEEE J. Selected Topics in Quantum Electronics*, 2006, 12, 148-155.

[163] Ksendzov, A., Lin, Y., *Optics Letters*, 2005, 30, 3344-3346.

[164] Krioukov, E., Klunder, D. J. W., Driessen, A., *et al.*, *Optics Letters*, 2002, 27, 512.

[165] Almeida, V. R., Xu, Q., Barrios, C. A., Lipson, M., *Optics Letters*, 2004, 29, 1209-1211.

[166] Jordana, E., Fedeli, J. M., El Melhaoui, L., Lyan, P., *et al.*, *Eur. Conf. on Integrated Optics Proc. (ECIO 2007)*, WB4, Copenhagen, 25-27 April 2007.

[167] Dell'Olio, F., Passaro, V. M. N., *Optics Express*, 2007, 15, 4977-4993.

[168] Barrios, C. A., Gylfason, K. B., Sánchez, B., *et al.*, *Optics Letters*, 2007, 32, 3080-3082.

[169] Veronis, G., Fan, S., J. Lightwave Technol. 2007, 25, 2511-2521.

[170] FIMMPROP by Photon Design, ver. 4.6, single license, 2007.

[171] Titterton, D., Weston, J., *Strapdown Inertial Navigation Technology*, The Institution of Electrical Engineers, 2004.

[172] Armenise, M.N., Passaro, V.M.N., De Leonardis, M., Armenise, M., patent n. EP1219926, European Patent Office, 2002.

[173] Sagnac, G., *R. Compete-renduz à l'Academie des Sciences*, 1913, 95, 708-710.

[174] Claps, R., Dimitropoulos, D., Raghunathan, V., Han, Y., Jalali, B., *Optics Express,* 2003, 11, 1731-1739.

[175] Passaro, V. M. N., De Leonardis, F., *J. Lightwave Technol.,* 2006, 24, 2920-2931.

[176] Boyraz, O., Jalali, B., *Optics Express,* 2004, 12, 5269-5273.

[177] [De Leonardis, F., Passaro, V. M. N., *New Journal of Physics*, 2007, 9, art. 25.

[178] [De Leonardis, F., Passaro, V. M. N., *J. Lightwave Technology*, 2007, 25, 2352-2366.

[179] Parriaux, O., Veldhuis, G. J., *J. Lightwave Technology*, 1998, 16, 573–582.

[180] Kogelnik, H., *Guided Wave Optoelectronics*, T. Tamir (Ed.), Springer–Verlag, 1988.

[181] Chao, C.–Y., Guo, L. J., *J. Lightwave Technology*, 2006, 24, 1395–1402.

[182] Protter, M. H., Morrey, C. B., *A First Course in Real Analysis*, Springer–Verlag, 1997.

In: Modeling of Photonic Devices
Editor: Vittorio M. N. Passaro
ISBN 978-1-60456-980-3

Chapter 2

SLOT AND HOLLOW WAVEGUIDES

Michele Lagioia*[*,1], Renzo Loiacono[2] and Vittorio M. N. Passaro[1]
[1]Photonics Research Group, Dipartimento di Elettrotecnica ed Elettronica,
Politecnico di Bari, via Edoardo Orabona n. 4, 70125 Bari, Italy
URL: http://dee.poliba.it/photonicsgroup
[2]Advanced Technology Institute, University of Surrey, Guildford, UK

ABSTRACT

In this chapter, an overview on low index core silicon waveguides is presented. In particular, the state-of-the-art for silicon slot waveguides is discussed. Numerical analysis for the optimization of SOI slot waveguides is described for both linear and nonlinear applications using a finite element method. The linear behaviour has been investigated at both 1.55 and 3.39μm, considering different values for geometrical parameters. For non linear applications, the waveguide dimensions and the nonlinear medium have been chosen in order to maximize the optical Kerr effect at 1.55μm using a novel iterative method. Possible fabrication tolerances have been included in the study. Afterwards, a brief overview on hollow-core waveguides is presented, and the theoretical analysis of a micron-sized device is described. An ideal and a real structure, inherently to the verticality of lateral sidewalls of the core, are considered. The dimensions chosen for the study are smaller than those recently presented in literature.

Keywords: Integrated Optics, SOI Slot Waveguides, Nonlinear Optical Waveguides, Finite-Element Method, Hollow-Core Waveguides.

1. INTRODUCTION

The basic operative principle of integrated optical waveguides is total internal refraction (TIR). According to Snell's law, a light beam is continuously reflected at the interface between a high index medium (core) surrounded by a lower index cladding. The efficiency of

[*] E-mail address: michele.lagioia@hotmail.it

this phenomenon mostly depends on the refractive index difference between core and cladding. Generally speaking, a greater index contrast results in a better light confinement, allowing a higher scale of integration in optoelectronics applications. However, in recent years a number of structures to guide and enhance light in low index material have been developed, too. Differently from TIR, the operative principle of these components is based on external reflections provided by interference effects in structures as, for example, multiple dielectric layers or photonic crystals. Although, the external reflection in these devices is never complete, often causing the propagation of leaky modes. Furthermore, these structures are considerably wavelength-dependent because of interference effects.

Recently, the possibility of guiding light in a low refractive index medium of a waveguide by total internal reflection has been proposed and demonstrated [1-3] thanks to an innovative structure, known in literature as a slot waveguide (see Figure 1). Slot waveguides are generally constituted by two silicon wires spaced by a nanometer low index region (called either slot or gap region), and confined by a silicon oxide layer using Silicon-on-Insulator technology (SOI substrate). The novelty of these devices is the possibility to guide light by confining the electromagnetic field in a narrow low-index gap between two high-index photonic wires. This effect enables the propagation of an eigenmode of the structure, which is fundamentally lossless, with very low wavelength sensitivity.

Another kind of low index device is the so-called hollow-core channel waveguide, consisting of a rectangular channel etched in a silicon substrate surrounded by alternating layers of high-index contrast materials. This kind of structure is reported to have low propagation losses [4].

In this chapter we present both slot and hollow-channel waveguides, focusing particular attention on the former. Operative principles, applications, and modeling are proposed for both structures, considering near and mid-infrared spectral windows. Moreover, an iterative method to evaluate the nonlinear performance of such waveguides is proposed and applied to SOI slot waveguides. By this method, optimal geometrical parameters to maximize nonlinear phase shift of the guided mode have been evaluated. Small variation in the device geometry are also simulated in order to account for technological tolerances. Figure 1 shows both vertical (a) and horizontal (b) configurations for slot waveguides, where w_s designates the gap region thickness, w_h the wire thickness and h the wire height. High index contrast is obtained between n_H and n_S refractive index regions.

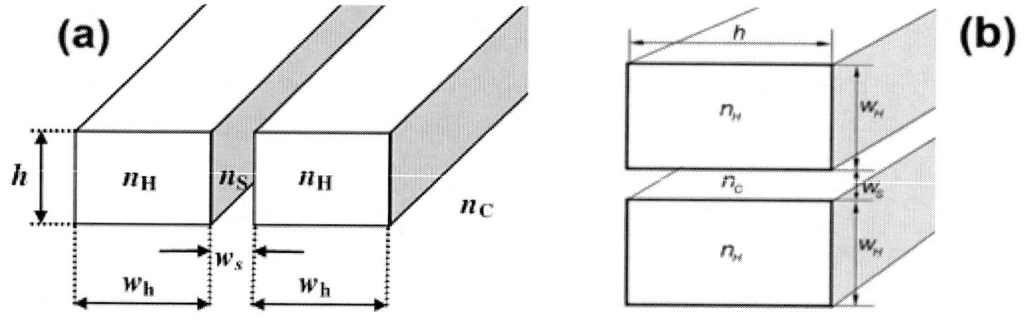

Figure 1. Schematic of slot waveguide structures: (a) vertical; (b) horizontal geometry.

2. Slot Waveguides: Principle of Operation

In the slot waveguide, the optical field can be confined in the low-index material of the gap region even if light is guided by TIR. The principle of operation of this structure is based on the continuity of normal component of electric flux density $D = \varepsilon_r \varepsilon_0 E = n^2 \varepsilon_0 E$ to the high-index contrast interface (where n, ε_r and ε_0 are refractive index, dielectric constant and vacuum permittivity, respectively), that enables a large discontinuity of the corresponding electric field components with a larger amplitude in the low-index medium. In fact, the amplitude ratio between the electric fields in the two regions is equal to $(n_H/n_S)^2$, where n_H and n_S are the refractive indices of wire and gap regions, respectively. The existence of this discontinuity is usually neglected in low-index contrast structures, while being very important for high-index contrast ones.

In detail, the principle of operation of the considered waveguide can be illustrated analyzing a 2D guiding structure with an infinite height, as that sketched in Figure 2, where a low-index region is surrounded by two high-index slabs. In this structure, the guided mode whose electric field dominant component is along the x axis is a TM-polarized mode, while, as it will be explained in the following, this is not true for a finite height structure. Rigorously, the analytic solution for electric field transversal component of TM mode is [1]:

$$E_x(x) = A \begin{cases} \dfrac{1}{n_S^2}\cosh(\gamma_s x) & |x| < a \\ \dfrac{1}{n_H^2}\cosh(\gamma_S a)\cos\left[\kappa_H(|x|-a)\right] + \dfrac{\gamma_s}{n_s^2\kappa_H}\sinh\left[\kappa_H(|x|-a)\right] & a < |x| < b \\ \dfrac{e^{\left[-\gamma_C(|x|-b)\right]}}{n_C^2}\left\{\cosh(\gamma_S a)\cos\left[\kappa_H(b-a)\right] + \dfrac{n_H^2\gamma_s}{n_s^2\kappa_H}\sinh(\gamma_S a)\sin\left[\kappa_H(b-a)\right]\right\} & |x| > b \end{cases} \quad (1)$$

where $A = A_0\sqrt{k_0^2 n_H^2 - \kappa_H^2}\,/k_0$, κ_H is the transverse wave number in high-index regions, γ_C is the field decay coefficient in the cladding, γ_S is the field decay coefficient in the slot region, $k_0 = 2\pi/\lambda_0$ is the vacuum wave-number and A_0 is an arbitrary constant. The transverse parameters κ_S, γ_H and γ_C obey the relations $k_0^2 n_H^2 - \kappa_H^2 = k_0^2 n_C^2 + \gamma_C^2 = k_0^2 n_S^2 + \gamma_S^2 = \beta^2$, where β is the eigenmode propagation constant, which can be calculated by solving the transcendental characteristic equation:

$$tan\left[\kappa_H(b-a) - arctan\frac{\gamma_C n_H^2}{\kappa_H n_S^2}\right] = \frac{\gamma_C n_H^2}{\kappa_H n_S^2} tanh(\gamma_S a) \quad (2)$$

From Eq. (1), the electric field immediately inside the slot ($|x| = a^+$) is n_H^2/n_S^2 times higher than that on the interface side into the gap region ($|x| = a^-$) for TM-polarized eigenmode. When the width of the gap region is much smaller than the decay length inside the slot ($a << 1/\gamma_S$), the optical field remains high all across the slot, because of the superposition in the slot region of the evanescent tails of the TM electric field x component of high index wires [1], as sketched in Figure 2(b).

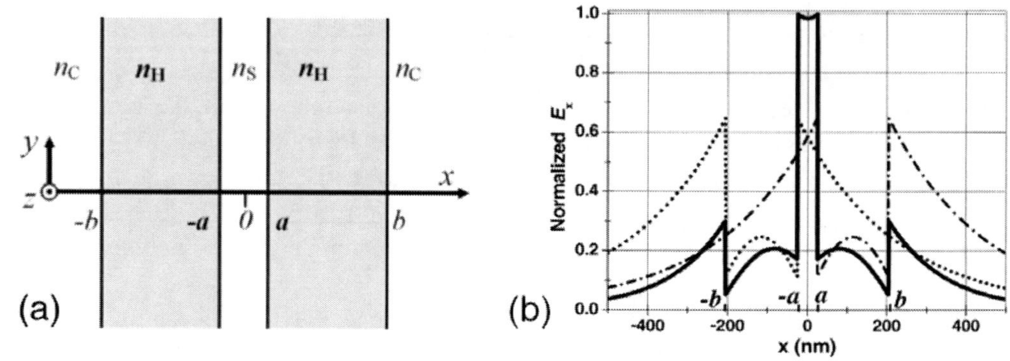

Figure 2. (a) Schematic of slot waveguide structure with infinite height and (b) relative normalized Ex distribution of fundamental TM eigenmode (solid curve) at $\lambda_0=1.55\mu m$, with $n_H=3.48$, $n_S=n_C=1.44$, $a=25nm$ and $b=205nm$. The individual slab waveguide TM modes are shown, too (dotted curves).

Now, considering a 3D slot waveguide structure with finite height, as that in Figure 1(a), the propagating mode, whose E-field dominant component along the direction orthogonal to the high index contrast interface designates a quasi-TE mode, is analogous to TM mode of previous 2D structure. In fact, the electric field along the x direction shows a strong discontinuity on the lateral interfaces and allows the quasi-TE mode to be confined in the gap region. As a result of the E-field enhancement in that region, the optical intensity along z is also much higher than that in the high index regions, as stated in the Poynting theorem, whereas the magnetic field of quasi-TE mode is continuous on the slot walls and is slowly variable across the structure. Quasi-TM mode, whose dominant component of the electric field is parallel to the slot sidewalls, the effect of the presence of the slot is negligible, since the electric field is continuous along the slot walls.

To experimentally demonstrate that quasi-TE optical power is really confined in the gap region, differently from what happens to quasi-TM mode, effective index measurements for both waveguide eigenmodes have been carried out and compared with the case without slot in the structure [3]. Thus, assuming that the field decays exponentially along the radial distance from the slot centre (in x-y plane), the effective index can be found from the decay coefficient γ_C of the optical field in the cladding regions through the relation:

$$n_{eff} = \sqrt{n_C^2 + \gamma_C^2 / k_0^2} \tag{3}$$

hence n_{eff} can be obtained experimentally by measuring γ_C. With this aim, directional couplers (Figure 3) have been fabricated and demonstrated in literature, each composed of two parallel slot waveguides separated by different distances d [3].

Since the coupling coefficient between the two parallel slot waveguides is proportional to the overlap of electric fields of both slot waveguides, the decay rate γ_C is related to the former, which is in turn related to power ratio K between output power from one slot waveguide and input power in the parallel waveguide.

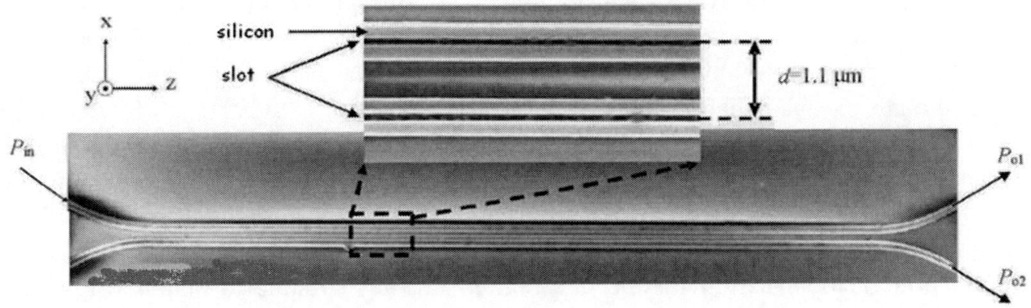

Figure 3. Picture of silicon slot directional coupler as proposed by Xu *et al.* [3].

Thus, γ_C can be obtained by means of [3]:

$$\gamma_C = -ln\left[arcsin\left(\sqrt{K(d)}\right)/L\right]/d + C \qquad (4)$$

where L is the directional coupler length and C a negative constant independent from d. Then, measuring K for different values of d, γ_C is calculated and, from Eq. (3), effective index n_{eff} too. By repeating this procedure for both quasi-TE and quasi-TM modes, and comparing the respective effective indices with that found without the slot region (i.e. $w_s=0$), it has been observed that the presence of gap region must cause a strong decrease for quasi-TE mode only, whereas the effective index of quasi-TM mode must be only marginally influenced [3]. This represents an experimental demonstration of quasi-TE mode high confinement (differently from quasi-TM) in low-index slot region.

2.1. SOI Slot Waveguides: Fabrication and Applications

Nowadays, a very promising solution to realize integrated optical devices is the well-known SOI technology, which allows fabrication of integrated photonic systems thanks to high index contrast between low-index buried oxide and high-index overlaying silicon, as well as low propagation losses and, above all, complete compatibility with widely developed and low-cost CMOS technology [5-6]. In traditional SOI waveguides, vertical confinement of light is obtained by exploiting the index contrast between the silicon core and the low index cladding and buried oxide layers, whereas lateral confinement can be obtained by etching. In such waveguide, the majority of light tends to be guided within the core and this can be problematic for some applications. In fact, at very high optical intensities, two-photon absorption in silicon may lead to significant optical losses. In addition, it is often desirable to maximize the field intensity overlap between the optical guided mode and the lower index cladding material [7]. Thus, a possible solution to these problems consists of employing a slot waveguide geometry, whose high-index wires are made by silicon, while the cladding and slot regions are in silicon oxide (SiO_2), air, or even other materials, depending on particular applications.

These waveguides have been fabricated in vertical and horizontal geometrical configurations (see Figure 1). In the former (see [3], [7] and [8]), the slot guiding structure is normally obtained on a SOI wafer depositing a silicon film on the top of buried oxide layer. After that, the two silicon wires are defined through electron beam (e-beam) lithography, which allows a submicron fabrication, followed by a deep plasma reactive ion etching (RIE). Finally, through plasma-enhanced chemical vapour deposition (PECVD) the SiO_2 cladding layer is deposed. A valid alternative to e-beam lithography, as well as widely used in components fabrication for Microelectronics, is the deep-ultra-violet (DUV) lithography. This low-cost technique is certainly promising for a future mass fabrication of slot waveguides. In particular, the 193nm DUV together with anisotropic plasma etching with HBr allows satisfactory fabrication margin for the gap region; making possible to obtain wires of the order of 100nm [8]. As regards the horizontal configuration, it is defined by thin film fabrication techniques and it does not require any high resolution lithographic process to pattern the narrow low-index region between the wires. By this way, both scattering losses due to Si/SiO_2 interface roughness are reduced. A gap region narrower than 100nm can be defined by employing ion implantation, lateral growth epitaxy and wafer bonding, that allow to obtain smoother interfaces than those produced by etching techniques used for vertical slots [9-10]. Thus, although an horizontal configuration allows to define more accurately both interfaces and slot thickness, a vertical configuration requires in principle a minor number of steps and a major simplicity of fabrication. As a result, the choice will be influenced by the particular application.

Another kind of SOI slot waveguide devices have been fabricated on a Si_3N_4/SiO_2 platform ([11]-[13]). In fact, even though the high index contrast between silicon and oxide allows a high power confinement in the gap region and a high integration level, a Si/SiO_2 waveguide is more sensitive to material interface roughness. Then, silicon nitride, whose refractive index is lower than silicon index but larger than SiO_2, allows to implement photonic devices less sensitive to fabrication tolerances. In particular, the use of Si_3N_4 for slot waveguide rails allows a larger slot region to be achieved, making easier to fill the slot volume with other materials (i.e. fluids) for specific applications. Fabrication of such a structure is based on a thermal growth of a few microns thick bottom SiO_2 cladding layer on silicon wafer to realize the SOI substrate; then a hundreds of manometers thick Si_3N_4 layer is deposited by low-pressure chemical vapour deposition (LPCVD) to define wires height and, finally, the device is patterned by using e-beam lithography and inductively coupled plasma (ICP) etching and covered by a SiO_2 layer deposited by LPCVD [12-13].

The opportunity of confining light in a slot nanostructure allows the introduction of new photonic devices in which, on one hand, the characteristics of some optical active materials can be more effectively exploited with respect to the traditional strip and rib waveguides, and on the other hand, several optical functions can be realized using particular architectural configurations where slot waveguide guarantees an high confinement of the guided light.

Important results have been obtained in realizing resonant structures with high quality factors based on SOI slot waveguides. As an example, micro-resonators have been designed and fabricated with an oval geometry [7]. This configuration is preferred to the conventional circular shape since displays larger coupling length. Through this method, a quality factor up to 27000 has been demonstrated. Alternatively, microdisk resonators with a horizontal slot structure (gap thickness 60nm), fabricated through deposition (instead of lithography) has allowed to reduce propagation losses caused by scattering phenomena at high-index sidewalls, and achieving satisfactory quality factors (about 1000) [9]. As previously mentioned, despite of lower confinement factor in the gap region, the use of a slot waveguide with Si_3N_4 wires presents some advantages. In fact, silicon nitride has been employed to realize both microring and Fabry-Perot resonators based on vertical slot waveguides with quality factors higher than 10000 and propagation losses lower than 20dB/cm [12]. Slot waveguide micro-resonators have found noticeable application in biochemical sensing. The first integrated device for such application based on a slot waveguide [13], and experimentally demonstrated, consisted of a microring resonator in Si_3N_4/SiO_2 with a slot width of 200nm, which is suitable to be filled with a liquid sample. The ring is fabricated by using the previously described techniques and before SiO_2 cladding deposition, a layer of Cr is thermally evaporated and patterned by lift-off technique on a rectangular region around the ring. Through the last step, a window is open over the ring sensor. A linear shift of 212.13 nm /RIU in the resonance wavelength has been demonstrated. This is more than twice than a conventional strip-waveguide ring resonator biochemical sensors [13].

Apart from sensing applications, SOI slot waveguide resonators have been used in fabrication of light emission devices compatible with CMOS technology, since an optical cavity enhances LED luminosity. In fact, using SiO_2 doped with erbium to fill the gap region, an electrically driven LED [10] has been realized working at 1.55μm; whereas, filling an horizontal slot with SiO_2 with erbium doped Si nanoclusters, a strong light enhancement of emission at 1.54μm has been experimentally demonstrated [14]. The use of materials doped with erbium in the slot region guarantees an high miniaturization level, because the carrier injection efficiency is related to the optical field confinement into the narrow low-index region, while this efficiency increases with the cross-section of active region for a traditional TIR-based waveguide.

Due to high index contrast between the materials, Si/SiO_2 waveguides are usually highly polarization dependent. Then, conventional optical devices as directional couplers need particular schemes employing splitters and rotators to align the polarization state, leading to an increase of both structure dimensions and complexity. Thus, thanks to the different sensitivities of quasi-TE and quasi-TM modes in slot waveguides (and their dependency on the waveguide parameters), novel polarization independent optical devices have been proposed [15-16]. One of these devices is a directional coupler composed by SOI slot waveguides [15], whose coupling length is independent from the particular field polarization. In fact, the coupling length shows an high sensitivity to refractive index n_s of the slot region only for the quasi-TE mode, which is highly confined in that region. On the other hand, quasi-TM mode has a profile distributed all over the waveguide cross-section, becoming almost independent from the index variation. As a result, a refractive index value that leads to the same coupling length for both modes can be found, and it decreases with the distance between the waveguides. The coupling length that guarantees polarization independence varies according to the waveguide dimensions. Another polarization independent optical device is the multimodal interference (MMI) beam splitter, shown in Figure 4 [16]. In this case the horizontal configuration is used, so the confined eigenmodes are quasi-TM. The input waveguides are connected with a larger slab waveguide, whose width is W. It is possible to demonstrate that there is a determined value of refractive index in the slot region n_s whose beat length is:

$$L_\pi = \frac{\pi}{\beta_0 - \beta_1} \quad (5)$$

L_π is the same for quasi-TE and quasi-TM modes because of the major sensitivity of quasi-TM modes to n_s variations, so it is important to appropriately choose both W and n_s. In Eq.(5), β_0 and β_1 are the propagation constants of fundamental and first order modes of the slab waveguides, respectively.

The fabrication of Si/SiO_2 slot waveguides could need integration with other kinds of waveguides. To this aim, a strip to slot (and vice versa) modal converter capable of converting Gaussian-like waveguide modes to non-Gaussian-like slot waveguide modes with a coupling efficiency of 99.8% and adopting several pairs of complementary tapers, has been presented [17], as well as a mode transformer between horizontal and vertical slot waveguides [18]. Furthermore, a 99.3% coupling efficiency was achieved for a 3μm long slot to channel waveguide (and vice versa) coupler [19].

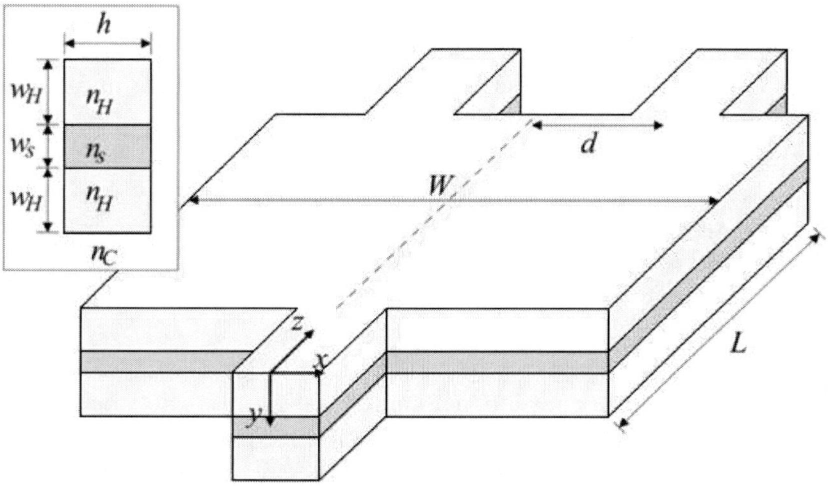

Figure 4. MMI beam splitter based on sandwiched structures (waveguide geometry in inset).

The innovative structure of slot waveguide makes possible to concentrate a large portion of the optical mode into the gap region and, because of the possibility to isolate the two silicon wires, it show a good predisposition to electro-optic interaction. In fact, a nonlinear $\chi^{(2)}$ polymeric material, deposed in cladding and gap regions, can exploit either nonlinear optical rectification to enable a current flux between two silicon electrodes [20], or Pockels effect (noticeably incremented thanks to device minimal sizes) to modulate the travelling field, obtaining a Mach-Zehnder modulator with $V_\pi \cdot L = 4$ mV/cm or a ring resonator with a tunability of 0.5THz/V [21]. Another proposed slot-based electro-optic modulator is characterized by integration with Bragg reflectors in order to vary the transmitted signal characteristics [22].

The best way to exploit the nonlinear Kerr effect in slot waveguides consists in filling the gap region with silicon nanocrystals (Si-nc), because this material exhibits a nonlinear index of orders of magnitude greater than both Si and SiO_2 ones, and it can be applied by using traditional CMOS fabrication techniques. Then, it has been possible to fabricate both a ring resonator optical switch with an operative frequency of 94.7GHz and pump power of only 6.2mW [23]. Other proposed devices using optical Kerr effect in slot waveguide directional couplers are all-optical logic gates [24], whose operative principle is based on a change of quasi-TE mode transmission characteristics when nonlinear effect is activated. For maximizing either electro-optic or Kerr effect, an increase of the superposition between the waveguide active material and light beam is needed. The best way to achieve it is slackening the group velocity of the beam integrating the slot waveguide with a periodic structure like a photonic crystal [25].

Another nonlinear application that uses slot waveguides is the frequency doubling due to birefringence and supporting cross-polarized fundamental and second-harmonic modes. In fact, using AlGaAs wires, the phase-matched second-harmonic generation could be achieved in a wide spectral range [26].

Recently, it has been shown that slot waveguides experience forces between the two sides, which are due to dipole moments induced by the guided light intensity. This force may be either attractive or repulsive, depending on the relative phase of the dipole moments, dictated by optical properties and geometrical parameters of the waveguide [27]. The combined optical, mechanical and electrostatic properties of Si/SiO_2 slot waveguides make them useful for Nano-Opto-Electro-Mechanical System (NOEMS). These devices are capable of achieving several functionalities, such as phase and/or amplitude modulation, polarization mode dispersion compensation, group velocity dispersion compensation, tuneable delay lines, polarization control, NOEMS tweezers, infrared near-field scanning optical microscopy tips, to mention a few [28]. A noticeable NOEMS application is a novel integrated optical sensor with ultrahigh sensitivity [11], consisting of a silicon-based disk resonator formed by a horizontal slot waveguide acting as a circular cantilever, whose sensitivity is four orders of magnitude larger than that of conventional microcantilever sensors. For realizing high optical mode confinement in nanometer thin low-index media, a multiple stacked horizontal slot waveguide structure has been proposed, achieving an increment on the confinement factor up to 85% than the traditional slot waveguide [29].

2.2. Slot Waveguides: Linear Response

In this section the most relevant numerical results about the linear response of SOI slot waveguides are reported and analyzed. The investigation has been carried out by a full-vectorial 2D finite-element method (FEM) [30] to calculate both modal profile and effective index of the eigenmodes supported by the structure. For this purpose, a modal analysis for the three components of the electric field (chosen as independent variable) has been carried out by solving the equation:

$$\nabla \times \left(\mu_r^{-1} \nabla \times \mathbf{E} \right) - k_0^2 \varepsilon_r \mathbf{E} = 0 \tag{6}$$

where $k_0 = \omega\sqrt{\varepsilon_0 \mu_0}$ and $\mathbf{E}(x,y,z) = \mathbf{E}(x,y)e^{-j\beta z}$, while ε_r and μ_r are relative complex permittivity and permeability, respectively, expressed as:

$$\varepsilon_r = \begin{bmatrix} \varepsilon_{rxx} & \varepsilon_{rxy} & 0 \\ \varepsilon_{ryx} & \varepsilon_{ryy} & 0 \\ 0 & 0 & \varepsilon_{rzz} \end{bmatrix} \qquad \mu_r = \begin{bmatrix} \mu_{rxx} & \mu_{rxy} & 0 \\ \mu_{ryx} & \mu_{ryy} & 0 \\ 0 & 0 & \mu_{rzz} \end{bmatrix} \tag{7}$$

in which the elements of the ε_r tensor are complex. The triangular mesh has been adopted as constituted by linear Lagrange elements (first order triangles). The maximum element size is fixed as equal to 3-3.5nm in the slot region, 4-5 nm in the wire regions, 7.5-9 nm in a rectangular region extending on cladding and buried oxide (1μm wide and 0.8μm high) and centered with the slot region, depending on the waveguide geometrical dimensions. However, as a general criterion, the total number of mesh elements has been chosen around 100,000 (about 200,000 degrees of freedom). By this way, a good trade-off between calculation time and accuracy can be achieved.

Considering a vertical slot waveguide whose cladding and slot regions are constituted by SiO_2, both effective indexes and confinement factors into the gap and cladding regions are reported in Figure 5 as a function of three dimensional parameters for both quasi-TE and quasi-TM modes at the telecommunication wavelength (1.55μm). Refractive indices of Si and SiO_2 are 3.48 and 1.444, respectively. The power confinement factor into a particular region Ω of the entire domain is defined as:

$$\Gamma_{\Omega} = \frac{\iint_{slot} \left|\mathbf{E}(x,y)\right|^2 dxdy}{\iint_{\mathbb{R}^2} \left|\mathbf{E}(x,y)\right|^2 dxdy} \tag{8}$$

and that the cladding region is the domain portion surrounding wires and gap regions (excluding buried oxide and silicon substrate).

It can be observed from graphs in Figure 5 that, for quasi-TE mode, a reduction of slot width w_s leads to an increase of both effective index and confinement factor into the slot region Γ_{slot} and, simultaneously, a reduction of confinement factor into the cladding region Γ_{cover}, because of the larger superposition of the evanescent tails of the field travelling into the wires. The quasi-TM confinement factors are only slightly affected by this variation, because the mode is principally distributed inside wires and cladding regions. However, it is worth observing that the quasi-TE effective index is lower than the quasi-TM one, since the former is more confined into the low index slot region, while the latter is mainly distributed into the high index wires. However, although reducing w_s improves Γ_{slot} for quasi-TE mode, there is a technological limit to this reduction [31]. Then, to avoid fabrication problems we have set the gap region width w_s=100nm for the next parametrical investigations about the vertical structure. Thus, considering w_s = 100nm and h=300nm, as it can be seen from Figure 5, an optimal waveguide height, w_h=220nm, can be found maximizing the quasi-TE field distribution into the gap region.

The presence of a maximum value for Γ_{slot} for the considered mode is due to two conflicting effects. In fact, fixing the waveguide height h, the field evanescent tails superposing into the gap region are short for lower values of the w_h/w_s ratio. Then, increasing the ratio, the tails are longer and Γ_{slot} rises, but, at the same time, the optical field is more localized within the wires and confinement factor decreases for much higher values of w_h/w_s. Then, the increase of the optical field into the high index wires is matched with the effective index that rises with w_h, but this trend is different for the two polarizations. Then, a w_h value can be found where a birefringence free condition occurs (quasi-TE and quasi-TM modes having the same effective index n_{eff}). Moreover, because of the increased width w_h of the wires, there is a simultaneous reduction of Γ_{cover} for both polarizations.

In order to complete the parametrical analysis, from the lowest-lying graphs in Figure 5, a linear increase of the effective indices with h is depicted if this parameter is comparable with w_h and w_s, otherwise a saturation behaviour for higher h values occurs.

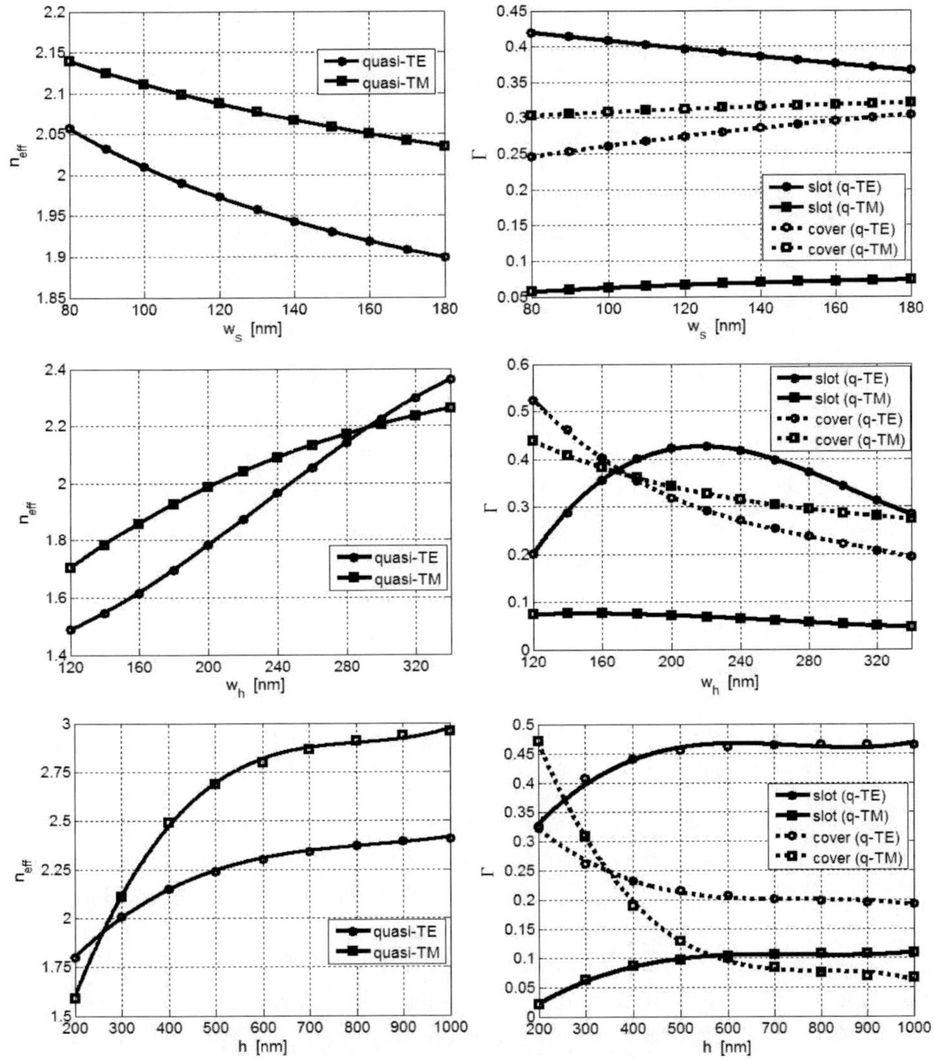

Figure 5. Effective indices n_{eff} and confinement factors Γ into the slot and cladding regions (constituted by SiO_2) for quasi-TE and quasi-TM modes at 1.55μm varying as a function of different dimensional parameters (where not specified, we assume w_s = 100nm, w_s = 250nm and h = 300nm).

This trend matches with Γ_{slot} and Γ_{cover} for both polarizations. Thus, as a general rule, the optimal height has to be chosen before reaching the top of the linear trend of Γ_{slot} (h=500nm for w_h = 250nm and w_s = 100nm) to avoid technological problems (as greater is the slot height, as more inaccurate the etching process is [31]). Another parameter used to optimize the linear performance of the structure is its effective area [32], defined as:

$$A_{eff} = \frac{\left(\iint\limits_{\square^2} \left| \mathbf{E}(x,y) \right|^2 dxdy \right)^2}{\iint\limits_{slot} \left| E(x,y) \right|^4 dxdy} \tag{9}$$

which determines the cross-section area where the confined mode within the gap region is distributed (quasi-TE in a vertical waveguide and quasi-TM in a horizontal one).

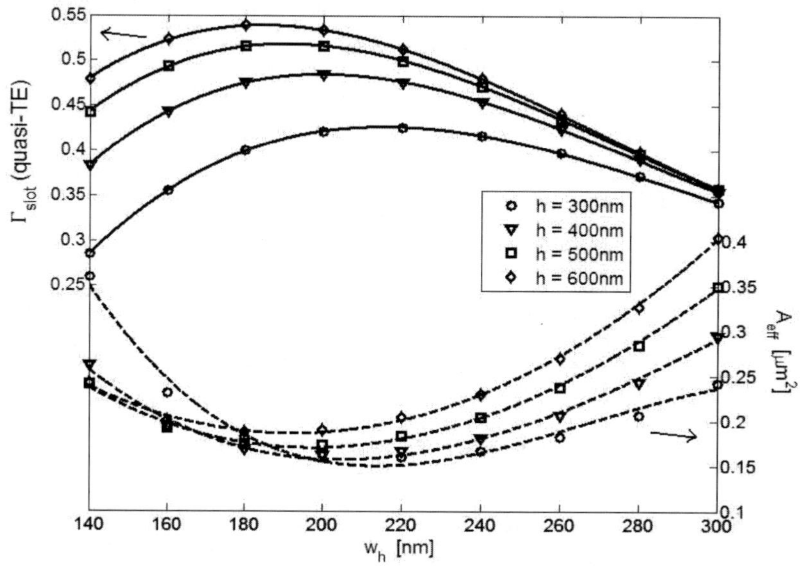

Figure 6. Quasi-TE Confinement factors Γ_{slot} and effective areas A_{eff} at 1.55μm mode for vertical slot with SiO_2 gap and cladding, and $w_s = 100$nm.

Taking into account both confinement factors and effective areas of guided modes and varying the wire dimensions, vertical and horizontal slot geometrical configurations can be optimized, considering w_s = 100nm for the former and w_s = 50nm for the latter (since in horizontal slot waveguide, the slot gap is defined by very accurate thin-layer deposition methods [8]). In particular, considering a vertical waveguide as the previous one and a horizontal waveguide with slot region composed by SiO_2 and covered by air, the mode confinement factors and effective areas have been evaluated as a function of the wire sizes and results are sketched in Figure 6 (TE) and Figure 7 (TM).

As it can be observed, the maximum (minimum) values of Γ_{slot} (A_{eff}) vary with both h and w_h. As an example, considering the graph in Figure 6 for h = 300nm, the maximum value of Γ_{slot} (≈43%) and the minimum of A_{eff} (≈ 0.17μm^2) is achieved for w_h = 220nm, while for h = 600nm the maximum value of Γ_{slot} (≈ 54%) and the minimum of A_{eff} (≈ 0.21μm^2) is achieved for w_h = 180nm. However, it is important to notice that, for small values of w_h (h), the effective area as a function of h (w_h) shows a different behaviour from the confinement factor. This means that, to achieve a good optimization, both Γ_{slot} and A_{eff} must be simultaneously taken into account. The optimal geometrical parameters to reach suitable values of Γ_{slot} and A_{eff} are summarized in Table 1.

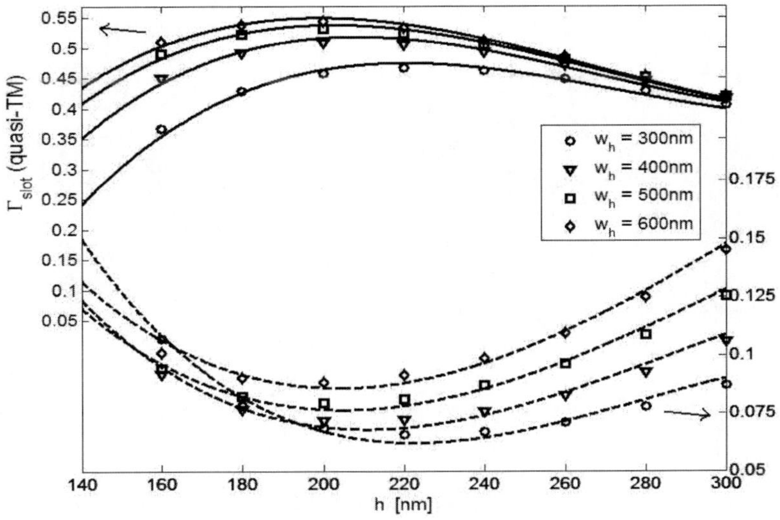

Figure 7. Quasi-TM Confinement factors Γ_{slot} and effective areas A_{eff} at 1.55μm mode for horizontal slot with SiO_2 gap and air cladding and w_s = 50nm.

Table 1. Dimensional parameters to reach suitable values of Γ_{slot} and A_{eff}, from graphs in Figure 6-7

Parameter	Vertical waveguide (quasi-TE) Γ_{slot}> 50%, A_{eff}< 0.2μm^2	Horizontal waveguide (quasi-TM) Γ_{slot}> 50%, A_{eff}< 0.1μm^2
w_s	100	50
w_h	170; 210	500; 600
h	500; 600	170; 240

As previously mentioned, for both vertical and horizontal configurations, the fabrication process includes a deep etching for horizontally defining the structure. Thus, the etching accuracy is very important. However, it is very hard to fabricate slot waveguides with perfectly vertical sidewalls, i.e. the resulting realistic structures are those represented in Figure 8. Thus, it is important to evaluate the influence on the field confinement of the variation of the slot sidewall angle [32].

In Table 2 the variations of effective index, confinement factor and effective area of slot guided mode are reported per each tilting degree of slot sidewalls. Thus, the vertical slot waveguide is much more sensitive to variations of θ angle with respect to horizontal configuration. This is due to the fact that, for the vertical configuration as in Figure 8(a), a variation of θ produces a variation of the total cross-section area of high index regions (wires) which in its turn influences the modal properties, while the total high index cross-section area of the horizontal waveguide remains constant with θ. Therefore, for a vertical waveguide in which the angle (θ') is considered at half height of the structure (as in Figure 9), the influence of the tilted sidewalls is almost negligible, because the cross-section area of the wires does not vary with θ'. However, larger is h, more sensitive to angle variations is the waveguide. It is important to notice that the dimensions of the wires considered in Figure 9 are the optimal values to maximize the confinement factor within the slot region, as shown in Figure 6.

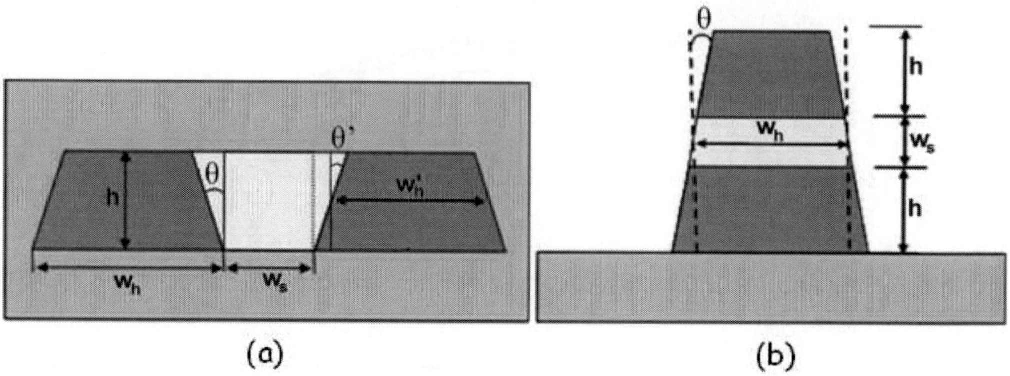

Figure 8. (a) Schematics of vertical and (b) horizontal slot waveguide structures with tilted sidewalls.

Table 2. Variations per degree of effective index, confinement factor and effective area of guided modes within the slot region, for vertical and horizontal slot structures

Structure	Δn_{eff}	$\Delta\Gamma_{slot}$	$\Delta A_{eff}\ [\mu m^2]$
vertical slot	- 0.025/°	- 0.08%/°	+ 0.015/°
horizontal slot	+ 0.002/°	- 0.04%/°	+ 0.002/°

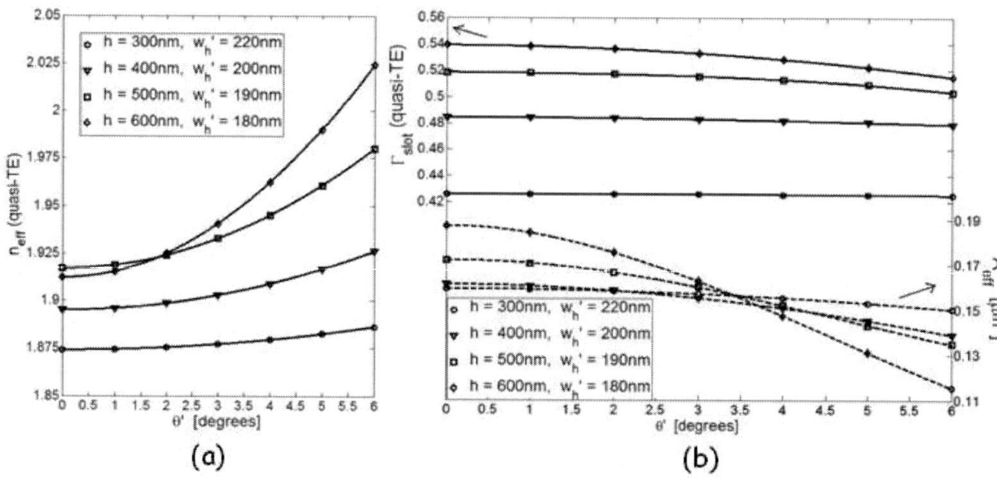

Figure 9. (a) Quasi-TE mode effective index, (b) confinement factor and effective area in a vertical slot waveguide as a function of sidewall tilting *θ'* angle, for a slot width of 100nm.

Besides the investigation at 1.55μm, the linear response of slot waveguide within the mid-infrared spectral region has been carried out too. In fact, recently the possibility of employing the mid-infrared region to achieve efficiently both emission [33-34] and amplification [35], i.e. avoiding nonlinear and free-carrier absorption, has been demonstrated. Therefore, due to the possibility of realizing a silicon amplifier at 3.39μm [35], the response of a slot waveguide at this wavelength has been investigated. For the physical definition of the structure, the linear effective indices of silicon oxide and silicon have been evaluated as 1.409 and 3.403, respectively, by considering the relevant Sellmeier equations at 300°K [36-

37]. Then, as it is shown in Figure 10, the optimal geometrical dimensions for reaching 50% confinement factor within the slot region for quasi-TE mode in vertical configuration (with air as cladding) can be chosen more than twice larger than optimal sizes at 1.55μm, reducing the influence of fabrication tolerances on structure behaviour and allowing to achieve a better structural accuracy. On the contrary, using same dimensions but at 1.55μm, the slot waveguide degenerates into a modal coupler [38]. In Figure 10 the single-mode condition for the quasi-TE mode in a vertical slot waveguide is also sketched, varying with w_h. It can be observed that single-mode condition (SMC) displays a nearly linear behaviour between slot width w_s and wire width w_h (see Figure 10(b)), as it happens for suitable dimensions at 1.55μm [39], while the condition originates a linear behaviour only for h values < 1.05μm, by fixing w_s= 400nm as in Figure 10(a). Furthermore, it can be noticed that, although w_s reduction increases the power confinement, it simultaneously reduces the available values of w_h providing SMC. Therefore, a trade-off is required between confinement factor and SMC, in order to optimize the linear response of the slot waveguide.

Silicon Photonics can be positively exploited for optical sensing applications; for example, silicon integrated lab-on-chip systems, in which photonic sensing functions are integrated with artificial intelligence and wireless communications, can be employed in environmental monitoring and medical diagnostic [5].

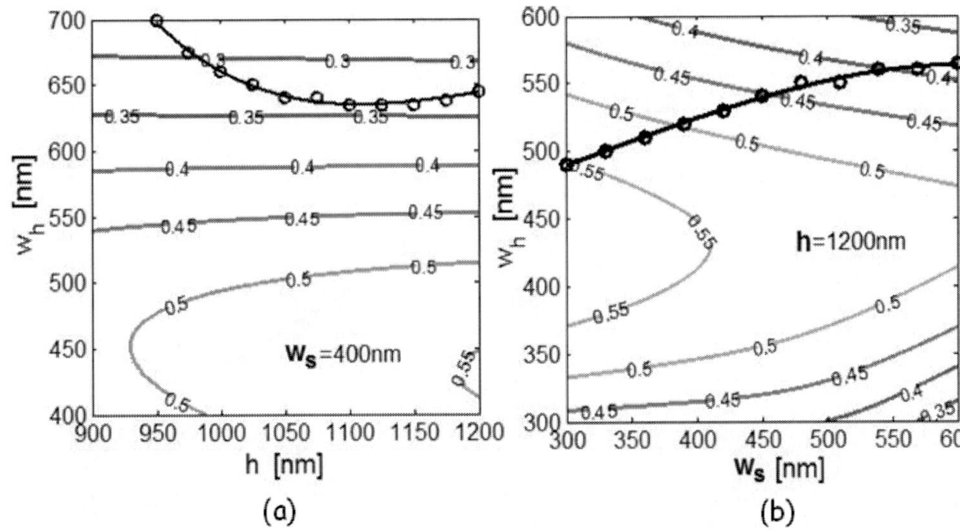

Figure 10. Confinement factors and single mode conditions (lines with circles) for quasi-TE mode in a vertical slot waveguide (covered by air) at wavelength 3.39μm, fixing (a) slot width w_s or (b) height h.

SOI slot waveguides for integrated optical sensors have been recently proposed [40], allowing to achieve an homogeneous sensitivity larger than one at 1.55μm using an optimized set of geometrical parameters. Homogeneous sensing (effective index change of guiding structure due to the refractive index change of material or aqueous solution inside the slot and cladding regions), can be evaluated according with variational theorem of dielectric waveguides, using the relation:

$$S_0 = \frac{n_{sol}}{Z_0 P} \iint_{\Omega} \left|\mathbf{E}(x,y)\right|^2 dxdy \tag{10}$$

where S_0 is the homogeneous sensitivity, Ω is the domain region in which the aqueous solution is distributed including slot and cladding regions, Z_0=120π is the free-space impedance, n_{sol} (for example 1.3 [41]) is the refractive index of the aqueous solution and P is the power transported along the propagation direction z, defined as:

$$P = \frac{1}{2}\iint_{\infty}\left[\left(\mathbf{E}\times\mathbf{H}^{*}+\mathbf{E}^{*}\times\mathbf{H}\right)\cdot\hat{z}\right]dxdy\,. \tag{11}$$

Considering a slot width of w_s = 400nm at 3.39μm, the sensitivity as a function of both wire dimensions, h and w_h, is represented in Figure 11. Although obtaining a sensitivity greater than one could seem strange and unexpected, it has been already demonstrated and physically explained [42].

The positioning of the gap region between the wires in a vertical slot waveguide depends on the mask alignment of the lithography step. It is clear how a small shift of the gap region is unavoidable during the fabrication process, causing a so-called asymmetric waveguide. However, it is useful to remember that the main advantage of employing a mid-infrared instead of near-infrared wavelength region is the relaxation of fabrication tolerance, because a 50% confinement factor can be achieved using larger gaps.

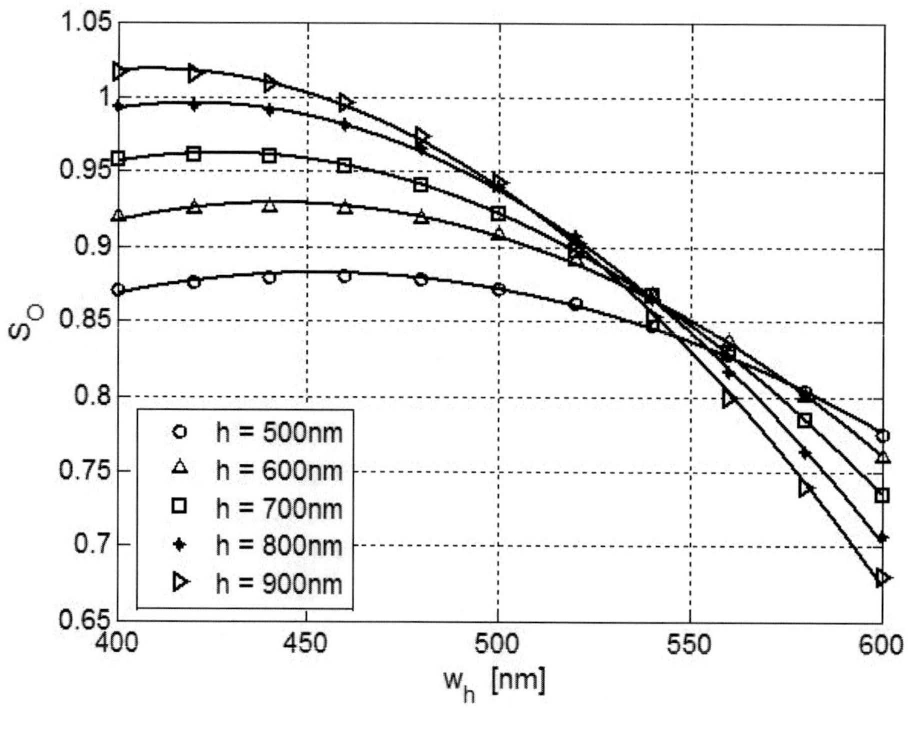

Figure 11. Homogeneous sensitivity of quasi-TE mode varying with wire width w_h and height h for a vertical slot at 3.39μm, with w_s = 400nm.

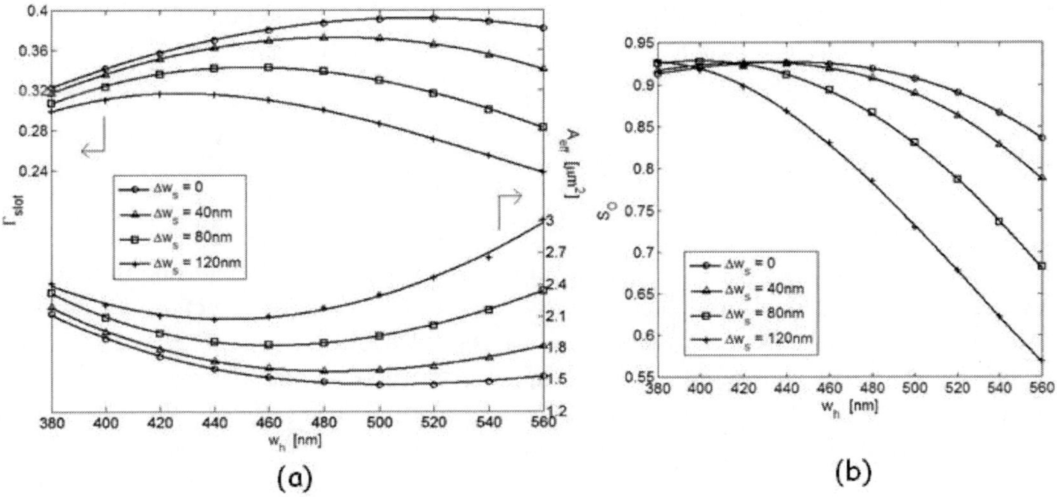

Figure 12. (a) Confinement factor, effective area and (b) sensitivity for quasi-TE mode in a vertical slot as a function of w_h and Δw_s with $w_s = 400$nm, $h = 600$nm, at 3.39μm.

Then, considering an horizontal shift Δw_s of the gap region as regards the central position, the wire widths should become w_h-Δw_s and w_h+Δw_s respectively, and the E_x contour lines should be more concentrated within the wider wire, determining an asymmetrical superposition of the field tales within the slot region. This asymmetrical field distribution causes a reduction of the total confined field within the slot region and an increase of the effective area of quasi-TE mode, as sketched in Figure 12(a), but also allows to decrease curvature losses in sharp-bended waveguides [43]. However, in Figure 12(b) it can be deduced that the maximum value of S_O achievable varying with Δw_s is always greater than 0.9, if w_h is appropriately chosen, because the homogeneous sensitivity does not directly depend on the confinement factor in the slot region but depends on the sum of the confinement factors within both regions, slot and cladding.

2.3. Slot Waveguides: Nonlinear Response

In such an ultra-small structure as the slot waveguide, the light intensity traveling along the propagation direction is considerably larger than that in conventional optical waveguides, giving a dramatic enhancement of nonlinear optical effects. Moreover, the waveguide nonlinearity is further enhanced [23-24, 44], due to the strong light confinement in the gap region for quasi-TE modes. Combining nonlinear optical materials with silicon photonic circuitry, both sizes, weight, and power consumption can be reduced by this potentially revolutionary new technology [45], achieving better performance in terms of speed and bandwidth. In this section, an optimization of silicon-on-insulator nanometer slot waveguides for nonlinear applications and nonlinear phase shifts induced by optical Kerr effect is described and proposed, adopting different materials as gap region medium and various input powers.

In general, the materials to take into account for all-optical signal processing belong to three groups: glasses, semiconductors and organic materials [46]. The starting point to make a choice among them depends on strong third order nonlinearity, i.e. optical Kerr effect. The ultrafast (<10 fs) nonlinear response of a material is given by the variation of its refractive index, depending on nonlinear coefficient n_2:

$$\Delta n = n_2 I \tag{12}$$

where I is the intensity of the input beam. Anyway, this evaluation of material efficiency for all-optical processing based only on n_2 strength is not enough. In fact, the influence of absorption phenomena plays a fundamental role too, because absorption reduces the intensity of the beam propagating into the medium. Thus, for sufficiently strong Kerr interactions in the active region of a waveguide at a certain wavelength λ_0, both linear and two photon absorption (TPA) should be weak enough compared with Kerr effect. Effectively, other conditions depend on the properties of the input optical pulse, i.e. laser characteristics, and not on medium ones [47]. Therefore, for establishing if the Kerr effect is dominant over absorption, two figures of merit are usually used [48]:

$$\begin{cases} W = \dfrac{n_2 I}{\alpha_0 \lambda} \\ T = \dfrac{\alpha_2 \lambda}{n_2} \end{cases} \tag{13}$$

where α_o is the linear absorption coefficient and α_2 is the TPA non linear coefficient. By this way, a good material for all-optical signal processing should have $W > 1$ and $T < 1$. Among the three classes of afore-mentioned materials, silica fibers are the best media as regards these figures of merit, but they also have a too low index n_2 to ensure practical use in integrated optic devices [46]. At the best of our knowledge, the materials that show simultaneously large nonlinear refractive indices, good figures of merit and not relatively difficult processing techniques include silicon nanocrystals [49-50], conjugated polymers MEH-PPV [51-52] and fullerene-containing polyurethane films [53].

Then, in order to consider the nonlinear response of the structure and its performance regards to Kerr effect, a geometrical optimization of SOI slot waveguides by means of an iterative method is needed. To this aim, a stationary modal analysis has been applied by full-vectorial 2D FEM [30] using the approach described at the beginning of the previous section, and achieving a good trade-off between calculation time and accuracy. Afterwards, the nonlinear effective index of slot quasi-TE mode has been calculated by means of an iterative method similar to that used by Fujisawa and Koshiba for chromatic dispersion in nonlinear holey fibers [54], but employing linear Lagrange elements instead of high order hybrid elements. The method consists of calculating the three electric field components from the solution of electromagnetic wave equation and, then, deriving the refractive index of each region as:

$$n = n_0 \left(1 + \frac{n_2 |\mathbf{E}|^2}{Z_0} \right)^{1/2} \tag{14}$$

where Z_0 is the free-space impedance, and n_0 and n_2 are the linear and non linear refractive indices for each region, respectively. Then, new values of effective index and field distributions are calculated and new refractive index evaluated again by Eq. (14). This iteration is repeated until a fixed effective index tolerance has been reached by numerical convergence. Briefly, the iterative method can be expressed on the grounds of the following steps:

1. Set up initial data in terms of linear refractive indices of whole domain regions, free-space wavelength λ_0 and input optical power P;
2. Set a tolerance value on effective index between two consecutive iterations.
3. Solve the linear problem by FEM, i.e. taking into account the linear refractive index n_0 for the entire domain, and derive linear effective index n_{eff}^{L}, electric field distribution $\mathbf{E}'(x,y,z)$ and optical power along propagation direction z, i.e. P';
4. Derive the normalization coefficient $\gamma = P / P'$ from which the square module of the new electric field is derived as $|\mathbf{E}|^2 = \gamma |\mathbf{E}'|^2$;
5. Substitute the parameters obtained by Eq. (14) to derive the refractive indices of nonlinear regions;
6. Solve the nonlinear wave equation with FEM, i.e. taking into account the refractive indices just obtained;
7. Derive the new values of nonlinear effective index n_{eff}^{NL}, electric field distribution $\mathbf{E}'(x,y,z)$ and power P';
8. Compare the difference between the new and previous effective indices with the tolerance value fixed at step 2. If the tolerance is bigger than the difference, then the last n_{eff}^{NL} is the most correct, otherwise return to step 4.

The correctness and applicability of the described method has been verified and proved by setting up the same analysis of nonlinear slot waveguides as carried out by Fujisawa and Koshiba [44] and comparing the results in terms of nonlinear effective index. The guiding structure used for the comparison is a SOI slot waveguide with silicon wires and SiO_2 cladding, while the slot (gap) region has been considered as filled with silicon nanocrystals, which allow a larger value of n_2 than SiO_2 to be obtained. The physical and geometrical parameters used for comparison are summarized in Table 3. As in [44], the nonlinear index has been considered only in the slot region, while all the other materials are assumed as linear. The extrapolation of $|\mathbf{E}'|$ on the space domain has been carried out by considering an uniform grid of 120,000 points, consisting of 300 points along x axis and 400 points along y axis. Moreover, the tolerance on the difference between effective index values derived from two successive iterative steps has been adopted as 10^{-4}, in order to ensure a sufficient accuracy for the effective index convergence without excessive time consuming. Using the parameters in Table 3 and the previously described simulation parameters, a linear effective index of 1.728703 has been obtained, in very good agreement with value of 1.729706, given in [44]. Moreover, calculation of quasi-TM mode shows an effective index of 1.755593, very similar to 1.755922 [44]. In Figure 13(a) the convergence of the iteration method before reaching the desired tolerance for the nonlinear effective index is sketched.

Table 3. Physical and geometrical parameters for validation of iterative method, as in [44]

Parameter	Definition	Value
λ_0	Operative free-space wavelength	1550 nm
n_0^{Si}	Silicon wire refractive index	3.48
$n_0^{SiO_2}$	Cladding and buried oxide refractive index	1.46
n_0^{nc}	Slot region refractive index	1.46
n_2^{nc}	Nonlinear index of silicon nanocrystals in slot region	10^{-16} m^2/W
w_s	Slot width	100 nm
w_h	Wire width	200 nm
h	Waveguide height	250 nm

The number of iterative cycles, required to reach the fixed tolerance, clearly depends on the input power. In fact, 6 cycles are needed for $n_2^{nc}P = 0.01\,\mu m^2$, while the cycles are increased up to 23 for $0.04\mu m^2$. When the definitive nonlinear effective index n_{eff}^{NL} is found, the nonlinear phase shift per unit length is calculated as:

$$\Delta\phi_{slot} = \frac{2\pi}{\lambda_0}\left(n_{eff}^{NL} - n_{eff}^{L}\right) \qquad (15)$$

where n_{eff}^{L} is the linear effective refractive index and $\Delta\phi_{slot}$ stands for a phase change depending only on nonlinear material in the slot region. Then, $\Delta\phi_{slot}$ versus product $n_2^{nc}P$ has been compared with that given in [44], as illustrated in Figure 13(b).

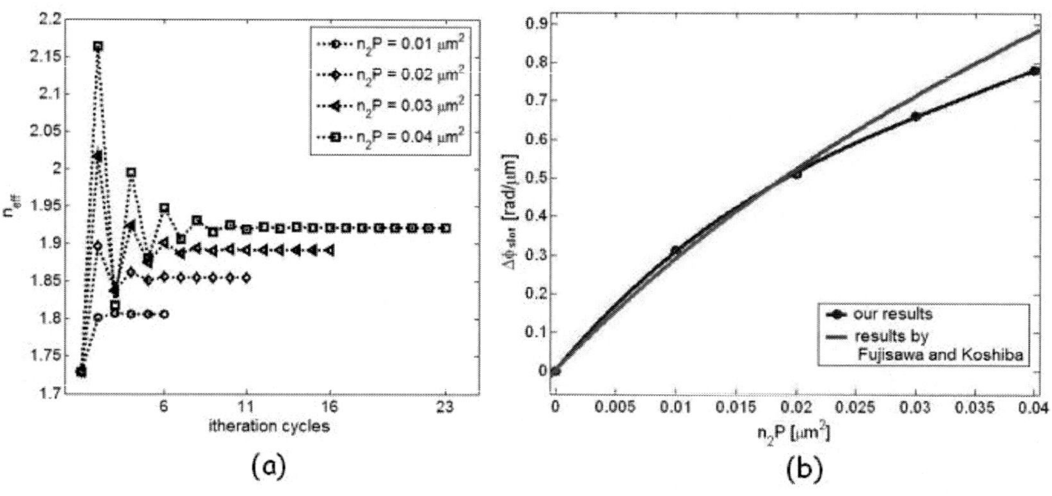

Figure 13. (a) Convergence of quasi-TE effective index versus iteration cycles for different values of $n_2^{nc}P$ (tolerance 10^{-4}); (b) comparison in terms of nonlinear phase shift per unit length as a function of $n_2^{nc}P$ between this chapter and Ref. [44]. Parameters are in Table 3.

It can be noted a remarkable correspondence at low input powers. In fact, the graphs are substantially identical for $n_2^{nc}P \leq 0.02$ μm^2, but the other method tends to overestimate $\Delta\phi_{slot}$ when $n_2^{nc}P$=0.02 μm^2. In fact, the phase difference is 13% smaller than that in [44] for $n_2^{nc}P$=0.04 μm^2. This result is clearly satisfactory, because larger input powers are unlikely due to possible optical damage and necessity of high quality and large intensity light sources.

The above results concern a medium nonlinearity only in the slot region, i.e. where the transversal electric field of quasi-TE mode is at least one order of magnitude larger than that in other regions. Although only a slight influence on the guided field is expected, nonlinearities of any material of the whole waveguide cross section should be considered, giving a more accurate and closer to real case estimation of nonlinear phase change $\Delta\phi_{glob}$. Therefore, the analysis has been carried out considering different cases, i.e. gap and cladding regions as filled with three materials for nonlinear applications available in literature and previously mentioned, whose linear refractive indexes are $n_0^{nc} = 1.46$ [50], $n_0^{C_{60}} = 2$ [53], $n_0^{PPV} = 1.63566$ [51-52], a tolerance of 10^{-4} and $n_2^{SiO_2}$, n_2^{Si}, n_2^{nc} values and other data as reported in Tables 3 and 4. In spite of low index contrast between slot and wire regions, the material good to achieve the largest value of nonlinear phase shift per unit length is demonstrated to be fullerene. In fact, results of the comparison between the considered materials are illustrated in Figure 14(a-b). The fitting coefficients used to draw the curves are reported in Table 5, using a second order polynomial function.

Table 4. Nonlinear refractive indices for different materials at λ_0=1.55 μm

Index	Material	Value	Reference
$n_2^{SiO_2}$	SiO_2	$0.026 \cdot 10^{-18}$ m^2/W	[46]
n_2^{Si}	Si	$4.5 \cdot 10^{-18}$ m^2/W	[55]
n_2^{nc}	Si nanocrystals embedded in SiO_2	$30 \cdot 10^{-18}$ m^2/W	[50]
$n_2^{C_{60}}$	Fullerene with polyurethane	$220 \cdot 10^{-18}$ m^2/W	[53]
n_2^{PPV}	Derivative polymer MEH-PPV	$1 \cdot 10^{-14}$ m^2/W	[52]

Table 5. Fitting coefficients of curves in Figure 14 (related equation $\Delta\phi = (a_2 P + a_1)P$)

Material	a_1	a_2
Fullerene with polyurethane	35681	-5.1987
Silicon nanocrystals	1450.9	-1.2778
MEH-PPV	56.0643	-0.0012

Using the iterative calculation method to find the optimal geometrical parameters, the nonlinear optical performance of slot waveguide can be maximized. The material assumed in the gap region has been silicon nanocrystals embedded in SiO_2, because of the easier and CMOS-compatible fabrication process compared with fullerene. Once the nonlinear material has been chosen, two kinds of simulations have been carried out.

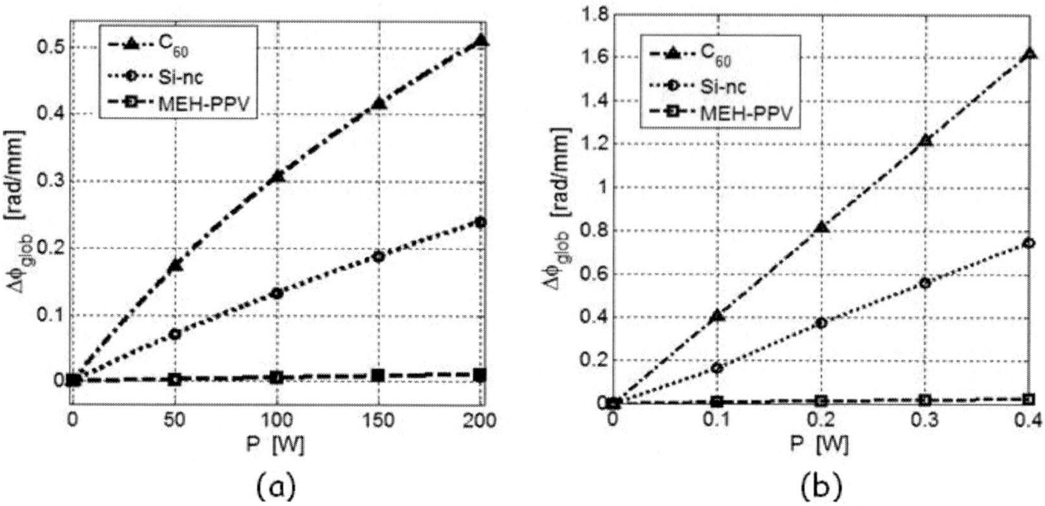

Figure 14. (a) Nonlinear shift per unit length for large and (b) low powers, with different materials in slot and cladding regions.

In the first of them, the nonlinear response of the waveguide has been estimated as a function of the input power for each of the three geometrical parameters, in order to verify if the dependence of $\Delta\phi_{glob}$ on geometrical parameters is similar to linear response in terms of quasi-TE mode effective area [32].

Quasi-TM modes, whose optical field distribution is confined in cladding oxide and silicon wire regions, show a small nonlinear phase shift, almost invariant with the input power. In the second analysis, we have fixed the input power and found the optimal geometrical configuration to have the maximum nonlinear phase shift, $\Delta\phi_{glob}$. The results obtained from the first analysis are illustrated in Figure 15. It can be observed a linear increase of $\Delta\phi_{glob}$ with increasing the input power, with a slope varying with the geometrical parameters, as expected. Moreover, the nonlinear phase shift and the effective area have a similar dependence from geometrical parameters, by comparing the graphs a and b in Figure 15 with the respective insets. However, the trend of $\Delta\phi$ as a function of waveguide height is surprisingly irregular and different from that of the effective area (9).

On the basis of different behaviour of nonlinear phase shift as a function of the waveguide height versus effective area, an optimization of the geometrical parameters as regards the structure nonlinear response has been achieved. With this aim, we have set the gap region width w_s = 100nm (to avoid fabrication problems related to excessive reduction of this parameter [31]) and varied both height and width of silicon wires. Then, a quite high input power (P = 100 W) has been chosen in order to obtain sufficiently accurate results with an effective index tolerance of 10^{-4} (being the accuracy of our FEM solver equal to 10^{-6} [30]). In fact, by decreasing either P or the tolerance, physically unexpected drops in the trend of $\Delta\phi_{glob}$ as a function of geometrical parameters are revealed. Anyway, the geometric dependence of $\Delta\phi_{glob}$ is the same for any power value, because of the linear trend of $\Delta\phi_{glob}$ with P (see Figure 15).

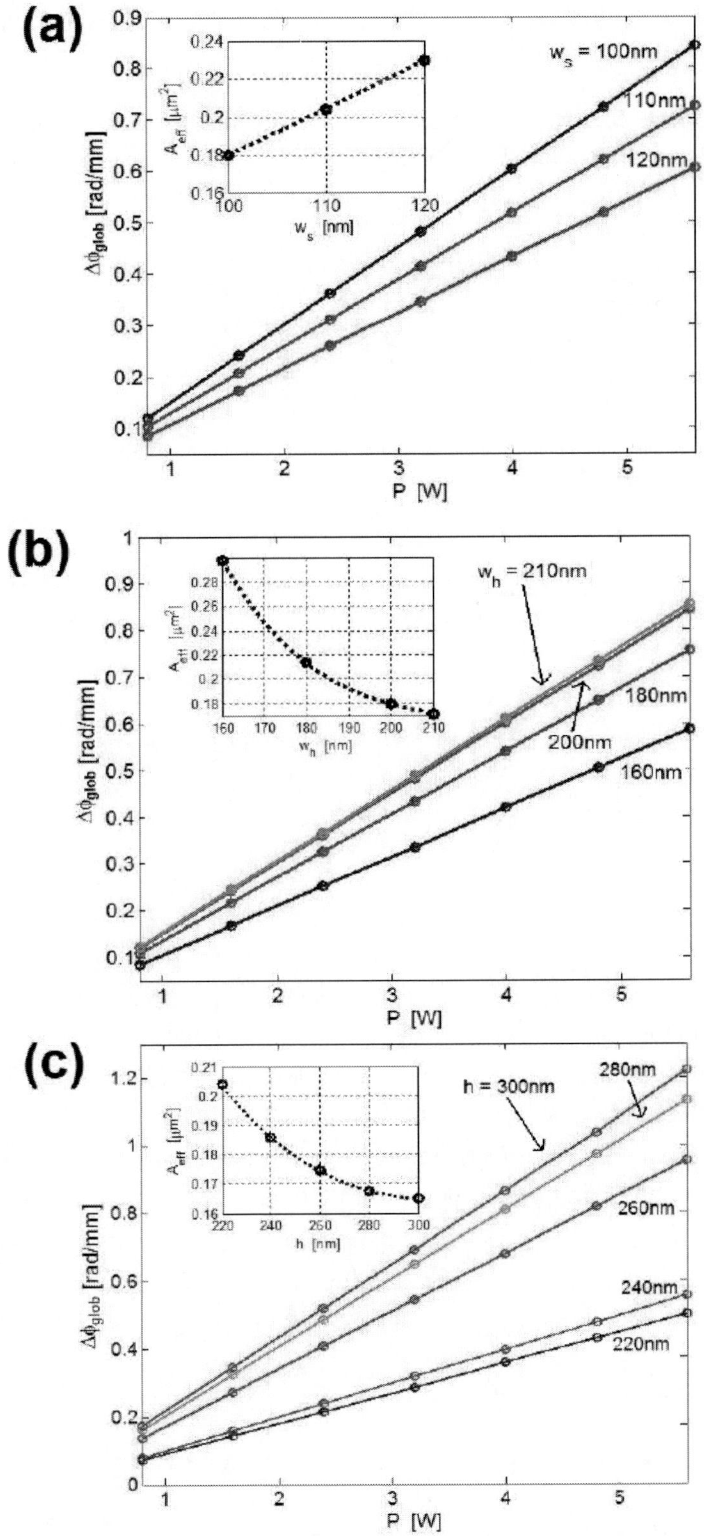

Figure 15. (a) Nonlinear shift as a function of input power for various w_s and h=250nm and w_h=200nm; (b) varying w_h, for w_s=100nm and h =250nm; (c) varying h, for w_s=100nm and w_h= 250nm.

The simulation results, sketched in Figure 16, have been fitted by a third order polynomial function, whose coefficients are summarized in Table 6.

Table 6. Fitting coefficients of curves in Figure 16 (fitting equation: $\Delta\phi = a_0 + a_1 w_h + a_2 w_h^2 + a_3 w_h^3$)

h [nm]	a_0	a_1	a_2	a_3
200	-0.0803	$-0.0342 \cdot 10^{-2}$	$1.2332 \cdot 10^{-5}$	$-3.3212 \cdot 10^{-8}$
300	-1.1206	$1.6879 \cdot 10^{-2}$	$-6.8264 \cdot 10^{-5}$	$8.3436 \cdot 10^{-8}$
400	-0.7274	$1.1262 \cdot 10^{-2}$	$-4.9305 \cdot 10^{-5}$	$6.8816 \cdot 10^{-8}$

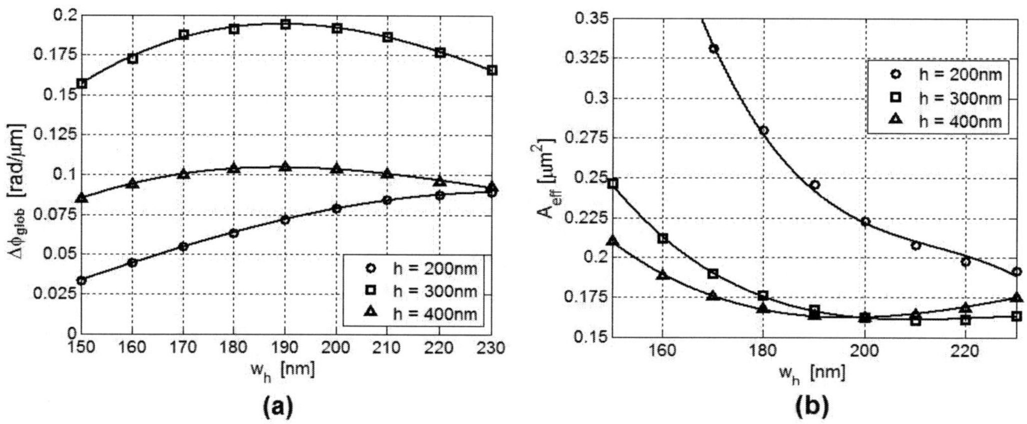

Figure 16. (a) Nonlinear shift of quasi-TE mode and (b) effective area as a function of silicon wire height and width, for w_s=100nm.

It can be observed the different trend of linear effective area and nonlinear phase shift versus both width and height of silicon wires. In fact, w_h values giving maxima of $\Delta\phi_{glob}$ do not correspond to those for A_{eff} maxima. For example, for h =300nm the minimum A_{eff} occurs for w_h=200nm but, for the same value of h, the maximum of $\Delta\phi_{glob}$ is obtained for w_h = 190nm. Furthermore, while the effective area decreases with h, nonlinear phase shift shows a maximum in correspondence of a wire height h = 300nm. Thus, it can be deduced that $\Delta\phi_{glob}$ is only partially influenced by the optical field confinement in the gap region. Then, these results indicate that the best nonlinear response of a SOI slot waveguide can be achieved with w_s= 100 nm, w_h= 190 nm ed h= 300 nm, taking into account the technological limitation on the gap width [31].

3. THE HOLLOW-CORE WAVEGUIDE

Another recently proposed low index core waveguide is the hollow-core waveguide, which has a fairly more complex fabrication procedure than a slot waveguide but enables a wider low-loss spectral range [4]. In fact, conventional optical waveguides based on total internal reflection cannot be used to guide light through a low-index core surrounded by cladding layers with higher refractive indices. Usually the sidewalls of a hollow-core waveguide consist of multi-layer coatings, i.e. alternating layers of material A and material B, where A has a higher refractive index than B.

Traditional waveguides are formed from highly transparent materials with index confinement. Such waveguide structures suffer the fundamental limitations associated with light propagation through solids, such as dispersion, absorption, scattering and nonlinear effects. On the contrary, hollow waveguides may minimize the dependence of light transmission on the optical properties of the waveguide materials, also avoiding the temperature dependence of the refractive index [56], because the hollow waveguide core strictly consists only of air or, more generally, other low index material.

Hollow waveguides fall in two categories: those based on semiconductor multilayer mirror [57] and those based on anti-resonant reflecting optical waveguide (ARROW) cladding [58]. In the first category, light is internally reflected in total at the multilayer. It has been theoretically demonstrated that the Bragg grating can be merged into the core of semiconductor-based hollow waveguides for ultrafast optical applications [59].

For the ARROW case, the layered wall is similar to a Fabry-Perot interferometer reflector and the waveguide loss strongly depends on the wavelength, making this structure useful for sensing applications [58]. ARROWs were first presented in 1986 by Duguay *et al.* [60]. These kind of structures relied both on total internal reflection and anti-resonance to achieve light confinement. Their particular configuration allows to design large single mode waveguides with low losses, and high fabrication tolerances, with a high reflectivity on the waveguide lower cladding. On the other hand, hollow core ARROWs apply the anti resonance principle on the whole waveguide perimeter, and could be extensively employed in sensor applications because of their capability to confine the optical field in a low index medium. The idea of confining light in a low refractive index medium was already pursued in the past [61], for example by employing materials whose refractive indices were lower than liquids (such as a particular type of Teflon, with refractive index 1.29). The basic hollow core ARROW structure is depicted in Figure 17.

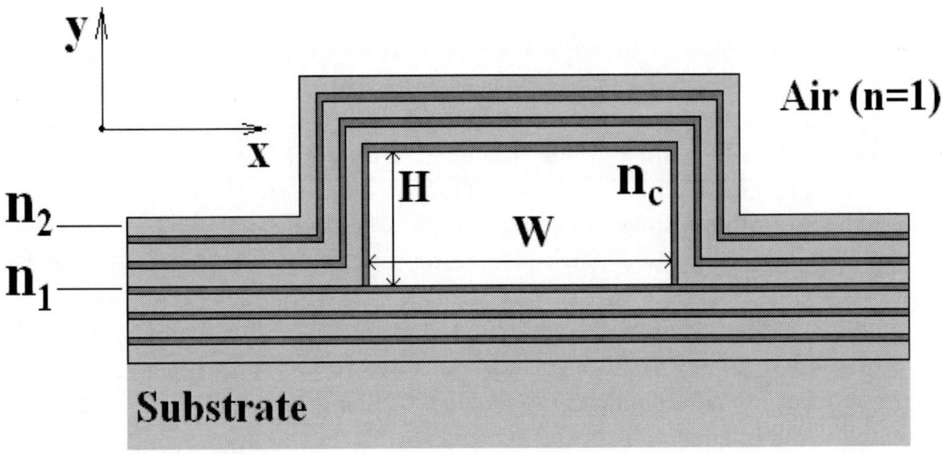

Figure 17. Hollow core ARROW waveguide with three periods.

The structure is composed of three materials: a low index core (whose refractive index is n_c), a high index material (with refractive index $n_1>n_c$) and a third material with refractive index in between, $n_c \leq n_2 < n_1$. Each reflecting period is constituted by a high index (n_1) layer followed by a lower index (n_2) layer. The low index core is surrounded by multiple periods. This structure produces high reflectivity outside the core when the phase Bragg condition is satisfied [62]. This condition can be written as:

$$\Phi = 2t_i k_T + \Phi_r = (2N-1)\pi \tag{16}$$

where N = 1,2..., k_T is the wave-vector transverse component experiencing high reflectivity and Φ_r is the total phase shift from reflection at the two interfaces with the adjacent layers. The phase condition (16) allows to evaluate the appropriate layer thickness leading to the anti-resonance condition:

$$t_i = \frac{\lambda}{4n_i}(2N-1) \Bigg/ \sqrt{1-\frac{n_c^2}{n_i^2}+\frac{\lambda^2}{4n_i^2 d_c^2}} \tag{17}$$

where n_i is the refractive index of the considered layer, d_c is the core height, and λ is the operating wavelength. Since Eq. (17) is largely dependent on λ, it can be generally approximated as:

$$t_i \approx \frac{\lambda}{4n_i} \tag{18}$$

As an example, for n=1 (air), Eq. (18) gives SiN layers 109nm thick (n_i = 2.1) and SiO_2 layers of 184nm (n_i=1.46) at the lowest order (N=0) and design wavelength of 785nm.

Hollow core ARROW waveguides are mainly built by using SiN/SiO_2 or Si/SiO_2 material systems [63]. Yin *et al* [64] presented non constant layer thickness waveguides for loss reduction purposes. The same authors also proposed the possibility of designing hollow waveguides operating by both principles, i.e. TIR (bottom layer) and anti-resonance (top layer). Hollow waveguides have generally been proposed in rectangular configuration, but less common shapes can be also employed. Bernini *et al.* [65] suggested the possibility of employing hexagonal waveguide, whereas Barber *et al.* [66] highlighted the use of semi-circular section profile. The latter configurations allow a good coupling between the waveguide device and the fiber to be achieved. The same authors also proposed the employment of hollow waveguide matrices by using several ARROWs intersected by rib TIR waveguides for sensor design.

The hollow-cladding silicon materials reported in literature also include SiGe/Si [4]. For long-wave infrared region (LWIR, from 1.55 to 100μm), which enables to realize optical components for sensing, communications, signal processing, imaging, the optical absorption in SiGe is significantly smaller than in SiO_2 and Si_3N_4. Furthermore, since the waveguide length can be of the order of a few millimetres, the loss introduced by the water vapour in the air core is negligible. Anyway, an estimation of propagation loss for these waveguides can be done looking at two fundamental sources of loss: absorption in the cladding and leakage due to defects in waveguide geometry. Considering the former, there is a penetration of roughly 10% of the guided mode into the lossy cladding, while the remaining loss contribution is negligible because the waveguide core is air. Considering a filling factor for the layer A (higher refractive index) of about 50%; the cladding absorption is about 10% of $0.5L_A+0.5L_B$, using the relation:

$$L_{clad} = \Gamma_{clad}\left[F_A L_A + (1 - F_A)L_B\right] \tag{19}$$

with F_A is filling factor of the layer A, Γ_{clad} is power confinement factor of guided mode into the cladding, L_A and L_B are A and B layer losses, respectively. Then, from relation (19) the knowledge of L_A and L_B is needed to estimate the power absorption into the cladding region, if both confinement Γ_{clad} and filling factors are known. In fact, with $F_A=50\%$ for hollow cladding-multilayers of SiGe and Si, a cladding-material loss < 1.65 dB/cm over the entire 1.5-100μm spectrum (apart from a 3dB/cm spike at 29μm) has been estimated. As stated before, even when the cladding material loss is zero, there is a mode leakage due to the improper waveguide geometry. Then, the overall leakage is proportional to λ^4/d^5 (where d is the hollow-core diameter) and it has been estimated that the layer thicknesses of SiGe and Si are 0.64 and 0.76 μm, respectively, at λ = 10 μm with d = 140 μm [4]. Since the thickness is beyond the critical value for stable strain, epitaxy of these layers is quite practical.

In general, there are fundamentally three fabrication methods for this kind of guiding structures: wafer bonding, self sealing, and sacrificial core. The latter is the most used, due to its higher yields. It employs plasma-enhanced chemical vapour deposition (PECVD) of sacrificial material [67].

One of the most severe limitations in hollow structure fabrication process is the fact that hollow micro channels tend to collapse due to the pressure of sidewalls. Moreover, these waveguides need accurate design because of their anti-resonance mechanism. For these reasons most authors prefer to focus their design on structures whose dimensions are much larger than those of common TIR waveguides (core width up to 15 μm). In this section, hollow waveguides with dimensions below 5 μm are investigated, thus focusing our attention on more compact devices with respect to other solutions [64]. Further, the choice of realizing a device by employing SOI technology should increase the optoelectronic integration possibility and the coupling mechanism with optical fibers.

3.1. Vertical Sidewall Hollow Waveguides

The waveguide analysis has been again developed using full-vectorial Finite Element Method (FV-FEM) [30]. Initially, a square waveguide with vertical sidewalls and core height (H) equal to core width, $W = H = 2$ μm, has been considered at operating wavelength 1.55 μm. Since mode confinement in this type of guiding structures depends on the number of periods covering the core, we have performed different analysis by varying this number, from one to six. Each reflecting period is assumed as formed by a layer of Si and a layer of SiO_2. As starting case, the waveguide core is assumed as filled with water (n_c = 1.33). The layer thickness can be calculated by using Eq. (19). For Si layer a refractive index of 3.477 is used, thus obtaining a thickness of t_{Si} = 111.4 nm. The relevant thickness for each SiO_2 layer is found as t_{oxide} = 268.3 nm. Calculations have shown that a hollow waveguide having these dimensions supports only the fundamental quasi TE and quasi TM modes. Figure 18 shows the modal distribution of quasi-TE mode in a hollow waveguide with 6 periods and $n_c = 1.33$.

The confinement factor Γ for both quasi-TE and quasi-TM modes is defined as:

$$\Gamma = \frac{\iint_{core} \left|E(x,y)\right|^2 dxdy}{\iint_{\infty} \left|E(x,y)\right|^2 dxdy} \tag{20}$$

where $\left|E(x,y)\right|^2$ is the squared electric field module of guided mode.

Figure 18. Modal distribution of hollow ARROW with 6 periods and vertical sidewalls (quasi-TE mode).

The optical field appears to be well confined inside the structure, confinement factor being reported in detail below for both polarizations. Figure 19 shows the effective index variation for quasi-TE polarized modes with a core index varying between 1.33 and 1.334. The variation exhibits a linear behavior as the core refractive index increases. A similar linear behavior can be observed with changing the number of reflecting periods.

Figure 20 shows the same analysis for quasi-TM modes, confirming a linear behavior. Moreover, it is possible to notice that the effective index decreases when the number of reflecting layers increases.

Figure 21 reports the confinement factor for both polarizations. It appears to be rather large, ranging from a minimum of 0.81 (for a single period waveguide) to a maximum of 0.97.

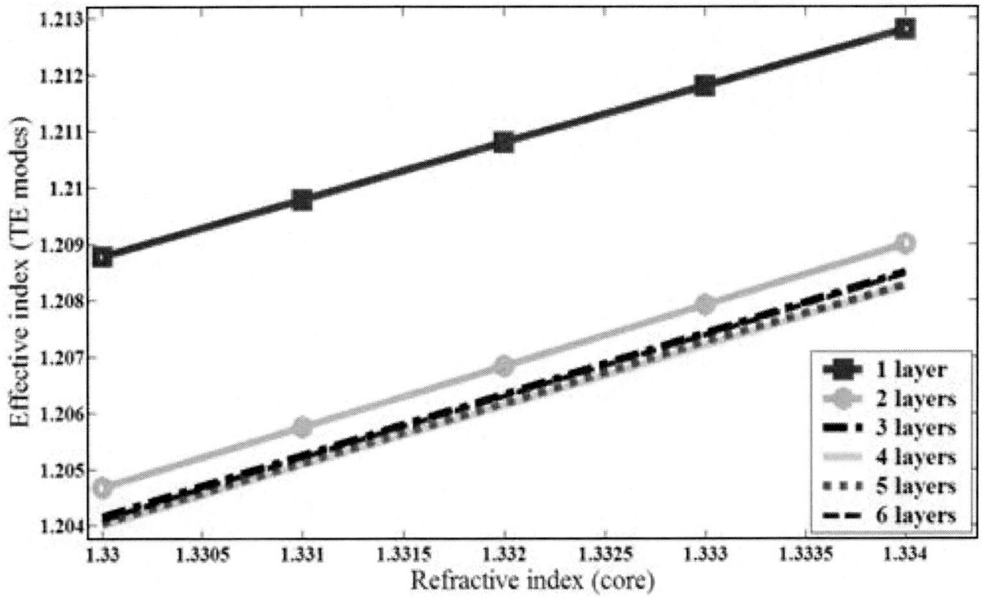

Figure 19. Effective index versus core refractive index for quasi-TE polarized mode.

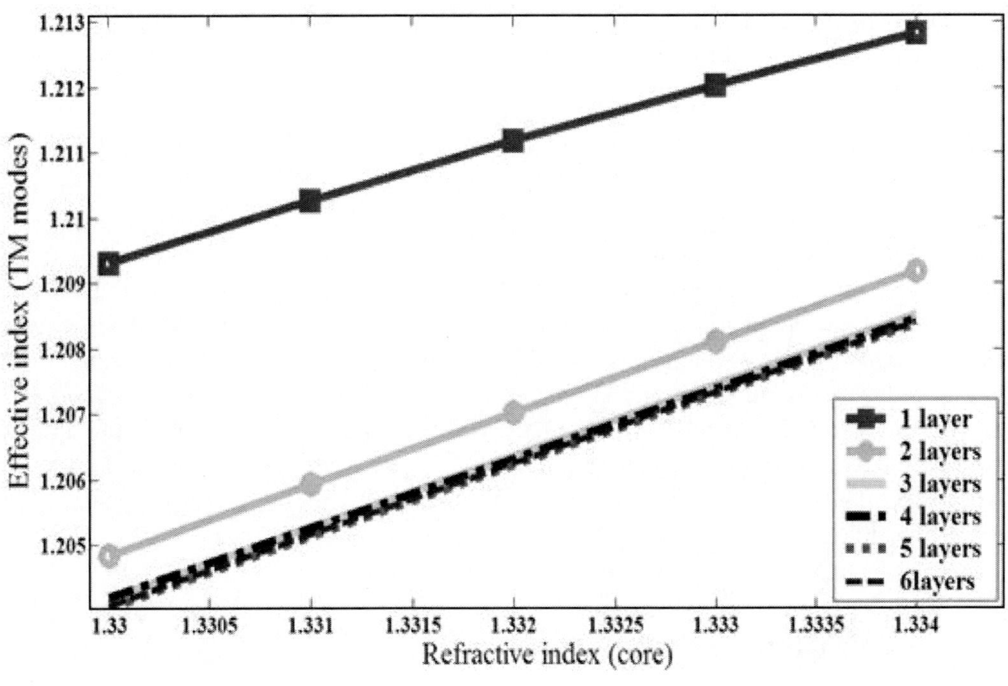

Figure 20. Effective index versus core refractive index for quasi-TM polarized mode.

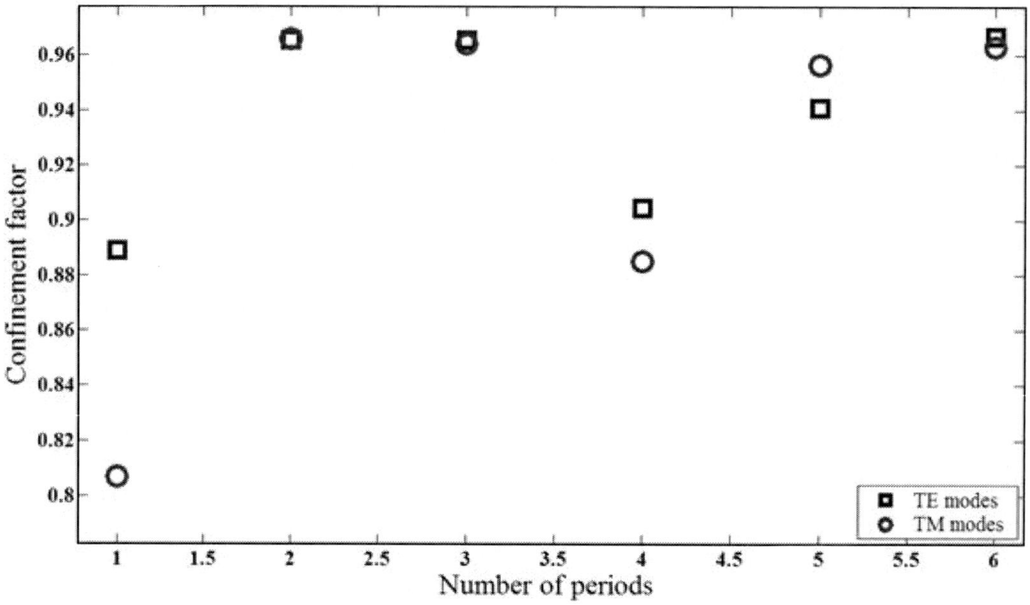

Figure 21. Confinement factor versus number of periods for quasi-TE (circles) and quasi-TM (squares) polarized modes.

To investigate the guiding structure sensitivity to index changes, we varied the core refractive index between 1.33 and 1.334 with a step equal to 0.001and evaluated the quantity:

$$S = \frac{\partial n_{eff}}{\partial n_c} \tag{21}$$

where n_{eff} is the effective index of guided mode. The analysis has been carried out for both quasi-TE and quasi-TM polarizations. As shown in Figure 22, the sensitivity slightly increases with the number of reflecting periods, ranging from 0.87 to 1.07 RIU (refractive index unit). This can be explained by considering that the confinement factor increases by adding more layers. Then, sensitivity significantly higher than unity (1.088 for quasi-TE with 6 periods and 1.085 for quasi-TM with 3 periods) can be obtained, well better than that allowed by any conventional waveguide structure (including rib, wire and slot). This result is not surprising, because these hollow waveguides allow an arbitrarily large low index core region to be achieved, so significantly improving their sensitivity.

3.2. Non Vertical Sidewall Influence

It must be remembered that all the results obtained so far have been obtained considering an ideal structure with vertical sidewalls, as that in Figure 18. These data do not include the fabrication tolerances that affect the device performance. In fact, accidental variations of layer thickness during deposition, etching related tolerances, and non vertical sidewalls can significantly influence the devices.

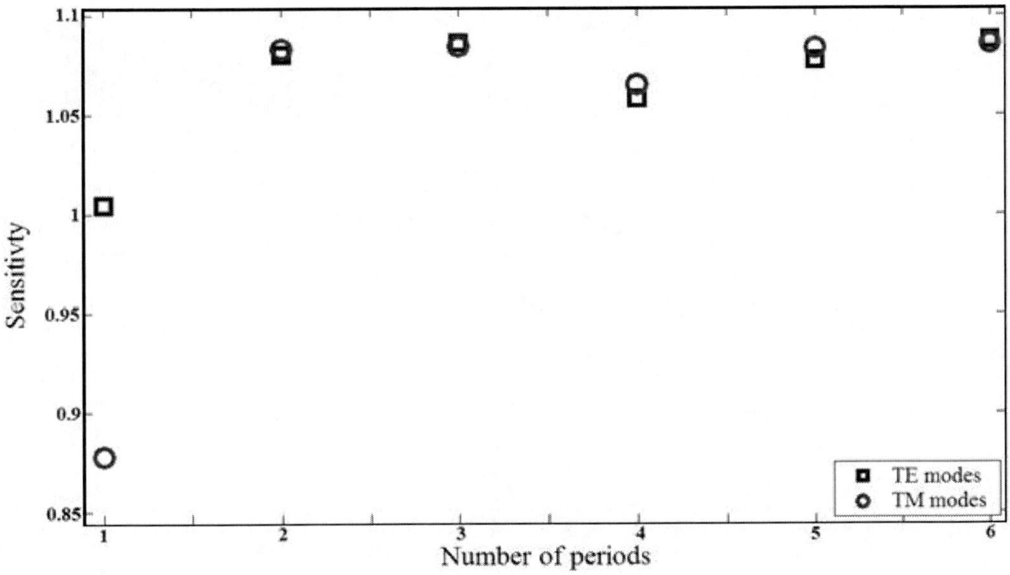

Figure 22. Sensitivity versus number of periods for quasi-TE (circles) and quasi-TM (squares) polarized modes.

We investigated the latter effect to determine its influence on waveguide sensitivity, by choosing a tilting angle of 8° with respect to the vertical. The resulting waveguide is shown in Figure 23. In Figures 24 and 25 the effective index variation due to n_c change in the range from 1.33 and 1.334 are sketched for quasi-TE and quasi-TM polarizations, respectively. Compared to the ideal case, the average effective index value is increased since the non vertical sidewalls produce a broader core cross section. The index variation tends to increase, in particular for quasi-TE polarization.

Figure 23. Modal profile in hollow ARROW waveguide with non vertical sidewalls and 6 periods (quasi-TE mode).

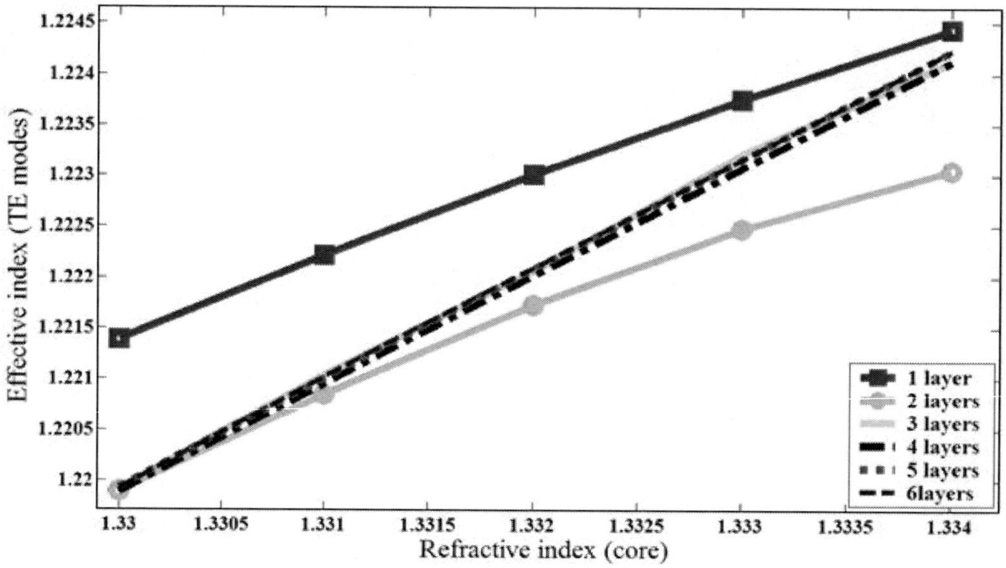

Figure 24. Effective index versus core refractive index for quasi-TE mode and non vertical sidewalls.

Figure 26 depicts the sensitivity variation for waveguides with non vertical sidewalls. Sensitivity values obtained in this case do not relevantly differ from those obtained for ideal case. In particular, these values are slightly lower and seem to vary more as the number of periods changes.

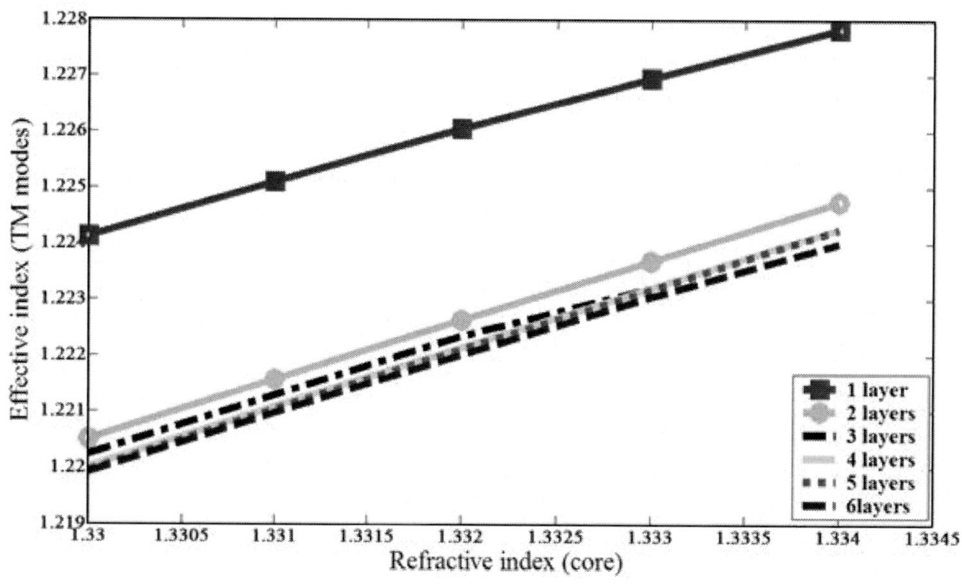

Figure 25. Effective index versus core refractive index for quasi-TM mode and non vertical sidewalls.

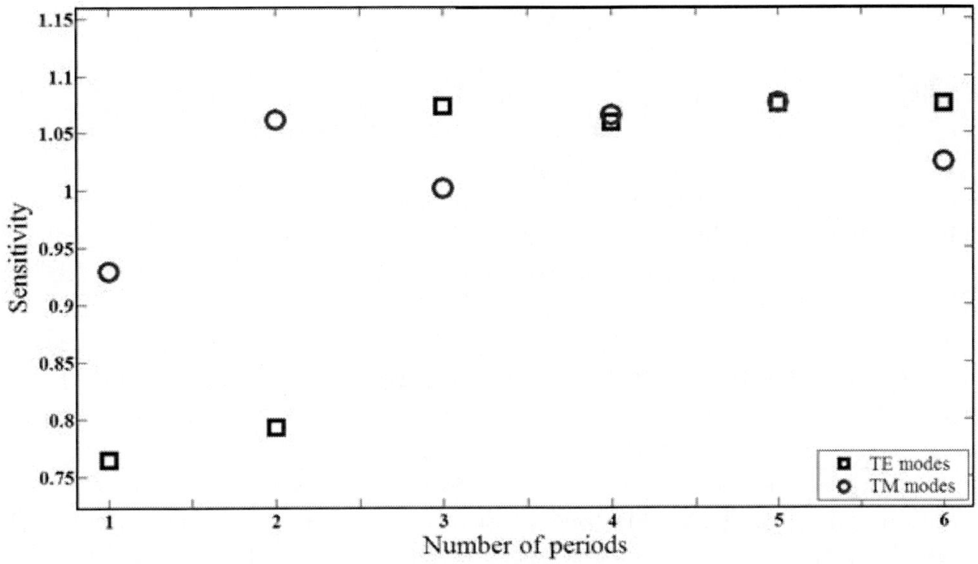

Figure 26. Sensitivity versus number of periods for quasi-TE (squares) and quasi-TM (circles) modes for non vertical sidewall waveguides.

Finally, Figure 27 shows the confinement factor versus the number of periods for both polarizations. As for sensitivity, confinement factor is slightly lower for non vertical sidewall cases, ranging from a 0.75 to 0.97.

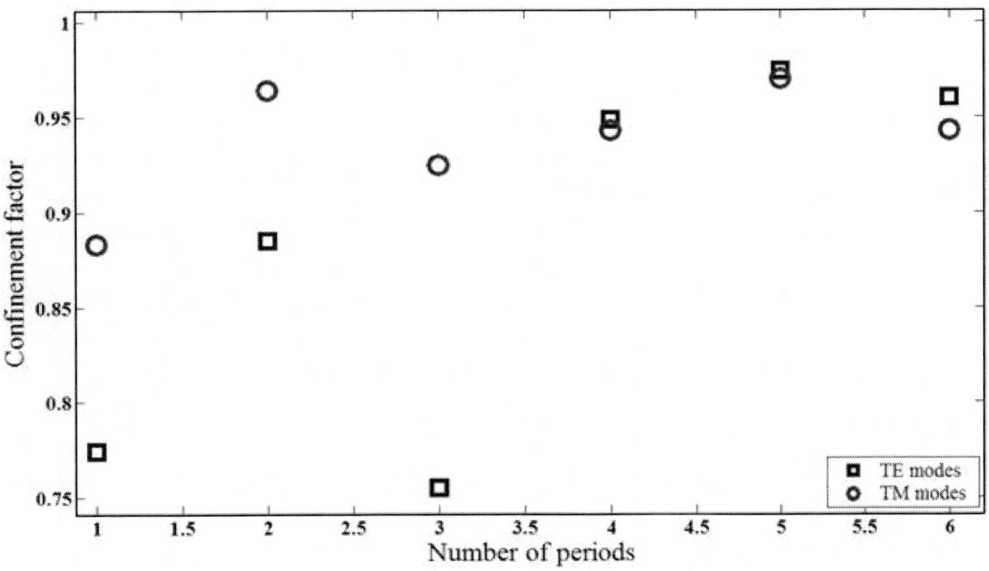

Figure 27. Confinement factor versus number of periods for quasi-TE (circles) and quasi-TM (squares) modes and non vertical sidewall waveguides.

CONCLUSION

In this chapter a description of the modal properties and performance of SOI slot waveguides for two infra-red wavelengths is carried out using FEM. In particular, both the linear and nonlinear behaviour of such a waveguide have been investigated, taking into account different nonlinear materials to fill the slot region, as well as the influence of technological limits for fabrication. Moreover, an iterative method for the calculation of nonlinear phase shift induced by optical Kerr effect has been proposed, allowing the optimization of geometrical parameters of the structure for nonlinear applications. In fact, slot waveguides are demonstrated to be very suitable for this kind of application, as well as for sensing. Finally, some results on hollow-core waveguides have been presented. Hollow waveguides still represent a quite unexplored topic in the field of integrated optics. The condition that allows these devices to work efficiently is not easily achieved, and for that reason these waveguides require great accuracy in the manufacturing process. On the other hand, hollow core ARROWs allow one to develop very sensitive devices that could be employed in several fields of metrology, especially those related to biochemistry and biology. In this chapter, we have analyzed a hollow core waveguide more compact than those recently proposed in literature. The simulation of the ideal structure (having vertical sidewalls) has shown a very high sensitivity. A non ideal structure with non vertical sidewalls has also been considered. Results obtained in the latter case show how fabrication tolerances affect waveguide performance and should be minimized in the manufacturing processes.

REFERENCES

[1] Almeida, V. R., Xu, Q., Barrios, C. A., Lipson, M., *Optics Letters*, 2004, 29, 1209-1211.

[2] Almeida, V. R., Xu, Q., Panepucci, R. R., Barrios, C. A., Lipson, M., *Materials Research Society Symposium*, 2003.

[3] Xu, Q., Almeida, V. R., Panepucci, R. R., Lipson, M., *Optics Letters*, 2004, 29, 1626-1628.

[4] Soref, R. A., Emelett, S. J., Buchwald, W. R., *J. Optics A: Pure and Applied Optics*, 2006, 8, 840-848.

[5] Jalali, B. and Fathpour, S., *J. Lightwave. Technology*, 2006, 40, 4600-4615.

[6] Bogaerts, W., Dumon, P., Jaenen, P., Wouters, J., Becks, S., Wiaux, V., Van Thourhout, D., Taillaert, D., Luyssaert, B., Baets, R., *SPIE Proceedings*, 2005, 5956, 1.

[7] Baehr-Jones, T., Hochberg, M., Walker, C., Scherer, A., *Applied Physics Letters*, 2005, 86, 081101.

[8] Jordana, E., Fedeli, J.-M., Lyan, P., Colonna, J. P., Gautier, P., Daldosso N., Pavesi, L., Lebour, Y., Pellegrino, P., Garrido, B., Blasco, J., Cuesta-Soto, F., Sanchis, P., *4th IEEE International. Conf. on Group IV Photonics*, 2007, 1.

[9] Barrios, C. A., Xu, Q., Shakya, J. , Monolatou, C., Lipson, M., in *Conference on Lasers and Electro-Optics*, 2006, CTuCC3.

[10] Barrios, C. A., Lipson, M., *Optics Express*, 2005, 13, 10092-10101.

[11] Barrios, C. A., *IEEE Photon. Technology. Letters*, 2006, 18, 2419-2421.

[12] Barrios, C. A., Sánchez, B., Gylfason, K., Griol, A., Sohlström, H., Holgado, M., Casquel, R., *Optics Express*, 2007, 15, 6846-6856.

[13] Barrios, C. A., Gylfason, K. B., Sánchez, B., Griol, A., Sohlström, H., Holgado, M., Casquel, R., *Optics Letters*, 2007, 32, 3080-3082.

[14] Galli, M., Gerace, D., Politi, A., Liscidini, M., Patrini, M., Andreani, L. C., Canino, A., Miritello, M., Lo Savio, R., Irrera, A., Priolo, F., *Applied Physics Letters*, 2006, 89, 241114-241116.

[15] Fujisawa, T., Koshiba, M., 2006, *Optics Letters*, 2006, 31, 56-58.

[16] Fujisawa, T., Koshiba, M., *IEEE Photon. Technology Letters*, 2006, 18, 1246-1248.

[17] Feng, N.-N., Sun, R., Kimerling, L. C., Michel, J., *Optics Letters*, 2007, 32, 1250-1252.

[18] Feng, N.-N., Sun, R., Michel, J., Kimerling, L. C., *Optics Letters*, 2007, 32, 2131-2133.

[19] Blasco, J., Barrios, C. A., *Conference on Lasers and Electro-Optics Europe*, 2005.

[20] Baher-Jones, T., Hochberg, M., Wang, G., Lawson, R., Liao, Y., Sullivan, P., Dalton, L., Jen, A., Scherer, A., *Optics Express*, 2005, 13, 5216-5226.

[21] Hochberg, M., Baher-Jones, T., Wang, G., Huang, J., Sullivan, P., Dalton, L., Scherer, A., *Optics Express*, 2007, 15, 8401-8410.

[22] McLauchlan, K. K., Dunham, S. T., *IEEE J. Selected Topics in Quantum Electronics*, 2006, 12, 1455-1460.

[23] Barrios, C. A., *Electronics Letters*, 2004, 40, 862-863.

[24] Fujisawa, T., Koshiba, M., *J. Optical Society America B*, 2006, 23, 684-691.

[25] Liu, T., Panepucci, R. R., *Optics Express*, 2007, 15, 4304-4309.

[26] Di Falco, A., Conti, C., Assanto, G., *Optics Letters*, 2006, 31, 3146-3148.

[27] Povinelli, M., Lončar, M., Ibanescu, M., Smythe, E., Johnson, S., Capasso, F., and Joannopoulos, J., *Optics Letters*, 2004, 29, 1209-1211.

[28] Almeida V. R., Panepucci, R. R., *Quantum Electronics and Laser Science Conf. (QELS)*, 2007, JThD104.

[29] Feng, N. N., Michel, J., Kimerling, L., *IEEE J. Quantum Electronics*, 2006, 42, 885-890.

[30] Comsol Multiphysics by COMSOL AB, ver. 3.2, *single license*, 2005.

[31] Jordana, E., Fedeli, J.-M., Lyan, P., Colonna, J. P., Gautier, P., Daldosso, N., Pavesi, L., Lebour, Y., Pellegrino, P., Garrido, B., Blasco, J., Cuesta-Soto, F. and Sanchis, P., Deep-UV, *4th IEEE International Conf. on Group IV Photonics*, 2007, 1.

[32] Sanchis, P., Blasco, J., Martínez, A., Martí, J., *J. Lightwave Technology*, 2007, 25, 1298-1305.

[33] Jalali, B., Raghunathan, V., Shori, R., Fathpour, S., Dimitropoulos, D., Stafsudd, *IEEE J. Selected Topics in Quantum Electronics*, 2006, 12, 1618-1627.

[34] Kier, A., *Mid-infrared semiconductor optoelectronics*, Springer Series in Optoelectronics, 2006.

[35] Raghunathan, V., Borlaug, D., Rice, R., Jalali, B., *Optics Express,* 2007, 15, 14355-14362.

[36] Leviton D., Frey, B., *Proceedings of SPIE*, 2006, 6273-2K.

[37] Frey, B., Leviton, D., Madison, T., *Proceedings of SPIE*, 2006, 6273-2J.

[38] Yamada, H., Chu, T., Ishida, S. and Arakawa, Y., *IEEE Photon. Technology Letters,* 2005, 17, 585-587.

[39] Müllner, P., Hainberger, R., *IEEE Photon. Technology Letters*, 2006, 18, 2557-2559.

[40] Dell'Olio, F., Passaro, V. M. N., *Optics Express*, 2007, 15, 4977-4993.

[41] Daly, D., Clarck, G., *The 49th European Study Group with Industry, OCIAM*, Oxford, 2004.

[42] Valduis, Parriaux, O., G., Hockstra, H., Lambeck, P., *J. Lightwave Technology*, 2000, 18, 677-682.

[43] Anderson, P., Schmidt, B., Lipson, M., *Optics Express*, 2006, 14, 9197-9202.

[44] Fujisawa, T., Koshiba, M., *IEEE Photon. Technology Letters*, 2006, 18, 1530-1532.

[45] Dalton, L., *SPIE News. on Communication and Networking*, 2007, 10.1117/2.1200709.0866.

[46] Boyd, R. W., *Nonlinear Optics*, Academic Press, San Diego, 2003.

[47] Kasap, S., Capper, P., *Handbook of Electronic and Photonic Materials*, Springer Science, New York, 2006, chap. 45.

[48] Gubler, U., Bosshard, C., Molecolar, in *Polymers for Photonics Applications II*, K.-S. Lee, Ed. Springer-Verlag, 2003.

[49] Gaburro, Z., Bettotti, P., Daldosso, N., Ghulinyan, M., in *Materials Science Foundations,* Trans Tech Publ. Zurich, 2006, vol. 27-28.

[50] Spano, R., Cazzanelli, M., Daldosso, N., Gaburro, Z., Ferraioli, L., Tartara, L., Yu, J., Degiorgio, V., Hernandez, S., Lebour, Y., Pellegrino, P., Garrido, B., Jordana, E., Fedeli, J., Pavesi, L., *Material Research Society Symposium Proceedings*, 2007, 958, L08-06.

[51] Bader, M. A., Marowski, G., Bahtiar, A., Koynov, K., Bubeck, C., Tillmann, H., Hörhold, H.-H., Pereira, S., *J. Optical Society America B*, 2002, 19, 2250-2262.

[52] Bahtiar, A., Ph.D. Thesis, Max-Planck Institute, 2004.

[53] Chen, Q., Kuang, L., Sargent, E. H., Wang, Z. Y., *Applied Physics Letters*, 2003, 83, 2115-2117.

[54] Fujisawa, T., Koshiba, M., *Optics Express*, 2003, 11, 1481-1489.

[55] Bristow, H., Rotenberg, A., van Driel, H. M., *Applied Physics Letters*, 2007, 90, 191104.

[56] Lo, S.-S., Wang, M.-S., Chen, C.-C., *Optics Express*, 2004, 12, 6589-6593.

[57] Miura, T., Koyama, F., *J. Japanese Applied Physics*, 2004, 43, L21-L23.

[58] Campopiano, S., Bernini, R., Zeni, L., Sarro, P. M., *Optics Letters*, 2004, 29, 1894-1896.

[59] Fedotov, A. B., Naumov, A. N., Sidorov-Biryukov, D. A., Chigarev, N. V., Zheltikov, A. M., Haus, J. W., Miles, R. B., *J. Optical Society of America B*, 2002, 19, 1162.

[60] Duguay, M. A., Kokubun, Y., Koch, T. L., *Applied Physics Letters*, 1986, 49, 13-15.

[61] Lowry, J. H., Mendelowitz, J. S., Subramanian, N. S., *Optical Engineering*, 1992, 31, 1982-1985.

[62] Schmidt, H., Yin, D., Barber, J. P., Hawkins, A. R., *IEEE J. Selected Topics in Quantum Electronics*, 2005, 11, 519-527.

[63] Bernini, R., Campopiano, S., Zeni, L., Sarro, P. M., *Sensors and Actuators B*, 2004, 100, 143-146.

[64] Yin, D., Barber, J. P., Hawkins, A. R., Schmidt, H., *Optics Express*, 2005, 13, 9331-9336.

[65] Bernini, R., Campopiano, S., Zeni, L., *IEEE J. Selected Topics in Quantum Electronics*, 2002, 8, 106-110.

[66] Barber, P. J., Lunt, E. J., George, Z. A.,Yin, D., Schmidt, H., Hawkins, A. R., *IEEE Photon. Technology Letters*, 2006, 18, 28-30.
[67] Barber, J. P., *et al.*, *IEEE Photon. Technology Letters*, 2005, 17, 363-365.

In: Modeling of Photonic Devices
Editor: Vittorio M. N. Passaro

ISBN 978-1-60456-980-3

Chapter 3

DIRECTIONAL COUPLERS

Vittorio M. N. Passaro*
Photonics Research Group, Dipartimento di Elettrotecnica ed Elettronica,
Politecnico di Bari, via Edoardo Orabona n. 4, 70125 Bari, Italy

ABSTRACT

In this chapter, an overview on the modeling of optical directional couplers for many applications of Photonics is presented. At first, a detailed investigation of silicon-on-insulator optical rib waveguides is carried out to find the best design parameters. Then, the modeling of the coupling mechanism both in coupled straight waveguides and in optical resonators is developed. Many numerical examples are presented and discussed.

Keywords: Directional Couplers, Silicon-on-Insulator, Coupled Mode Theory, Waveguide Modal Analysis, modeling

1. INTRODUCTION

Optical directional coupling is one of the most important tasks of Photonics in a nextremely large number of photonic devices, including those based on fibers or planar or two-dimensional (2D) confined waveguides. Some examples of passive coupling devices are polarization converters, directional couplers, mode converters, beam splitters, wavelength filters or multiplexers [1]. For many years, Coupled Mode Theory (CMT) [2] has been conceived with the aim to model and design these kinds of devices, and its fundamental details are briefly described in chapter 6 of this book. In this chapter, some novel aspects of the modeling of directional couplers, coupling to optical resonators and coupling among multi-stage optical resonators are presented, based on the use of a mixed modeling procedure, i.e. CMT with full-vectorial finite element method (FV-FEM) [3]. It will be shown as this

---

* E-mail address: passaro@deemail.poliba.it

approach is rather general and powerful for many different coupling mechanisms. However, before describing the modeling of couplers, a detailed modal analysis of rib waveguides in Silicon-on-Insulator (SOI) technology has been achieved by FV-FEM. This is needed to have a number of geometrical and optical parameters to be used in examples of coupling devices.

2. SOI Waveguide Modal Analysis

In this section, a modal analysis of SOI rib waveguides has been carried out by FV-FEM [3], using about 100,000 mesh elements. The waveguide with vertical sidewalls and silicon oxide cover, sketched in Figure 1, has total rib height WGH, etch depth ED, rib width WGW. The following data were considered: $n_{core} = 3.476409$, $n_{oxide} = 1.444$ (wavelength 1.55 μm), WGH = 1.35μm, and ED = 675 nm, 800 nm, 925 nm.

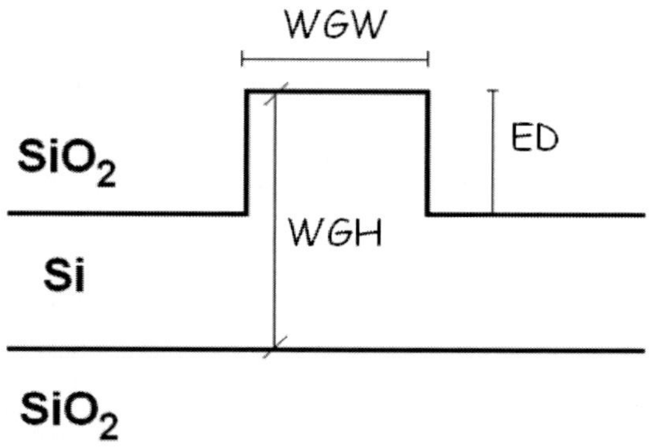

Figure 1. Cross section of SOI rib waveguide with vertical sidewalls.

An effective index has been found for both quasi-TE and quasi-TM modes and the results are shown in Figure 2 as a function of rib width for ED =675 nm. The second mode is clearly excited for rib widths larger than 1000 nm.

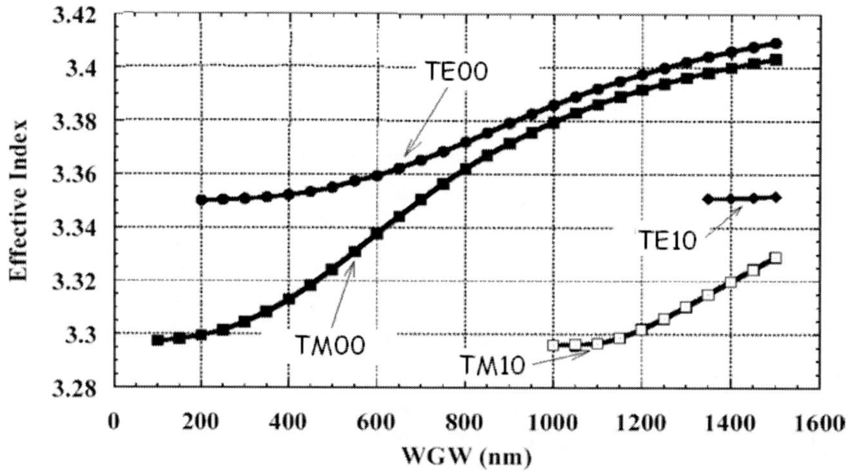

Figure 2. Effective index versus rib width for both polarizations (ED = 675nm).

Similar results can be seen in Figures 3 and 4 for ED = 800nm and 925nm, respectively.

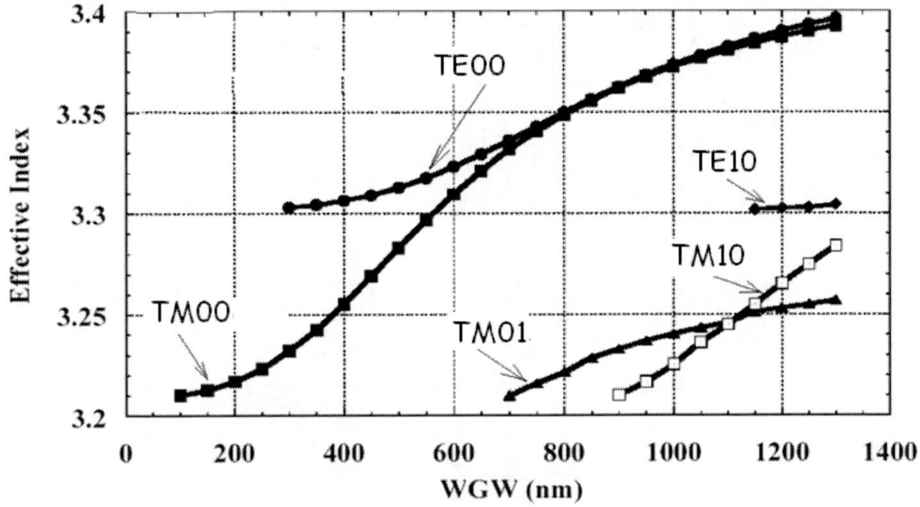

Figure 3. Effective index versus rib width for both polarizations (ED = 800nm).

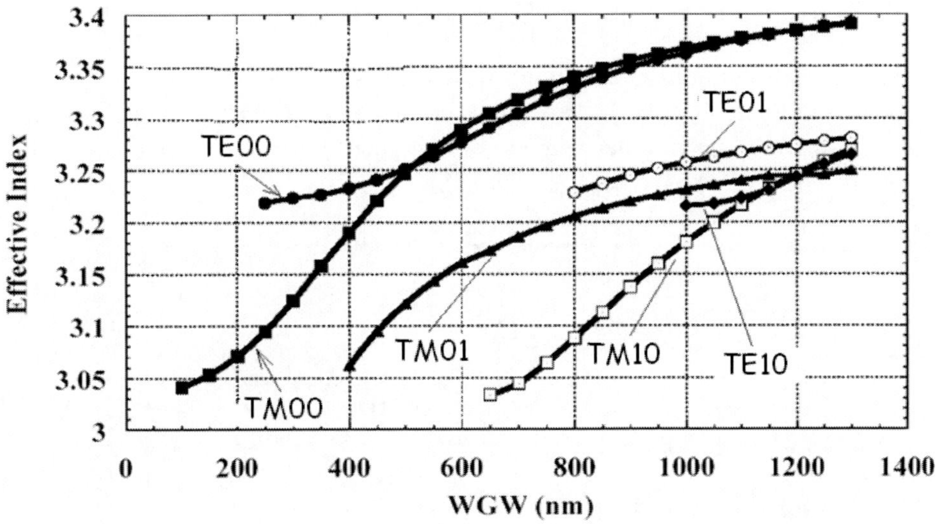

Figure 4. Effective index versus rib width for both polarizations (ED = 925nm).

Excitation of higher order modes clearly occurs for smaller rib widths as large as the etch depth, because the field confinement is enhanced in the rib region. Waveguide birefringence has been investigated for ED = 825 nm, 850 nm, 875 nm, 900 nm, 925 nm, 950 nm and WGW in the range from 175 to 1175 nm. Results are summarized in Figure 5. In this chapter birefringence is defined as $B = n_{eff}^{TE} - n_{eff}^{TM}$. Zero-Birefringence (ZBR) condition has been calculated as a function of waveguide dimensions. Figure 5 shows that two values of WGW exist achieving ZBR for each ED. The first value (WGW ~400nm for ED = 900nm) is more difficult to fabricate, while the second one (WGW ~1000nm) allows a more relaxed technological condition. Further, first order quasi-TE and quasi-TM modes cut-off conditions have been evaluated as a function of waveguide dimensions.

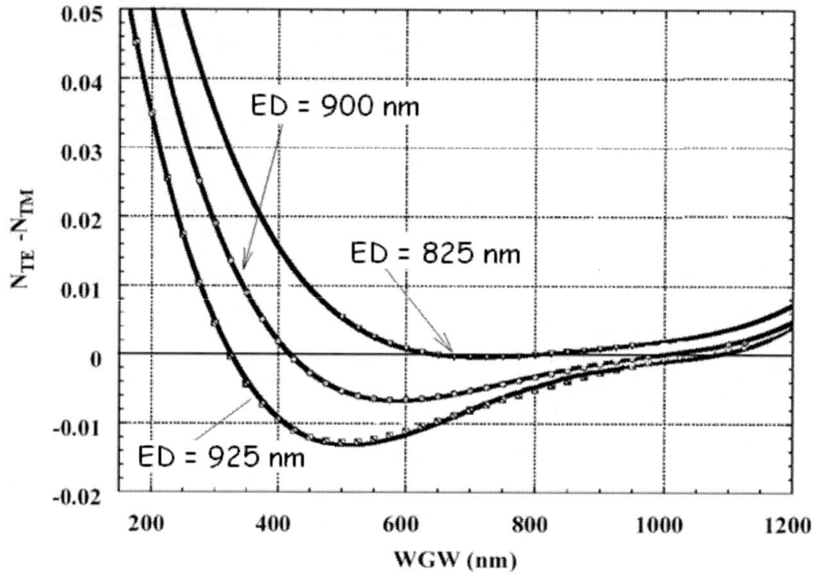

Figure 5. Birefringence versus SOI rib width, for various etch depths (WGH = 1.35 μm).

As a result, next plot in Figure 6 shows the rib sizes to simultaneously achieve ZBR and single mode condition (SMC). This design criteria are obtained by using only FEM predictions, differently from other papers in literature where other methods like beam propagation method (BPM) have been employed [4-5].

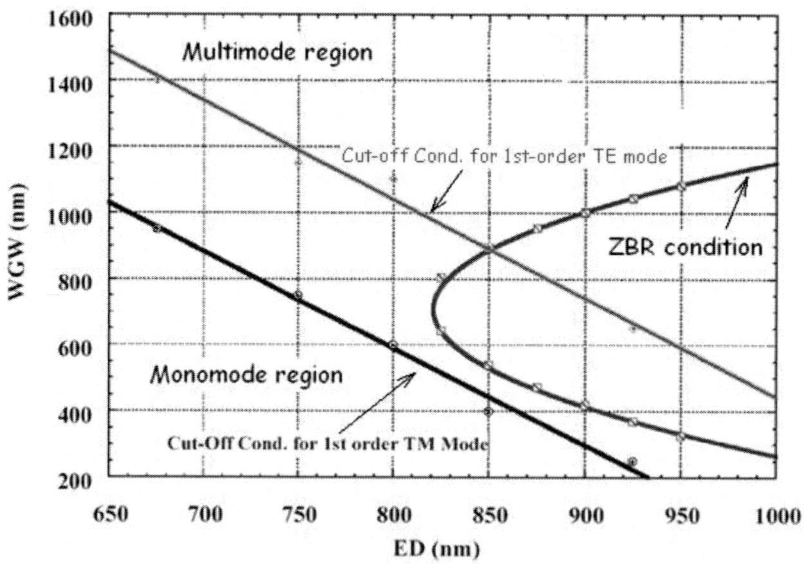

Figure 6. SOI rib width versus etch depth to simultaneously achieve ZBR and SMC.

Moreover, effect of SiO_2 cladding finite thickness (500 nm) has been investigated assuming vertical side walls, as sketched in Figure 7. A maximum shift in quasi-TE and quasi-TM mode effective indices of 0.1% has been observed over rib waveguide with oxide infinite cladding. A small change in waveguide birefringence (B) dependence on waveguide dimensions is obtained, showing a negligible effect on both ZBR and SMC.

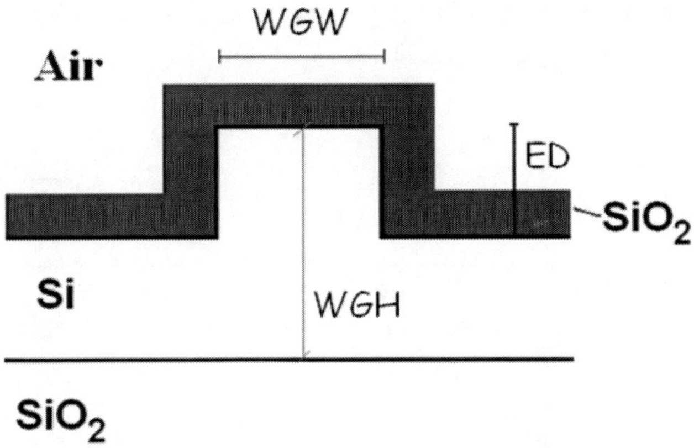

Figure 7. SOI rib waveguide with a finite thickness oxide cladding (500 nm).

Moreover, effect of rib waveguide non-vertical sidewalls (8° or 10° tilting) has been studied, observing some changes in zero-birefringence and cut-off conditions. Both symmetric and asymmetric waveguide cross sections, as in Figure 8, have been considered. Tilting can be a secondary effect of fabrication tolerances, but also a further design parameter.

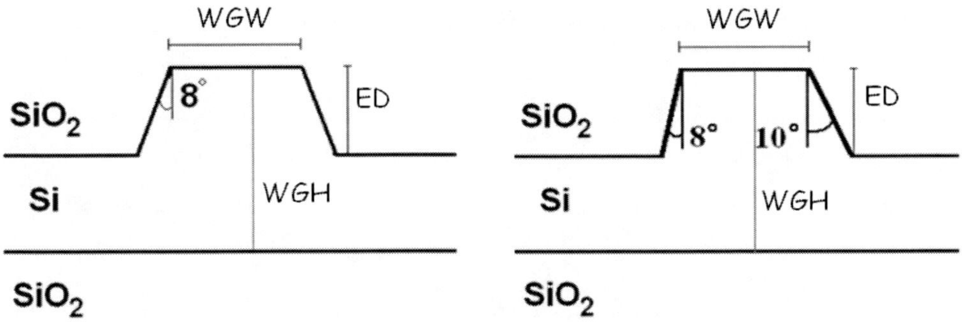

Figure 8. Cross section of rib waveguide with tilted sidewalls in symmetric and asymmetric cases.

Similar results as before are plotted in Figure 9, 10 and 11 in terms of effective index versus rib width, to find SMC. We can note as SMC occurs for smaller widths as large as the etch depth, as expected. No significant changes (lower than 5%) in zero-birefringence and cut-off conditions have been observed for symmetrically tilted sidewalls.

An analysis of ZBR condition has been also performed in these structures. Waveguide (symmetric 8°/8°) birefringence has been investigated for ED=825 nm, 850 nm, 875 nm, 900 nm, 925 nm, 950 nm and for WGW in the range from 175 to 1175 nm.

Waveguide (symmetric 8°/8°, asymmetric 8°/10°, symmetric 10°/10°) birefringence has been investigated for ED=925 nm and WGW in the range from 300 nm to 1100 nm, as sketched in Figure 12.

We note that asymmetry in tilting angles (8° and 10° tilting) produces very small shifts in zero-birefringence and cut-off conditions. For instance, ZBR condition is obtained with WGW = 369.563 nm and 1031.68 nm for symmetric structure, (8°/8°), or with WGW= 354.155 nm and WGW = 1018.32 nm for asymmetric case (8°/10°), giving a maximum shift with over symmetric 8°/8° case equal to 4%.

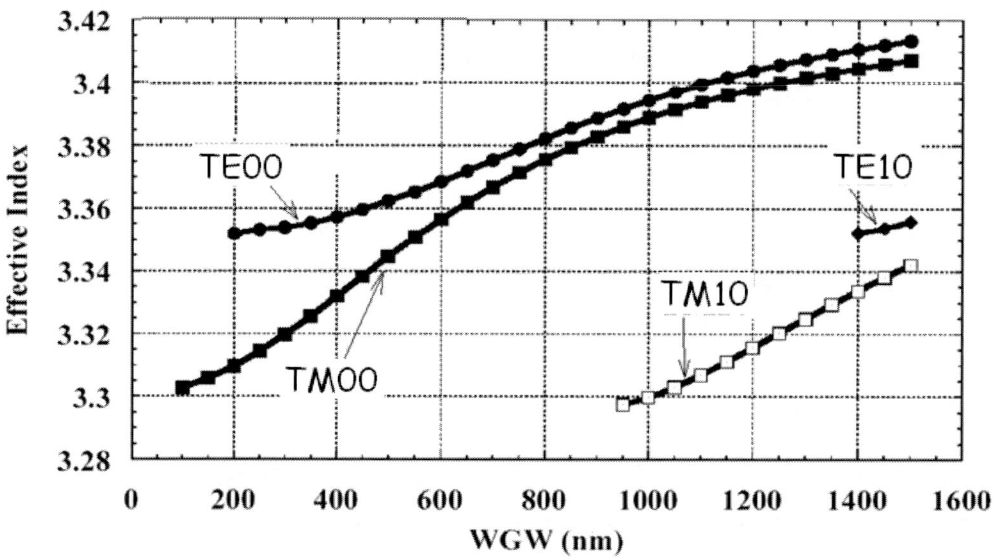

Figure 9. Effective index versus rib width for symmetrically tilted sidewalls (ED = 675nm).

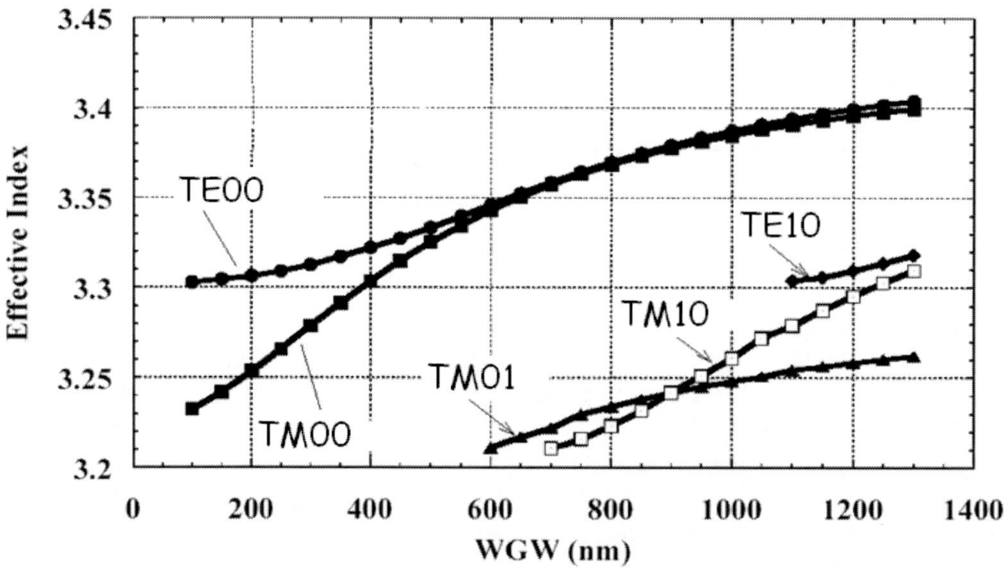

Figure 10. Effective index versus rib width for symmetrically tilted sidewalls (ED = 800nm).

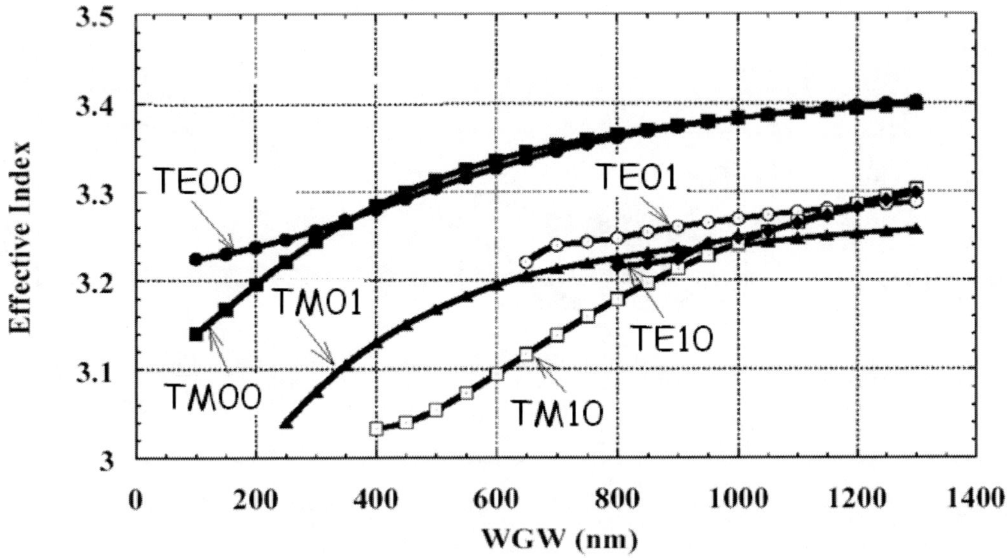

Figure 11. Effective index versus rib width for asymmetrically tilted sidewalls (ED = 925nm).

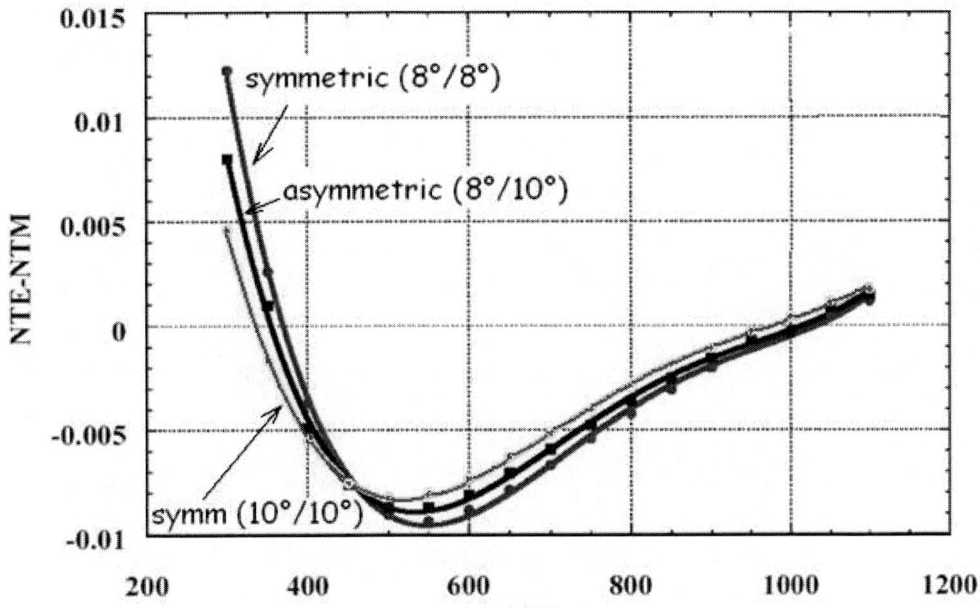

Figure 12. Birefringence versus rib width for ED =925 nm and different structures.

Finally, for symmetric structure (10°/10°) we obtain WGW = 335.254 nm and WGW= 973.745 nm, with a maximum shift with respect to asymmetric 8°/10° or symmetric 8°/8° case equal to 5% and 9%, respectively. Under all these conditions, we cannot simultaneously obtain SMC and ZBR in the same SOI rib waveguide.

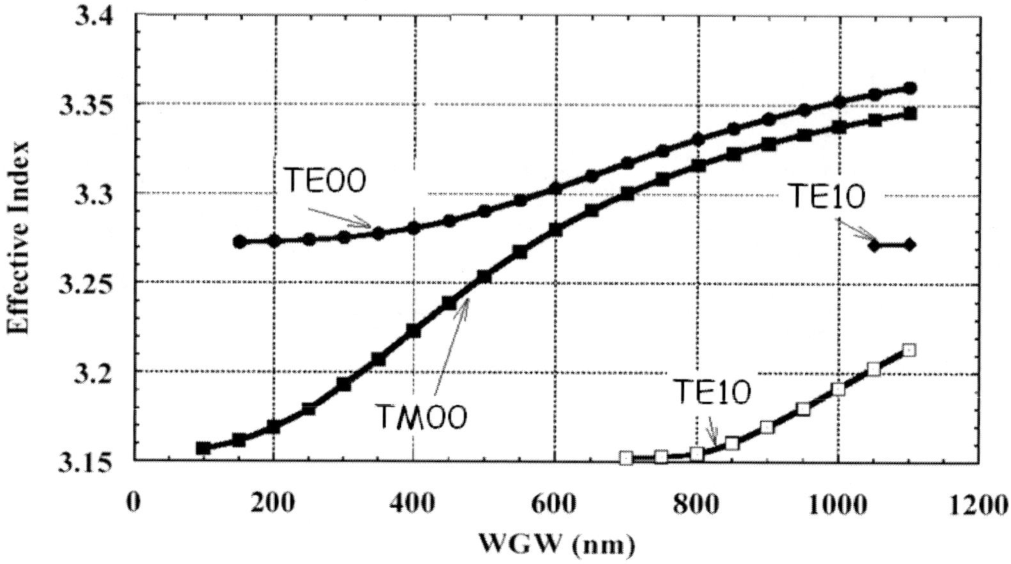

Figure 13. Effective index versus rib width for WGH =1000 nm and ED= 500nm.

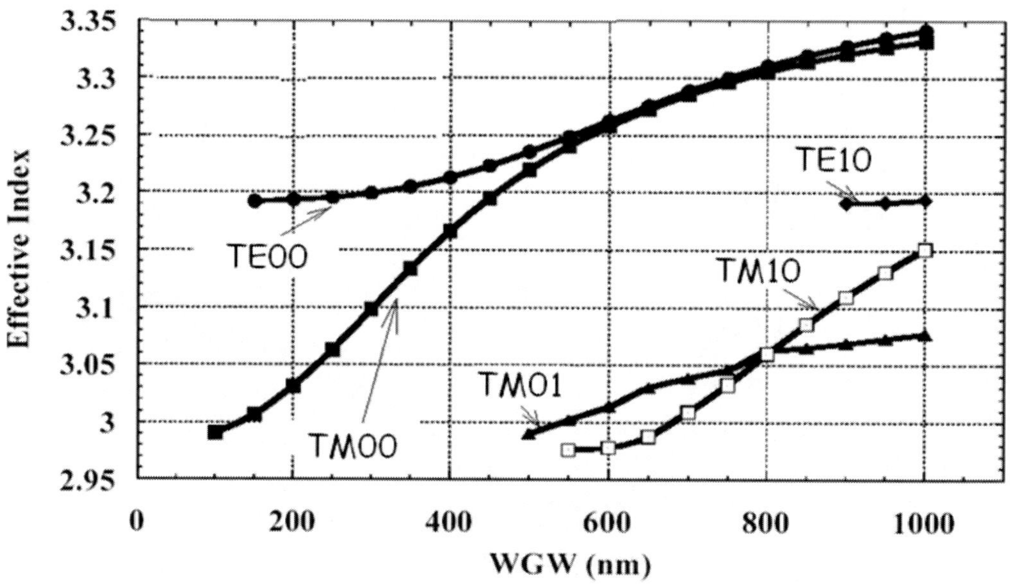

Figure 14. Effective index versus rib width for WGH = 1000 nm and ED = 600nm.

Therefore, the parameters have been changed as in Figure 13 and 14 for vertical sidewalls, WGH = 1000 nm and ED =500 and 600nm, respectively. Best result is sketched in Figure 15 for ED= 700 nm and WGH =1000 nm, proving the possibility to obtain a single-mode and zero-birefringence waveguide for WGW = 380nm.

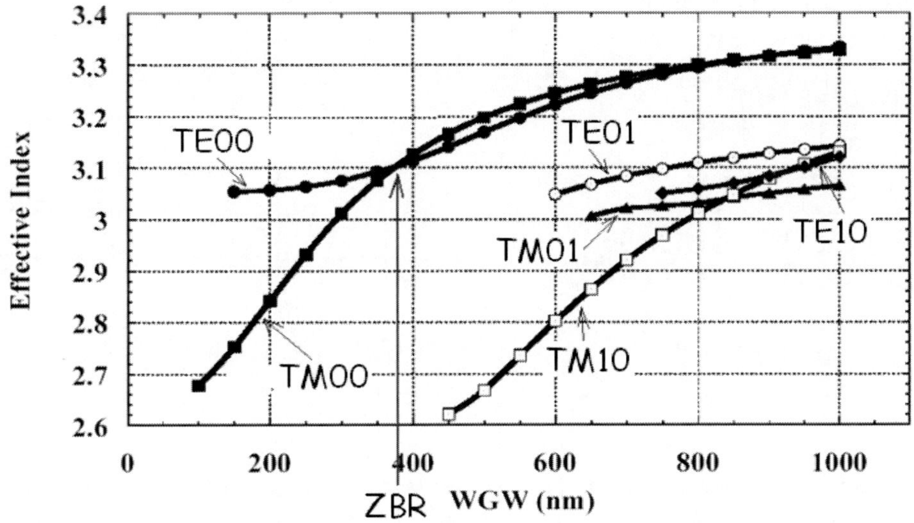

Figure 15. Effective index versus rib width for WGH = 1000nm and ED = 700nm and vertical sidewalls.

3. STRAIGHT DIRECTIONAL COUPLERS

In this section a lossless directional coupler (see Figure 16) formed by two monomodal waveguides is considered, modelled by Coupled Mode Theory (CMT), as usually formulated in literature [6]. Assuming input power in waveguide b as $P_{in}^{b} = 0$, it is well known that:

$$\frac{P^a_{out}}{P^a_{in}} = \left[\frac{1}{\Delta^2 + \kappa_{ab}\kappa_{ba}} \left(\kappa_{ab}\kappa_{ba} cos^2 \left(L_c \sqrt{\Delta^2 + \kappa_{ab}\kappa_{ba}} \right) + \Delta^2 \right) \right] \tag{1}$$

with

$$\kappa_{ab} = \frac{\tilde{\kappa}_{ab} + C_{ab}\left[\beta_a - \beta_b - \tilde{\kappa}_{bb}\right]}{1 - C_{ab}C_{ba}}; \quad \kappa_{ba} = \frac{\tilde{\kappa}_{ba} + C_{ba}\left[\beta_b - \beta_a - \tilde{\kappa}_{aa}\right]}{1 - C_{ab}C_{ba}}$$

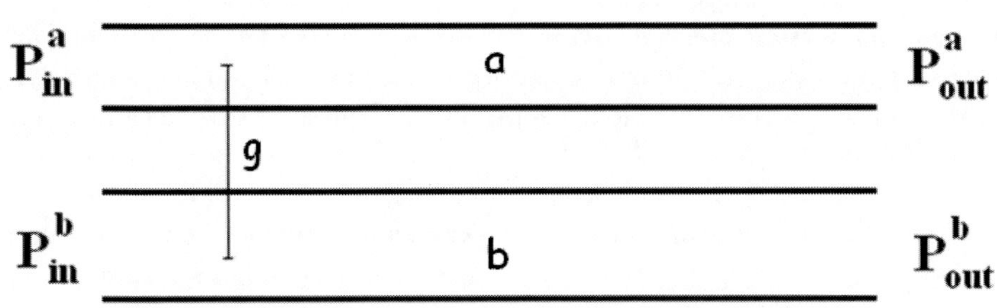

Figure 16. Geometric scheme of directional coupler.

Moreover, in Eq.(1) it results:

$$\kappa_{pq} = \omega \iint_{\infty} \Delta\varepsilon^{p} \left(\mathrm{E}_t^p \cdot \mathrm{E}_t^q - \frac{\varepsilon^q}{\varepsilon_0 n^2} E_z^p E_z^q \right) dxdy \qquad p,q = a,b$$

$$C_{pq} = 2 \iint_{\infty} \left(\mathrm{E}_t^q \times \mathrm{H}_t^p \right) \cdot \hat{z}\, dxdy \qquad p,q = a,b$$

$$\Delta\varepsilon^a(x,y) = \varepsilon_0 n^2(x,y) - \varepsilon^a(x,y)$$

$$\Delta\varepsilon^b(x,y) = \varepsilon_0 n^2(x,y) - \varepsilon^b(x,y)$$

$$\Delta = \frac{1}{2}\left[\beta_b - \beta_a + \frac{\left(\kappa_{bb} - \kappa_{aa}\right) + \left(C_{ab}\kappa_{ba} - C_{ba}\kappa_{ab}\right) + 2C_{ab}C_{ba}\left(\beta_b - \beta_a\right)}{1 - C_{ab}C_{ba}} \right]$$

where L_c is the coupler length, β_a and β_b are propagation constants of modes propagating in isolated waveguides (a and b in Figure 16) constituting the coupler, $n(x,y)$ is coupler refractive index profile, $\varepsilon^a(x,y)$ and $\varepsilon^b(x,y)$ are single waveguide refractive index profiles, respectively. If the coupler is perfectly symmetric, we have:

$$\varepsilon^a(x,y) = \varepsilon^b(x,y) \;\Rightarrow\; \Delta\varepsilon^a(x,y) = \Delta\varepsilon^b(x,y)$$

$$\beta_b = \beta_a$$

$$\mathrm{E}_t^a = \mathrm{E}_t^b \qquad \mathrm{H}_t^a = \mathrm{H}_t^b \qquad E_z^a = E_z^b$$

$$C_{ab} = C_{ba}$$

$$\kappa_{ab} = \kappa_{ba} \qquad \kappa_{aa} = \kappa_{bb}$$

$$\Delta = 0 \qquad \kappa_{ab} = \kappa_{ba} = k$$

and, under this assumption, we can write in terms of coupling coefficient k:

$$\frac{P_{out}^a}{P_{in}^a} = cos^2\left(L_c k\right) \qquad (2)$$

Calculations included in this CMT formulation are again carried out by FV-FEM. This mixed CMT+FEM approach for modelling of directional couplers has been validated by considering a coupler realized using a SOI ridge waveguides [7], whose geometrical dimensions are WGW=300 nm and distance between waveguide centres (gap) g=300 nm. Dependence of optical power exiting from both coupler ports on coupler length L_c has been investigated for quasi-TE (E parallel to the substrate) and quasi-TM mode (E perpendicular to the substrate). Results are summarized in Figure 17, showing a very good agreement with 3D FDTD predictions and experiments presented in that paper.

In general, losses are present in the coupler. They can be taken into account by a coefficient Ψ so that $r^2 + t^2 = \Psi \le 1$, being r and t the reflected and transmitted normalized power, respectively. Directional couplers have been simulated by 3D FEM assuming SOI waveguides with geometrical dimensions WGH = 1350 nm, ED = 800nm, and WGW= 800 nm, 1000 nm, 1200 nm.

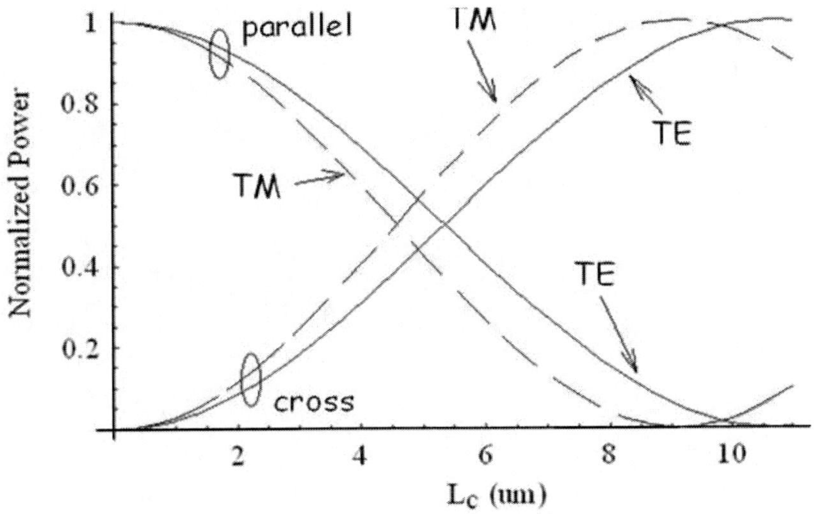

Figure 17. Normalized powers versus coupler length for both polarizations (for structure in [7]).

Gap (*g*) between coupled waveguides has been varied in the range from 200 nm to 700 nm. From simulations it has been observed that it holds:

$$r^2 \cong \Psi cos^2\left(\frac{\pi}{2}\frac{L_c}{L^*}\right) \tag{3}$$

where L_c is coupler length and L^* is the coupler length allowing to obtain a complete power transfer between waveguides. L^* is related to coupling coefficient *k* by:

$$k = \frac{\pi}{2L^*} \tag{4}$$

that permits to estimate *k* knowing L^* values. L^* dependence on *g* has been investigated, for different values of WGW and quasi-TM modes (see Figure 18). Coupling is clearly stronger for narrower waveguides. Coupling coefficients are listed in Table 1 as a function of gap.

Similar investigation has been performed for a directional coupler realized using ridge SOI waveguides (WGW = 380 nm, WGH = 290 nm), as sketched in Figure 19. Plot of coupling length is shown in Figure 20 as a function of gap for quasi-TE polarization. Starting from L^* values, coupling coefficient *k* has been calculated also in this case, as in Table 2. General dependence of L^* on *g* can be expressed by a cubic polynomial interpolation:

$$L^* \cong c_0 + c_1 g + c_2 g^2 + c_3 g^3 \tag{5}$$

where c_0, c_1, c_2, c_3 are fitting parameters.

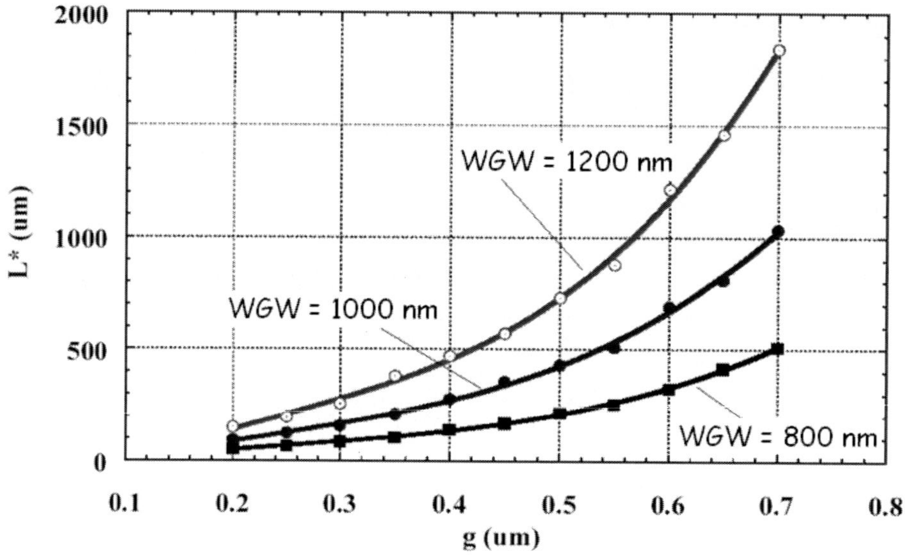

Figure. 18. Coupling length versus gap for various rib widths (WGH = 1350 nm, ED = 800nm).

Table 1. Coupling coefficients versus gap

g (μm)	k (μm^{-1}) (WGW = 800 nm)	k (μm^{-1}) (WGW = 1000 nm)	k (μm^{-1}) (WGW = 1200 nm)
0.20	0.0302	0.0170	0.0105
0.25	0.0242	0.0127	0.0081
0.30	0.0185	0.0100	0.0062
0.35	0.0150	0.0075	0.0041
0.40	0.0112	0.0057	0.0033
0.45	0.0092	0.0044	0.0028
0.50	0.0073	0.0037	0.0022
0.55	0.0062	0.0031	0.0018
0.60	0.0048	0.0023	0.0013
0.65	0.0038	0.0019	0.0011
0.70	0.0031	0.0015	0.0009

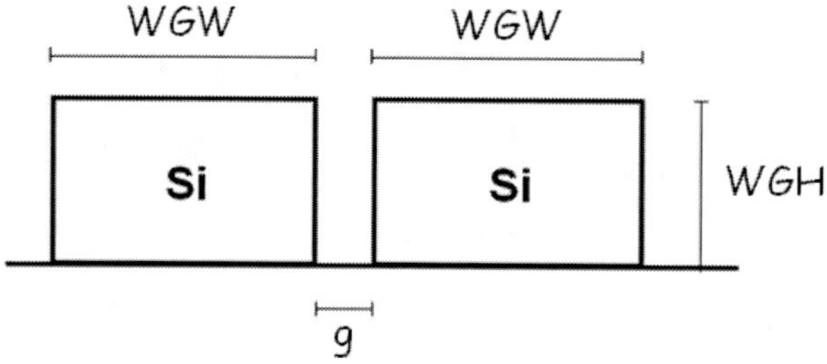

Figure 19. Scheme of directional coupler using SOI ridge waveguides.

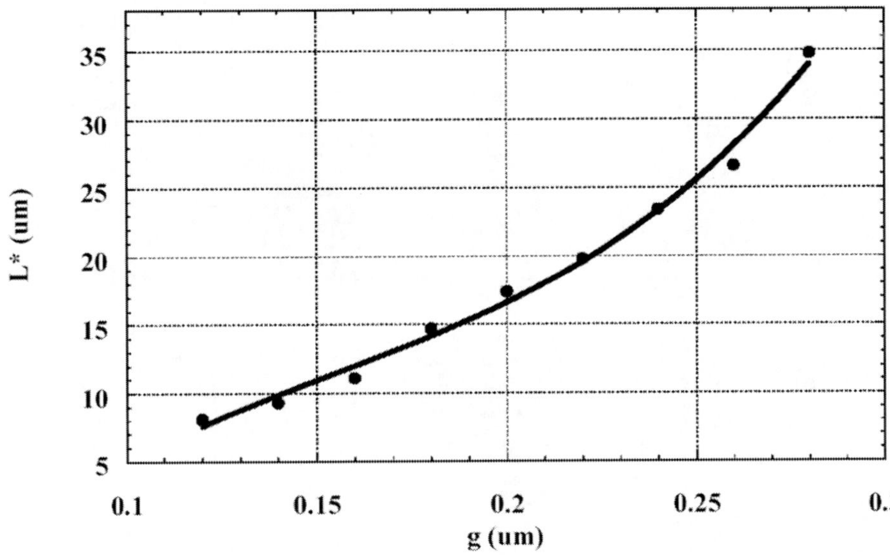

Figure 20. Coupling length versus gap (WGH = 290nm, WGW = 380nm, quasi-TE).

Table 2. Coupling coefficient versus g

g (nm)	k (μm^{-1}) (WGW = 380 nm)
120	0.1939
140	0.1689
160	0.1415
180	0.1069
200	0.0903
220	0.0793
240	0.0671
260	0.0591
280	0.0451

Combining (1), (3) and (5), we can write:

$$r^2 = \Psi - t^2 = \Psi cos^2\left(\frac{\pi}{2}\frac{L_c}{c_0 + c_1 g + c_2 g^2 + c_3 g^3}\right)$$
$$t^2 = \Psi\left[1 - cos^2\left(\frac{\pi}{2}\frac{L_c}{c_0 + c_1 g + c_2 g^2 + c_3 g^3}\right)\right] \qquad (6)$$

Assuming $\Psi = 0.75$, t dependence on g and L has been investigated as sketched in Figures 21-23, using 3D plots for WGH =1350 nm, ED =800 nm, WGW =800, 1000 and 1200nm, respectively (quasi-TM polarization). A similar 3D plot is presented in Figure 24 for WGW =390nm, WGH = 380nm and quasi-TE polarization.

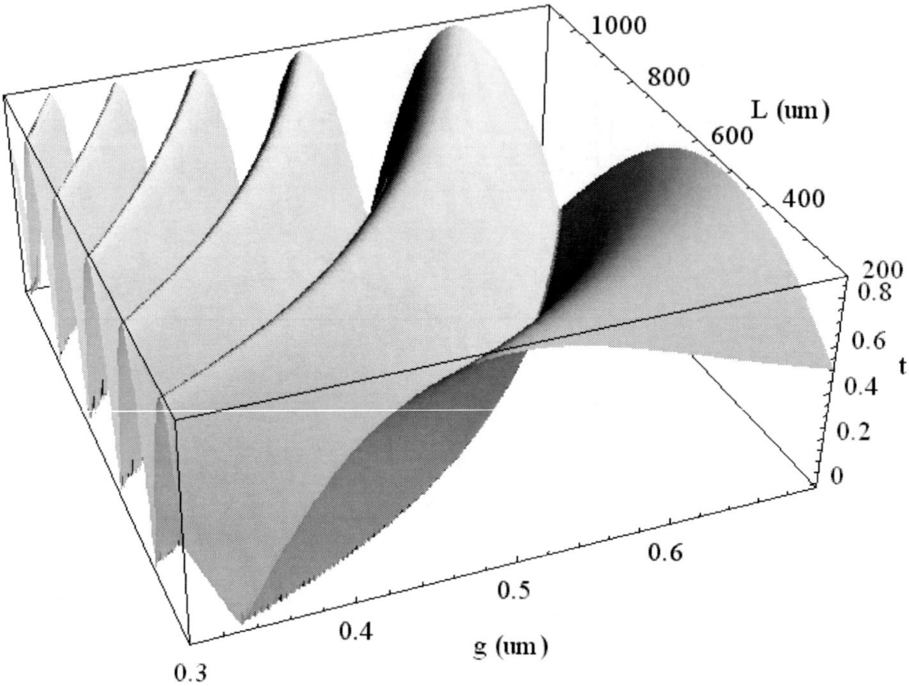

Figure 21. Transmitted power versus *g* and *L* (WGW = 800nm, quasi-TM).

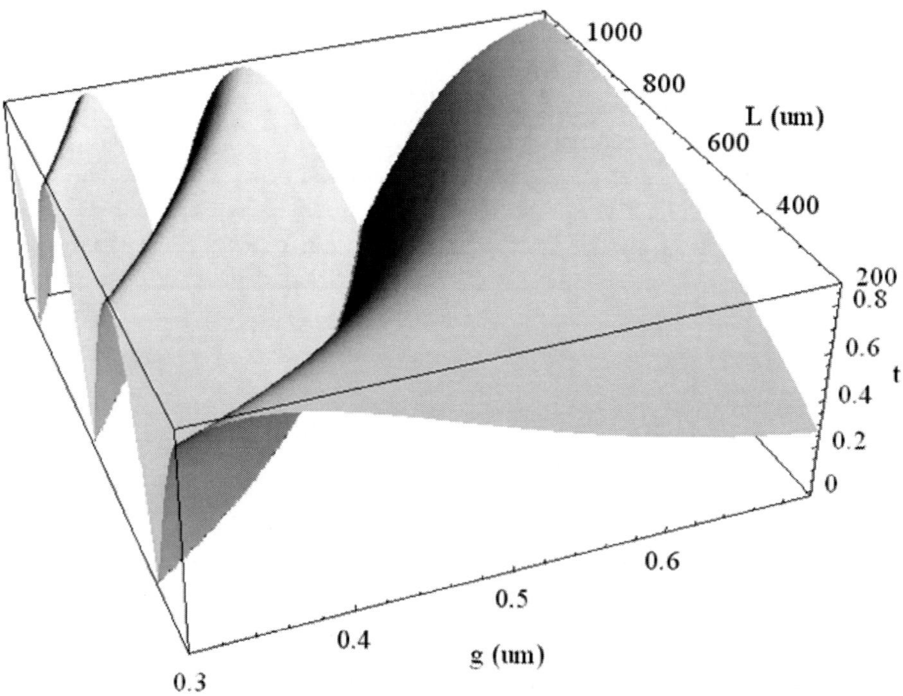

Figure 22. Transmitted power versus *g* and *L* (WGW = 1000nm, quasi-TM).

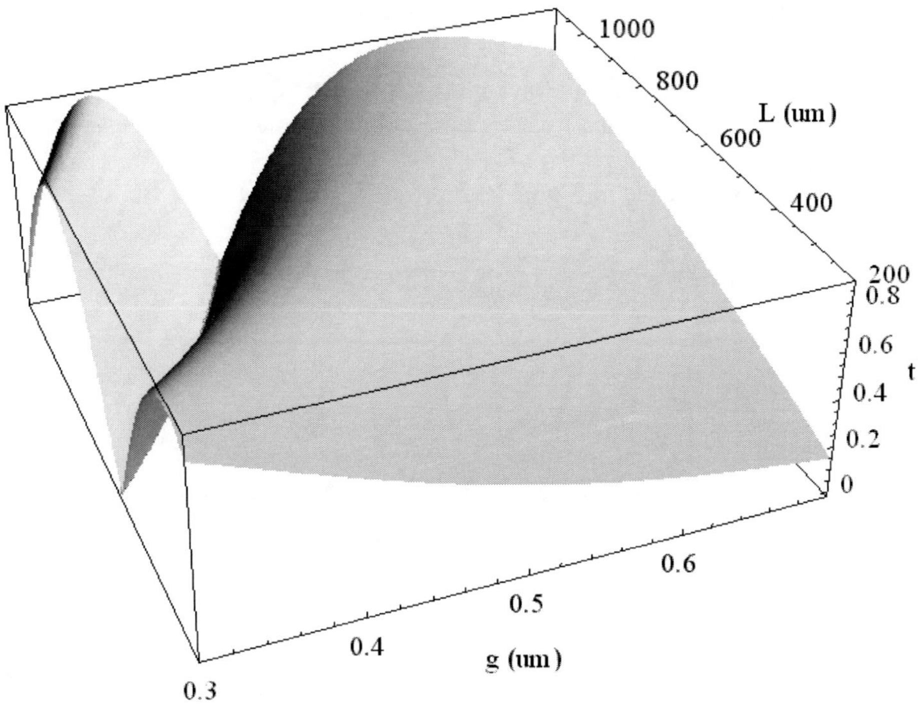

Figure 23. Transmitted power versus *g* and *L* (WGW =1200nm, quasi-TM).

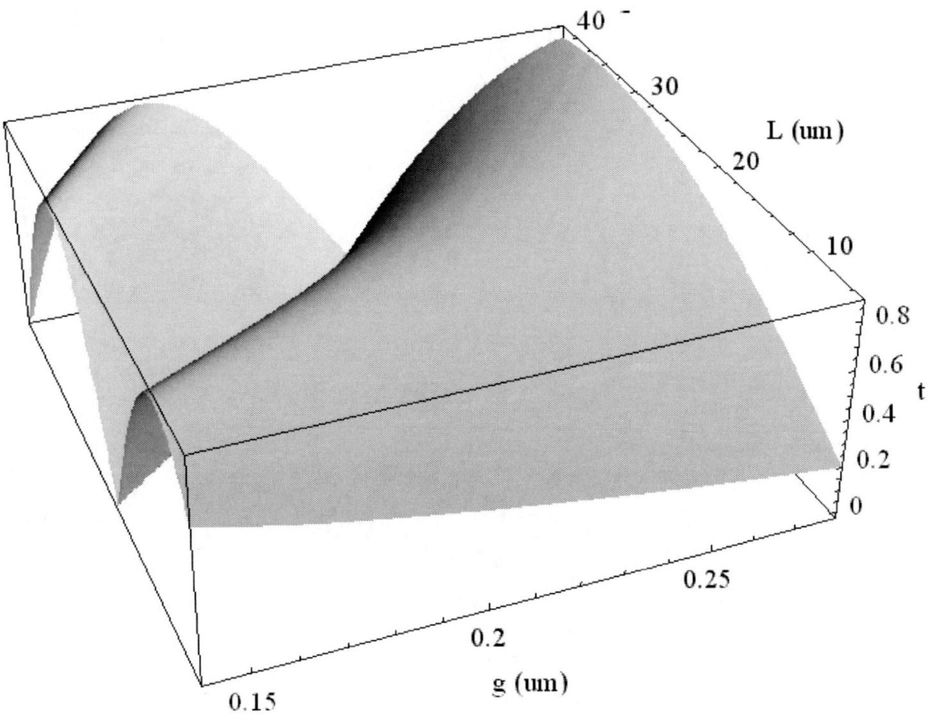

Figure 24. Transmitted power versus *g* and *L* (WGW = 380nm, WGH = 290nm, quasi-TE).

3.1. Effect of Tilted Sidewalls

Using again the FEM-based approach, the parameter L^* has been investigated as a function of SOI rib width WGW and *g*, for a directional coupler realized using a SOI ridge waveguides (WGW = 300 nm, 340 nm, 380 nm and WGH = 290 nm) with 8° tilted symmetric sidewalls, as sketched in Figure 25.

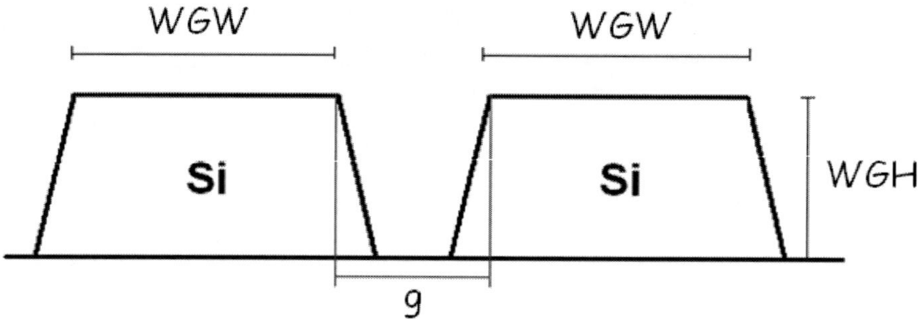

Figure. 25. Cross section of coupled SOI ridge waveguides with tilted sidewalls.

The following plots in Figures 26, 27 and 28 show the coupling length as a function of waveguide gap. Coupling length is always smaller for quasi-TM than quasi-TE modes. For quasi-TM polarization, practical influence of tilting angle cannot be revealed, while for quasi-TE coupling this influence is rather significant, since the mode is well confined in the waveguide rib, where tilting effect occurs. Then, vertical sidewalls should be fabricated to reduce the coupling length. Of course, these considerations are less important with increasing the rib width (see Figure 28), because the effect of tilting angle tends to be reduced.

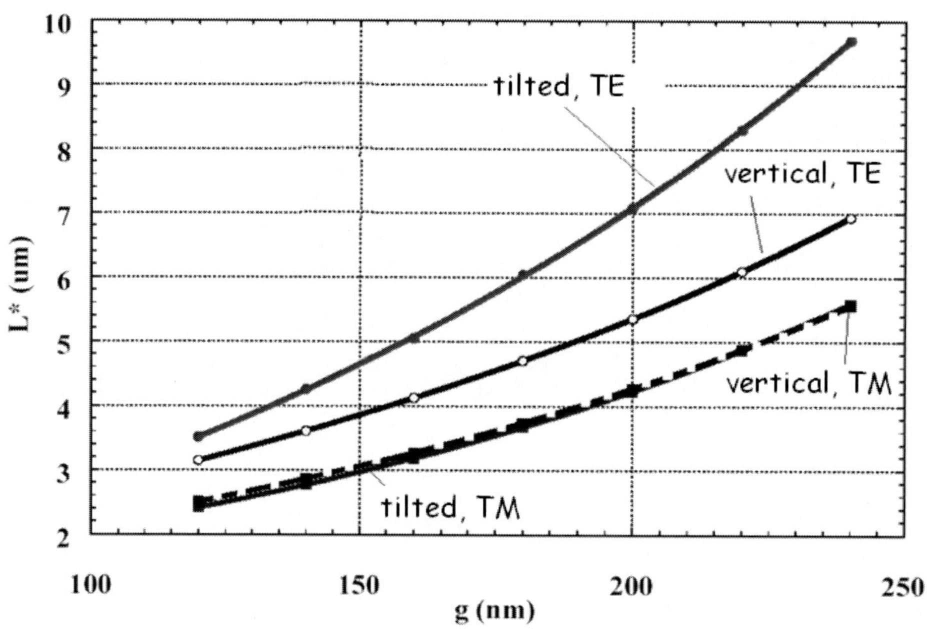

Figure 26. Coupling length versus gap for various structures and both polarizations (WGW = 300 nm).

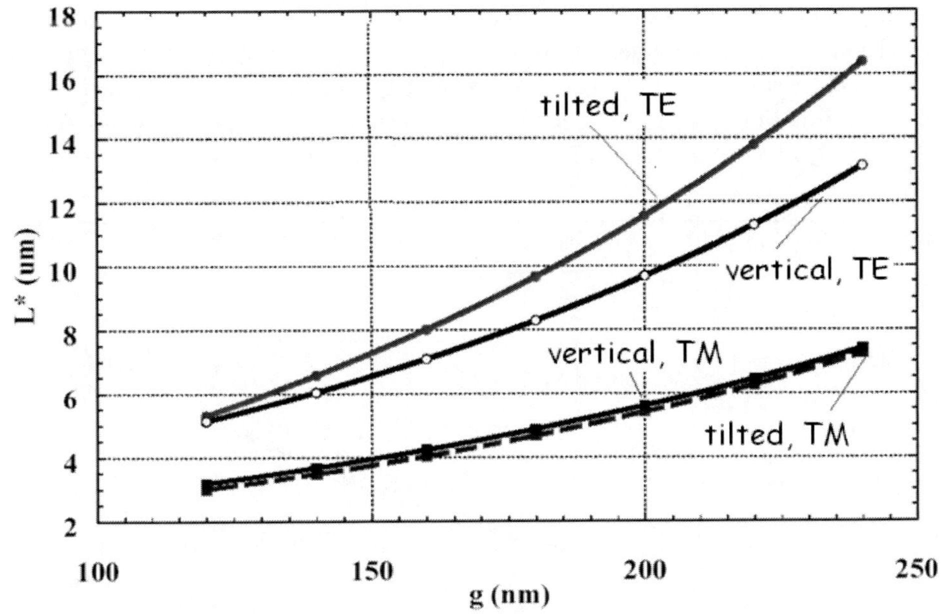

Figure 27. Coupling length versus gap for various structures and both polarizations (WGW = 340 nm).

Table 3. Coupling coefficients versus waveguide gap for tilted cross sections

	WGW =300 nm (8° tilted)		WGW =340 nm (8° tilted)		WGW =380 nm (8° tilted)	
g	k_TE	k_TM	k_TE	k_TM	k_TE	k_TM
120	0.447340	0.651139	0.296566	0.52178	0.20557	0.424642
140	0.369571	0.567469	0.239095	0.451459	0.162572	0.364818
160	0.311508	0.495226	0.19624	0.390028	0.131306	0.313203
180	0.259877	0.426788	0.162787	0.336833	0.107402	0.268878
200	0.221691	0.371521	0.135964	0.290818	0.088536	0.230844
220	0.189331	0.322992	0.11405	0.251153	0.073361	0.198257
240	0.162005	0.280834	0.095944	0.21695	0.060987	0.170317

Table 4. Coupling coefficients versus waveguide gap

	WGW =300 nm (vertical)		WGW =340 nm (vertical)		WGW =380 nm (vertical)	
g	k_TE	k_TM	k_TE	k_TM	k_TE	k_TM
120	0.499069	0.627307	0.306252	0.492553	0.193820	0.390674
140	0.435502	0.548913	0.260104	0.427656	0.161433	0.336981
160	0.381170	0.480218	0.221843	0.371328	0.135190	0.290759
180	0.334207	0.420177	0.189718	0.322514	0.113606	0.250963
200	0.293329	0.367826	0.162544	0.280248	0.095699	0.216715
220	0.257634	0.322183	0.139426	0.243646	0.080737	0.187233
240	0.226371	0.282399	0.119713	0.211958	0.068199	0.161847

Coupling coefficient k has been calculated in all considered cases, as summarized in following Table 3. For comparison, k values obtained for a coupler realized by SOI ridge waveguides with vertical sidewalls are reported in Table 4. In quite all considered cases, we can observe a coupling coefficient decrease for quasi-TE and a simultaneous increase for quasi-TM modes, when tilted waveguide sidewalls are considered.

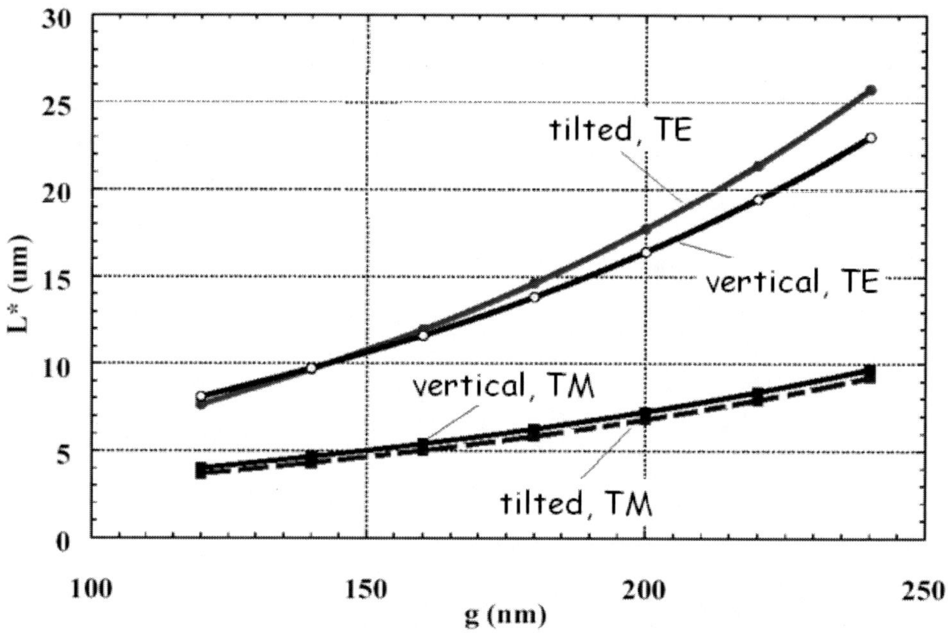

Figure 28. Coupling length versus gap for various structures and both polarizations (WGW = 380 nm).

4. Coupling of Straight Waveguide with Ring/Racetrack Resonator

According with literature [2], a directional coupler with weighted coupling can be again modelled adopting CMT. Referring to Figure 29 where two generic waveguides (WG1 and WG2) are shown, we can write:

$$\begin{aligned} \frac{da_1(z)}{dz} - i\delta a_1(z) &= -i\kappa(z)e^{-i\vartheta(z)}a_2(z) \\ \frac{da_2(z)}{dz} + i\delta a_2(z) &= -i\kappa(z)e^{i\vartheta(z)}a_1(z) \end{aligned} \tag{7}$$

where

$$\vartheta(z) = \int_0^z \left[\Delta\beta(\zeta) - \Delta\beta(0)\right] d\zeta \tag{8}$$

$$\delta = \frac{\Delta\beta(0)}{2} \tag{9}$$

$$\Delta\beta(z) = \frac{2\pi}{\lambda}\left[n_{eff,2}(z) - n_{eff,1}(z)\right] \tag{10}$$

$a_1(z)$ and $a_2(z)$ represent the complex amplitudes of optical modes propagating in the two isolated waveguides, $n_{eff,1}(z)$ and $n_{eff,2}(z)$ are the relevant effective indices and $\kappa(z)$ is the z-varying coupling coefficient.

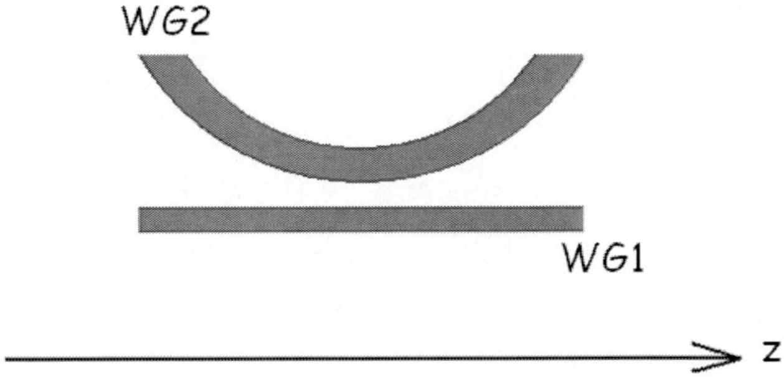

Figure 29. Directional coupler with weighted coupling along z.

This kind of formalism can be successfully used to investigate the generic coupling between a straight waveguide and a ring (or racetrack) in a microring (micro-racetrack) resonator. In a multistage microring (micro-racetrack) resonator, the same formalism can be used to investigate the coupling between two different rings (or racetracks). In this section we investigate the coupling between the straight waveguide and the racetrack in a micro-racetrack resonator, whose partial section is in Figure 29 and complete geometry is in Figure 30. WG1 (straight waveguide) has a constant effective index N_S ($n_{eff,1}(z) = N_S$), whereas WG2 has a z-varying effective index given by:

$$n_{eff,2}(z) = \begin{cases} N_B & z \in \left[0, \frac{3}{4}R\right] \cup \left[\frac{3}{4}R + L, \frac{3}{2}R + L\right] \\ N_S & z \in \left]\frac{3}{4}R, \frac{3}{4}R + L\right[\end{cases} \tag{11}$$

where N_B is the effective index of mode propagating in the bent portion of the racetrack. Effective indices N_B and N_S have been calculated by 2D full-vectorial FEM. According with (10), it results:

$$\Delta\beta(z)=\begin{cases}\dfrac{2\pi}{\lambda}(N_B\text{-}N_S) & z\in\left[0,\dfrac{3}{4}R\right]\cup\left[\dfrac{3}{4}R+L,\dfrac{3}{2}R+L\right]\\[2ex] 0 & z\in\left]\dfrac{3}{4}R,\dfrac{3}{4}R+L\right[\end{cases} \tag{12}$$

where R is the generic resonator radius. According with (8) and (9), we have:

$$\delta=\frac{\pi}{\lambda}(N_B\text{-}N_S) \tag{13}$$

$$\vartheta(z)=\begin{cases}0 & z\in\left[0,\dfrac{3}{4}R\right]\\[2ex] -2\delta L & z\in\left]\dfrac{3}{4}R,\dfrac{3}{4}R+L\right[\\[2ex] -2\delta\left(z-\dfrac{3}{4}R\right) & z\in\left[\dfrac{3}{4}R+L,\dfrac{3}{2}R+L\right]\end{cases} \tag{14}$$

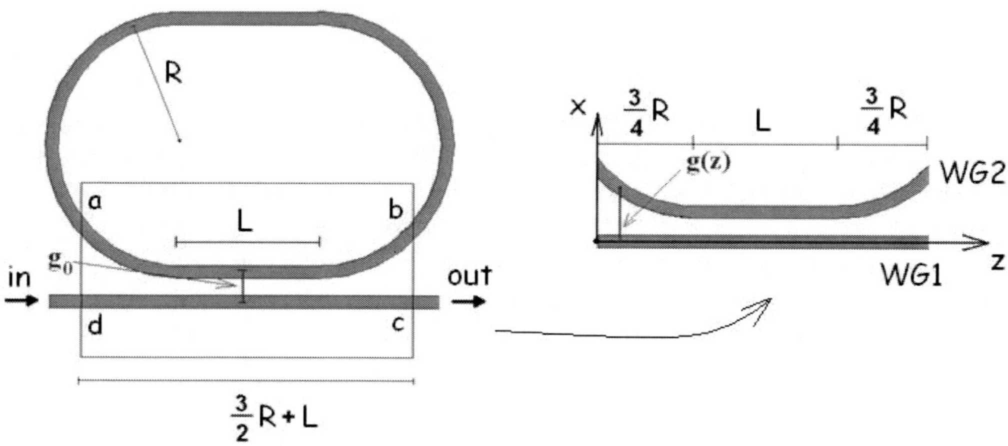

Figure 30. Racetrack resonator geometry, coupled to a straight waveguide.

From geometrical considerations, we can write the distance *g(z)* between straight waveguide and resonator as:

$$g(z)=\begin{cases}g_0+R-\dfrac{1}{4}\sqrt{7R^2+24Rz-16z^2} & z\in\left[0,\dfrac{3}{4}R\right]\\[2ex] g_0 & z\in\left]\dfrac{3}{4}R,\dfrac{3}{4}R+L\right[\\[2ex] g_0+R-\dfrac{1}{4}\sqrt{-16L^2-24LR+7R^2+32Lz+24Rz-16z^2} & z\in\left[\dfrac{3}{4}R+L,\dfrac{3}{2}R+L\right]\end{cases} \tag{15}$$

where g_0 is the minimal distance between the coupled waveguides. Then, coupling coefficient dependence on distance between the waveguides can be numerically investigated by FEM, calculating symmetric and anti-symmetric super-mode effective indices. Finally, coupling coefficient dependence on $g(z)$ can be expressed by:

$$\kappa(z) = Ue^{Vg(z)} \tag{16}$$

where U and V are fitting parameters obtainable again by FEM calculations. Knowing U, V, N_S and N_B, it is possible to solve the equation system (7) assuming two initial conditions:

$$\begin{bmatrix} a_1(0) \\ a_2(0) \end{bmatrix} = \begin{bmatrix} 0 \\ 1 \end{bmatrix} \text{ and } \begin{bmatrix} a_1(0) \\ a_2(0) \end{bmatrix} = \begin{bmatrix} 1 \\ 0 \end{bmatrix}$$

Finally, the coupler behaviour is described by the following matrix:

$$\begin{bmatrix} b \\ c \end{bmatrix} = \begin{bmatrix} re^{i\phi} & ite^{i\varphi} \\ ite^{-i\varphi} & re^{i\phi} \end{bmatrix} \begin{bmatrix} a \\ d \end{bmatrix}$$

where a, b, c and d represent the complex amplitudes of optical modes at coupler output (b, c) and input (a, d) ports (see Figure 30) [8].

5. Coupled Multi-Stage Racetrack Resonators

In this section we consider a multi-stage racetrack resonator (n is the number of racetrack resonators), as a generalized case of previous analysis. Figs 31 and 32 show the multi-stage racetrack resonator geometry for n odd and even, respectively. The electromagnetic interactions in coupling regions can be described using the following matrices:

$$\begin{bmatrix} b_k \\ c_k \end{bmatrix} = \begin{bmatrix} r_k & it_k \\ it_k & r_k \end{bmatrix} e^{i\beta_k L_c} \begin{bmatrix} a_k \\ d_k \end{bmatrix} \qquad (k = 1,\dots,n+1) \tag{17}$$

where a_k, b_k, c_k, d_k are the output and the input electric fields in the k-th coupling region, L_c is the coupling length (equal for each region), r_k and t_k are self- and cross-coupling coefficients and β_k is the propagation constant of optical signal in the k-th coupling region.

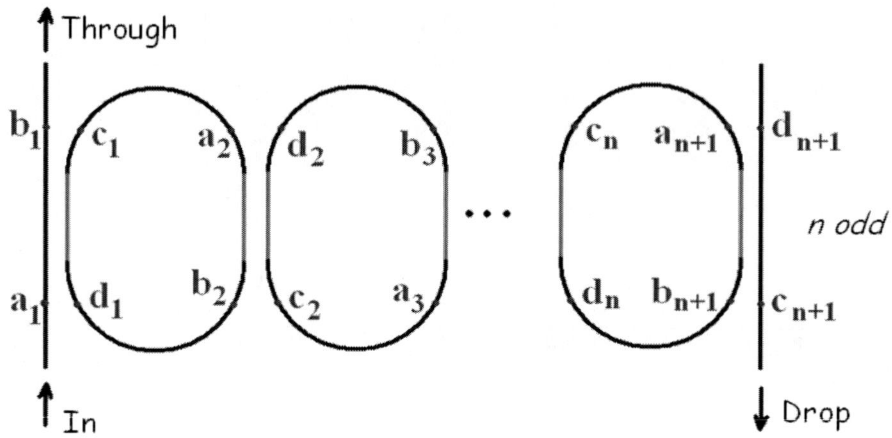

Figure 31. Scheme of n coupled racetrack resonators (n odd).

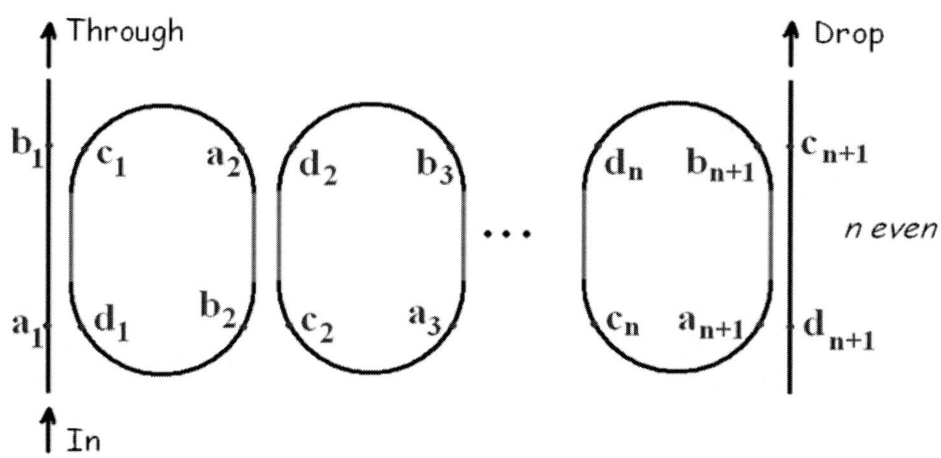

Figure 32. Scheme of n coupled racetrack resonators (n even).

Moreover, we have:

$$a_{k+1} = c_k \sqrt{A_k} e^{i\beta_{Bk}\frac{L_p}{2}} \qquad (k = 1, 2, ..., n) \tag{18}$$

$$d_k = b_{k+1}\sqrt{A_k} e^{i\beta_{Bk}\frac{L_p}{2}} \Rightarrow b_{k+1} = \frac{1}{\sqrt{A_k}} e^{-i\beta_{Bk}\frac{L_p}{2}} d_k \qquad (k = 1, 2, ..., n) \tag{19}$$

where L_p is the optical path length of light travelling inside each racetrack resonator except for the coupling region, A_k is the optical field loss factor resulting from the light propagating over $L_p/2$ (in k-th racetrack resonator) and β_{Bk} is the propagation constant of optical signal in the k-th racetrack resonator bend. We can write in matrix form:

$$\begin{bmatrix} a_{k+1} \\ b_{k+1} \end{bmatrix} = U_k \begin{bmatrix} c_k \\ d_k \end{bmatrix} \qquad U_k = \begin{bmatrix} \sqrt{A_k} e^{i\beta_{Bk}\frac{L_p}{2}} & 0 \\ 0 & \frac{1}{\sqrt{A_k}} e^{-i\beta_{Bk}\frac{L_p}{2}} \end{bmatrix} \tag{20}$$

and rewrite in a different form:

$$\begin{bmatrix} c_k \\ d_k \end{bmatrix} = V_k \begin{bmatrix} a_k \\ b_k \end{bmatrix} \qquad V_k = \frac{1}{it_k} \begin{bmatrix} -\left(r_k^{\,2} + t_k^{\,2}\right) e^{i\beta_k L_c} & r_k \\ -r_k & e^{-i\beta_k L_c} \end{bmatrix} \tag{21}$$

Combining (20) and (21), we obtain:

$$\begin{bmatrix} a_{k+1} \\ b_{k+1} \end{bmatrix} = U_k V_k \begin{bmatrix} a_k \\ b_k \end{bmatrix} \quad \Rightarrow \quad \begin{bmatrix} a_{n+1} \\ b_{n+1} \end{bmatrix} = P_1 P_2 \ldots P_n \begin{bmatrix} a_1 \\ b_1 \end{bmatrix} \tag{22}$$

where $P_k = U_k V_k$. Writing (21) for $k=n+1$ and applying (22), we have:

$$\begin{bmatrix} c_{n+1} \\ d_{n+1} \end{bmatrix} = V_{n+1} \begin{bmatrix} a_{n+1} \\ b_{n+1} \end{bmatrix} = V_{n+1} P_1 P_2 \ldots P_n \begin{bmatrix} a_1 \\ b_1 \end{bmatrix} = W \begin{bmatrix} a_1 \\ b_1 \end{bmatrix} \qquad W = V_{n+1} P_1 P_2 \ldots P_n$$

which can be rewritten in a more convenient form:

$$\begin{bmatrix} c_{n+1} \\ b_1 \end{bmatrix} = \frac{1}{w_{22}} \begin{bmatrix} w_{12} & det(W) \\ 1 & -w_{21} \end{bmatrix} \begin{bmatrix} d_{n+1} \\ a_1 \end{bmatrix} \tag{23}$$

where w_{12}, w_{21}, w_{22} and w_{11} are W matrix elements. Assuming that $d_{n+1} = 0$, we obtain:

$$b_1 = -\frac{w_{21}}{w_{22}} a_1 \tag{13}$$

$$c_{n+1} = \frac{det(W)}{w_{22}} a_1 \tag{14}$$

$$T = \left| \frac{b_1}{a_1} \right|^2 = \left| -\frac{w_{21}}{w_{22}} \right|^2 \tag{15}$$

$$D = \left| \frac{c_{n+1}}{a_1} \right|^2 = \left| \frac{det(W)}{w_{22}} \right|^2 \tag{24}$$

For n =1, single racetrack resonator (SRR):

$$T = \left| \frac{r_1 - A_1 r_2 \left(r_1^2 + t_1^2 \right) e^{i\delta}}{1 - A_1 r_1 r_2 e^{i\delta}} \right|^2 \qquad \delta = L_c \left(\beta_1 + \beta_2 \right) + L_p \beta_{B1} \tag{25}$$

For n =2, double racetrack resonator (DRR):

$$T = \left| \frac{r_2 - r_1 A_2 \left(r_2^2 - t_2^2 \right) e^{i\delta_1} - A_1 r_1 r_2 r_3 e^{i\delta_2} + A_2 A_1 r_3 \left(r_1^2 + t_1^2 \right)\left(r_2^2 + t_2^2 \right) e^{i(\delta_1+\delta_2)}}{1 - A_2 r_1 r_2 e^{i\delta_1} - A_1 r_1 r_3 e^{i\delta_2} + A_2 A_1 r_3 r_2 \left(r_1^2 + t_1^2 \right) e^{i(\delta_1+\delta_2)}} \right|^2 \tag{26}$$

$$\delta_1 = L_c \left(\beta_1 + \beta_2 \right) + L_p \beta_{B2}$$
$$\delta_2 = L_c \left(\beta_1 + \beta_3 \right) + L_p \beta_{B1}$$

For example, simulating by 3D FEM a directional coupler with L_c = 1030 μm and g = 0.58 μm, realized using SOI waveguides having geometrical sizes WGH=1350 nm, WGW= 1000 nm, ED= 800 nm, we obtain r^2 = 0.5821 and t^2 = 0.1679 (with Ψ = 0.75). Effective indices of fundamental quasi-TM modes propagating in straight and bent (R=100 μm) waveguides are: N_{TM}^S = 3.371529 and N_{TM}^B = 3.405307, respectively.

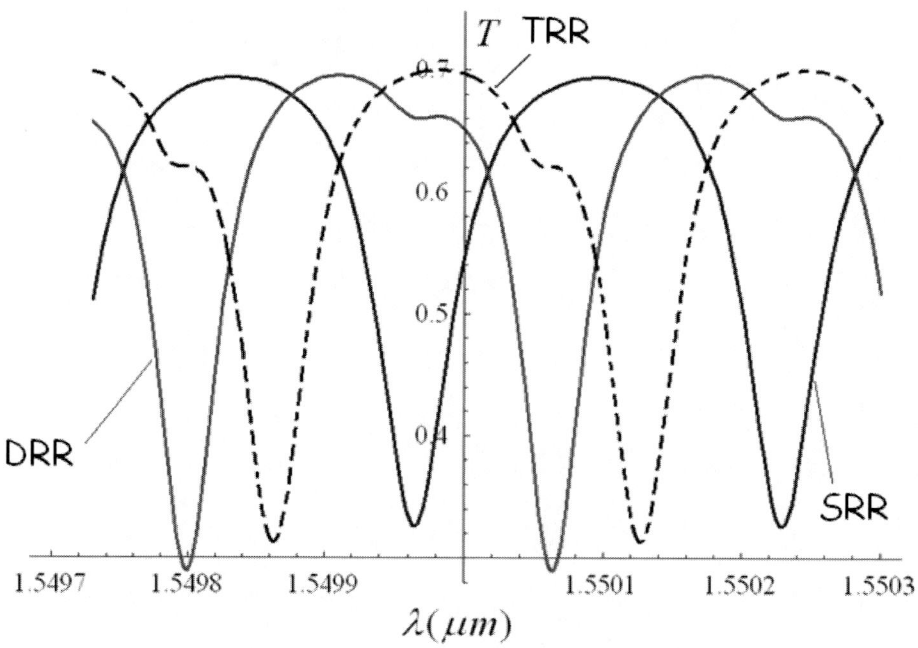

Figure 33. Transmission spectra for multi-stage racetrack resonators (SRR, DRR, TRR).

To model a multi-stage racetrack resonator employing an above-described directional coupler, we assume: R =100 μm, $A_1 = A_2 = A_3 = 0.8$, $r_1^2 = r_2^2 = r_3^2 = r_4^2 = 0.5821$, $t_1^2 = t_2^2 = t_3^2 = t_4^2 = 0.1679$, $N_{B1} = \beta_{B1}\lambda / 2\pi = 3.405307$, $N_{B2} = \beta_{B2}\lambda / 2\pi = 3.405307 + 9{\cdot}10^{-4}$, where the additive term is used to simulate fabrication imperfections between two close racetrack resonators, for example an increase of 10 nm in waveguide width (1%), and so on. Moreover, we have $N_{B3} = \beta_{B3}\lambda / 2\pi = 3.405307$, $N_1 = \beta_1\lambda / 2\pi = 3.371529$, $N_2 = \beta_2\lambda / 2\pi = 3.371529 + 9{\cdot}10^{-4}$, $N_3 = \beta_3\lambda / 2\pi = 3.371529 + 9{\cdot}10^{-4}$, $N_4 = \beta_4\lambda / 2\pi = 3.371529 + 3{\cdot}10^{-4}$. Transmission spectra calculated by Eq. (25) and (26) have been obtained for SRR, DRR and triple racetrack resonator (TRR), and are sketched in Figure 33.

Similar results were obtained in literature [9] by experimental transmission spectra of SRR, DRR and TRR realized using the same geometrical parameters used in our modeling. Other significant results on this approach can be found in literature [10-11].

Conclusion

In this chapter, an extensive modal analysis of SOI rib waveguides is presented in order to meet single-mode and birefringence free conditions. Finite element method is demonstrated as a powerful approach to study both rib and wire waveguides. Moreover, a mixed calculation procedure based on coupled mode theory and FEM is presented and applied to both straight directional couplers and coupling between a straight waveguide and an optical resonator. A generalization of this approach is derived for coupling among multi-stage racetrack resonators.

References

[1] Liu, J.-M., *Photonic Devices*, Cambridge University Press, 2005.
[2] Kogelnik, H., "Theory of dielectric waveguides," in *Integrated Optics,* T. Tamir (Ed.), Springer, 1975.
[3] Comsol Multiphysics by COMSOL AB, ver. 3.2, *single license*, 2005.
[4] Chan, S. P., Passaro, V. M. N., Reed, G. T., *Electronics Letters*, 2005, 41, 528-529.
[5] Chan, S. P., Png, C. E., Lim, S. T., Reed, G. T., Passaro, V.M.N., *J. Lightwave Technology*, 2005, 23, 2103-2111.
[6] Hardy, A., Streifer, W., *J. Lightwave Technology*, 1985, LT-3, 1135-1146.
[7] Yamada, K., *et al.*, *IEEE Photon. Technology Letters*, 2005, 17, 2376-2378.
[8] Popović, M. A., Manolatou, C., Watts, M. R., *Optics Express*, 2006, 14, 1208-1222.
[9] Timotijevic, B. D., *et al., J. Optics A: Pure and Applied Optics*, 2006, 8, S473-S476.
[10] Dell'Olio, F., Passaro, V. M. N., *et al.*, *J. Optics A: Pure and Applied Optics*, 2008, 10, 064003.
[11] Passaro, V. M. N., Dell'Olio, F., Timotijevic, B. D., Mashanovich, G., Reed, G. T., *The Open Optics Journal*, 2008, 2, 6-9.

In: Modeling of Photonic Devices
Editor: Vittorio M. N. Passaro

ISBN 978-1-60456-980-3

Chapter 4

RING RESONATORS

***Francesco De Leonardis**[*1] **and Vittorio M. N. Passaro**[2]*

[1]Photonics Research Group, Dipartimento di Ingegneria dell'Ambiente e per lo Sviluppo Sostenibile, Politecnico di Bari, viale del Turismo n. 8, 74100 Taranto, Italy

[2]Photonics Research Group, Dipartimento di Elettrotecnica ed Elettronica, Politecnico di Bari, via Edoardo Orabona n. 4, 70125 Bari, Italy

ABSTRACT

In this chapter, nonlinear optical propagation of optical pulses in silicon-on-insulator passive ring resonator is theoretically investigated. Two photon absorption and free carrier absorption are taken into account to investigate the performance limits of passive resonators under pulsed excitation. Moreover, a model including the fundamental physical effects of active ring resonators (ring lasers) for understanding the optical mode behavior and control of the oscillating mode in semiconductor injection ring lasers is briefly reviewed. The fully physical model is derived from a quantum-mechanical approach and do not depend on any semi-empirical approximation. The ring laser behavior is related to its physical and technological constraints, as the backscattering effect and ring radius.

Keywords: Integrated Optics, Nonlinear Optics, Silicon-on-Insulator Technology, Modelling, Ring Laser, Optical resonator.

1. INTRODUCTION

Channel dropping filters that access one channel of a wavelength division multiplexed (WDM) signal without disturbing the other channels, are useful elements for any WDM telecommunication system. Ring or disk resonators are attractive candidates for this application, because they can potentially realize the narrowest linewidth for a given device

[*] E-mail address: f.deleonardis@poliba.it

size. In addition, a cascade of coupled resonators can further modify the simple Lorentzian response of a single resonator, giving desirable higher order filter characteristics [1]. Potentials of microring resonators are related to the fact of supporting the traveling wave resonant modes. Unlike the standing wave resonators, any ring, disk or racetrack resonator may completely extract a particular wavelength from the input bus waveguide, offering superior performance over other optical approaches for the same task. In applications such as WDM systems, the ring resonators have to be small enough so that the spacing between resonant frequencies accommodates the set of WDM channels within a telecommunications window (ITU standard). The telecommunication window supported by erbium amplifiers is 30 nm. Rings with free spectral range FSR > 30 nm require radii $R \leq 5$ μm, which is possible to achieve by the state-of-the-art fabrication technologies. Hence, these microring structures require high index contrast to make the bending losses negligible. In that case, microring performance is limited by scattering due to sidewall roughness. Such scattering leads to two detrimental effects: energy loss due to the radiation continuum and scattering into the counter-propagating mode. Contra-directional coupling induced by surface roughness can be more problematic than the radiation loss, because the periodic nature of the ring leads to a natural phase matching between forward and backward modes, due to any possible perturbation effect [2-3].

Recent research activities have focused on achieving better filter functionality based on microrings. Generally speaking, the configuration of vertically coupled microring resonators has many features, such as the independent optimum design of lower waveguide from upper ring, as well as the possibility of dense integration resulting from cross-grid configuration [4]. In addition, the coupling strength between the ring and the bus waveguides can be controlled more precisely than the lateral coupling, because the vertical separation is obtained by well-controlled deposition techniques, rather than etching fine gaps. However, flatness of passband, sharp roll-off from passband to stop band, and large out-of-band rejection are necessary to minimize the pulse broadening and the packing efficiency of wavelength channels. The ideal response shape is the box-like function with a unity shape factor, which is defined by the ratio of -1 dB bandwidth to -10 dB bandwidth. However, since the response shape of the single microring resonator filter is expressed by the Lorentzian function, the shape factor is as small as 0.17 [5]. In the arrayed waveguide grating (AWG) filters, where the response can be approximated by the Gaussian function, the theoretical value of the shape factor is 0.32. In addition, the crosstalk is limited to -20 dB due to the large Lorentzian wing. Therefore, the tailoring of the filter response shape is required to improve the performance of microring resonator filters.

The series-coupled microring resonator is one solution [1], and a triple-coupled microring based on planar structure has been reported [6]. However, the crosstalk was about -10 dB and the FSR remained smaller than 20 nm. In addition, very fine etching was required to realise a gap less than 500 nm., because each microring and bus waveguide were fabricated in the same layer. In [7] a vertically triple-coupled microring resonator filter has been proposed and demonstrated. With this configuration, a box-like filter response shape and an expansion of FSR larger than 20 nm have been demonstrated.

The high-quality factors achievable in microring resonators have also attracted considerable interest for realizing microring-based active components and nonlinear switching devices. In fact, semiconductor ring lasers are important for monolithic and integrated optoelectronic applications. Since these lasers do not require the use of cleaved facets or difficult Bragg gratings for optical feedback, they can be conveniently integrated so reducing the occupation area. Moreover, the semiconductor ring lasers offer new capabilities in travelling-wave unidirectional oscillation. In general, unidirectional operation is desired because it offers the advantages of enhanced mode purity, reduced sensitivity to feedback and higher single pump. Driven by these advantages, both experimental and theoretical studies have been proposed for understanding the physical behaviour of bidirectional and unidirectional regimes in active ring resonators. For example, the influence of injection current on either unidirectional or bidirectional behaviour in Multi-Quantum-Well (MQW) ring lasers based on GaAs/AlGaAs material system has been experimentally demonstrated [8]. More recently, a fully physical model derived from a quantum-mechanical approach and not depending on any semi-empirical approximation has been proposed, with the aim to relate the different operation regimes to laser physical and technological constraints, such as ring radius and ring sidewall roughness [9].

Another approach to use microring resonators in active devices has been proposed [10], where a wide tunable double ring resonator-coupled laser has been designed. This structure consists of four main regions: gain region, two passive microring resonators, passive waveguides and absorption regions. The two passive microring resonators have slightly different radii providing a mode selection and wavelength tuning mechanism. This configuration offers many promising advantages over conventional tunable lasers, including ultra wide wavelength tuning range, high side mode suppression ratio, uniform threshold and efficiency, narrow linewidth, low frequency chirp, and simple fabrication.

Further, all optical nonlinear switches in compact GaAs-AlGaAs microring resonators has been demonstrated at 1.55 μm wavelength [11-12]. Switching is accomplished in pump and probe configuration, in which both pump and probe signals are tuned to different microring resonance wavelengths. Refractive index change in the microring due to free carriers generated by two photon absorption (TPA) is used to switch the probe beam in and off resonance. However, the switching speed is currently limited by carrier life time, then the performance of this type of device may be further improved by dc biasing to sweep out the carriers.

Thus, generally speaking two different technologies are very promising to realise optical resonators: III/V semiconductor compounds and silicon-on-insulator (SOI). The former is mainly used to implement active functionality such as ring lasers and nonlinear devices. The latter can represent the best candidate to include ring resonators as building blocks in the context of optical interconnects. In fact, coupled resonator optical waveguide (CROW) structures [13], consisting of a chain of coupled integrated optical micro-resonators, are emerging as very attractive components to be included in delay lines [14], buffers [15], add-drop multiplexers [16] and nonlinear devices [17]. To fabricate this type of photonic structures, SOI technological platform is very adequate because of SOI waveguide low losses and high index contrast, low cost of silicon and wide production infrastructure available for silicon based integrated device fabrication. Recently, high quality factor microring resonators and polarization independent coupled racetrack resonators have been fabricated using SOI technology [18-19].

In this chapter, various aspects of ring resonator modeling and design are briefly described. In section 2 the mathematical modeling used to investigate TPA effects in passive SOI ring resonator, and the influence of some technological parameters (ring sidewall roughness, ring radius) over the backscattering coefficient influencing the operation regime and performance of active ring resonators, are described. In section 3 some numerical results and simulations applied to passive SOI ring resonators and active GaAs-AlGaAs MQW semiconductor ring lasers are presented.

2. THEORY

2.1. Passive Ring Resonator

The most popular approach to model single ring and racetrack micro-resonators is based on Coupled Mode Theory (CMT) [20-21]. Both 2D and 3D Finite Difference Time Domain (FDTD) algorithms have been also used to investigate the electromagnetic propagation in these devices. Very recently, circuit-based [22] and matrix-based [23-24] approaches have been introduced to analyze and design coupled ring resonators.

In addition, filters based on SOI technology have been considered, where high index contrast between waveguide cladding and core facilitates light guiding in nanometer-scale structures, thus making the bending loss in SOI microring negligible. However, when the ring cross section becomes very small, some effects induced by TPA cannot be considered negligible. In this sense, we believe that the Coupled Mode Theory in time domain [1] could be considered the best candidate to model the physical features of ring resonators in presence of nonlinear effect such as the TPA effect.

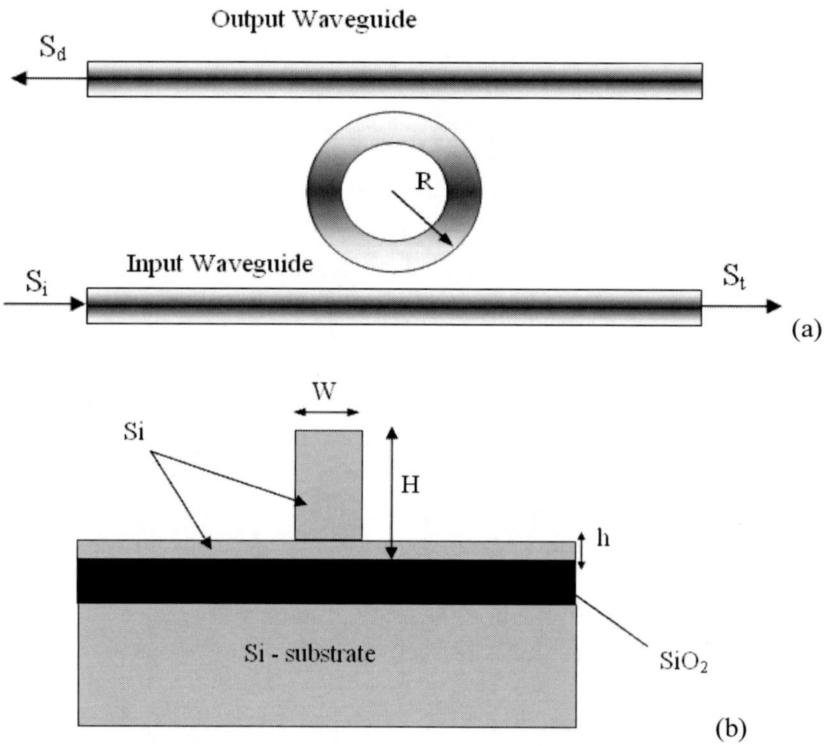

Figure 1. (a) Evanescently coupled microring resonator; (b) Cross section of the ring waveguide.

In fact, in this case it is possible to couple the rate equation governing the generation of free carriers as induced by TPA, with the time domain differential equation governing the ring resonator dynamics.

In Figures 1(a) and (b) the single ring resonator, evanescently coupled to two external waveguides, and the ring cross section are depicted, respectively.

We assume that the ring supports a travelling wave of amplitude $A(t)$, such that $|A(t)|^2$ represents the total power flowing through each ring cross section at each time instant t. With this assumption, the time evolution of the field envelope amplitude is governed by means of the following differential equation [1]:

$$\frac{dA(t)}{dt} = j(\omega_0 - \omega)A(t) - \frac{1}{\tau}A(t) - v_g \alpha^{(FCA)} A(t) - j v_g \frac{2\pi}{\lambda}\Delta n A(t) + \\ - v_g \frac{\beta^{(TPA)}}{A_{eff}} |A(t)|^2 A(t) - j\sqrt{\frac{2}{\tau_e}}\sqrt{\frac{v_g}{2\pi R}} S_i \qquad (1)$$

where ω_0 is the angular frequency at the resonance, ω is the angular frequency of the input beam, v_g is the group velocity and S_i denotes the input wave amplitude. In Eq. (1) τ is the ring cavity photon lifetime, defined as:

$$1/\tau = 1/\tau_e + 1/\tau_d + 1/\tau_l \qquad (2)$$

where $1/\tau_e$ and $1/\tau_d$ are related to the power leaving the ring due to external coupling with input and output waveguides, respectively. As described in chapter 8 of this book, the coupling factor κ_e^2 (κ_d^2) is related to the coupling time constant τ_e (τ_d) by means of $\kappa_{e,(d)}^2 = 2\pi R/(\tau_{e,(d)} v_g)$. The coefficient $1/\tau_l$ represents the optical power loss due to other intrinsic effects, such as propagation, bending and scattering losses. The last term is given by $1/\tau_l = \alpha_{loss} v_g$, where α_{loss} is the linear loss coefficient and v_g is the group velocity of the optical mode inside the ring resonator. Eq. (1) has been obtained from wave equation under a slowly-varying amplitude approximation. In particular, the terms including $\alpha^{(FCA)}$ (free carrier absorption), Δn (effective index change due to plasma dispersion) and $\beta^{(TPA)}$ (two photon absorption) have been derived by introducing linear and third order polarization vectors in the wave equation. We evaluate $\alpha^{(FCA)}$ and Δn by the following expressions:

$$\alpha^{(FCA)} = \sigma_0 \cdot \left(\frac{\lambda}{1.55}\right)^2 N_c \qquad (3)$$

$$\Delta n = \left(-8.8 \times 10^{-22} N_c - 8.5 \times 10^{-18} N_c^{0.8}\right) \qquad (4)$$

where N_c is the density of electron-hole pairs generated by TPA, $\sigma_0 = 1.45\times10^{-17}$ cm^{-2} is the FCA cross section measured at $\lambda = 1.55$ μm, and λ is the input signal wavelength. To obtain the complete mathematical model, Eq. (1) has been coupled to the rate equation governing the free carrier dynamics into the waveguide core [25]:

$$\frac{dN_c}{dt} = -\frac{N_c}{\tau_{eff}} + \frac{\beta^{(TPA)}}{2\hbar\omega}\left(\left|A(z,t)\right|^2 / A_{eff}\right)^2 \tag{5}$$

where τ_{eff} is the effective recombination lifetime of free carriers and A_{eff} is the effective area of optical mode inside the ring.

2.2. Active Ring Resonator

The theory for active ring resonator (see Figure 2) is based on the semi-classical interaction between radiation and matter. Then, the atomic systems can be modeled as quantum phenomena while the electromagnetic (e.m.) field is classically described by Maxwell's equations. In particular, the electric dipole $\mathrm{e_r}$ operator relates the system quantum-mechanical description, being the polarization P of medium as the source of e.m. field.

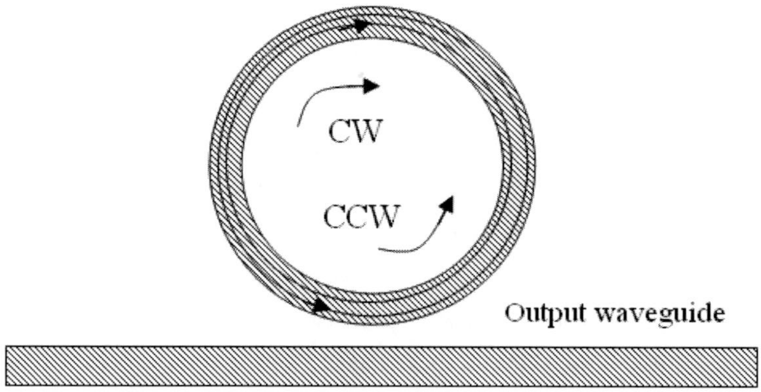

Figure 2. Ring laser with output coupler and grating to force the laser unidirectional regime.

Assuming a predominant single transverse mode as the electric field inside the ring laser, we can write:

$$\mathrm{E}(\mathbf{r},t) = \sum_n E_n(t) e^{j(\omega_n t + \phi_n(t))} \mathrm{F}(\mathbf{r}) + c.c. \tag{6}$$

where c.c. indicates the conjugate complex terms, $E_n(t)$ is the electric field amplitude, ω_n is the angular pulsation of ring cavity optical mode and $\phi_n(t)$ is the time-dependent phase of the electric field. In general, the field function $F(r)$ can be written as $F(r) = G(x,y)U(z)$, where $G(x,y)$ and $U(z)$ indicate the transverse and longitudinal electric field profiles, respectively, z representing the propagation direction (curvilinear coordinate). The subscript n takes into account all possible longitudinal modes in the ring cavity. With this representation for the field, the polarization P induced by the gain medium is given by:

$$\mathrm{P}(\mathbf{r},t) = \sum_n P_n(t) e^{j(\omega_n t + \phi_n(t))} \mathrm{F}(\mathbf{r}) + cc. \tag{7}$$

where $P_n(t)$ is the complex, slowly-varying component of the polarization of n-th longitudinal mode. The wave equation for the electric field time evolution is:

$$-\nabla^2 \mathbf{E} + \mu_0 \frac{\partial \mathbf{J}}{\partial t} + \frac{1}{v^2} \frac{\partial^2 \mathbf{E}}{\partial t^2} = -\mu_0 \frac{\partial^2 \mathbf{P}}{\partial t^2} \tag{8}$$

where v is the electric wave velocity in the ring resonator. The second term is included to take into account the cavity losses. In particular the current density is expressed in terms of the fictional conductivity $\mathbf{J} = \sigma \mathbf{E}$, where the conductivity σ is assumed as the sum of two contributions, σ_1 and σ_2. The former includes all optical losses (propagation loss, radiation loss, leakage loss, etc..) of the n-th longitudinal mode. It is related to the quality factor Q_n of the cold cavity by means of $\sigma_1 = (\varepsilon \omega_n)/Q_n$, where ε is the cavity permittivity. The latter contribution is included to take into account the effect of backscattering induced by any cavity sidewall roughness. According to [9], the backscattered wave induces a contribution to fictional conductivity as $\sigma_2 = \varepsilon b_{scat}$, where b_{scat} is the backscattering rate. This rate coefficient is related to amplitude reflectivity R_{amp} as $b_{scat} = cR_{amp} / (\pi n_{eff} R_{eff})$, being c the free-space light velocity, n_{eff} the effective index of optical wave in the ring cavity and R_{eff} the ring effective radius.

With the aim to study the dynamics of two counter-propagating modes inside the ring cavity, we can particularise Eqs. (6)-(7) to only two modes, one clockwise (CW, mode 1) and the other counter-clockwise (CCW, mode 2). Under slowly-varying amplitude and phase approximation (extensively used in laser dynamics modelling), the following set of equations can be derived from wave equation (8):

$$\dot{E}_1 = -\frac{\omega_1}{2Q_1} E_1 - \frac{1}{2} \frac{\omega_2}{\omega_1} b_{scat} E_2 \cos(\psi + \delta_2) + \frac{1}{2} \frac{\omega_1}{\varepsilon} \mathrm{Im}\{P_1\} \tag{9}$$

$$\dot{E}_2 = -\frac{\omega_2}{2Q_2} E_2 - \frac{1}{2} \frac{\omega_1}{\omega_2} b_{scat} E_1 \cos(-\psi + \delta_1) + \frac{1}{2} \frac{\omega_2}{\varepsilon} \mathrm{Im}\{P_2\} \tag{10}$$

$$\dot{\phi}_1 = -\frac{\omega_1}{2\varepsilon} \frac{1}{E_1} \mathrm{Re}\{P_1\} - \frac{1}{2} \frac{\omega_2}{\omega_1} b_{scat} \frac{E_2}{E_1} \sin(\psi + \delta_2) + \Omega_1 - \omega_1 \tag{11}$$

$$\dot{\phi}_2 = -\frac{\omega_2}{2\varepsilon}\frac{1}{E_2}\mathrm{Re}\{P_2\} - \frac{1}{2}\frac{\omega_1}{\omega_2} b_{scat}\frac{E_1}{E_2}\sin(-\psi + \delta_1) + \Omega_2 - \omega_2 \quad (12)$$

where $\psi = \phi_2 - \phi_1$ and Ω_n are the eigen-frequencies of the cold cavity eigen-modes. Eqs. (9)-(12) are the Lamb's self-consistency equations and take into account the effects related to both gain medium and backscattering. Finally, the system (9)-(12) can be solved when the polarization vector is known.

The model proposed in this section leads to evaluate the linear and nonlinear terms of P by performing quantum calculations applied to a generic semiconductor structure. By this way, it is possible to calculate all physical parameters, involved in the ring laser dynamics, starting from the physical description of the energy band without any semi-empirical approximation. To evaluate the P vector, we use the density matrix formalism [26] to write:

$$P = \overline{n}Tr(\rho M) = \overline{n}\sum_{b,a}(\rho_{ba}M_{ab} + \rho_{ab}M_{ba}) \quad (13)$$

where ρ is the density matrix operator, given by:

$$\rho = \begin{bmatrix} \rho_{aa} & \rho_{ab} \\ \rho_{ba} & \rho_{bb} \end{bmatrix}$$

being ρ_{aa} the probability to have an electron in $|a\rangle$ state of a valence band (VB) sub-band, ρ_{bb} the probability to have an electron in $|b\rangle$ state of a conduction band (CB) sub-band, ρ_{ab} and ρ_{ba} the probabilities to have a transition between a and b levels or b and a levels, respectively (it holds $\rho_{ba} = \rho_{ab}^*$). M is the dipole moment operator (formed by the electron-hole pair relevant to two sublevels of the same order, one in CB and the other in VB) in the form of a 2x2 matrix as:

$$M = \begin{bmatrix} 0 & M_{ab} \\ M_{ba} & 0 \end{bmatrix}$$

where M_{ab}, M_{ba} can be calculated as in [26-27]. In Eq. (13), $\overline{n}$ denotes the electron total density included in the conduction and valence sub-bands, satisfying the following relationship for semiconductor lasers:

$$\overline{n}\sum_{b,a}\left[\rho_{bb}^{(0)} - \rho_{aa}^{(0)}\right] = \int_{W_g}^{\infty} g_{cv}(W_{ba})\left[f_c(W_{ba}) - f_v(W_{ba})\right]dW_{ba} \quad (14)$$

being $g_{cv}(W_{ba})$ the state density, $W_{ba} = W_b - W_a$ the energy difference between level b and a, f_c (f_v) the Fermi-Dirac distribution function at level $|b\rangle$ and $|a\rangle$, respectively. By solving the continuity equation as proposed in a previous work [9], we derive the elements of density matrix and, then, the polarization vector as a sum of two terms, one linear and one non linear. By substituting the components of polarization vector obtained by the quantum-mechanical analysis in Eq. (9)-(10), considering only two modes, one clockwise (CW, 1) and the other counter-clockwise (CCW, 2), rearranging the equations in terms of the mode intensities $I_1 = |E_1|^2$ and $I_2 = |E_2|^2$, and assuming $\omega_1 = \omega_2$, we obtain the rate equations for any semiconductor ring laser in the form:

$$\frac{dI_1}{dt} = 2I_1\left(\alpha_1 - \tilde{\beta}_1 I_1 - \tilde{\theta}_{12} I_2\right) - 2\sqrt{I_1 I_2}\left(\xi_1 I_1 + \eta_{12} I_2\right) + b_{scat}\sqrt{I_1 I_2}\cos\left(\Psi_{tot} + \delta\right) \quad (15)$$

$$\frac{dI_2}{dt} = 2I_2\left(\alpha_2 - \tilde{\beta}_2 I_1 - \tilde{\theta}_{21} I_1\right) - 2\sqrt{I_1 I_2}\left(\xi_2 I_2 + \eta_{21} I_1\right) + b_{scat}\sqrt{I_1 I_2}\cos\left(\Psi_{tot} - \delta\right) \quad (16)$$

with $\psi_{tot} = \psi - \pi + \frac{1}{2}(\delta_2 - \delta_1)$, $\delta = \frac{1}{2}(\delta_1 + \delta_2)$ and

$$\tilde{\beta}_i = \frac{\omega_i}{2\varepsilon}|M|^4\, \tilde{\mathrm{N}}_{iiii}\mathrm{I}_{iiii}$$

$$\tilde{\theta}_{ij} = \frac{\omega_i}{2\varepsilon}|M|^4\left[\tilde{\mathrm{N}}_{iijj}\mathrm{I}_{iijj} + \tilde{\mathrm{N}}_{ijji}\mathrm{I}_{ijji} + \tilde{\mathrm{N}}_{ijij}\mathrm{I}_{ijij}\right]$$

$$\xi_i = \frac{\omega_i}{2\varepsilon}|M|^4\left[\tilde{\mathrm{N}}_{iiij}\mathrm{I}_{iiij} + \tilde{\mathrm{N}}_{iiji}\mathrm{I}_{iiji} + \tilde{\mathrm{N}}_{ijii}\mathrm{I}_{ijii}\right]$$

$$\eta_{ij} = \frac{\omega_i}{2\varepsilon}|M|^4\, \tilde{\mathrm{N}}_{ijjj}\mathrm{I}_{ijjj} \quad i = 1,2 \text{ and } j \neq i$$

where

$$\mathrm{I}_{nqkm} = \int_{W_g}^{\infty} g_{cv}(W_{ba})\left[f_c(W_{ba}) - f_v(W_{ba})\right]\mathrm{Im}\left\{je^{-j\Psi_{nqkm}}\mathrm{C}_{qkm}\right\}dW_{ba}$$

$\tilde{\mathrm{N}}_{nqkm} = \dfrac{\int_{V_l} F_n(r)^* F_q F_k^* F_m dr}{\mathrm{N}_n}$ is a real quantity (field overlapping integral, being $\mathrm{N}_n = \int F_n(r) F_n(r)^* dr$), which is not zero just in the active region volume, and $\alpha_n = g_n - \omega_n/2Q_n$ is the net linear gain, being g_n the gain of active structure. The parameters Ψ_{nqkm} and C_{qkm} are appropriately defined in Ref. [9].

The introduction of variable δ in Eq. (15)-(16) avoids the explicit dependence on the individual scattering phases δ_1 and δ_2, and, then, all physical quantities will depend only on an average value δ. The equation system (15)-(16) has a stable solution as $\psi_{tot} = 0$ when $\psi = \pi$ and $\delta_1 = \delta_2$. The coefficients in the model, namely $\tilde{\beta}_i$, $\tilde{\theta}_{ij}$, ξ_i and η_{ij}, represent self-saturation, cross-saturation, self interference and cross interference coefficients, respectively. Among various effects, they are also responsible of the mode competition phenomenon [9]. Then, by manipulating Eq. (11)-(12), we obtain:

$$\begin{aligned}\frac{d\psi_{tot}}{dt} &= (\Omega_2-\Omega_1)-(\omega_2-\omega_1)-(\gamma_2-\gamma_1)+(\nu_2 I_2-\nu_1 I_1)+(\nu_{21}I_1-\nu_{12}I_2)\\ &+\sqrt{I_1I_2}(\chi_{21}-\chi_{12})+\left(\sigma_{21}I_1\sqrt{\frac{I_1}{I_2}}-\sigma_{12}I_2\sqrt{\frac{I_2}{I_1}}\right)-\frac{b_{scat}}{2}\left(\sqrt{\frac{I_2}{I_1}}\sin(\psi_{tot}+\delta)+\sqrt{\frac{I_1}{I_2}}\sin(\psi_{tot}-\delta)\right)\end{aligned} \tag{17}$$

with

$$\nu_i = \frac{\omega_i}{2\varepsilon}\tilde{N}_{iiii}R_{iiii}$$

$$\nu_{ij} = \frac{\omega_i}{2\varepsilon}\left[\tilde{N}_{iijj}R_{iijj}+\tilde{N}_{ijji}R_{ijji}+\tilde{N}_{ijij}R_{ijij}\right]$$

$$\chi_{ij} = \frac{\omega_i}{2\varepsilon}\left[\tilde{N}_{iiij}R_{iiij}+\tilde{N}_{iiji}R_{iiji}+\tilde{N}_{ijii}R_{ijii}\right]$$

$$\sigma_{ij} = \frac{\omega_i}{2\varepsilon}\tilde{N}_{ijjj}R_{ijjj}$$

where Ω_1 and Ω_2 are the cold cavity frequencies for CW and CCW modes, respectively, ω_1, ω_2 are the laser beam frequencies, γ_1, γ_2 are the pulling effect coefficients, ν_1, ν_2 are the self-pushing effect coefficients, ν_{12} and ν_{21} are the cross-pushing effect coefficients. The terms χ_{12}, χ_{21}, σ_{12} and σ_{21} are nonlinear coefficients depending on nonlinear part of polarization vector. Moreover, it holds:

$$R_{nqkm} = \int_{W_g}^{\infty}|M|^4 g_{cv}(W_{ba})\left[f_c(W_{ba})-f_v(W_{ba})\right]\mathrm{Re}\left\{je^{-j\Psi_{nqkm}}\mathrm{C}_{qkm}\right\}dW_{ba}$$

and

$$\gamma_i = \frac{\omega_i}{2\varepsilon}\frac{\int_{V_I}|F(r)|^2dr}{\mathrm{N}_m}\int_{W_g}^{\infty}|M|^2 g_{cv}(W_{ba})\left[f_c(W_{ba})-f_v(W_{ba})\right]\frac{(W_i-W_{ba})}{(W_i-W_{ba})^2+\left(\frac{\hbar}{\tau_{in}}\right)^2}dW_{ba}$$

being τ_{in} the electron average relaxation time. In order to complete the ring laser model, the classical rate equation for injected carriers has been used as:

$$\frac{dN}{dt}=\frac{J}{ed}-AN-BN^2-CN^3-\frac{2\varepsilon_0 n_{eff}^2}{h\omega_1}\left[I_1\left(g_1-\tilde{\beta}_1 I_1-\tilde{\theta}_{12}I_2\right)-\sqrt{I_1 I_2}\left(\xi_1 I_1+\eta_{12}I_2\right)\right]$$
$$-\frac{2\varepsilon_0 n_{eff}^2}{h\omega_2}\left[I_2\left(g_2-\tilde{\beta}_2 I_2-\tilde{\theta}_{21}I_1\right)-\sqrt{I_1 I_2}\left(\xi_2 I_2+\eta_{21}I_1\right)\right] \quad (18)$$

where ε_0 is the vacuum dielectric permittivity, d is the active region thickness, J is the laser current density, n is the active region refractive index, A, B, C are leakage recombination, bimolecular recombination and Auger coefficients, respectively. Eqs. (15)-(18) are the coupled differential equations for two counter-propagating modes inside any ring laser based on III/V semiconductor material system (such as, for example, GaAs-based MQW).

It is worth to note that self and cross interference coefficients are equal to zero only in absence of any backscattering (i.e. $R_{amp}=0$) inside the laser cavity. In this ideal case Eqs (15)-(16) include only self- and cross-saturation coefficients and, then, the only stable regime is the unidirectional one. In next section, we will show how the backscattering effect can be also responsible of a stable bidirectional regime, depending on R_{amp} values. The goal of this model is also to evaluate the operating regime of a quantum well (QW) ring laser related to the statistical information on sidewall roughness. Differently from [8], where the backscattering coefficient is empirically evaluated, we have calculated R_{amp} coefficient by means of self–correlation function of sidewall roughness, as given in [2]:

$$R_{amp}^2=2\pi R_{eff}\left(\varepsilon_0\omega\delta n^2 t_{rib}F_\phi/4\right)^2\sqrt{\pi}\left|\sigma_c\right|^2\exp\left[-\left(k_0 n_{eff}L_c\right)^2\right] \quad (19)$$

where δn^2 is the ring-air relative permittivity, t_{rib} is the ridge overall height, F_ϕ is the azimuth component of normalized electric field, σ_c and L_c are the standard deviation and correlation length of sidewall roughness self–correlation function, assumed as a Gaussian function.

3. NUMERICAL RESULTS

3.1. Passive Ring Resonator

In this section, we show the TPA influence on passive microring resonator time evolution. It is worth noting that TPA could be an undesirable effect in the design of a ring resonator filter or, on the other side, an advantage in the design of some nonlinear all-optical devices. Hence, sometimes TPA has to be avoided, while in other cases it can be controlled and excited. In this section, we focus our analysis on the limits induced by TPA on the performance of filters based on small ring resonators. It is evident from Eq. (5) that TPA effect can be reduced by decreasing the optical power and the effective recombination lifetime of free carriers. It is not trivial to find the maximum input power in order to neglect the TPA, because it depends on the ring enhancement factor. This factor also depends on a number of different parameters such as ring radius, cavity photon lifetime and input pulse FWHM. On the contrary, the reduction of τ_{eff} can be carried out by an appropriate design of the ring cross section. To obtain an estimation for τ_{eff}, free carrier diffusion needs to be considered in addition to the recombination lifetime. If diffusion carriers move out from the modal area, this results in an effective lifetime that can be shorter than the recombination lifetime in SOI structures, τ_r. If τ_t is the transit time, then $1/\tau_{eff} = 1/\tau_r + 1/\tau_t$. It has been demonstrated [14] that τ_r depends on the surface-recombination velocity at the interface between the top silicon and buried oxide, with a typical value of about 100 ns. Transit time τ_t also depends on the optical mode size, total height of SOI rib waveguide and ambipolar diffusion coefficient D, as described in chapter 8 on Raman effect. Thus, a small cross section is necessary to obtain a low value for τ_{eff}. To enhance the scaling effect, we consider a ring cross section of few hundreds of nanometres. We have fixed the rib total height to H = 500 nm (~ λ/3), $r = h/H = 0.1$ and W = 300 nm, in order to achieve very high confinement in the rib and negligible bending losses. For these values, τ_{eff} has an estimation of about 1 ns (large lifetime). The parameters used in the simulations are $R = 5$ μm, $\alpha_{loss} = 2$ dB/cm, $\kappa_e^2 = 2\%$ (power fraction coupled from the input waveguide to the ring), The effective area, calculated by full-vectorial finite element method (FV-FEM) [28], is $A_{eff} = 0.16$ μm^2 and $\beta^{(TPA)} = 0.7$ cm/GW [25].

Firstly, we investigate the TPA effects in continuous wave (CW) regime. Figure 3 shows the spectrum of normalized power at the end of output waveguide (Reflectivity) and input waveguide (Transmittivity) versus wavelength detuning for 10 mW of input power ($|S_i|^2$). The plot has been obtained by solving the system (1)-(5) for each wavelength. The wavelength detuning is referred to the value of $\lambda_0 = 1.5450$ μm (TPA-free resonance condition). For comparison reason, solution without the TPA is also plotted by solid line. The structure has been designed such to obtain optimum traveling behavior without the TPA. By setting the matching condition to [1]:

$$\frac{2}{\tau_e} = \frac{\kappa^2 v_g}{2\pi R} \tag{20}$$

The plot shows that the total signal power could be extracted from the resonator at the resonance even in the presence of ring losses (see the solid curve in second subplot) [1]. Thus, Eqs. (2) and (20) require that input and output waveguides are no longer symmetrically coupled to the ring ($\tau_d \neq \tau_e$). Therefore, Figure 2 shows that, due to FCA induced by nonlinear TPA effect, the power at output waveguide decreases over case in absence of TPA.

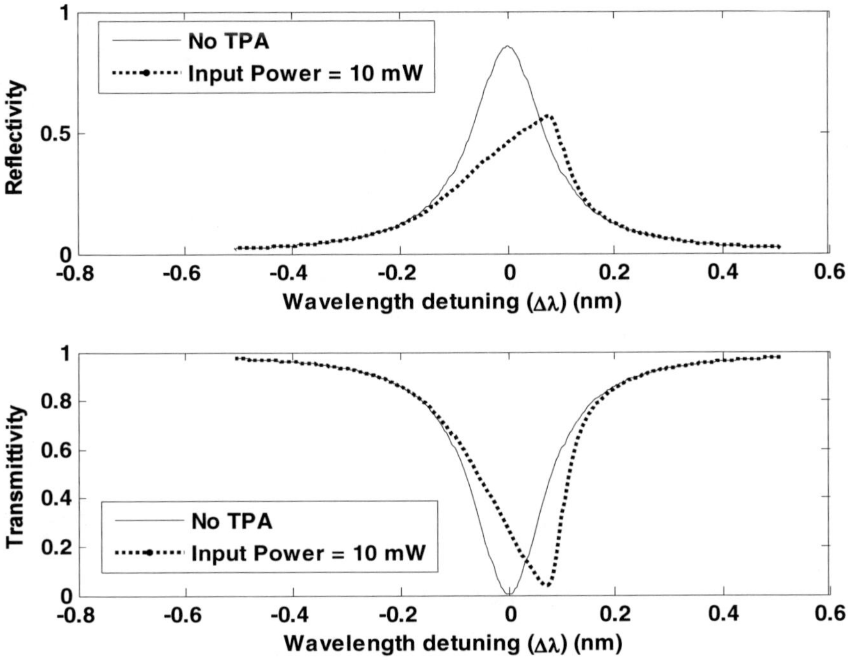

Figure 3. Reflectivity and transmittivity versus wavelength detuning for different input powers.

In fact, the nonlinear absorption induces a decrease in the cavity photon lifetime. In this condition, Eq. (6) does not hold and the microring is unmatched with the waveguide, i.e. the transmittivity at the resonance is different from zero. Moreover, it can be seen that for these relatively high levels of input power, both broadening of transmittivity and reflectivity curves, and shift of peak position towards larger wavelengths will occur. These effects mainly depend on the plasma dispersion effect induced by free carrier excess and other nonlinear terms in Eq. (1). Our calculations show that the considered microring could work as a TPA-free CW optical filter for input powers lower than 2.5 mW (15.6 mW/μm^2).

In several nonlinear applications, it could be interesting to use the ring resonator under pulsed regime, as the Raman resonant amplifier described in chapter 8. In such applications, it is crucial to hold the cavity enhancement factor larger than 1. In the following simulations we will show as the cavity enhancement factor can be influenced by the pulsed regime operation. To maximize the evaluation of the cavity enhancement factor Γ, we remove the output bus waveguide and solve the Eq. (1) by neglecting the TPA effect. Figure 4 shows the enhancement factor Γ versus ring radius for different pulse widths (T_{FWHM}), assuming κ_e^2 = 3 %. As expected, Γ assumes a monotonically decreasing shape versus R for each value of T_{FWHM}.

Anywhere, a very small ring microcavity requires large pulse widths to realise a high enhancement factor. In fact, as the input pulse is narrower than the time constant of the evanescent coupling process (τ_e), the field inside the ring cannot be enhanced. This is evident for $T_{FWHM} = 1$ ps and $R = 5$ μm, where the power enhancement factor falls to 0.66, while it is $\Gamma \geq 10$ for $T_{FWHM} = 100$ ps and $R < 12$ μm (here $R = 5$ μm can be chosen as an optimal case).

Thus, pulsed excitation of microring resonators should require $T_{FWHM} >> \tau_e$ to hold the cavity enhancement factor to high level. Further improvements can be obtained by reducing κ_e^2. Figure 5 shows the enhancement factor versus κ_e^2 for different pulse widths, assuming $R = 5$ μm (optimal case).

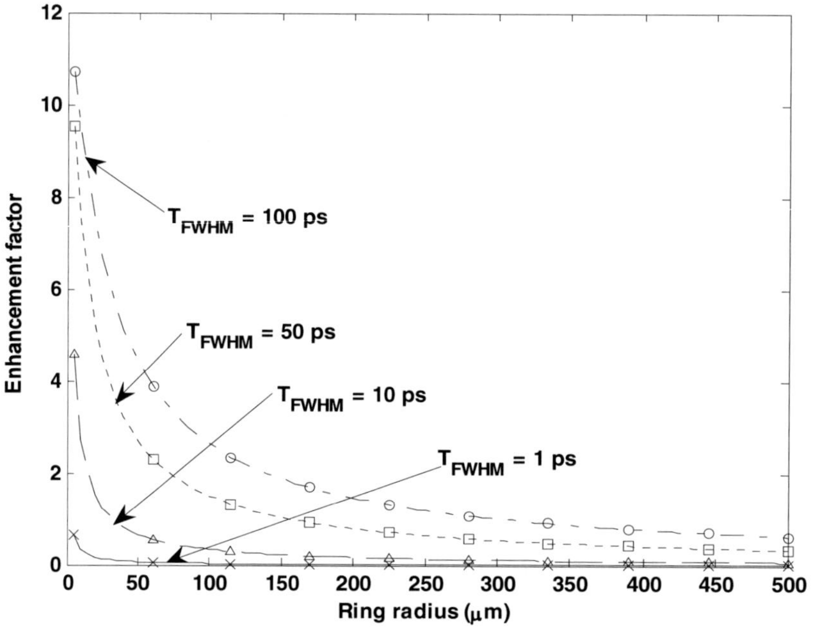

Figure 4. Pump enhancement factor versus ring radius for various pump pulse widths.

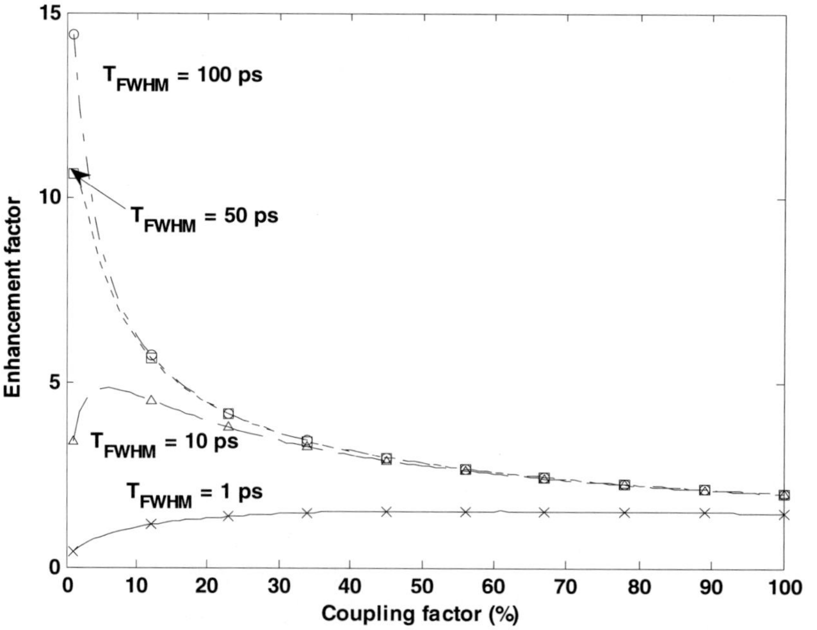

Figure 5. Pump enhancement factor versus pump coupling factor for various pulse widths.

The figure shows that, in case of T_{FWHM} = 1 ps, the maximum value of Γ is only 1.52, corresponding to κ_e^2 = 60%. In addition, we can observe a decrease of enhancement factor for $\kappa_p^2 \geq$ 20% and T_{FWHM} in the range 10 ÷ 100ps. It is worth noting that the shape of Γ versus κ_e^2 presents an absolute maximum (see for T_{FWHM} = 10ps), and this peak shifts towards smaller values of κ_e^2 with increasing T_{FWHM}. Thus, Figures. 4 and 5 indicate the best operation region for micro-scale ring resonators (R < 12 μm) for pump pulses with $T_{FWHM} \geq$ 100 ps and coupling factors κ_e^2 in the range 1% ÷ 5%. The previous curves can be considered a good approximation of the cavity enhancement factor when T_{FWHM} □ τ_{eff} (FCA negligible). However, the condition $T_{FWHM} \geq$ 100 ps gives an increasing of Γ even in the presence of FCA effect. Following the previous discussion, now it is possible to investigate the behaviour of architecture of Figure 1 in the pulsed regime. We have first simulated the nonlinear propagation in SOI microring assuming a Gaussian input pulse with $T_{FWHM} = 300$ps $>> \tau_e = 48.33$ ps (evaluated without TPA). The input pulse has been applied with a delay time T_d = 600 ps. Figure 6 shows three subplots: time evolution of normalized power inside the ring, reflectivity and transmittivity, respectively. Each normalization is referred to the peak input power ($|S_i|^2_{peak} = 250$ mW).

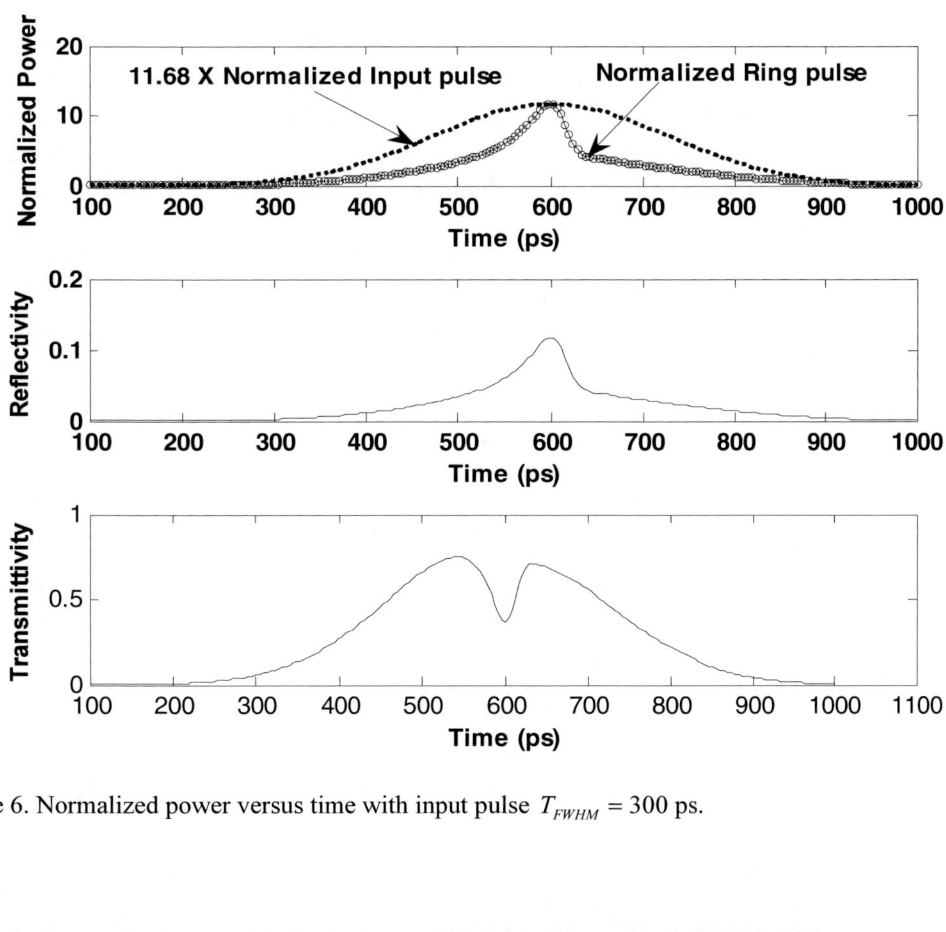

Figure 6. Normalized power versus time with input pulse T_{FWHM} = 300 ps.

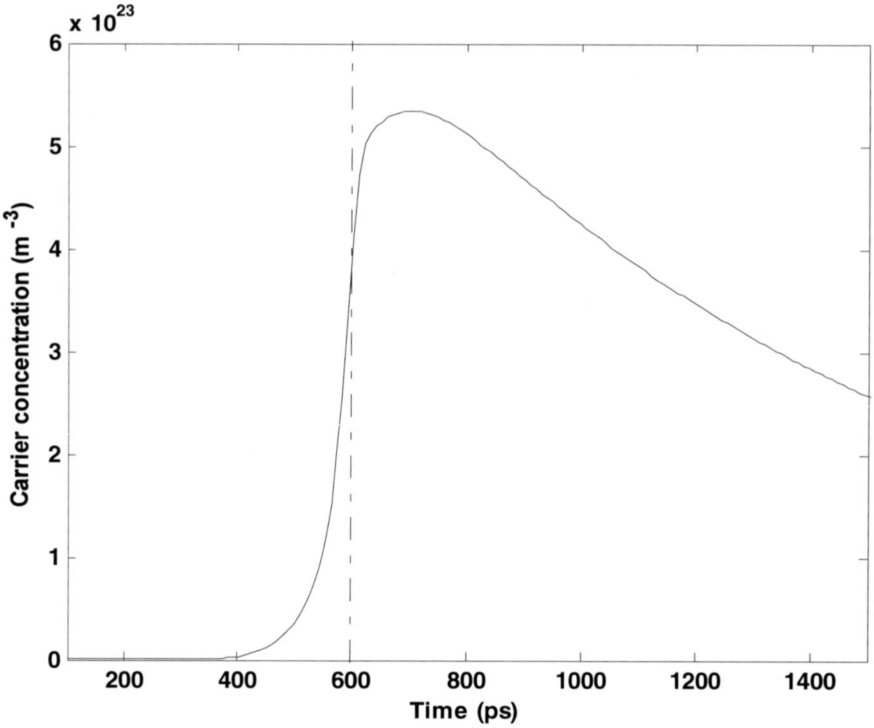

Figure 7. Carrier concentration versus time for input pulse $T_{FWHM} = 300$ ps.

The wavelength of the input pulse was detuned of 0.2253 nm with respect to the cavity resonance mode. Several features of Figure 6 can be noticed. The power cavity enhancement factor is 11.68 (see the peak of the first subplot). The power at the input waveguide end initially rises with the ring power, but then drops to a minimum as the microring resonance becomes closer to the input pulse wavelength. We note that this dip exactly corresponds to the power peak in the ring and at the end of output waveguide. The dip occurs at time instant t = 600 ps, where the net decrease of refractive index induced by carrier concentration (i.e., $3.6{\times}10^{17}$ cm^{-3}) generated by TPA (see Figure 7), causes a resonance mode wavelength shift by precisely 0.2253 nm (i.e. the same amount of initial detuning). Thus, the shifted wavelength is locked to the resonance condition and the power can exit from ring and be detected at the output waveguide. In addition, at the falling edge of the input pulse, the power in the ring begins to decrease, and the power at the input waveguide end rises again as the cavity resonance mode returns to its initial position.

In Figure 8, time evolution of the normalized ring power is shown, for two Gaussian input pulses with $T_{FWHM} = 50$ ps and 5 ps, applied at $T_d = 80$ ps and 10 ps, respectively. The plot shows that the power enhancement decreases into the ring, with decreasing the pulse width. In addition, we can observe from Figure 8 a reshape of the optical pulse as T_{FWHM} value move down the time constant of the evanescent coupling process ($\tau_e = 48.33$ps in this case). Thus, a stringent condition for pulsed excitation of nano-section microrings to work as filters with a high enhancement factor and without any pulse distortion, is to have the input pulse FWHM significantly larger than τ_e.

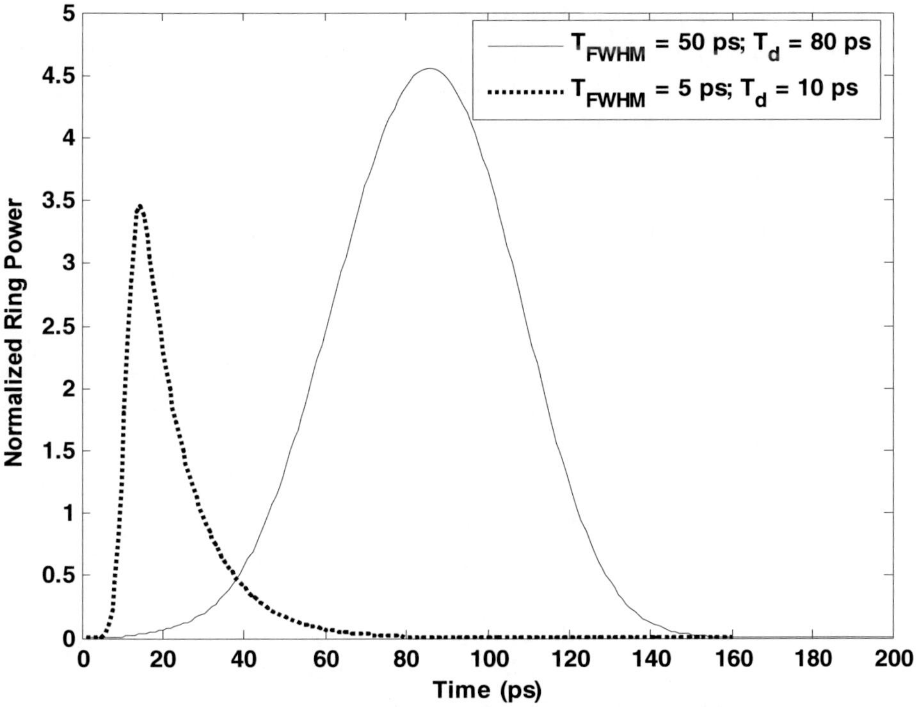

Figure 8. Normalized ring power versus time for different values of T_{FWHM} .

Finally, importance of FEM approach in simulations of photonic devices can be demonstrated even for ring resonators. To this aim, we have compared a 2D simulation obtained by well known finite difference time domain (FDTD) [29] with 2D FEM [28]. Results are sketched in following Figure 9 (a) and (b), in case of refractive index contrast 2 (core 3, cladding 1).

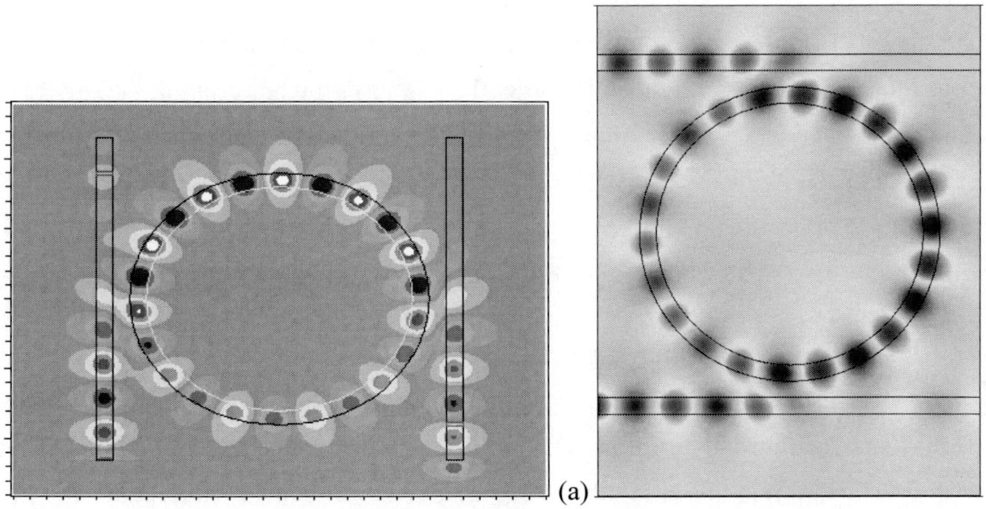
(a) (b)

Figure 9. (a) Electric field simulation under CW excitation by 2D FDTD; (b) FEM.

Other simulation parameters include resonant wavelength of 1.983 μm, outer radius of 1.8 μm, waveguide and ring width of 0.2 μm, gap between waveguides and ring of 0.2 μm. The agreement is proved to be very good, and FEM predictions allow to simulate very well the radiation loss outside the resonator.

3.2. Active Ring Resonator

The numerical simulations for active ring resonators have been performed by considering a standard GaAs-AlGaAs MQW laser structure. The active region is constituted from one (or more) GaAs wells sandwiched between two $Al_{0.2}Ga_{0.8}As$ waveguide regions, each 100 nm thick. The p-type and n-type $Al_{0.4}Ga_{0.6}As$ cladding layers are 1.0 and 1.5 μm thick, respectively. A top GaAs cap has been also included. A total optical loss of 25 cm^{-1} has been taken into account. We have assumed R = 200 μm, rib width of 2 μm and considered the backscattering coefficient b_{scat} as a parameter. Thus, we can show as the physical operation of MQW ring laser depends on the relationship between injection current and backscattering coefficient. Anyway, it is possible to calculate the backscattering coefficient starting from the statistical information on ring sidewall roughness taken from experimental data, as explained by Eq. (19).

Analyzing the stationary solution of Eqs. (15)-(16), the relationship $\delta_1 + \delta_2 = 0, \pi$ must be satisfied. In particular $\delta_1 + \delta_2 = 0$ allows a condition of minimum stimulated energy, whereas $\delta_1 + \delta_2 = \pi$ gives an instable condition of maximum stimulated energy (it can be discarded).

In the following, we investigate the impact of statistic information on ring sidewall roughness on the operating characteristics of semiconductor MQW ring laser. Figure 10 shows the intensities of CW (mode 1) and CCW (mode 2) beams versus injection current in the stationary condition, for different values of the statistical parameters of sidewall roughness, standard deviation and correlation length.

If the current is close to the threshold, i.e. $I \approx I_{th}$, the beams hold the same intensity (bidirectional condition). For currents well larger than I_{th}, a dominant beam in the ring cavity grows inducing the unidirectional regime. In Figure 10 we have assumed σ_c=12 nm and used different values of L_c. We can observe that the range of injection currents where the MQW ring laser works in bidirectional regime increases by increasing L_c.

The previous discussion is also confirmed in Figure 11, which shows the stationary regimes of I_1 and I_2 versus correlation length L_c for different values of standard deviation σ_c, by assuming an injection current value of I=100mA (MQW with one well) and $\delta_1 + \delta_2 = 0, \pi$. As usual, mode 1 designates the CW solution and mode 2 the CCW one, respectively. The plot shows that for σ_c smaller than a critical value, depending of injection current ($\sigma_{c,th}$=0.0047 μm in this case), the MQW ring laser works in an unidirectional regime without any dependence on L_c. This means that, for each L_c, the backscattering coefficient is too small to compensate the mode competition effect induced by self- and cross- saturation coefficients $\tilde{\beta}_i$, $\tilde{\theta}_{ij}$ (see Eqs. (15)-(16)). Therefore, a dominant beam in the ring cavity grows while the backscattering effect will always induce to excite a very weak wave traveling in the opposite direction, orders of magnitude below the dominant beam.

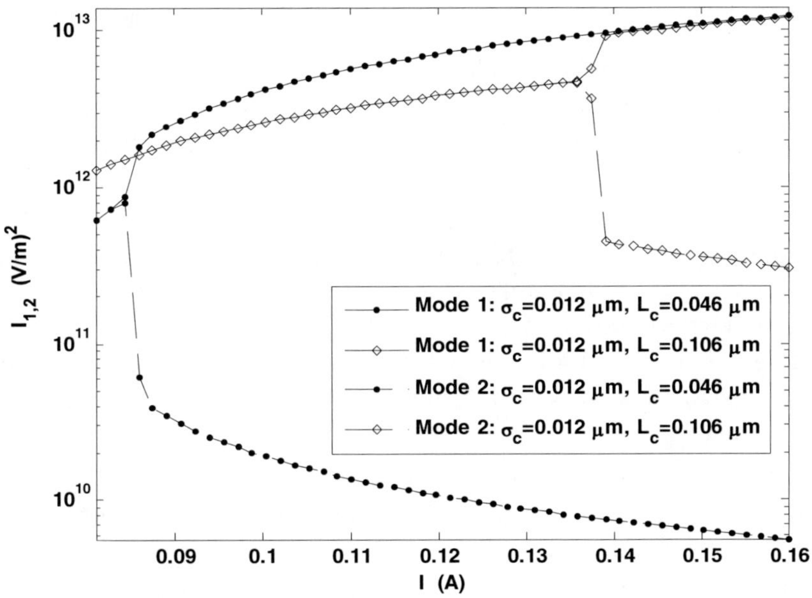

Figure 10. Intensities of both beams versus the laser current for various roughness correlation lengths.

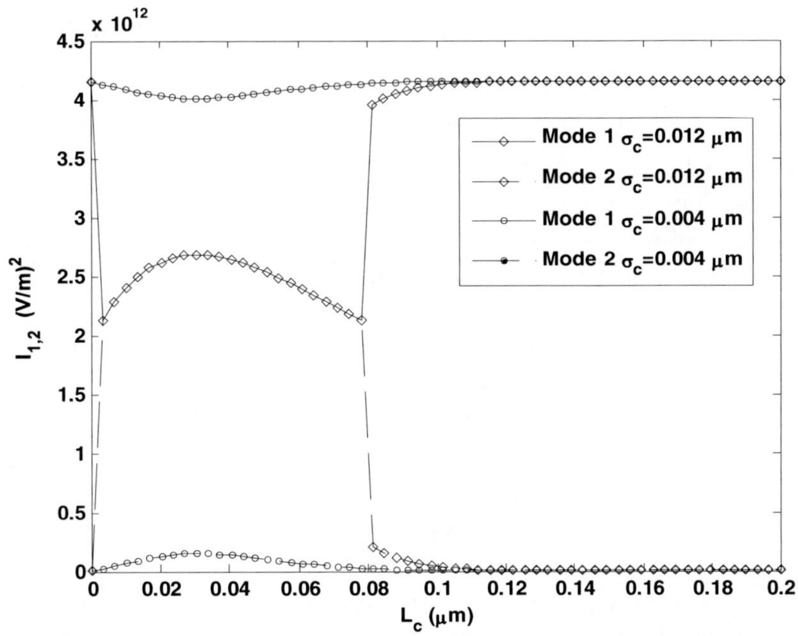

Figure 11. Intensities of both beams versus correlation length for various values of ring sidewall roughness function standard deviation.

The dominant beam can be randomly either CW or CCW beam, depending mathematically on the initial conditions or, physically, on the local optical losses inside the ring cavity. For values of roughness standard deviation $\sigma_c > \sigma_{c,th}$, there exists a range for L_c where the MQW ring laser shows a bidirectional operating regime. In particular, this range increases with increasing σ_c. Since the self-correlation function of the sidewall roughness is described as a Gaussian function (see [2]), this means that exists a range of L_c, close to the peak of the Gaussian shape, where the backscattering coefficient compensates the mode competition effect. In the range of L_c where the regime is bidirectional, it is possible to observe a maximum in the plot. This maximum occurs where the peak of Gaussian self-correlation function is situated.

Figures 10 and 11 are particularly important because they lead to an estimation of the MQW ring laser operation regime related to the etching step of the fabrication process. In fact, by performing a number of measurements of the etching profile of ring sidewalls obtained on different samples, it is possible to extract the statistical (Gaussian) information on the sidewall roughness function and, then, give theoretical predictions (by our model) on the laser working regime.

Moreover, our numerical investigation indicate that the stationary regimes of I_1 and I_2 can be unidirectional o bidirectional depending of the coupling factor between the ring laser and the output bus waveguide (see Figure 2). In fact, solving Eqs. (15)-(18) for given values of L_c, σ_c and injection current I, it is possible to observe that the laser regime becomes bidirectional by increasing the coupling factor, starting from an unidirectional condition. In fact, for relatively low coupling factors the quality factor of the ring resonator [9] assumes relatively large values, so inducing the injection current I to be well larger than the threshold, I_{th}. A dominant beam in the ring cavity grows due to the mode competition effect, as induced by self- and cross-saturation coefficients $\tilde{\beta}_i$, $\tilde{\theta}_{ij}$. On the contrary, for a coupling factor larger than an appropriate value (depending on L_c, σ_c and I), a significant part of the optical power leaves the ring resonator and this induces the threshold current to approach the fixed value of I. In this case, it is not possible for only one of counter-propagating beams to be completely extinguished. In fact, the backscattering effect between the beams always induces radiation travelling in the opposite direction. On the basis of these results, one additional component has to be included in the architecture of any MQW ring laser to favor only one circulating direction over the other, i.e. to achieve an unidirectional regime also for injection current values where the ring laser should be bidirectional. The solution could consist of an output coupler including a grating [9].

CONCLUSION

The microring resonant filters are very interesting devices that can contribute to increase the integration level in photonic systems. In this chapter we present a theoretical investigation of the influence of input power levels and FWHM widths in pulsed excitation of SOI microring resonators with nanometer-scale cross section. Moreover, the active ring resonator (ring laser) behavior related to its physical and technological constraints, as the backscattering effect and ring radius, has been briefly investigated.

REFERENCES

[1] Little, B. E., Chu, S. T., Haus, H. A., Foresi, J., Laine, J. P., *J. Lightwave Technol.,* 1997, 15, 998-1005.

[2] Armenise, M. N., Passaro, V. M. N., De Leonardis, F., Armenise, M., *J. Lightwave Technol.*, 2001, 19, 1476-1494.

[3] Van, V., Absil, P. P., Hryniewicz, J. V., Ho, P. T., *J. Lightwave Technology,* 2001, 19, 1734-1739.

[4] Chu, S. T., Little, B. E., Pan, W., Kaneko, T., Sato, S., Kokubun, Y., *IEEE Photon. Technology Letters,* 1999, 11, 691-693.

[5] Chu, S. T., Pan, W., Suzuki, S., Little, B. E., Sato, S., Kokubun, Y., *IEEE Photon. Technology Letters,* 1999, 11, 1423-1425.

[6] Hryniewicz, J. V., Absil, P. P., Little, B. E., Wilson, R. A., Ho, P. T., *IEEE Photon. Technology Letters,* 2000, 12, 320-322.

[7] Yanagase, Y., Suzuki, S., Kokubun, Y., Chu, S. T., *J. Lightwave Technology,* 2002, 20, 1525-1529.

[8] Soref, M., Giuliani, G., Scirè, A., Miglierina, R., Donati, S., Laybourn, P. J. R., *IEEE J. Quantum Electronics,* 2003, QE-39, 1187-1195.

[9] De Leonardis, F., Passaro, V. M. N., *Laser Physics Letters,* 2005, 2, 59-70.

[10] Liu, B., Shakouri, A., Bowers, J. E., *IEEE Photon. Technology Letters,* 2002, 14, 600-602.

[11] Absil, P. P., Hryniewicz, J. V., Little, B. E., Cho, P. S., Wilson, R. A., Joneckies, L. G., Ho, P. T., *Optics Letters,* 2000, 25, 554-556.

[12] Van, V., Ibrahim, T. A., Absil, P. P., Johnson, F. G., Grover, R., Ho, P. T., *IEEE J. Selected Topics in Quantum Electronics,* 2002, 8, 705-713.

[13] Yariv A., Xu, Y., Lee, R. K., Scherer, A., *Optics Letters,* 1999, 24, 711-713.

[14] Poon, J. K. S., Scheuer, J., Xu, Y., Yariv, A., *J. Optical Society of America B,* 2004, 21, 1665-1673.

[15] Xia, F., Sekaric, L., Vlasov, Y., *Nature Photonics,* 2007, 1, 65-71.

[16] Barwicz T., Popović, M. A., Watts, M. R., Rakich, P. T., Ippen, E. P., Smith, H. I., *J. Lightwave Technology*, 2006, 24, 2207-2218.

[17] Heebner, J. E., Boyd, R., Park, Q., *Physical Review E,* 2002, 65, 036619.

[18] Niehusmann, J., Vörckel, A., Bolivar, P. H., Wahlbrink, T., Henschel, W., Kurz, H., *Optics Letters,* 2004, 29, 2861-2863.

[19] Timotijevic, B. D., Gardes, F. Y., Headley, W. R., Reed, G. T., Paniccia, M. J., Cohen, O., Hak, D., Masanovic, G. Z., *J. Optics A: Pure and Applied Optics*, 2006, 8, S473-S476.

[20] Hiremath, K. R., Stoffer, R., Hammer, M., *Optics Communications,* 2006, 257, 277-297.

[21] Yariv, A., *Electronics Letters,* 2000, 36, 321-322.

[22] Van, V., *J. Lightwave Technology,* 2006, 24, 2912-2919.

[23] Darmawan, S., Chin, M. K., *J. Optical Society of America B,* 2006, 23, 834-841.
[24] Landobasa, Y. M., Darmawan, S., Chin, M. K., *IEEE J. Quantum Electronics,* 2005, 41, 1410-1418.
[25] Liu, A., Rong, H., Paniccia, M., Cohen, O., Hak, D., *Optics Express*, 2004, 12, 4261-4268.
[26] Yamada, M., Suematsu, Y., *J. Applied Physics*, 1981, 52, 2653-2664.
[27] Asada, M., Kameyama, A., Suematsu, Y., *IEEE J. Quantum Electronics,* 1984, QE-20, 745-753.
[28] Comsol Multiphysics by COMSOL AB, ver. 3.2, *single license*, 2005.
[29] Fullwave by RSoft Design Group, *trial license*, 2005.

In: Modeling of Photonic Devices
Editor: Vittorio M. N. Passaro
ISBN 978-1-60456-980-3

Chapter 5

SILICON MODULATORS

Vittorio M. N. Passaro[1*] and Francesco De Leonardis[2]
[1]Photonics Research Group, Dipartimento di Elettrotecnica ed Elettronica, Politecnico di Bari, via Edoardo Orabona n. 4, 70125 Bari, Italy
[2]Photonics Research Group, Dipartimento di Ingegneria dell'Ambiente e per lo Sviluppo Sostenibile, Politecnico di Bari, viale del Turismo n. 8, 74100 Taranto, Italy

ABSTRACT

In this chapter, an overview on silicon optical modulators is presented. In particular, some aspects of mathematical modeling of these devices are reviewed. In the first part of the chapter, we have focused on a fully integrated multiphysics approach for the rigorous design of metal-oxide-semiconductor modulators. Then, modeling techniques for silicon modulators based on p-n junction are shown by optimizing structures already demonstrated in literature. Finally, a novel approach to all-optical modulation in silicon waveguides is briefly presented as based on Raman effect.

Keywords: Silicon Modulators, Plasma Dispersion Effect, Metal-Oxide-Semiconductor, Modeling, Raman Modulator.

1. INTRODUCTION

For decades, silicon has been considered the optimal material for Electronics mass production. In the last few years, the possibility to realize highly performing active and passive photonic integrated devices using a Silicon-on-Insulator (SOI) technological platform has been widely proved [1]. One of the most important requirements of any integrated optical technology is the ability to perform optical modulation, which permits one to encode a bit stream onto the optical carrier provided by the laser source. Light can be modulated using

---

* E-mail address: passaro@deemail.poliba.it

different physical effects, depending on materials and technologies used to fabricate modulators. For high-speed modulation, electro-optic effect is commonly employed. Unfortunately, unstrained silicon does not exhibit this physical effect, thus to solve the problem of optical modulation in silicon the plasma dispersion effect is commonly adopted, in order to electrically change the silicon refractive index by varying electron and hole distribution within the material [2]. In general, the free-carrier concentration in silicon electro-optic devices can be varied by charge carrier injection, accumulation or depletion. Recently, *p-i-n* [3], *p-n* [4] and metal-oxide-semiconductor (MOS) [5-6] structures have been used to this purpose.

To study this kind of device, two different models are usually developed. The first allows one to investigate the modulator electrical behaviour, whereas the second simulates the device from an optical point of view, taking into account the electrically induced silicon refractive index change [7]. These two models are usually developed in literature in two different simulation environments, the former purely electrical (often through commercial CAD tools [8]) and the latter purely optical, usually by beam propagation method (BPM), giving some consequent problems and inaccuracies in using the electrical simulation results in the environment where optical simulation is performed. In the next section a different, completely integrated multiphysics approach is explained and some significant results are given. In another section, the model of an improved concept of silicon modulator achieving low power consumption and low losses is presented. Finally, an architecture of a SOI waveguide optical modulator based on Raman effect is shown.

2. MOS Silicon Modulators

In this section the use of full-vectorial finite element method (FEM) [9] is demonstrated to obtain a fully integrated optoelectronic model of a SOI optical modulator, assuming a typical metal-oxide-semiconductor (MOS) structure as that sketched in Figure 1.

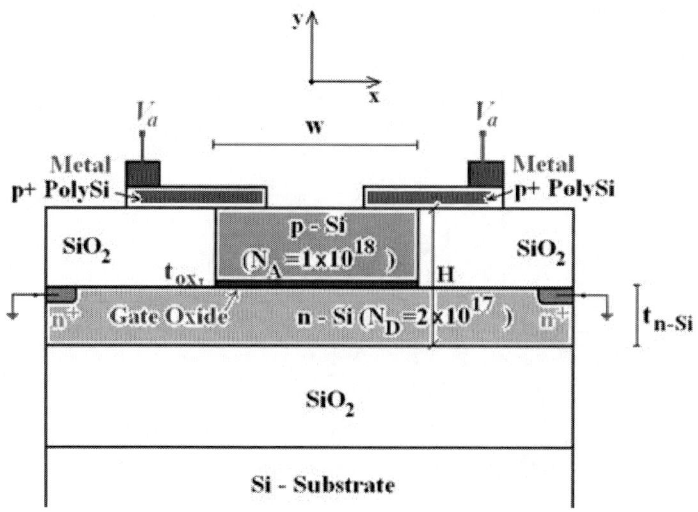

Figure 1. SOI modulator cross section (dopant concentrations in cm-3).

Assuming to ground n-type silicon layer, a free carrier thin layer is accumulated on both sides of the gate oxide when a positive voltage V_a, larger than the MOS flat band voltage, is applied to the device. According to plasma dispersion effect, this voltage-induced free carrier accumulation locally changes the silicon refractive index. This operating principle allows high modulation speeds to be achieved, because no slow phenomena are involved in the accumulation process. The electrically induced localized change of silicon refractive index affects the effective index of optical mode propagating within the modulator and, finally, the phase of optical signal passing through it. Then, this phase shift is usually converted as an amplitude modulation using the typical Mach-Zehnder interferometer architecture. In our modeling we have assumed device geometrical sizes and simulation parameters as follows: p-Si region width w =1600nm, modulator total height H = 1550nm, n-Si region thickness $t_{n\text{-}Si}$ = 900nm, gate oxide thickness t_{ox} = 10.5 nm, operating wavelength λ = 1550nm.

2.1. Electrical Investigation

Electron and hole distribution dependence on applied voltage (V_a) has been investigated using 2D static drift-diffusion model, solved by FEM. This model consists of three coupled partial differential equations (PDEs) in the form:

$$\varepsilon_0\varepsilon_r\left(\nabla\Psi\right)=q\left(N_e-N_h-C\right) \tag{1}$$

$$\nabla\cdot\left(\mu_p v_T\nabla N_h+\mu_p N_h\nabla\Psi\right)=R_{SRH} \tag{2}$$

$$\nabla\cdot\left(\mu_n v_T\nabla N_e-\mu_n N_e\nabla\Psi\right)=R_{SRH} \tag{3}$$

where R_{SHR} is the Shockley-Hall-Read recombination rate:

$$R_{SHR}=\frac{N_eN_h-N_i^2}{\tau_p\left(N_e+N_i\right)+\tau_n\left(N_h+N_i\right)} \tag{4}$$

and ε_0 is the vacuum permittivity, ε_r is the relative permittivity, q is the elementary charge, N_e is the electron distribution, N_h is the hole distribution, N_i is the silicon free-carrier intrinsic concentration, C is the doping profile, μ_n is the electron mobility, μ_p is the hole mobility, v_T is the thermal voltage, ψ is the electro-static potential, τ_n is the electron life time in silicon and τ_p is the hole life time in silicon. For boundaries in contact with air, normal components of electric displacement and current density vectors are imposed to be equal to zero. At boundaries in contact with a metal, both neutrality condition and action mass law have been applied. Then, hole and electron concentrations and electro-static potential have been imposed as:

$$\Psi=V_a+\frac{kT}{q}ln\left[\frac{1}{N_i}\left(\frac{C}{2}+\sqrt{\left(\frac{C}{2}\right)^2+N_i^2}\right)\right] \tag{5}$$

$$N_e = \frac{C}{2} + \sqrt{\left(\frac{C}{2}\right)^2 + N_i^2} \quad (6)$$

$$N_h = -\frac{C}{2} + \sqrt{\left(\frac{C}{2}\right)^2 + N_i^2} \quad (7)$$

where k is the Boltzmann constant and T = 300K is the operating absolute temperature. By solving coupled equations (1)-(3) with above defined boundary conditions, both electro-static potential and free carrier distribution can be evaluated in modulator cross section for any value of applied voltage, V_a. For instance, electro-static potential $\psi(x,y)$ for $V_a = 2$ V is shown in Figure 2. Free carrier distribution along the center of modulator cross section is shown in Figure 3 for V_a ranging from 0 to 3 V. From simulations, flat band voltage has been estimated as 1.45 V.

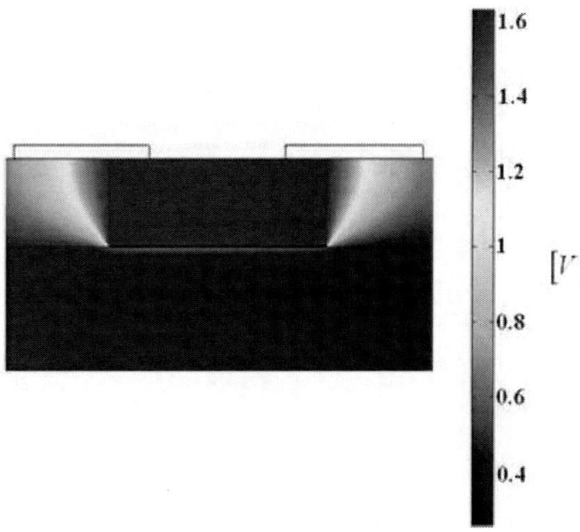

Figure 2. Electro-static potential for Va = 2 V.

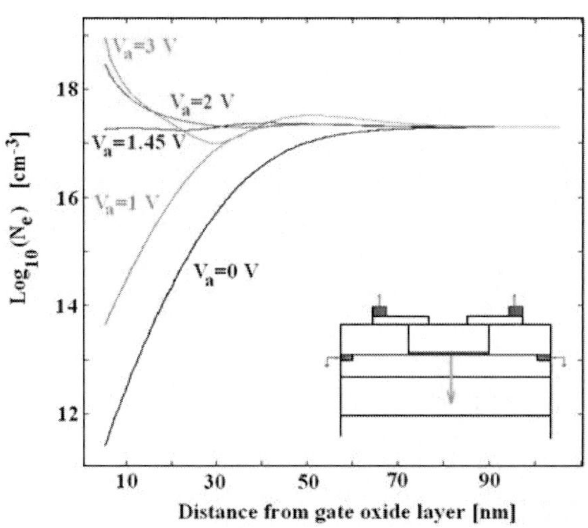

(a)

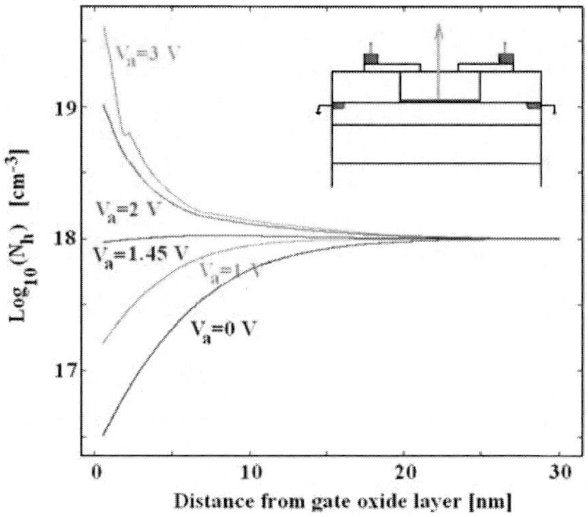

(b)

Figure 3.(a) Electron and (b) hole distributions along the modulator cross section center for V_a ranging from 0 V to 3 V.

2.2. Optical Investigation

Starting from 2D free carrier distribution, silicon refractive index change Δn_{Si} as due to applied voltage can be calculated at each point of the modulator cross section.

Refractive index change is related to free carrier distribution by the following, well-known semi-empirical Soref's relation at operating wavelength of 1550 nm [2]:

$$\Delta n_{Si} = -8.8\times10^{-22}\Delta N_e - 8.5\times10^{-18}\left[\Delta N_h\right]^{0.8} \tag{8}$$

where ΔN_e and ΔN_h are the changes in electron and hole concentration (cm^{-3}), respectively. Using this approach, refractive index distribution $n(x,y)$ can be calculated for any V_a value. In Figure 4 the refractive index distribution along the center of modulator cross section is shown for V_a ranging from 0 to 3 V.

Once knowing $n(x,y,V_a)$, optical mode field distributions propagating within the device and their effective indices n_{eff} can be calculated as eigenfunctions and eigenvalues of the following wave equation:

$$\nabla^2 \mathrm{E}_t + \nabla\left(\frac{\nabla n^2(x,y)\cdot \mathrm{E}_t}{n^2(x,y)}\right) + \left(\frac{2\pi}{\lambda}\right)^2 n^2(x,y)\mathrm{E}_t = \left(\frac{2\pi}{\lambda}\right)^2 n_{eff}\mathrm{E}_t \tag{9}$$

where E_t is the transverse electric field. For all optical modes supported by the modulator, it is possible to calculate the electromagnetic field distribution. For $V_a = 2$ V, quasi-TE and quasi-TM mode field distributions have been calculated and results are sketched in Figure 5. By comparing both figures, a considerable difference in quasi-TE and quasi-TM mode profiles can be recognized. Quasi-TE is more confined within the waveguide as compared to quasi-TM one.

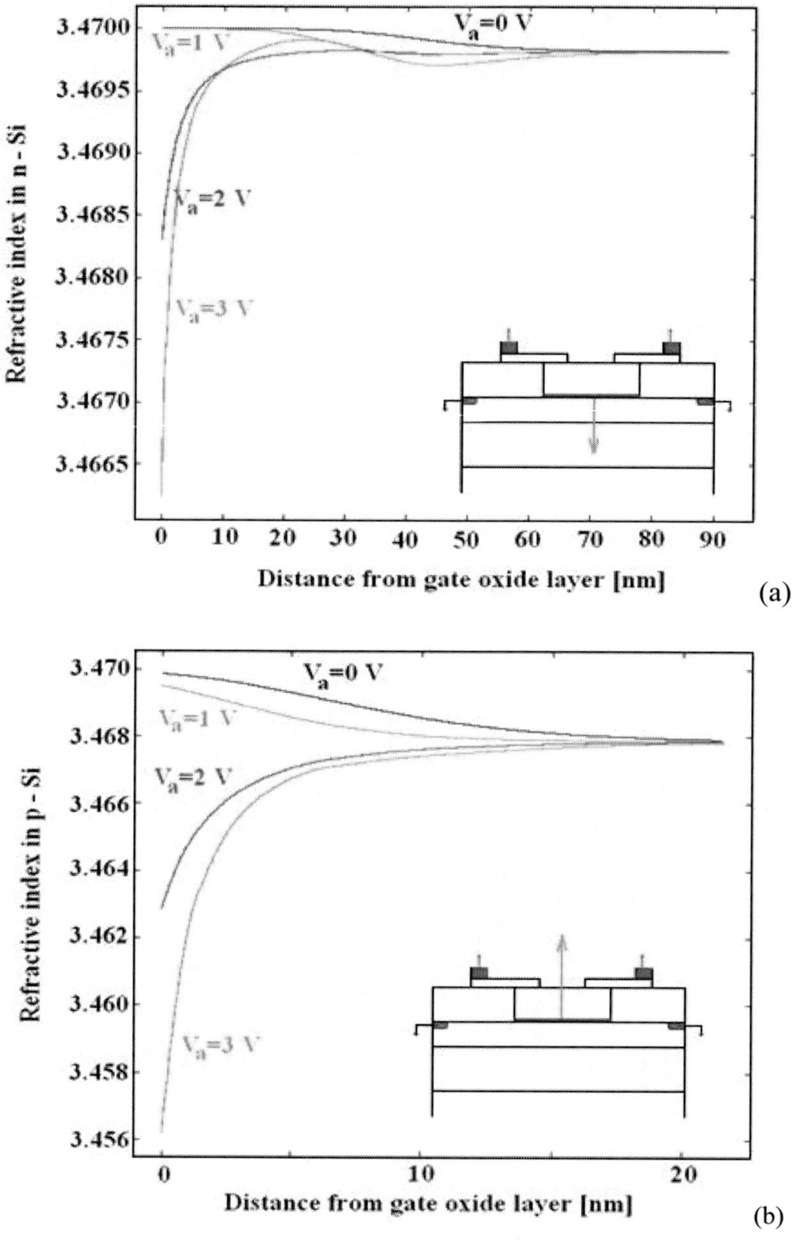

Figure 4. Refractive index distribution along the modulator cross section center in (a) n-Si and (b) p-Si, for Va ranging from 0 V to 3 V.

This difference is due to the continuity condition of electric field x component, E_x, and discontinuity of electric field y component, E_y, across the interface between silicon and gate oxide. For quasi-TE mode, E_y is the minor electric field component, whereas for quasi-TM mode E_y is the major electric field component, so practically E_y discontinuity produces effects only on quasi-TM polarized light. Therefore, the MOS capacitor based phase-shifter performance is strongly dependent on the incoming light polarization.

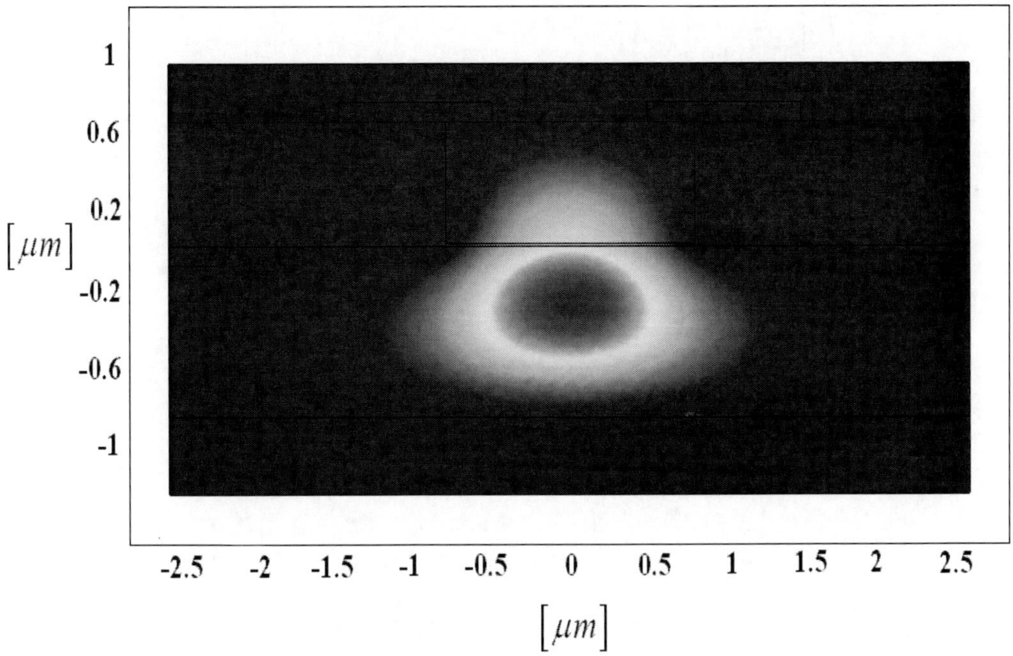

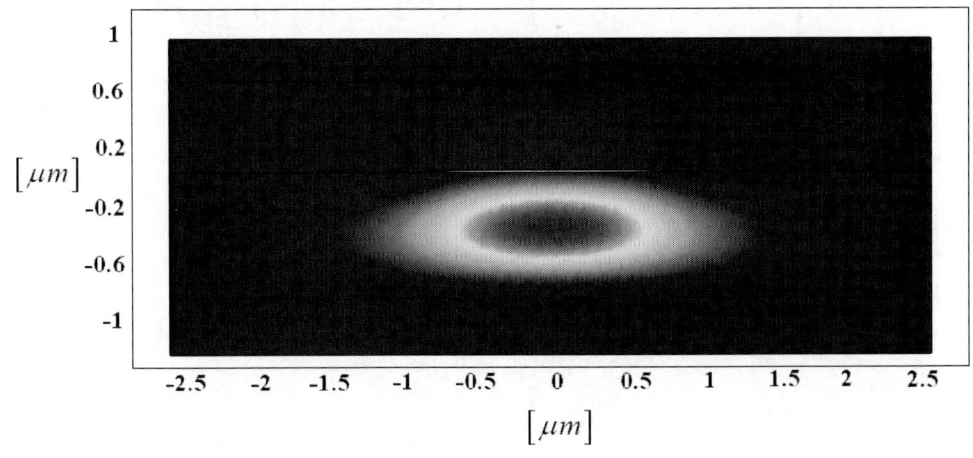

Figure 5. (a) Quasi-TE and (b) quasi-TM mode electric field distributions for Va = 2 V.

Phase shift induced in the optical signal propagating within the modulator, and its dependence on applied voltage V_a, have been investigated either for quasi-TE or quasi-TM mode, assuming a modulator length L = 8 mm, as in Figure 6. As expected, phase shift induced by an applied voltage change is significantly lower (about 66% for L=8 mm) for quasi-TM polarized light than for quasi-TE one.

A very detailed analysis of scaled and optimized structure of MOS silicon modulator has been recently published, based on the modelling technique summarized above [10].

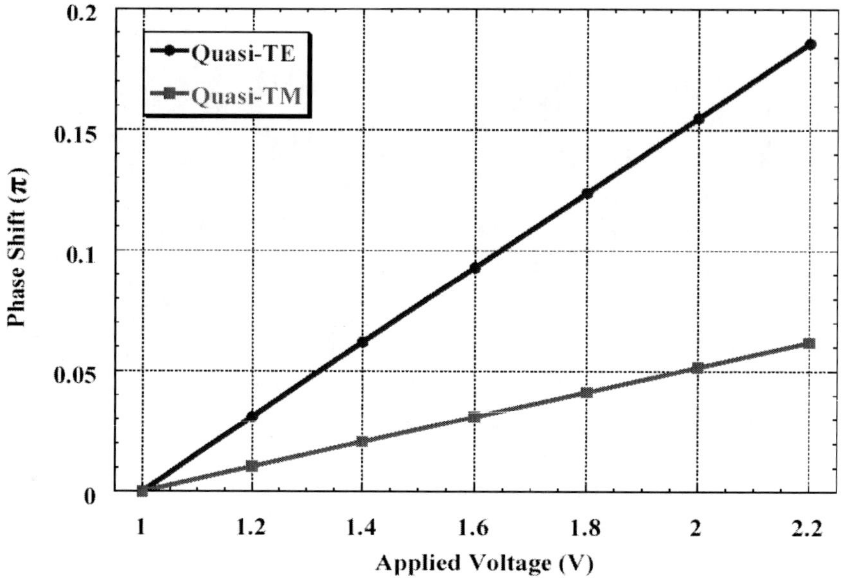

Figure 6. Phase shift induced in TE and TM polarized light by applied voltage.

3. Dual-Channel Photonic Modulator

It is well known that the main methods to modulate the refractive index in silicon-based waveguides are thermo-optic and plasma dispersion effects [11]. However, the latter is much faster, allowing modulation frequencies of the order of GHz. It is based on changes of both real refractive index and optical absorption coefficient due to modulation of free carrier concentrations by biasing a MOS, *p-i-n*, or *p-n* junction. Recently, an efficient depletion-type *p-n* device for varying the refractive index of high index contrast SOI waveguides has been presented in literature by using a four terminal p-n junction diode [4]. In this section, an optimized architecture of that structure is studied and a dual-channel approach for a fast Mach-Zehnder (MZ) SOI modulator is theoretically demonstrated.

3.1. Device Structure

The optical device structure originally presented by Gardes *et al.* [4] is illustrated in Figure 7. It is a depletion-type vertical phase modulator integrated in a low loss SOI rib waveguide. The device has an asymmetrical *p-n* structure, where two slab n^+ silicon regions work as a common cathode and two p^+ poly-silicon regions are joined as a common anode (four terminals). Both n^+ and p^+ regions are modeled as highly doped regions with peak doping concentrations of 1×10^{19} ions/cm^3. The designed structure includes a silicon thickness of 450 nm, etched rib waveguide 415 nm wide and slab thickness of 100 nm. The silicon slab and the bottom part of the rib have an *n*-type background doping concentration of 4×10^{17} cm^{-3} and the top part of the rib has a *p*-type uniform doping concentration of 2×10^{17} cm^{-3} , named uniform profile in Figure 8.

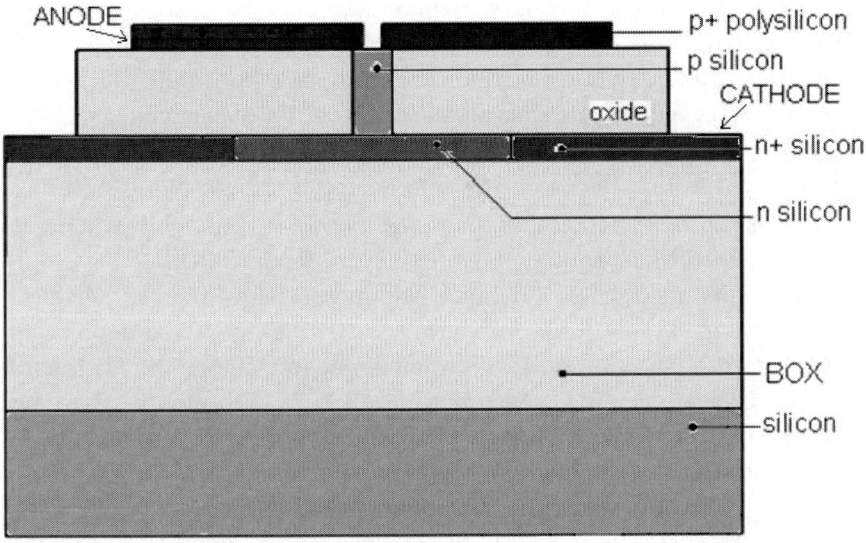

Figure 7. Cross section of depletion-type p-n structure.

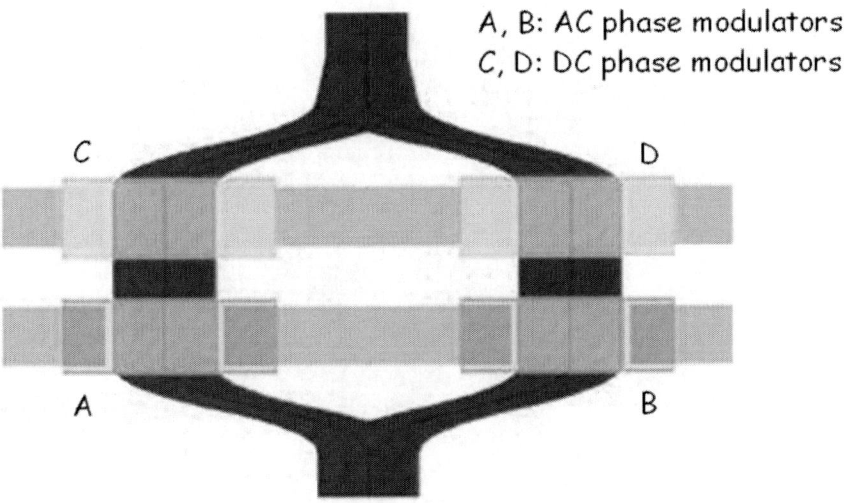

Figure 8. Mach-Zehnder interferometer using original structure [4].

The oxide thickness is chosen to be 1000 nm, ensuring good optical confinement, while a top silicon oxide cladding layer covers the whole structure. The n^+ doped regions are situated on both sides of the wave-guiding region, in the slab, 1.5 μm away from the waveguide centre. In general, the design was optimized in order to achieve single mode and birefringence free propagation (equal effective index for quasi-TM and quasi-TE waves in the rib), optimum value and efficient positioning of *n*-type and *p*-type doping to have minimum losses and large free carrier concentration changes, as well as fast dynamics and maximum bandwidth. In particular, the rib width has a direct influence on the device capacitance, since *p-n* junction is situated inside the rib waveguide. By this structure, very low switching time (7 ps) and DC power (~0 mW) and small length (2.5 mm) of single phase shifter were theoretically demonstrated [4].

By using the original depletion device of Figure 7, a push-pull Mach-Zehnder (MZ) interferometer was achieved, having four phase shifters of the same geometry, two of them as AC depletion devices (sections A and B in Figure 8), and the other two as DC phase shifters (C and D in Figure 8) to bias the device by introducing a static π-phase shift in one arm. If each depletion phase shifter was 2.5 mm long to achieve a π-phase shift when a reverse bias of 10 V is applied, a total MZ length of about 5 mm was obtained [4].

Starting from this modulator structure, we have changed some characteristics, in particular optimizing its doping profile. Figure 9 shows the optimized doping profile (b) in the rib (logarithmic scale) as compared with original one (a). The modified structure includes a peak *p*-type doping at rib center (1×10^{18} ions/cm^3 over 75 nm), a reduced *p*-type doping in the rest of the rib (1×10^{16} ions/cm^3 along 110 nm at the rib top and along 165 nm at rib bottom), and an increased *n*-type doping inside the silicon slab (3×10^{18} ions/cm^3). By this designed profile, a reduction of carrier absorption inside the rib, an increase of refractive index modulation due to stronger changes of free carrier concentrations, and an increase of breakdown voltage larger than 10 V can be simultaneously obtained, having a capacitance of about 1.31 pF.

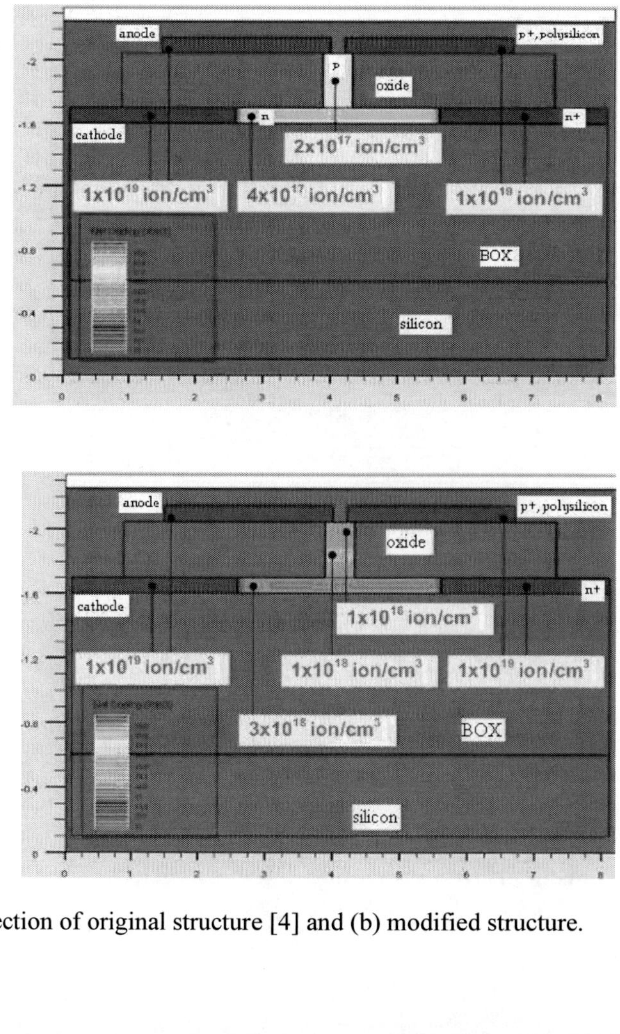

Figure 9. (a) Cross section of original structure [4] and (b) modified structure.

Moreover, by keeping the same four terminal structure and sizes as before, the other advantages already mentioned (single mode and birefringence free) are again satisfied. Moreover, the importance of the optimized structure is clear when we consider the mode effective index versus applied voltage for both structures. In fact, the structures change the effective index as a function of applied voltage covering a different range, from 2.7615 to 2.7618 for uniform profile and from 2.76185 to 2.7621 for optimized profile (quasi-TE mode). This opens the possibility to use both structures to modulate a Mach-Zehnder interferometer without any DC phase shifter, if both structures are used in different arms (dual-channel approach). In Figure 10 such solution is sketched by including the optimized structure in one arm and the original structure in the other one. Thus, the "OFF state" is achieved by driving the "uniform" AC modulator at a reverse bias of 3 V and the optimized AC modulator at 4 V bias, enabling destructive interference at the output. The "ON state" corresponds to AC modulators having a bias of 10V and 0V, respectively, so allowing constructive interference. Since the interferometer can be achieved by using only two AC modulators, the MZ total length is significantly reduced by about half, each phase shifter being 2.52 mm long. Moreover, the control of driving voltages would allow very high modulation depth to be achieved. Similar behavior for both polarizations assure birefringence free performance, important for coupling with fibers.

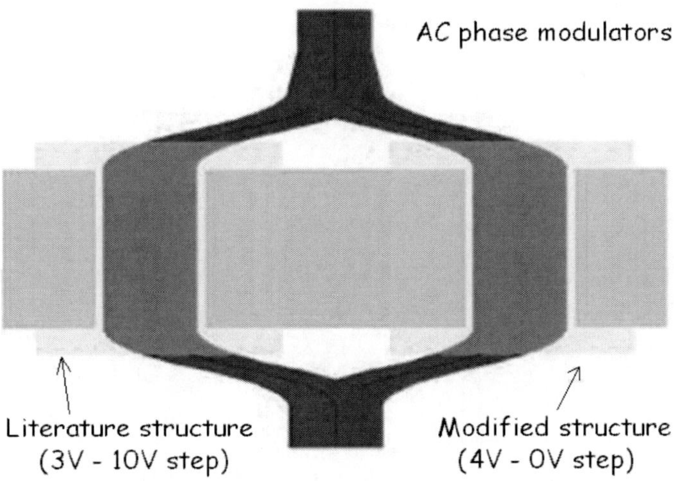

Figure 10. Dual-channel interferometer architecture.

3.2. Dual-Channel Device Modeling

The device has been electrically modeled for both its static and dynamic behavior using a commercial simulation CAD package [8], by solving the equations which describe the semiconductor physics, such as Poisson's equation and charge continuity equations for holes and electrons. This method has been used to find the free carrier concentrations in the wave-guiding region for both DC and transient biasing conditions. From those concentrations, both refractive index and absorption coefficient changes have been evaluated by using the well known Soref's equations [2]. Figure 11 and 12 show the refractive index and absorption coefficient changes versus position along the rib as induced from free carrier concentration changes by biasing the structures from 0 to 10V for uniform and optimized doping profiles, respectively. An index change up to six times larger than that originally provided [4] is so demonstrated, as in Figure 11. Moreover, we note from Figure 12 a moderate increase of absorption over the most part of rib, so keeping low the losses, about 2 dB/cm for bias of 0V and 1 dB/cm for 10V. These losses are lower than in literature [4].

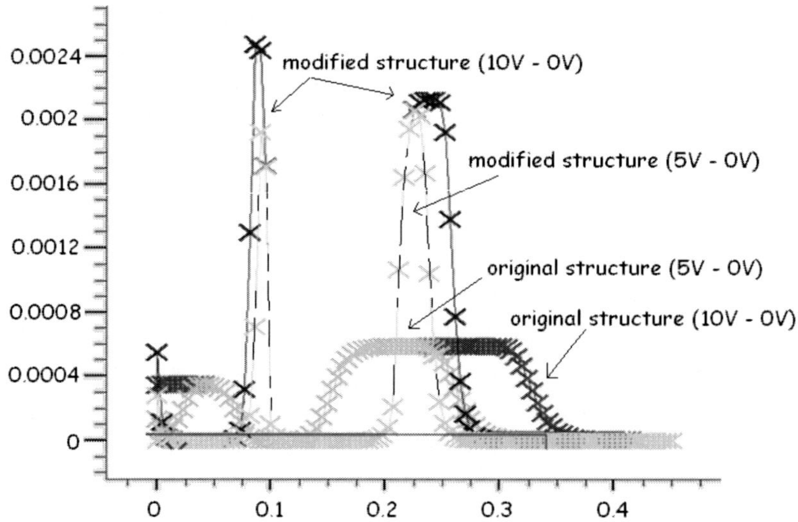

Figure 11. Refractive index absolute change versus rib position for both structures.

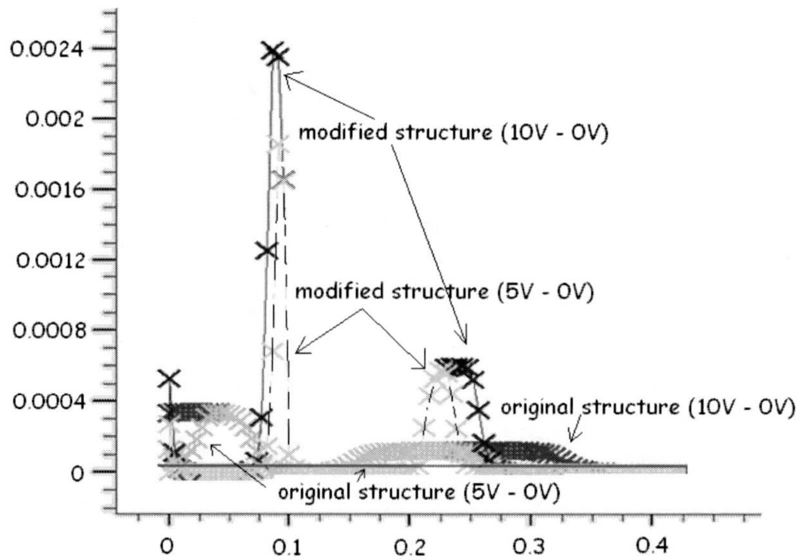

Figure 12. Absorption coefficient absolute change versus rib position for both structures.

Moreover, the device has been optically modeled using the full-vectorial beam propagation method [12] at optical wavelength of 1.55μm. By considering the refractive index and absorption change provided by electrical simulation of free carrier distributions in the *p-n* device, it is possible to derive the relevant refractive index 2D distribution and simulate the mode effective index change for different applied voltages, as well as the optical attenuation, as in Figure 13 and 14 for original and modified structure, respectively.

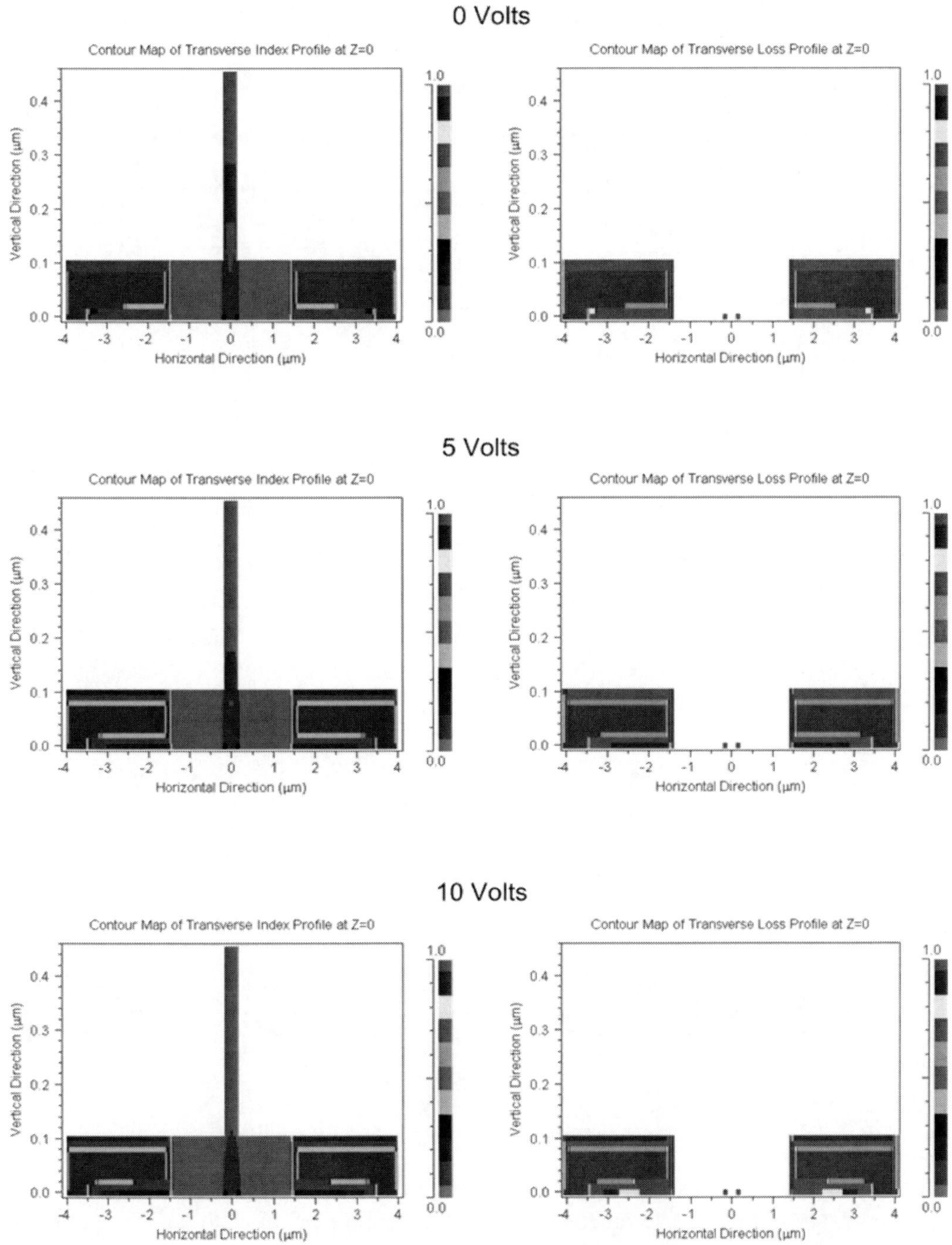

Figure 13. Refractive index profile and attenuation versus position for original structure.

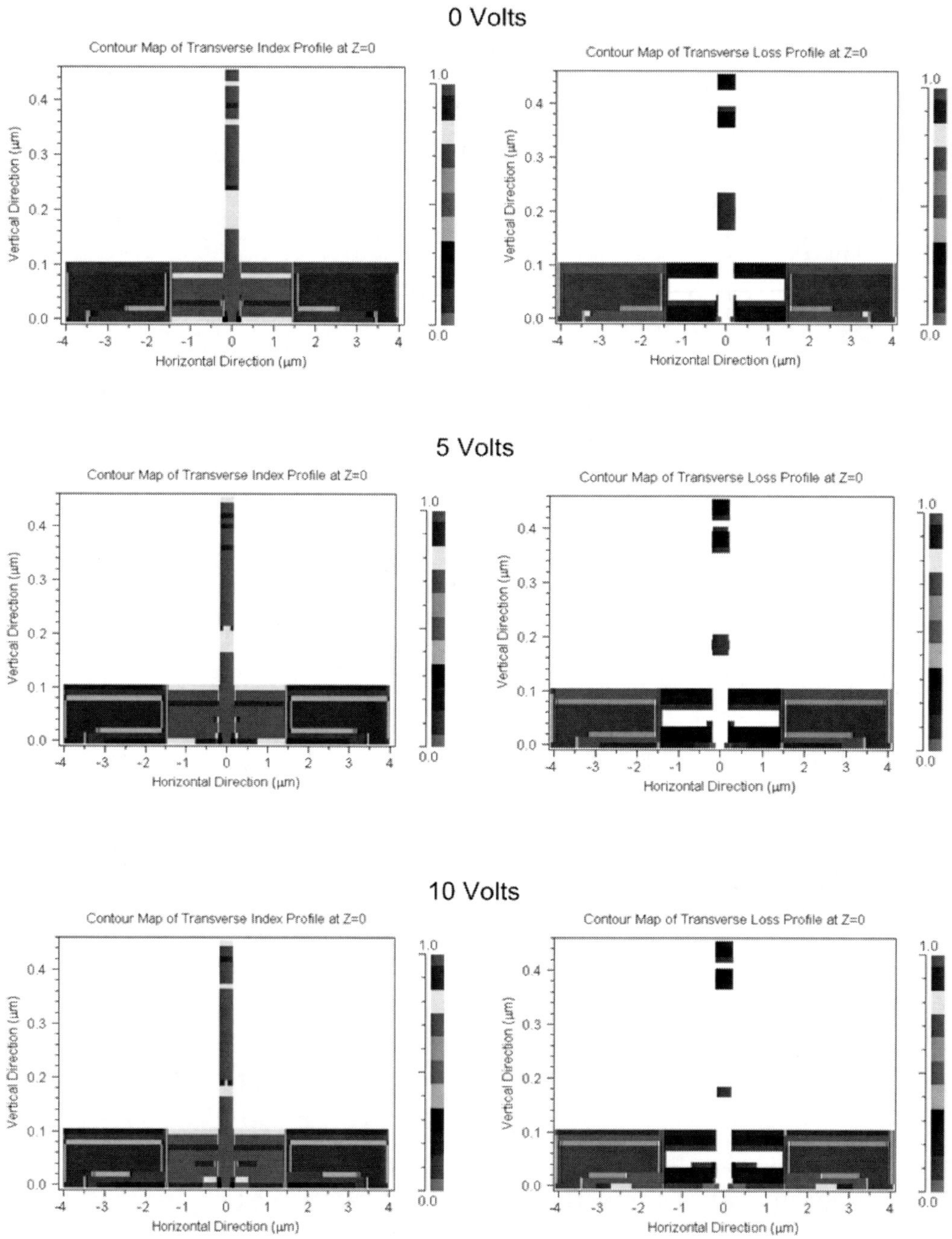

Figure 14. Refractive index profile and attenuation versus position for modified structure.

The same approach has been used to find the static phase shift allowed by the simple phase modulator or MZ interferometer, and their dynamic performance, i.e. the phase shift against transient time.

Figure 15 shows the static phase shift (degrees) as a function of applied voltage (volts) achieved for both polarizations, quasi-TE and quasi-TM, in case of single phase shifter with optimized structure.

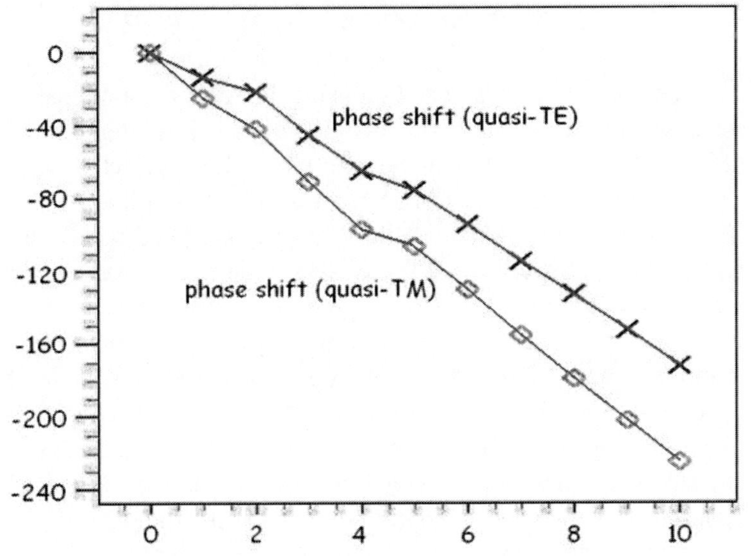

Figure 15. Static phase shift (degrees) versus applied voltage (volts) for both polarizations.

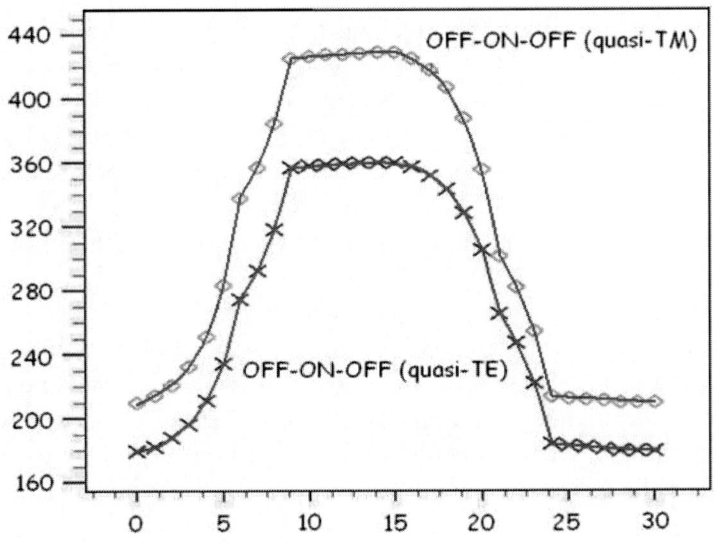

Figure 16. Dual-channel Mach-Zehnder phase switch versus transient time (ps) for both polarizations.

A phase shift of 172° for quasi-TE and 225° for quasi-TM excitation can be obtained by applying 10 V over a device 2.5 mm long. The DC power is negligible, well lower than in [4]. Moreover, in Figure 16 the dynamic phase switch (degrees) is sketched versus transient time (ps) for dual-channel MZ device, 2.52 mm long as in Figure 10, biased in OFF-ON-OFF switching state, for both polarizations. This dynamic study clearly demonstrates a switching (rise or fall) time of 8 ps, giving a device bandwidth of 59.7 GHz. The switching power required for ON-OFF or OFF-ON switching state is well 38% down that in previous structure [4], because both switching currents and voltages are smaller.

4. Raman Silicon Modulator

Very recent activities have been focused on achieving active functionality in Silicon-on-Insulator (SOI) waveguides, mostly wavelength conversion, light amplification and generation, based on Stimulated Raman Scattering (SRS) effect [13]. In fact, as a transmission medium silicon has much higher nonlinear effects than the commonly used silicon dioxide (optical fibre), in particular the Raman gain. The high Raman gain and the small mode area in SOI waveguide make the SRS observable over the millimetre-scale interaction length, usually encountered in integrated optical devices. Starting from the approach developed and experimentally demonstrated in literature for optical fibres [14], in this section the model of an all-optical intensity modulator based on Raman effect in SOI rib waveguide is briefly presented. The study is motivated by to main strategic reasons. First, an all-optical signal processing should be highly suitable for high speed systems in order to avoid the bottlenecks associated with electrical-to-optical and optical-to-electrical signal conversion and, secondly, Raman physical effect should allow to realise different functionalities on the same platform, i.e. light generation and amplification, Stokes and anti-Stokes conversion, continuum generation, all-optical switching, and all-optical modulation.

The architecture of the integrated all-optical intensity modulator is shown in Figure 17 and includes two different blocks, the generation block (GB), where a pump signal is launched into the SOI rib waveguide travelling over a length L_1, and a cancellation block (CB).

If the pump power is large enough, a Stokes pulse is generated via Stimulated Raman Scattering (SRS). When a data stream is introduced at the optical beam combiner, the power transfer from the data to the Stokes pulses occurs under quasi velocity-matching condition. Since the power transfer occurs if a space-time overlap between data and Stokes pulses is verified, the Stokes pulse will erase only the matched data pulse, selected in the cancellation block (CB) of length L_2.

Finally, a selective Bragg grating or micro-ring filter is used to extract at the output only the modulated data signal. Although this principle of operation is similar to that already presented in literature [14], it is here investigated in a totally integrated optical silicon architecture. It could constitute an approach for ultra-fast all-optical modulation or data encoding. In the following section, details of modelling aspects and some numerical results are presented and briefly discussed.

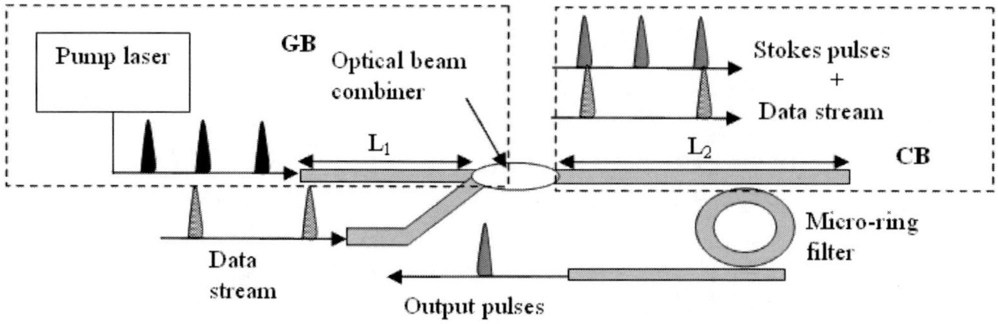

Figure 17. Architecture of all-optical Raman intensity modulator in SOI rib waveguide.

4.1. Modelling and Simulations

Since the cancellation is due to an all-optical power transfer mechanism between different optical pulses, the usual hypothesis of quasi continuous wave (CW) operation cannot be used. In fact, it applies for experimental investigation in SOI waveguides where the pump pulses have a full wave half maximum (FWHM) time width $T_{FWHM} \geq 1$ ns, the walk-off length L_w generally exceeding the waveguide length L. However, for ultrashort pulses with $T_{FWHM} \leq 100$ ps, typically it holds $L_w \leq L$ (see chapters 7 and 8 of this book, too).

SRS is then limited by the group-velocity mismatch and occurs only over distances z □ L_w, even if the actual waveguide length L is considerable larger than L_w. At the same time, the nonlinear effects such as self phase modulation (SPM) and cross phase modulation (XPM), can become significant and affect the evolution of both pump and Raman pulses. Thus, in general the time derivative cannot be neglected and appropriate numerical techniques are necessary to solve the partial differential equation system governing the mutual interaction among Stokes, pump and data pulses, including all nonlinear effects in a general and self-consistent formalism.

In particular, in both generation and cancellation blocks, Raman effect (generation of the fundamental, first and second order Stokes waves), walk-off effect, SPM and XPM induced by Kerr effect, two photon absorption (TPA), plasma dispersion and free carrier absorption (FCA) induced by TPA have been simultaneously considered.

It is worth to note that the TPA effect is generated only by pump pulse in GB, while TPA is produced in CB by the combination of residual pump, data stream pulse and fundamental Stokes wave. Therefore, the model is constituted by a system of partial differential equations in both time and space domains, together with a rate equation for the time dynamics of TPA generated free carriers [15]. In particular, the system involves four partial differential equations in GB for pump pulse, fundamental Stokes wave, first and second order Stokes waves, respectively, as:

$$\begin{cases} \frac{\partial A_p}{\partial z} + \beta_{1p}\frac{\partial A_p}{\partial t} + j\frac{1}{2}\beta_{2p}\frac{\partial^2 A_p}{\partial t^2} = -\frac{\left(\alpha_p^{(prop)} + \alpha_p^{(FCA)}\right)}{2}A_p - 0.5\beta^{(TPA)} f_{p,p}\left|A_p\right|^2 A_p + j\gamma_{p,p}\left|A_p\right|^2 A_p + \\ + j2\gamma_{p,s0}\left|A_{s0}\right|^2 A_p + j\frac{2\pi}{\lambda_p}\Delta n_p A_p - \frac{1}{2}g_R\frac{\omega_p}{\omega_{s0}} f_{p,s0}\left|A_{s0}\right|^2 A_p \\ \frac{\partial A_{s0}}{\partial z} + \beta_{1,s0}\frac{\partial A_{s0}}{\partial t} + j\frac{1}{2}\beta_{2,s0}\frac{\partial^2 A_{s0}}{\partial t^2} = -\frac{\alpha_{s0}^{(prop)}}{2}A_{s0} - \frac{1}{2}\alpha_{s0}^{(FCA)} A_{s0} - \beta^{(TPA)} f_{s0,p}\left|A_p\right|^2 A_{s0} + \\ + j2\gamma_{s0,p}\left|A_p\right|^2 A_{s0} + j\gamma_{s0,s0}\left|A_{s0}\right|^2 A_{s0} + j\frac{2\pi}{\lambda_{s0}}\Delta n_{s0} A_{s0} + \frac{1}{2}g_{R,s0} f_{s0,p}\left|A_p\right|^2 A_{s0} - \frac{1}{2}g_{R,s0} f_{s0,s1}\left|A_{s1}\right|^2 A_{s0} \\ \frac{\partial A_{s1}}{\partial z} + \beta_{1,s1}\frac{\partial A_{s1}}{\partial t} + j\frac{1}{2}\beta_{2,s1}\frac{\partial^2 A_{s1}}{\partial t^2} = -\frac{\alpha_{s1}^{(prop)}}{2}A_{s1} - \frac{1}{2}\alpha_{s1}^{(FCA)} A_{s1} - \beta^{(TPA)} f_{s1,p}\left|A_p\right|^2 A_{s1} + j2\gamma_{s1,s0}\left|A_{s0}\right|^2 A_{s1} + \\ + j2\gamma_{s1,p}\left|A_p\right|^2 A_{s1} + j\gamma_{s1,s1}\left|A_{s1}\right|^2 A_{s1} + j\frac{2\pi}{\lambda_{s1}}\Delta n_{s1} A_{s1} + \frac{1}{2}g_{R,s1} f_{s1,s0}\left|A_{s0}\right|^2 A_{s1} - \frac{1}{2}g_{R,s1} f_{s1,s2}\left|A_{s2}\right|^2 A_{s0} \\ \frac{\partial A_{s2}}{\partial z} + \beta_{1,s2}\frac{\partial A_{s2}}{\partial t} + j\frac{1}{2}\beta_{2,s2}\frac{\partial^2 A_{s2}}{\partial t^2} = -\frac{\left(\alpha_{s2}^{(pt)} + \alpha_{s2}^{(FCA)}\right)}{2}A_{s2} - \beta^{(TPA)} f_{s2,p}\left|A_p\right|^2 A_{s2} + j2\gamma_{s2,s1}\left|A_{s1}\right|^2 A_{s2} + \\ + j\gamma_{s2,s2}\left|A_{s2}\right|^2 A_{s2} + j2\gamma_{s2,p}\left|A_p\right|^2 A_{s2} + j2\gamma_{s2,s0}\left|A_{s0}\right|^2 A_{s2} + j\frac{2\pi}{\lambda_{s2}}\Delta n_{s2} A_{s2} + \frac{1}{2}g_{R,s2} f_{s2,s1}\left|A_{s1}\right|^2 A_{s2} \end{cases} \quad (10)$$

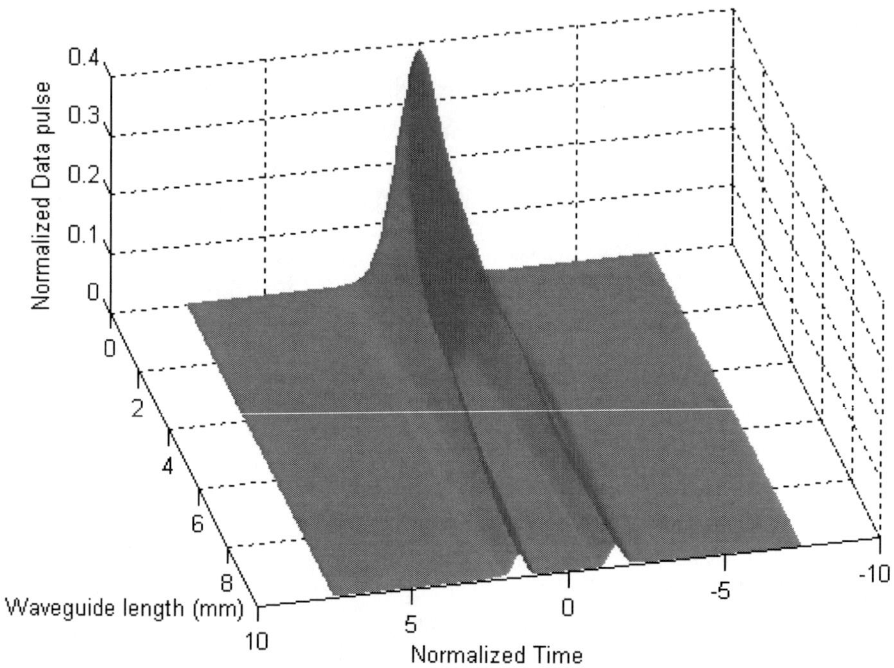

Figure 18. Time-space evolution of data pulse in CB.

Moreover, it is clear that the model involves five partial differential equations in CB, for residual pump, data pulse, fundamental Stokes pulse, first and second order Stokes waves. They are similar to the previous ones and are not reported here. The main design criteria selected for our device have been absence of any birefringence effect in nano-scale SOI rib waveguide for the fundamental Stokes wave, and absence of any excitation of higher order Stokes waves in both GB and CB sections. As an example, by assuming a pump pulse at λ_p = 1.4332 μm, the condition of free polarization for the fundamental Stokes wave (λ_{s0} = 1.5487 μm) can be obtained using a SOI waveguide with a rib total height H = 500 nm, a slab height H_s=150 nm and rib width W = 396 nm, as in literature [15]. Other numerical parameters were $T_{FWHM} = 100\,\text{ps}$ for both pump and data pulses (1.67 Gb/s pulse rate in this case), and Raman gain g_R = 10.5 cm/GW. Then, the excitation of higher order Stokes waves is avoided in GB by assuming a peak pump power $P_0 = 1.5$ W and a waveguide length $L_1 = 10$ mm. Figure 18 shows the time-space evolution of data pulse in CB. It is evident the cancellation operation induced by the fundamental Stokes pulse (see also Figure 19).

The high cancellation rate essentially depends on the relative high level of the Stokes wave at the input of CB, as a result of the amplification process in GB. In Figure 19 a depletion effect of the fundamental Stokes wave as induced by the building up of the first order Stokes wave (given in Figure 20) can be clearly seen. As it is confirmed by pulse evolution in Figure 20, it is important to set the waveguide length $L_2 = 4$ mm in order to avoid the excitation of higher order Stokes waves in CB. Finally, under this condition, Figure 21 shows an example of signature operation on the data stream. In this case, the cancellation induces secondary lobes which are 5 dB down with respect to data pulse level.

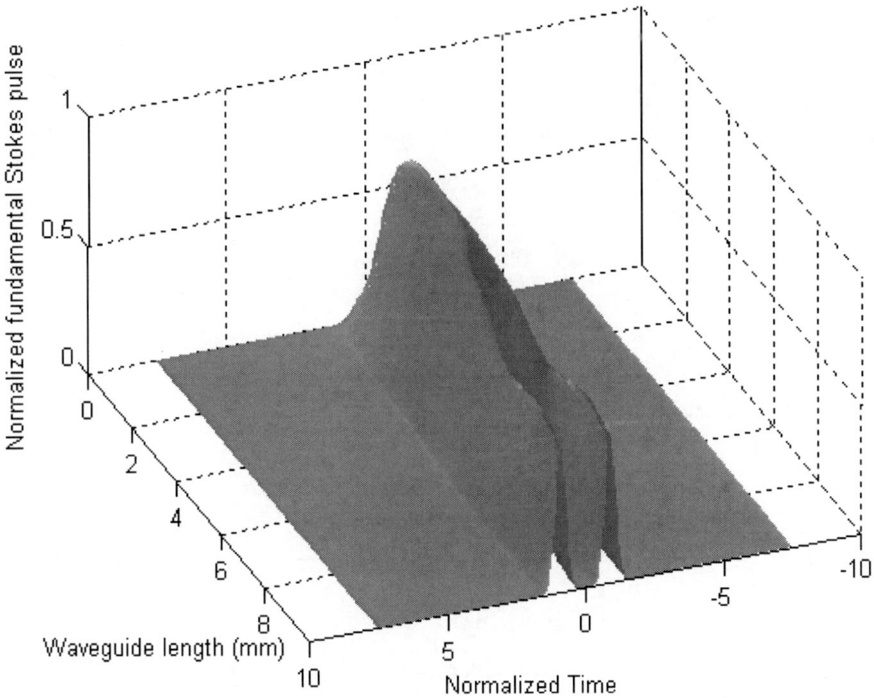

Figure 19. Evolution of fundamental Stokes pulse.

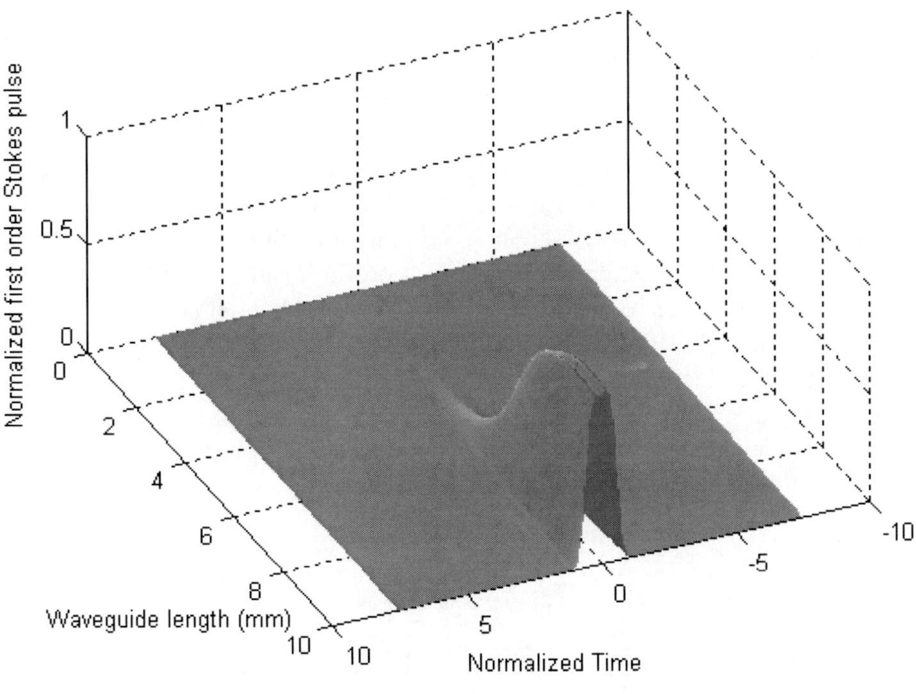

Figure 20. Space-time evolution of first order Stokes pulse in CB.

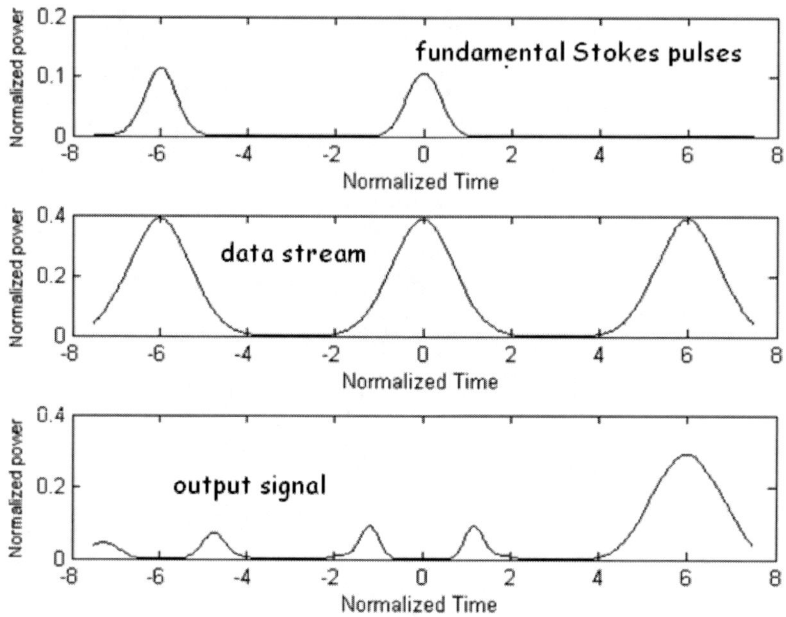

Figure 21. Result of interaction between data stream and fundamental Stokes pulses.

Conclusion

In this chapter, some modeling techniques of silicon modulators are briefly described. In particular, a fully integrated multiphysics model for SOI modulators having a MOS structure is developed. Electro-static potential and free carrier distribution dependences on applied voltage are investigated, and relevant phase shifts induced on quasi-TE and quasi-TM optical modes are calculated for each value of applied voltage. Through this calculation, a $V_\pi \times L$ product has been estimated as 3.4 V·cm for quasi-TE and 10.2 V·cm for quasi-TM polarized light. Moreover, an optimized architecture of a 2.52 mm long SOI optical modulator based on a depletion p-n junction is investigated. The doping profile inside the rib has been optimized in order to achieve very large refractive index change through plasma dispersion effect and, then, large phase shifts while keeping low the relevant losses. By using a novel concept of dual-channel Mach-Zehnder modulator, high performance compact device is theoretically demonstrated, including small length, low switching time and very low power consumption.

Finally, the model and results of an all-optical intensity modulator based on Raman effect in SOI rib waveguide are presented.

References

[1] Passaro, V. M. N. (Ed.), *Silicon Photonics*. Research Signpost Publ., India, 2006.
[2] Soref, R. A., Bennett, B. R., *IEEE J. Quantum Electronics*, 1987, QE-23, 123-129.
[3] Hewitt, P. D., Reed, G. T., *J. Lightwave Technology*, 2001, 19, 387-390.
[4] Gardes, F. Y., Reed, G. T., Emerson, N.G., *Optics Express*, 2005, 13, 8845-8854.

[5] Liu, A., Jones, R., Liao, L., Samara-Rubio, D., Rubin, D., Cohen, O., Nicolaescu, R., Paniccia, M., *Nature*, 2004, 427, 615-618.

[6] Liao, L., Samara-Rubio, D., Morse, M., Liu, A., Hodge, D., *Optics Express*, 2005, 13, 3129-3135.

[7] Barrios, C. A., Lipson, M., *J. Applied Physics*, 2004, 96, 6008-6015.

[8] TCAD by Silvaco Corp., *single license*, 2008.

[9] Comsol Multiphysics by COMSOL AB, ver. 3.2, *single license*, 2005.

[10] Passaro, V.M.N., Dell'Olio, F., *IEEE Trans. on Nanotechnology*, 2008, 7.

[11] Reed, G. T., Png, C. E., *Materials Today*, 2005, 8, 40-50.

[12] OptiBPM by Optiwave Corp., Ottawa, Canada, *ver. 7.0, single license*, 2005.

[13] Claps, R., Dimitropoulos, D., Raghunathan, V., Han, Y., Jalali, B., *Optics Express*, 2003, 11, 1731-1739.

[14] Burdge, G., *et al.*, *ECOC Proceedings*, 1997, 61-64.

[15] Passaro, V. M. N., De Leonardis, F., *J. Lightwave Technology.*, 2006, 24, 2920-2931.

In: Modeling of Photonic Devices
Editor: Vittorio M. N. Passaro

ISBN 978-1-60456-980-3

Chapter 6

DIFFRACTION GRATINGS

Giuseppe Tandoi*[*,1] and Vittorio M. N. Passaro[2]

[1]Optoelectronics Group, Department of Electronics and Electrical Engineering, Rankine Building, Oakfield Avenue, Glasgow G12 8LT, UK

[2]Photonics Research Group, Dipartimento di Elettrotecnica ed Elettronica, Politecnico di Bari, via Edoardo Orabona n. 4, 70125 Bari, Italy

ABSTRACT

In this chapter, an overview of diffraction gratings is presented. In particular, some aspects of mathematical modelling of these photonic components are reviewed. In the first part we have focused our attention on the main grating structures, configurations, technologies and applications. Then, a description of the grating behaviour is presented by using the Coupled Mode Theory. Finally, the main numerical methods to solve Maxwell's equations in periodic media are presented, i.e., Finite-Difference Time-Domain (FDTD), Finite Element Method (FEM), Transfer Matrix Method (TMM), and Floquet-Bloch Theory (FBT), with some design examples of grating-based devices.

Keywords: Diffraction Gratings, Fiber Gratings, Coupled Mode Theory, Finite-Difference Time-Domain, Finite Element Method, Transfer Matrix Method, Floquet-Bloch Theory, Modeling, Grating Devices

1. INTRODUCTION

Diffraction gratings are photonic components used in several optic and optoelectronic applications because of their simple structure and flexible behaviour. Indeed, the structure of gratings is based on refractive index periodicity along one or more directions in a propagation medium. This characteristic can make easier the power exchange between co-directional or

[*] E-mail address: peppe@elec.gla.ac.uk

contra-directional modes of a waveguide or two or more different waveguides. This general property of the gratings is the main principle on which many important devices are based and is the reason for all the efforts that have been done in modeling, design and fabrication of these fundamental photonic components. The study of grating characteristics has also brought the research to explore other more complex periodic structures, named *photonic crystals*, that have a refractive index periodicity in one, two or three dimensions. These photonic components are more powerful but their study and realization present more complex problems than simple gratings which, in many applications, are efficient enough.

For these reasons, diffraction gratings have been extensively studied and used in a number of configurations, technologies and applications. Many powerful theories and numerical methods have been proposed in literature to understand and control their physical behaviour for design of specific photonic devices. In this chapter the main characteristics and configurations of gratings are explained in section 2, and main technologies and applications described in section 3. The physical behaviour of gratings is investigated by coupled-mode theory (CMT) in section 4 as a powerful method, widely used in the design of grating-based devices such as grating couplers and diffraction Bragg reflectors (DBR). In the following sections, the main numerical methods used in grating modeling are presented, including the finite-difference time-domain (FDTD) method in section 5 as a powerful approach to solve Maxwell's equations in complex optical structures, i.e., to find optical field distributions and propagation constants. Moreover, the finite element method (FEM) is described in section 6, where a mixed procedure based on both FEM and CMT is presented for the design of selection gratings in distributed feedback quantum cascade laser (DFB QCL). Moreover, the transfer matrix method (TMM) is a method used in grating analysis and is explained in section 7. In section 8, the Floquet-Bloch theory (FBT) is presented as a rigorous method to describe the optical field in a periodic medium as a superposition of space harmonics. Examples of FBT application to design a DBR and a vertical grating-assisted waveguide coupler are also briefly reviewed.

2. GRATING STRUCTURES

The structure of a simple grating consists of a one-dimensional (1D) periodic repetition of a well defined refractive index profile. The most common shapes are rectangular, triangular and sinusoidal ones, as represented in Figure 1, but each refractive index profile, periodic or aperiodic, could be considered as well in the definition of a diffraction grating.

The most important geometrical parameter of a grating is the period Λ, or its grating constant $K = 2\pi/\Lambda$, because of the grating operating wavelength dependence on it, as expressed by:

$$\lambda_B = 2n_{eff}\Lambda / m \qquad (1)$$

where λ_B is the characteristic Bragg wavelength of the grating, n_{eff} is the effective index of a mode propagating in the guiding medium, and m is an integer indicating the grating order. The Bragg wavelength derives from the Bragg condition, that consists of the matching between the mode propagation constant in the optical medium β and the resonant one β_0, as:

$$\beta = 2\pi n_{eff} / \lambda_B = \beta_0 = m\pi / \Lambda \quad (2)$$

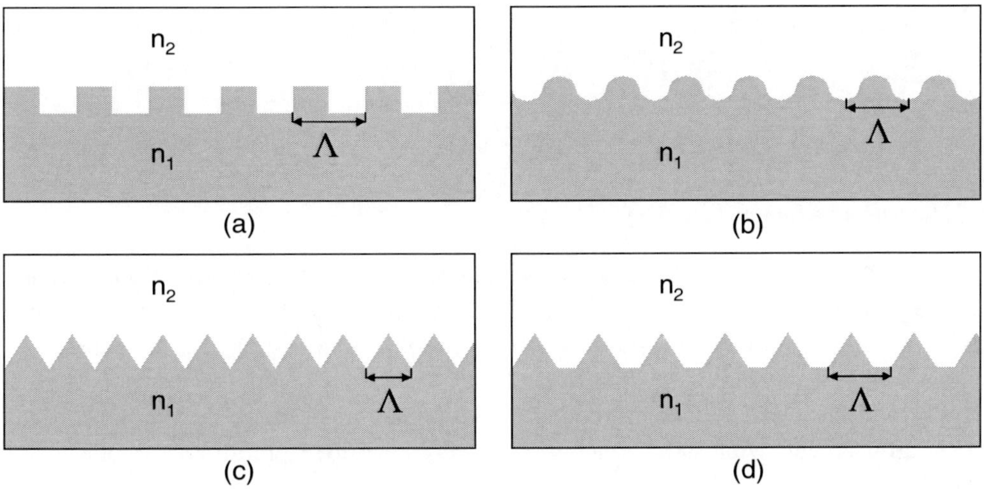

Figure 1. Scheme of (a) rectangular, (b) sinusoidal, (c) triangular and (d) generic periodic refractive index profiles of gratings.

The refractive index contrast Δn determines interaction (e.g., reflection) of an incident wave at Bragg wavelength. This behaviour can be used in all applications that need a selective behaviour. In general, as it will be seen in the next sections, gratings allow the interaction and the power exchange between two co-directional or contra-directional modes in an optical waveguide or two or more different waveguides.

The periodic structures described above are examples of simple uniform gratings. However, different kinds of gratings still exist. Gratings that do not have a uniform index profile, or equivalently, a constant value of Δn along their length, are called *apodized gratings*, while gratings having changeable period Λ along their length are called *chirped gratings*. These particular gratings are represented in Figure 2.

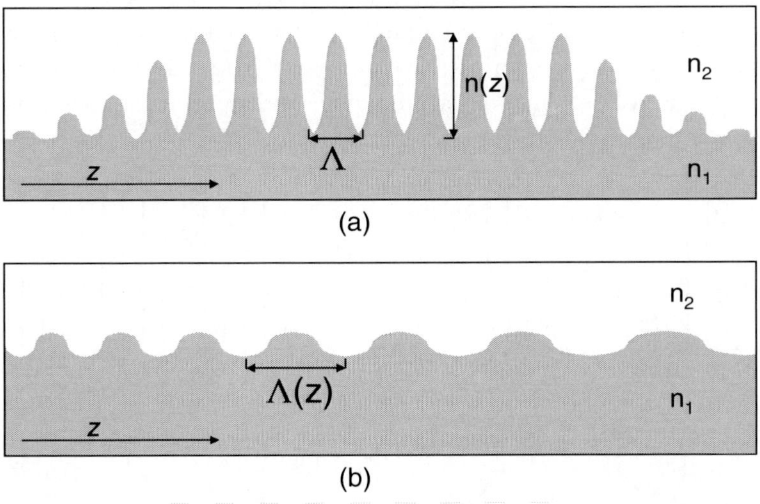

Figure 2. Scheme of (a) apodized and (b) chirped refractive index grating profiles.

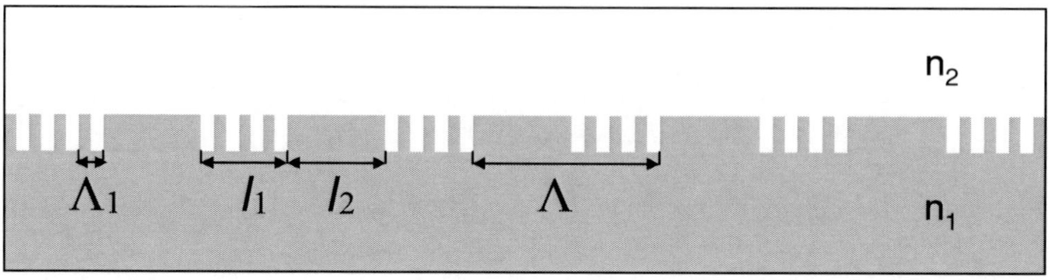

Figure 3. Scheme of a sampled grating.

A chirped grating is a grating with varying spatial period and, therefore, changeable spatial Bragg wavelength. For this reason, its reflection spectrum is the superposition of the spectra of gratings with different periods. This kind of grating can successfully reflect more than one resonant wavelength. An apodized grating has a more selective spectrum, in terms of suppression of higher order resonant modes. The modulation of the refractive index along the grating length, indeed, yields the reduction of secondary lobe amplitudes in the reflection spectrum.

Sampled gratings are another important kind of gratings, obtained by a periodic repetition of a certain number of sections, each made by two parts, one containing a grating of period Λ_1 and length l_1, and the other of length l_2 and constant refractive index profile. Therefore, the period of such gratings is $\Lambda = l_1 + l_2$, as represented in Figure 3.

The reflection spectrum of sampled gratings is the periodic repetition (period $f_s = 1/\Lambda$) of the spectrum of the inner repeated gratings with period Λ_1.

Gratings can also be used in a cascade configuration, where more gratings of different period are positioned in cascade, in order to reflect more than one wavelength. Gratings in this configuration are called *superstructured gratings* and show a reflection spectrum given by the superposition of the spectra of different cascaded gratings.

Diffraction gratings are also classified by their use, i.e. as *transmission grating*, when the output wave is the transmitted one, or as *reflection grating*, when the output wave is the reflected one, as represented in Figure 4.

Depending on its application and technology, the grating can be opportunely positioned in free space or, more commonly, be integrated into the medium in which the input and the output waves propagate, such as optical waveguides or fibers. In both cases, the input and output waves are modes of propagation medium and, under particular conditions, can interact and exchange power to each other.

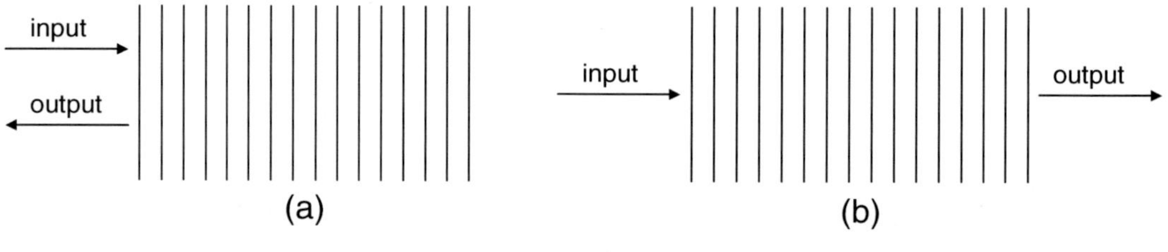

Figure 4. Scheme of (a) reflection grating and (b) transmission grating.

3. Grating Technologies and Applications

Gratings are photonic components useful in many fields, such as fibre optic telecommunications systems and optical sensing applications. In fibre optic devices, gratings are used to improve the performance of some optoelectronic devices such as monolithic or non-monolithic lasers, and as part of fibre optic components, such as filters, interferometers, multiplexers, demultiplexers, fibre-dispersion compensators and modulators. This is due to relatively simple and flexible use of gratings. Indeed, depending on the particular application, diffraction gratings have been fabricated in many different technologies, including III-V semiconductor compounds, silicon compounds (Si, SiO_2, SiGe, silicon oxynitride, silicon-on-insulator) and ferroelectric materials like lithium niobate ($LiNbO_3$).

The first diffraction gratings [1] made for commercial use were mechanically ruled, manufactured by burnishing grooves individually with a diamond tool against a thin coating of evaporated metal applied to a plane or concave surface. Such *ruled gratings* [1] included the majority of diffraction gratings used in spectroscopic instrumentation. Since late 1960s, a method different from mechanical ruling has also been used to manufacture diffraction gratings, involving the photographic recording of a stationary interference fringe field. Thus, in 1901 Aimé Cotton produced experimental interference gratings, fifty years before the concepts of holography were developed by Gabor. A few decades later, Michelson considered the interferometric generation of diffraction gratings, but recognized that an intense monochromatic light source and a photosensitive material of sufficiently fine granularity were not available. In the mid 1960s, ion lasers and photoresists (grainless photosensitive materials) became available: the former provided a strong monochromatic line, and the latter was photoactive at molecular level, rather than at crystalline level (unlike, for example, photographic film). In 1967 *D. Rudolph* and *G. Schmahl* at University of Göttingen and *A. Labeyrie* and *J. Flamand* in France independently produced the first holographic diffraction gratings of spectroscopic quality. Decades of research and development at Thermo RGL have contributed to the process for manufacturing replicated diffraction gratings (*replicas*) [1]. This process is capable of producing thousands of duplicates of master gratings with the same quality and performance as the master. The replication process has reduced the price of a typical diffraction grating by a factor of 100 or more, compared with the cost of acquiring a master grating.

Gratings have been realized in many different technologies that depend, according to the particular application, on the material system used for the wave-guiding medium. The main material systems used for this purpose are semiconductors and glasses. III-V semiconductor compounds as InP and GaAs compounds are mainly used for active optical telecommunications devices (as receivers, transmitters, modulators, amplifiers, active couplers), operating at wavelengths 1.5 or 1.3 µm. Silicon and SiGe are also used for active optical devices (optical receivers), while silica (SiO_2) and silicon oxynitride (SiON) waveguides are usually used for passive photonic grating-based or waveguide coupler-based components. Ferroelectric materials, as $LiNbO_3$, are used to realize passive components, but, thanks to their significant electro-optic and piezoelectric properties, they are also used for active components as optical modulators, tunable filters, active couplers and multiplexers, as well as for optical amplifiers and lasers because of their light amplification properties (doping with rare earths like erbium). Other important media for optical communications and sensing applications include silica or polymeric optic fibres [2]. Fiber gratings will be better described later in this section.

Using above mentioned material systems, the main types of gratings are those realized by selective-doping, selective etching, lithographic or holographic techniques, as well as $LiNbO_3$ proton exchange. In all these techniques, the fabrication process yields the selective change of refractive index along the surface of optical medium exposed to that particular process.

According to fabrication procedure, gratings are classified by two general typologies, i.e. *index modulation* and *structural corrugation*. In case of structural corrugation, the periodic (or almost periodic) refractive index profile is a permanent structure of the waveguide, realized by corrugating it by means of selective etching, deposition or material growth. This corrugation is usually located at the interface between layers or between guiding and cover layers of a planar waveguide. It can also be placed away from the interfaces, close to the guiding layer so long as the mode fields have sufficient penetration into the neighbouring layers to see the corrugation. In case of index modulation, the periodic (or almost periodic) refractive index profile can be permanently written in a waveguide medium or created by any effect, electro-optic, acousto-optic, thermo-optic or nonlinear. In the latter case, the grating can be time dependent if the modulation is time varying, or a moving grating if the modulation signal is a travelling wave. $LiNbO_3$ and other ferroelectric materials can be used for electro-optically and acousto-optically induced gratings, while polymers for electro-optically and thermo-optically induced gratings and SiO_2 for thermo-optically induced ones. These effects are not described in this chapter, that focus the attention only on modelling of permanent gratings. In this section, the main characteristics and applications of III-V semiconductor compounds and fiber diffraction gratings, realized by typical fabrication rules for these two technologies combined with interferometer techniques, will be considered. In next sections, the main grating modelling rules will be described, with general considerations applied to particular III-V semiconductor and fiber gratings, but extensible also to other grating technologies.

3.1. DBR and DFB Semiconductor Lasers

III-V semiconductor compounds have been used to fabricate many different grating-based optical devices like filters, two-modes or multiple-modes couplers and sensors, but, certainly, the most important application in this material system is the wavelength selection in Diffraction Bragg Reflector (DBR) and Distributed Feedback (DFB) lasers. In both cases, the grating is used to induce the optical feedback and improve the single mode operation of the laser. In DBR lasers, the grating is fabricated on one facet or both facets of the laser cavity, in order to partially reflect the desired wavelength. By this way, the reflected mode is continuously amplified in the active region at the expenses of the other longitudinal modes of the cavity, giving the emission of the desired wavelength. On the contrary, the grating is fabricated in DFB lasers on the active region and optical feedback is distributed along the whole cavity. Then, the selected mode has multiple reflections on each refractive index discontinuity. Figure 5 shows a scheme of both DFB and DBR lasers, respectively. In most cases, the Bragg mirror is more specifically a quarter-wave mirror, providing the maximum amount of reflection for given number of layers. DBR lasers are usually laser diodes, but sometimes they also designate fiber lasers containing fiber Bragg gratings (FBG). Both laser types are described below.

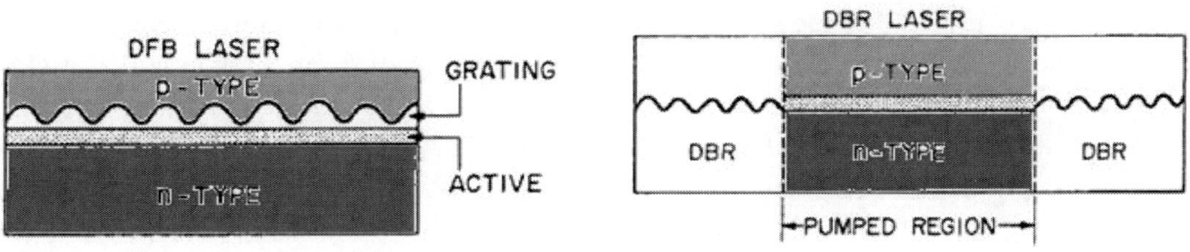

Figure 5. Scheme of (a) DFB and (b) DBR laser.

Most solid state bulk lasers actually use laser mirrors which are Bragg mirrors. Nevertheless, such lasers are not called DBR lasers. A DBR laser diode contains some corrugated waveguide structures (grating section) providing wavelength-dependent feedback to define the emission wavelength. Another section of the laser waveguide acts as amplifying medium (active region), and last section of the resonator may have another DBR.

DBR laser diodes are usually single-frequency lasers with diffraction-limited output, and often they are wavelength-tunable (tunable lasers). Tuning within the free spectral range (FSR) of the laser resonator may be accomplished through a separate phase section, which can be electrically heated simply by varying the temperature of the gain region via the drive current. If the temperature of the whole device is varied, the wavelength response is significantly smaller than for an ordinary single-mode laser diode, since the grating reflection band is shifted less than the gain maximum. Mode-hop free tuning over a larger wavelength region is possible by coordinated tuning of the Bragg grating and gain structure. More sophisticated device designs are also possible, exploiting a kind of Vernier effect with sampled gratings, that offer a tuning range as wide as 40 nm, although not without mode hops. The linewidth of DBR diodes is typically a few megahertz. The actual DBR laser is the *seed laser*, which can achieve output powers above 100 mW. Applications of DBR laser diodes include optical fiber telecommunications, free-space optical communications, laser cooling, optical metrology, sensors, and high resolution spectroscopy. DBR lasers are actually competing with *external cavity diode lasers* (ECDLs), which also offer wavelength-tunable single-frequency output, with potentially better performance in terms of noise, but also requiring a significantly more complex set-up. Chips containing DBR laser arrays can serve as very compact sources for use in wavelength-division-multiplexing (WDM) systems. Other important types of DBR lasers are *surface-emitting lasers* (SEL) and *vertical-cavity surface-emitting lasers* (VCSEL). SELs have two DBRs as Bragg mirrors at the end facets, that yield both the optical feedback and the surface-coupling of the emitted light. VCSELs have a vertical active cavity with the two planar DBRs sections made of epitaxial layers of alternate materials.

DFB semiconductor lasers have been realized by using many different grating configurations. The most important of them include top *surface gratings* [3], *buried gratings* [4] and *lateral gratings* [5]. Moreover, DFB laser gratings can have a stronger effect if the grooves are etched up to the active region of the laser. In this case, the periodicity of the grating will influence the dielectric constant, whose complex part depends on the laser modal gain. In this case, the coupling of modes is defined as complex. These so-called *complex-*

coupled DFB lasers show better performance in terms of single-mode operation and tunability.

External cavity diode lasers (ECDL) represent another type of grating-assisted lasers [6-8]. They are non-monolithic diode lasers where the laser cavity (resonator) is completed with external optical elements. These diode lasers are semiconductor lasers based on a laser diode chip which typically has one anti-reflection coated edge, and the laser resonator includes a collimating lens and an external mirror. The longer resonator increases the damping time of intracavity light and, thus, allowing for lower phase noise and smaller emission linewidth in single-frequency operation. An intracavity filter such as the diffraction grating can further reduce the linewidth. Typical ECDL linewidths are below 1 MHz. Wavelength tuning is possible by including some adjustable optical filters as tuning elements, using gratings. In this case, the external resonator also adds important features for mode locking (mode-locked laser diodes) with a lower (possibly tunable) pulse repetition rate and a higher pulse energy, compared with monolithic laser diodes. A saturable absorber can be included for passive mode locking. Mode locking may also be achieved with an absorber section integrated into the gain chip, or by modulation of drive current (active mode locking), or with both methods simultaneously (hybrid mode locking). A special type of external-cavity laser uses a resonator based on optical fiber rather than free-space optics. Narrowband optical feedback can then be obtained by a fiber Bragg grating.

Tunable external-cavity diode lasers usually use a diffraction grating as the wavelength-selective element in the external resonator. They are also called grating-stabilized diode lasers. The common Littrow configuration contains a collimating lens and a diffraction grating as the end mirror. The first-order diffracted beam provides optical feedback in the laser diode chip, which has an anti-reflection coating on the right side. The emission wavelength can be tuned by rotating the diffraction grating. A disadvantage consists of the change of output beam direction, not convenient for many applications.

In the Littman-Metcalf configuration [7], the grating orientation is fixed, and an additional mirror is used to reflect the first-order beam back to the laser diode. The wavelength can be tuned by rotating that mirror. This configuration offers a fixed direction of the output beam, and also tends to exhibit a smaller linewidth as the wavelength selectivity is stronger. In fact, the wavelength-dependent diffraction occurs twice per resonator round trip. A disadvantage is that the zero-order reflection of the beam reflected by the tuning mirror is lost, so that the output power is somewhat lower than for a Littrow laser. Typical applications of ECDLs are in optical telecommunications systems and spectroscopy. Competing types of lasers are DBR laser diodes and small fiber lasers.

3.2. Fiber Bragg Gratings

Another important type of grating is the *fiber Bragg grating* (FBG). These are optical fibers with a periodic or aperiodic perturbation of refractive index in the core (see Figure 6), or fiber devices containing such a perturbation. Typically, the perturbation is approximately periodic over a certain length of a few millimeters or centimeters, and the period is of the order of hundreds of nanometers, or much longer for long-period fiber gratings (LPFG).

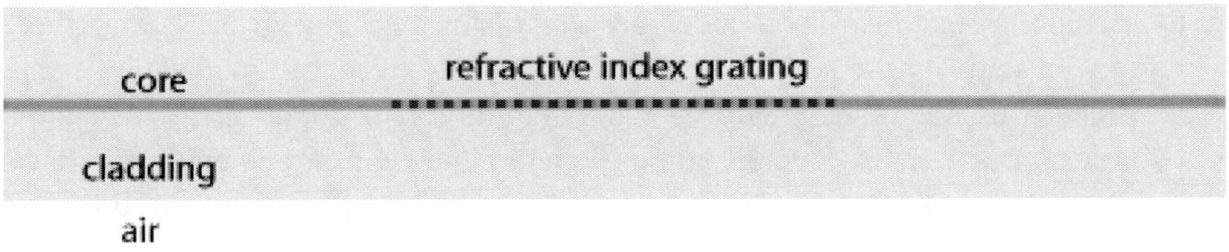

Figure 6. Schematic structure of a fiber Bragg grating (FBG). The fiber core has a periodically varying refractive index. Typical sizes are 125 μm cladding diameter and 8 μm core diameter; grating period in the range of hundreds of nanometers.

The refractive index perturbation leads to light reflection (propagating along the fiber) in a narrow range of wavelengths, where the Bragg condition holds. Even a weak index modulation (amplitude 10^{-4}) is enough for achieving nearly total reflection, if the grating is sufficiently long (e.g. a few millimeters). Light at other wavelengths, not satisfying the Bragg condition, is nearly not affected by the Bragg grating, except for some side lobes which frequently occur in the reflection spectrum (suppressed by grating apodization). The reflection bandwidth of a fiber grating, which is typically well below 1 nm, depends on both length and strength of the refractive index modulation. The narrowest bandwidth values, desirable for single-frequency fiber lasers or certain optical filters, are obtained from long gratings with weak index modulation. Large bandwidths may be achieved with short and strong gratings, as well as with aperiodic designs.

As the wavelength of maximum reflectivity depends not only on the Bragg grating period but also on physical effects like temperature and/or mechanical strain, Bragg gratings can be used as efficient temperature and/or strain sensors. Transverse stress, as generated by squeezing a fiber grating between two flat plates, induces birefringence and thus polarization-dependent Bragg wavelengths.

Most fiber Bragg gratings are used in single-mode fibers, and in that case the physical modelling is often relatively simple. The same formalism as for dielectric mirrors can in principle be used, assuming for example a rectangular index modulation. At least the main reflection band can be properly calculated with that approximation. However, the number of periods of index modulation can become rather large for long gratings, significantly increasing the computation time. It may then be preferable to use a model based on mode coupling, leading to a couple of differential equations with a coupling term, the magnitude of which is related to local strength of index modulation. Then, the coupling is effectively assumed to be smoothly distributed, and the numerical integration is carried out with a step size much larger than the grating period. Such methods can be used for calculating the frequency-dependent complex amplitudes for light transmission and reflection. Thus, not only the fractions of reflected and transmitted power, but also the chromatic dispersion can be obtained. Numerical models become substantially more complicated if many propagation modes are involved. Even for a single-mode fiber, it may be necessary to consider four modes (rather than just two counter-propagating modes) if birefringence is relevant, or even a larger number of modes if coupling to cladding modes occurs. For multimode fibres, a multitude of core modes has to be accounted for. In such cases, the coupling coefficients depend not only on index modulation amplitude, but also on grating three-dimensional profile. If the strength of the index modulation in a grating is constant over some length, and suddenly drops to zero outside that range, the reflection spectrum exhibits sidelobes, in particular if the peak reflectivity is high. Sometimes these sidelobes are disturbing, e.g. in applications of fiber Bragg gratings as optical filters. They can be largely removed by apodization technique, where the strength of the index modulation is smoothly ramped up and down along the grating. Of course, an increased overall grating length is needed for achieving a certain peak reflectivity. For optimal profile of index modulation in apodized fiber Bragg gratings, a trade-off must be usually met between optimum sidelobe suppression and maximum reflectivity for limited grating length and given maximum strength of index modulation.

Telecom applications of FBGs often involve wavelength filtering, e.g. for combining or separating multiple wavelength channels in WDM systems. Extremely narrow-band filters can be realized with rather long FBGs (with lengths of tens of centimeters). FBGs can be used as end mirrors of fiber lasers, then typically restricting the optical emission to a very narrow spectral range. Even single-frequency operation can be achieved by having the whole laser resonator formed by a FBG with a phase shift in the middle. Outside a laser resonator, an FBG can serve as a wavelength reference, e.g. for stabilization of the laser wavelength. This method can also be applied for wavelength-stabilized laser diodes.

A DBR fiber laser usually has a linear laser resonator formed by an active rare-earth doped fiber between two fiber Bragg gratings, as in Figure 7. Compared with a fiber DFB laser, which has the gain within single grating structure, a DBR fiber laser has a longer laser resonator and thus the potential for higher output power, higher power efficiency, and narrower linewidth. On the other hand, this can also lead to less robust single-frequency operation, or multimode operation with a correspondingly much larger emission bandwidth. Single-frequency DBR fiber lasers offer similar output powers as DBR laser diodes, tens of milliwatts or sometimes >100 mW. Sampled grating designs can also be used in DBR fiber lasers, obtaining again a tuning range which is tens of nanometers wide.

If the polarization of the writing beams is perpendicular to fiber axis, a significant deviation between Bragg wavelengths for both polarization directions can occur, i.e. a birefringence. This may be used for example for fabricating rocking filters. The range of interesting phenomena in FBGs is further enriched by the occurrence of optical nonlinearities.

Fiber gratings with aperiodic index modulation can have interesting properties, such as reflectivity curves without sidelobes, multiple tailored reflection bands, or special dispersion profiles. Particularly for dispersion compensation, so-called *chirped fiber gratings* are used [9], where the Bragg wavelength varies with position. Then, it is possible to achieve very large group delay dispersion in a short fiber length, sufficient for dispersion compensation of a long span of transmission fiber. Chirped fiber gratings are also interesting for application as distributed fiber-optic sensors with *intragrating sensing*, i.e. monitoring the temperature along the device length.

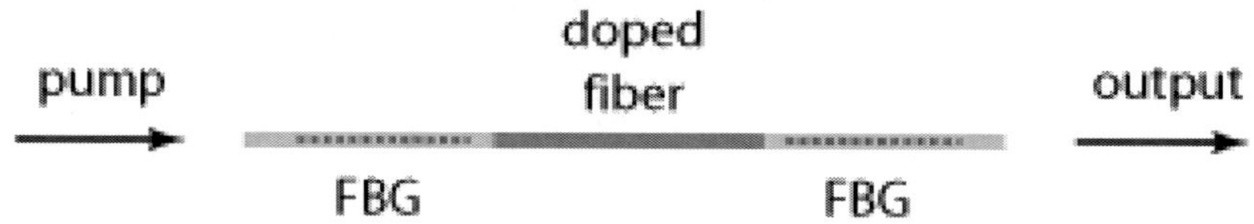

Figure 7. Short DBR fiber laser for narrow-linewidth emission.

Typical FBGs have grating periods of a few hundreds of nanometers, coupling counter-propagating waves in the core. A second possibility is to use *long-period Bragg gratings* (LPBG) [10] with periods of the order of hundreds of microns (often with tilted grating planes) and a length of a few of centimeters. Such gratings can couple modes with the same propagation direction. For example, the fundamental mode of a multimode fiber can be coupled to a certain higher-order mode, or a core mode can be coupled to cladding modes propagating in the same direction. In the latter case, the coupling effectively introduces propagation losses, because light in cladding modes normally experiences strong losses in the fiber coating. Such gratings are used for introducing carefully controlled wavelength-dependent losses, e.g. for gain equalization in erbium-doped fiber amplifiers, but are also used for fiber-optic sensing.

It is also possible to write FBGs in *polymer optical fibers*. As with silica fibers, ultraviolet (UV) light is usually used, but the physical mechanisms are somewhat different. An advantage of Bragg gratings in polymer fibers is the larger wavelength tunability. Polymer fibers can be stretched more strongly, and they react more strongly to temperature changes. The fabrication of fiber Bragg gratings typically involves the illumination of fiber core material with ultraviolet laser light (using a KrF or ArF excimer laser or other type of deep ultraviolet laser), which induces some structural changes and thus a permanent modification of refractive index. The photosensitivity of the core glass is actually strongly dependent on chemical composition and UV wavelength. Silica glass has a very weak photosensitivity, while germane-silicate glass exhibits a much stronger effect, making possible a refractive index contrast up to $\approx 10^{-3}$. A significant further increase of photosensitivity is possible by loading the fiber with hydrogen (*hydrogenated fibers*). For that purpose, the fiber is kept in a high-pressure hydrogen atmosphere for some time. Phosphate glasses are not normally regarded as suitable for FBG fabrication, but special methods make this possible [11].

The first fiber Bragg gratings [12] have been fabricated with a visible laser beam propagating along the fiber core, but a more versatile technique has been demonstrated in 1989 by G. Meltz [13], using the interferometer superposition of ultraviolet beams coming from the fiber side (transverse holographic technique). The angle between ultraviolet beams determines the period of light pattern in the fiber core and thus the Bragg wavelength. The two ultraviolet beams are often generated by exposing a periodic phase mask (photomask) with a single UV beam [14] (phase mask technique), using the two first-order diffracted beams. Non-periodic phase masks can be used to obtain more complicated patterns.

Another fabrication method is the point-by-point technique, where the regions with increased refractive index are written point by point with a small focused laser beam. This is an appropriate and very flexible technique particularly for long-period Bragg gratings. Instead of ultraviolet light, infrared light in the form of intense femtosecond pulses can also be used for writing Bragg gratings [15] in various kinds of glasses. In that case, two-photon absorption occurs near the laser beam focus, but not in regions outside the focus. Then, it is possible to write such gratings through the polymer coating of fiber, since the intensity in the coating is much lower when the beam is focused to the fiber core [16].

A totally different method also using infrared light consists of the fabrication of long-period FBGs in photonic crystal fibers by irradiation with CO_2 laser beam. Depending on the writing conditions, actually different physical mechanisms are involved in the Bragg grating formation, and different types of gratings can be distinguished. Type I gratings are written with moderate intensities and exhibit an index grating right across the core. Type II gratings

[17] can be written with much higher intensities within very short times, often with a single nanosecond pulse from an excimer laser (single-shot damage gratings). They can be written on the drawing tower [18] just before the fiber is coated, so that one avoids the process of removing an already fabricated coating, and obtains a grating with the full mechanical strength of ordinary fiber. Fiber Bragg gratings are quite durable, but the degree of durability, e.g. the temperature at which the grating may be erased, strongly depends on the fiber material and details of grating fabrication. The optical properties may change during some time after fabrication, before they settle at their final values. To reach a stable state more quickly, an annealing procedure can be applied, which typically means that the fiber is kept at some elevated temperature for a few hours.

Optical fiber sensors are fiber-based devices for sensing some physical or chemical quantity, typically temperature or mechanical strain, but sometimes also vibrations, pressure, acceleration, or concentrations of chemical species. The general principle of such devices is that light from a single-frequency fiber laser or super-luminescent source is sent through an optical fiber, experiences some changes of its parameters either in fiber or in one or several fiber Bragg gratings, and then reaches a detector arrangement which measures these changes. Compared with other types of sensors, fiber-optic sensors exhibit a number of advantages. First, they consist of electrically insulating materials, i.e. electric cables are not required, which makes possible the use in high voltage environments. Moreover, they can be safely used in explosive environments, because there is no risk of electrical sparks, even in case of defects. They are immune to electromagnetic interference (EMI), and do not cause such effects themselves. Their materials can be chemically passive, i.e. they do not contaminate their surroundings and not subject to corrosion. They have a very wide operation temperature range, much wider than possible for many electronic devices. Finally, they have multiplexing capabilities, i.e. multiple sensors in a single fiber line can be interrogated with a single optical source.

The operation principle of many fiber-optic sensors based on FBG is that the FBG Bragg wavelength (i.e., the wavelength of maximum reflectivity) depends not only on the Bragg grating period, but also on temperature and/or mechanical strain. For silica fibers, the fractional response of the Bragg wavelength to strain is roughly 20% smaller than the strain itself, since the direct effect of strain is to some extent reduced by a decrease of refractive index. The temperature effect is close to that expected from thermal expansion alone. The effects of strain and temperature can be distinguished with various techniques, e.g. by using reference gratings which are not subject to the strain, or by combining different types of fiber gratings, so that both quantities are obtained at the same time. For pure strain sensing, the resolution can be in the range of a few $\mu\varepsilon$, i.e. relative length changes of a few times 10^{-6}, and the accuracy may not be much lower. For dynamic measurements as for acoustic phenomena, sensitivities better than 1 nε in a 1-Hz bandwidth are achievable. A single fiber may contain many grating sensors in a series to monitor the temperature and strain distribution along the whole fiber. This is called quasi-distributed sensing. There are different techniques to address the single gratings (and thus certain locations along the fiber).

In one technique, called wavelength division multiplexing (WDM) or optical frequency domain reflectometry (OFDR), the gratings have slightly different Bragg wavelengths. A wavelength-tunable laser in the interrogation unit can be tuned to the wavelength belonging to a particular grating, and the wavelength of maximum reflectivity indicates the influences of strain or temperature, for example. Alternatively, a broadband light source as a super-luminescent source may be used together with a wavelength-swept photodetector or a CCD-based spectrometer. In any case, the maximum number of gratings is typically between 10 and 50, limited by the tuning range or bandwidth of the light source and the required wavelength interval per fiber grating. Another technique, called time division multiplexing (TDM), uses identical weakly reflecting gratings, interrogated with short pulses. The reflections from different gratings are then distinguished via their arrival times. Time division multiplexing is actually often combined with WDM in order to multiply the number of different channels to hundreds or even thousands. An optical switch allows to select between different fiber lines, further multiplying the possible number of sensors.

Apart from the approaches described above, there are many alternative techniques. Fiber Bragg gratings may be used in interferometer fiber sensors as reflectors, and the measured phase shift results from fiber spans between them. There are also Bragg grating laser sensors, where a sensor grating forms the end mirror of a fiber laser cavity, containing e.g. an erbium-doped fiber, which receives pump light (e.g. 980nm) via the fiber line. The Bragg wavelength, depending on either temperature or strain, determines the lasing wavelength. This approach, which has many further variations, promises very high resolution due to small linewidth of such fiber lasers, and very high sensitivity. In some cases, pairs of Bragg gratings are used as fiber Fabry-Perot interferometers, which can react with high sensitivity to external influences. The Fabry-Perot interferometer can also be made with other means, e.g. with a variable air gap in the fiber. Long-period fiber gratings are particularly interesting for multi-parameter sensing (temperature and strain), and alternatively for strain sensing with very low sensitivity to temperature changes.

Even after quite a few years of development, fiber-optic sensors have not yet received a great commercial success, since it is difficult to replace other well established technologies, although they can exhibit certain limitations. However, for some application areas, fiber-optic sensors are increasingly recognized as a technology with very interesting possibilities. This holds particularly for harsh environments, such as sensing in high voltage and high power machinery, or in microwave ovens. Bragg grating sensors can also be used to monitor very specific environmental conditions, such as within the wings of airplanes, wind turbines, bridges, large dams, oil wells, oil and gas pipelines, seismic stations, or along electrical power lines. Sometimes, buildings with integrated fiber-optic sensors are called "smart structures", since they allow to monitor inside conditions and gain important information on the strain to which different parts of the structure are subject, such as aging phenomena, vibrations, etc. Smart structures are a main driver for the further development of fiber-optic sensors.

4. Modelling Techniques for Gratings

4.1. Coupled Mode Theory

Coupled-mode theory [19] deals with mode coupling of different spatial distributions or different polarizations, or both [2]. Although the theory described in this section is specifically formulated in terms of coupling of waveguide modes, it can be easily extended to other kind of spatial modes, such as Gaussian. The spatial dependence of the mode fields at a single frequency ω, can be written in a lossless waveguide as:

$$\mathrm{E}(r) = A_\nu \mathrm{E}_\nu(x,y)\exp(\mathrm{i}\beta_\nu \mathrm{z}) \quad (3)$$

$$\mathrm{H}(r) = A_\nu \mathrm{H}_\nu(x,y)\exp(\mathrm{i}\beta_\nu \mathrm{z}) \quad (4)$$

being the two Maxwell's equations:

$$\nabla \times \mathrm{E} = j\omega\mu_0 \mathrm{H} \quad (5)$$

$$\nabla \times \mathrm{H} = -j\omega\varepsilon \mathrm{E} \quad (6)$$

The normal modes with fields given by (3)-(4) are characteristic solutions of Maxwell's equations (5)-(6). These modes are orthogonal and normalized, and form a basis for linear expansion of any optical field at a given frequency ω in the waveguide, as:

$$\mathrm{E}(r) = \sum_\nu A_\nu E_\nu(x,y)\exp(\mathrm{i}\beta_\nu \mathrm{z}), \quad (7)$$

$$\mathrm{H}(r) = \sum_\nu A_\nu H_\nu(x,y)\exp(\mathrm{i}\beta_\nu \mathrm{z}), \quad (8)$$

where E_ν and H_ν are normalized mode fields satisfying the orthonormality relation, and the summation extends over all discrete indices of the guided modes and integrates over all continuous indices of both radiation and evanescent modes. In an ideal waveguide where these modes are defined, normal modes are not coupled. Then, the expansion coefficients A_ν are constants and independent from x, y and z. When a spatial perturbation is present in a waveguide (i.e., a grating), the modes defined by the unperturbed ideal waveguide are no longer exact normal modes of the perturbed waveguide. They could now be coupled by the perturbation as they propagate along the waveguide. As a result, if the fields are still expanded in terms of normal modes of unperturbed waveguide, the expansion coefficients are no longer constants but vary with propagation direction z, as the fields propagate down the waveguide, becoming $A_\nu(z)$. We first consider the coupling between normal modes in a single waveguide that is subject to some perturbation. The spatially dependent perturbation of the waveguide can be represented by a perturbing polarization $\Delta P(r)$ at frequency ω. Then, the Lorenz reciprocity theorem can be applied, which holds for any pair of arbitrary sets of fields, (E_1, H_1) and (E_2, H_2). If we take (E_1, H_1) and (E_2, H_2) to be the normal modes of the perturbed and unperturbed waveguide, respectively, it holds $\Delta P_1 = \Delta P$ and $\Delta P_2 = 0$. Integrating both sides of the resultant equation over the waveguide cross section, we have:

$$\sum_\nu \frac{\mathrm{d}}{\mathrm{dz}} A_\nu(z)\mathrm{e}^{\mathrm{i}(\beta_\nu - \beta_\mu)z} \int_{-\infty}^{\infty} (E_\nu \times H_\mu^* + E_\mu^* \times H_\nu)\cdot z\mathrm{d}x\mathrm{d}y = i\omega \mathrm{e}^{\mathrm{i}\,\beta_\mu z} \int_{-\infty}^{\infty}\int_{-\infty}^{\infty} E_\nu^* \cdot \Delta\mathrm{P}\,\mathrm{d}x\mathrm{d}y \quad (9)$$

and, applying the orthonormality relation, the coupled-mode equation is found:

$$\pm\frac{dA_v}{dz} = i\omega e^{i\beta_\mu z}\int_{-\infty}^{\infty}\int_{-\infty}^{\infty} E_v^* \cdot \Delta P \, dx dy \tag{10}$$

where the plus sign is for $\beta_v > 0$ (forward-propagating v mode), and the minus sign is for $\beta_v < 0$ (backward-propagating v mode). The coupled-mode equation can be used for mode coupling caused by any kind of spatially dependent perturbation on the characteristics of the waveguide, such as a change due to nonlinear optical interactions of the fields at frequency ω in the waveguide, written in linear polarization as:

$$\Delta P = \Delta\varepsilon E = \Delta\varepsilon \sum_v A_v E_v e^{i\beta_v z} \tag{11}$$

In this case we have:

$$\pm\frac{dA_v}{dz} = \sum_\mu i\kappa_{v\mu} A_\mu e^{i(\beta_\mu - \beta_v)z} \tag{12}$$

where

$$\kappa_{v\mu} = \omega\int_{-\infty}^{\infty}\int_{-\infty}^{\infty} \Delta\varepsilon E_v^* \cdot E_\mu \, dx dy \tag{13}$$

is the *coupling coefficient* between modes v and μ. This result can also be extended to anisotropic waveguides by simply considering $\Delta P = \Delta\varepsilon \cdot E$, where $\Delta\varepsilon$ is a tensor. By this way, the coupling coefficient is written as:

$$\kappa_{v\mu} = \omega\int_{-\infty}^{\infty}\int_{-\infty}^{\infty} E_v^* \cdot \Delta\varepsilon \cdot E_\mu \, dx dy \tag{14}$$

In a lossless waveguide, the dielectric tensor is a Hermitian matrix and, then, $\Delta\varepsilon_{ij} = \Delta\varepsilon^*_{ji}$ and $\Delta\kappa_{v\mu} = \Delta\kappa^*_{\mu v}$. In an optical structure with more than one waveguide, Maxwell's equations are directly solved with the boundary conditions defined by the entire structure, in order to find its normal modes. Alternatively, the structure can be divided into separate individual waveguides, expanding the fields in terms of normal modes of individual waveguides, and treating the problem with a coupled-mode approach. The first procedure can yield exact solutions and is sometimes desirable. However, it is not generally possible to obtain the exact solutions for complicated structures, then the second procedure yields approximate solutions, but it can be applied to most structures without difficulty. In addition, it gives an intuitive insight of how optical waves interact in a multiple-waveguide structure. The multiple-waveguide is described by $\varepsilon(x, y)$, while the individual ones are described by $\varepsilon_a(x, y)$, $\varepsilon_b(x, y)$, $\varepsilon_c(x, y)$ and so on. The normal modes are solved for each individual waveguide. The fields in the entire structure can be expanded in terms of these normal modes as before, but with index v representing modes of different waveguides. The coupled-mode equation for the multiple-waveguide structure can be obtained by using the reciprocity theorem as:

$$\pm\frac{dA_\nu}{dz}=\sum_\mu i\kappa_{\nu\mu}A_\mu e^{i(\beta_\mu-\beta_\nu)z} \tag{15}$$

where "+" sign is for $\beta_\nu > 0$ (forward-propagating ν mode) and "-" one for $\beta_\nu < 0$ (backward-propagating ν mode). It is noted that the summation over index μ runs through the modes of every individual waveguide, not just the modes of one single waveguide. In addition, the coupling coefficients $\kappa_{\nu\mu}$ have a complicated form and are better expressed in terms of matrix elements:

$$\kappa_{\nu\mu} = c_{\nu\nu}\,[\, c^{-1}\cdot \underline{\kappa}\,]_{\mu\nu} \tag{16}$$

where the elements of the matrices c = $[c_{\nu\mu}]$ and κ = $[\underline{\kappa}_{\nu\mu}]$ are given by:

$$\begin{aligned} c_{\nu\mu} &= \int_{-\infty}^{\infty}\int_{-\infty}^{\infty}(E_\nu\times H_\mu^* + E_\mu^*\times H_\nu)\cdot z\,dxdy = c_{\mu\nu}^* \\ \underline{\kappa}_{\nu\mu} &= \omega\int_{-\infty}^{\infty}\int_{-\infty}^{\infty} E_\nu^*\cdot\Delta\varepsilon\cdot E_\mu\ dxdy \end{aligned} \tag{17}$$

Therefore, $c_{\nu\nu}$= 1 for ν forward-propagating mode, and $c_{\nu\nu}$ = -1 for backward-propagating mode. The coefficient $c_{\nu\mu}$ represents the *overlap coefficient* of fields (E_μ, H_μ) and (E_ν, H_ν), i.e. the mode fields of different individual waveguides. Note that $c_{\nu\mu} \neq 0$ in general, because modes of different waveguides are not necessarily orthogonal to each other. Because the fields (E_μ, H_μ) and (E_ν, H_ν) are normalized, we have $c_{\nu\nu}= \pm 1$ and $|c_{\nu\mu}| \leq 1$. Note also that, in general, $\underline{\kappa}_{\nu\mu} \neq \underline{\kappa}^*_{\nu\mu}$ and $\kappa_{\nu\mu} \neq \kappa^*_{\nu\mu}$, where ν and μ refer to modes of different waveguides. Therefore, $\Delta\kappa_{\nu\mu} = \Delta\kappa^*_{\mu\nu}$ is not always valid for coupling between waveguides. Indeed, by using the reciprocity theorem it can be shown that:

$$\underline{\kappa}_{\nu\mu} - \underline{\kappa}^*_{\nu\mu} = \frac{c_{\nu\mu}+c^*_{\nu\mu}}{2}(\beta_\nu-\beta_\mu) = c_{\nu\mu}(\beta_\nu-\beta_\mu) \tag{18}$$

where a direct relationship between coupling coefficients and propagation constants is evident. This is an important relation that has four fundamental implications; the first one is that the coupling between two modes is not symmetric, $\kappa_{\nu\mu} \neq \kappa^*_{\nu\mu}$ unless $\beta_\nu = \beta_\mu$ or $c_{\nu\mu} = c^*_{\nu\mu}$ = 0, since the normal modes of different waveguides are not necessary orthogonal to each other. The second one is that the coupling of modes of the same order between two identical waveguides is always symmetric, resulting $\underline{\kappa}_{\nu\mu} = \underline{\kappa}^*_{\nu\mu}$ and $\kappa_{\nu\mu} = \kappa^*_{\nu\mu}$. The third one is that the relation (18) is valid also for modes in a single waveguide, where the normal modes are always orthogonal to each other, and therefore $c_{\nu\mu} = c^*_{\nu\mu} = 0$ if $\nu \neq \mu$, but $\underline{\kappa}_{\nu\mu} = \kappa_{\mu\nu}$. The last one is that it is not possible to change the coupling between two modes without simultaneously changing their overlap coefficients or propagation constants.

In most applications, coupling between two modes is of interest. This includes coupling between two modes in the same waveguide, such as in a periodic waveguide, or coupling between two parallel waveguides, as in a directional coupler. For coupling between two modes, the coupled-mode equation can be written in a simple form that can be analytically solved. It is clear that both coupling among modes in the same waveguide and among multiple waveguides can be described by coupled-mode equations, with the only difference in coupling coefficients. This is convenient because general solutions of the coupled-mode equations can be applied to both cases. For a particular problem, we only have to calculate the specific coupling coefficients either in a single waveguide or between two separate waveguides. Then, the field expansion is referred to only two modes of amplitudes A and B, and therefore coupled-mode equations simply become:

$$\pm\frac{\mathrm{d}A}{\mathrm{d}z} = i\kappa_{aa}A + i\kappa_{ab}B\mathrm{e}^{\mathrm{i}(\beta_b - \beta_a)z}$$
$$\pm\frac{\mathrm{d}B}{\mathrm{d}z} = i\kappa_{bb}B + i\kappa_{ba}A\mathrm{e}^{\mathrm{i}(\beta_a - \beta_b)z} \tag{19}$$

For coupling in a single waveguide, the coupling coefficients are simply given by (13) in case of isotropic waveguide or by (14) for anisotropic waveguide. If the waveguide is lossless, we have also $\kappa_{ab} = \kappa^*_{ba}$. For coupling between two waveguides, the coupling coefficients are given by (16), which can be explicitly expressed as:

$$\kappa_{aa} = \frac{\underline{K}_{aa} - c_{ab}\underline{K}_{ba} / c_{bb}}{1 - c_{ab}c_{ba} / c_{aa}c_{bb}}, \kappa_{ab} = \frac{\underline{K}_{ab} - c_{ab}\underline{K}_{bb} / c_{bb}}{1 - c_{ab}c_{ba} / c_{aa}c_{bb}}$$
$$\kappa_{ba} = \frac{\underline{K}_{ba} - c_{ba}\underline{K}_{aa} / c_{aa}}{1 - c_{ab}c_{ba} / c_{aa}c_{bb}}, \kappa_{bb} = \frac{\underline{K}_{bb} - c_{ba}\underline{K}_{ab} / c_{bb}}{1 - c_{ab}c_{ba} / c_{aa}c_{bb}} \tag{20}$$

As discussed earlier, in general it holds $\kappa_{ab} \neq \kappa^*_{ba}$ for coupling between two waveguides. Moreover, a self-coupling term is present in each of the coupled equations (19), because normal modes "see" an index profile in the perturbed waveguide different from that of the original waveguide, where the modes are defined. They can be removed from these equations by expressing the normal-mode expansion coefficient as follows:

$$A(z) = \underline{A}(z)\exp\left[\pm\int_{-\infty}^{\infty}(\kappa_{aa}(z)\mathrm{d}z\right],$$
$$B(z) = \underline{B}(z)\exp\left[\pm\int_{-\infty}^{\infty}(\kappa_{bb}(z)\mathrm{d}z\right]. \tag{21}$$

where "+" or "-" sign is chosen for forward-propagating or backward-propagating mode, respectively. We can also note that if Δε is a function of z, the coupling coefficients are also functions of z and the coupled-mode equation can be only numerically solved. However, for waveguide equations of practical interest for two-mode coupling, both Δε and the coupling coefficients are constants or periodic with z. In either cases, the coupled-mode equations can be reduced to the following general form:

$$\pm\frac{\mathrm{d}\underline{A}}{\mathrm{d}z} = i\kappa_{ab}\underline{B}\mathrm{e}^{\mathrm{i}\,2\delta z}$$
$$\pm\frac{\mathrm{d}\underline{B}}{\mathrm{d}z} = i\kappa_{ba}\underline{A}\mathrm{e}^{-\mathrm{i}\,2\delta z} \qquad (22)$$

in terms of $\underline{A}$, $\underline{B}$, κ_{ab} and κ_{ba} being constants independent from z. The parameter 2δ is the *phase mismatch* between the two modes being coupled. Phase-matched coupling with $\delta = 0$ is always symmetric with $\kappa_{ab} = \kappa^{*}_{ba}$, irrespective of whether these two modes belong to the same waveguide or different waveguides. This general form of coupled-mode equations applies to both cases of constant and periodic perturbations, but the detailed parameters in these two equations will vary. In case of constant perturbation, we find:

$$A(z) = \underline{A}(z)\mathrm{e}^{\pm\mathrm{i}\kappa_{aa}z}, \quad B(z) = \underline{B}(z)\mathrm{e}^{\pm\mathrm{i}\kappa_{bb}z} \qquad (23)$$

and

$$2\delta = \left(\beta_b \pm \kappa_{bb}\right) - \left(\beta_a \pm \kappa_{aa}\right) \qquad (24)$$

The choice of sign ± here is consistent with that in equations (22). The physical meaning of self-coupling coefficient is a change in the propagation constant of relevant normal mode. While the propagation constants of the normal modes in the original waveguide are $\beta_a = \beta_b$, their values are changed because of the perturbation on the waveguide. These modes now propagate with modified propagation constants, $\beta_a \pm \kappa_{aa}$ and $\beta_b \pm \kappa_{bb}$, respectively, which take into account the effect of the perturbation. In addition, they couple to each other through κ_{ba} and κ_{ab}. In the case of periodic perturbation, with period Λ and wave-number $K = 2\pi/\Lambda$, $\Delta\varepsilon$ and coupling coefficients κ_{aa}, κ_{ab}, κ_{ba} and κ_{bb} are periodic functions of z, with period Λ and wave-number $K = 2\pi/\Lambda$. For this reason, the coefficients κ_{ba} and κ_{ab} can be expanded in a Fourier series with $\kappa_{ba}(q)$ and $\kappa_{ab}(q)$ constant coefficients and phase factor qK, where q is an integer. Because $\kappa_{aa}(z)$ and $\kappa_{bb}(z)$ are periodic in z, we find that:

$$\left|\int_0^z (\kappa_{aa}(z)\mathrm{d}z\right| \ll Kz$$
$$\left|\int_0^z (\kappa_{bb}(z)\mathrm{d}z\right| \ll Kz \qquad (25)$$

Therefore, the contribution to the phase-mismatch parameter by $\kappa_{aa}(z)$ and $\kappa_{bb}(z)$ is negligible compared to the contribution by qK. As a result, the coupled-mode equations can be also given in the form (22), but including constants $\kappa_{ab} = \kappa_{ab}(q)$ and $\kappa_{ba} = \kappa_{ba}(q)$ independent from z, and:

$$2\delta = \Delta\beta + qK = \beta_b - \beta_a + qK \qquad (26)$$

where q is an integer minimizing δ. With these general considerations, Eq. (22) represents the most general coupled equations for two-mode coupling in waveguide structures of practical interest. They can be analytically solved and their solutions apply to several different two-mode coupling schemes. First, we consider the co-directional coupling of two modes propagating in the same direction, forward direction in z, over a length l, as in Figure 8. In this case $\beta_a > 0$ and $\beta_b > 0$, and the coupled equations are written:

$$\frac{\mathrm{d}\underline{A}}{\mathrm{d}z} = i\kappa_{ab}\underline{B}\mathrm{e}^{\mathrm{i}\,2\delta z}$$
$$\frac{\mathrm{d}\underline{B}}{\mathrm{d}z} = i\kappa_{ba}\underline{A}\mathrm{e}^{-\mathrm{i}\,2\delta z} \tag{27}$$

These equations for co-directional coupling are generally solved as an initial-value problem with initial values $\underline{A}(z_0)$ and $\underline{B}(z_0)$ given at $z = z_0$, to find the values of $\underline{A}(z)$ and $\underline{B}(z)$ at any other location z.

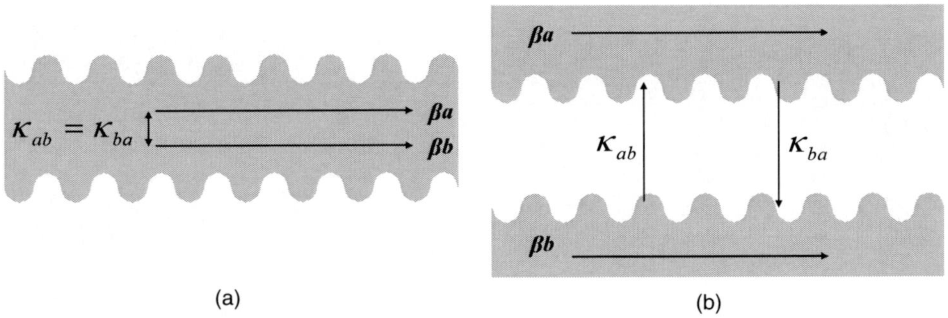

Figure 8. Co-directional coupling between two modes of propagation constants β_a and β_b: (a) in same waveguide and (b) in two parallel waveguides.

The general solution can be expressed in matrix form:

$$\begin{bmatrix}\underline{A}(z)\\ \underline{B}(z)\end{bmatrix} = \mathrm{F}(z;z_0)\begin{bmatrix}\underline{A}(z_0)\\ \underline{B}(z_0)\end{bmatrix} \tag{28}$$

where the *forward-coupling matrix* $\mathrm{F}(z;\, z_0)$ relates the field amplitudes at location z_0 to those at generic location z. In the simple case of power launched only into mode a at $z = 0$, with initial values $\underline{A}(0) \neq 0$ and $\underline{B}(0) = 0$, the *coupling efficiency* for a length l is:

$$\eta = \frac{P_b(l)}{P_a(0)} = \frac{\left|\kappa_{ba}\right|^2}{\beta_c^2}\sin^2 \beta_c z \tag{29}$$

Thus, power is periodically exchanged between two modes with a *coupling length:*

$$l_c = \frac{\pi}{2\beta_c} \tag{30}$$

where maximum power transfer occurs. One can see that the complete power transfer can occur only under phase-matching condition, $\delta = 0$.

Contra-directional coupling of two modes propagating in opposite directions over a length l, is shown in Figure 9, with forward-propagating mode a and backward-propagating mode b.

In this case, $\beta_a > 0$ and $\beta_b < 0$. Thus, the coupled equations are:

$$\begin{aligned} \frac{\mathrm{d}\underline{A}}{\mathrm{d}z} &= i\kappa_{ab}\underline{B}\mathrm{e}^{\mathrm{i}\,2\delta z} \\ -\frac{\mathrm{d}\underline{B}}{\mathrm{d}z} &= i\kappa_{ba}\underline{A}\mathrm{e}^{-\mathrm{i}\,2\delta z} \end{aligned} \tag{31}$$

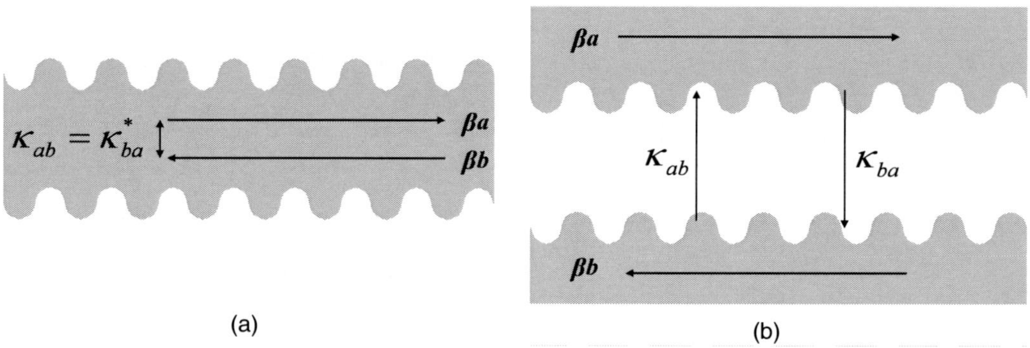

Figure 9. Contra-directional coupling between two modes of propagation constants β_a and β_b (a) in the same waveguide and (b) in two parallel waveguides.

These contra-directional coupling equations are generally solved as a boundary-value problem with boundary values for $\underline{A}(0)$ at one end and $\underline{B}(l)$ at the other end. The general solution can be expressed in matrix form:

$$\begin{bmatrix} \underline{A}(z) \\ \underline{B}(z) \end{bmatrix} = \mathrm{R}(z;0,l) \begin{bmatrix} \underline{A}(0) \\ \underline{B}(l) \end{bmatrix} \tag{32}$$

where the *reverse-coupling matrix* R(z; 0, l) relates the field amplitudes, $\underline{A}(0)$ at $z = 0$ and $\underline{B}(l)$ at $z = l$, to those at location z. When the boundary values are $\underline{A}(0) \neq 0$ at one end and $\underline{B}(l) = 0$, because mode b is propagating backward with no input at $z = l$ but with output at $z = 0$, the coupling efficiency for a length l is:

$$\eta = \frac{P_b(0)}{P_a(0)} = \frac{\kappa_{ba}^{*}}{\kappa_{ab}} \frac{\sinh^2 \alpha_c l}{\cosh^2 \alpha_c l - \delta^2 / \kappa_{ba}\kappa_{ab}} \tag{33}$$

where $\alpha_c = \sqrt{\kappa_{ab}\kappa_{ba} + \delta^2}$. Complete power transfer occurs as $l \to \infty$ if $\delta^2 < \kappa_{ab}\ \kappa_{ba}$. In the case when $\underline{A}(0) \neq 0$ and $\underline{B}(l) = 0$, contra-directional coupling can be seen as a reflection of the field amplitude $\underline{A}(0)$ at $z = 0$, with a reflection coefficient:

$$r = |r|\, e^{i\varphi_{DBR}} = \frac{\underline{B}(0)}{\underline{A}(0)} = \frac{\kappa_{ba}\sinh\alpha_c l}{\alpha_c\cosh\alpha_c l + i\delta\sinh\alpha_c l} \tag{34}$$

and reflectivity $R = |r|^2 = \eta$. The phase shift is:

$$\varphi_{DBR} = \varphi_B - \tan^{-1}\left(\frac{\delta}{\alpha_c}\tanh\alpha_c l\right) \tag{35}$$

The coupling between two modes a and b is symmetric with $\kappa_{ab} = \kappa^*_{ba}$ if these two modes are orthonormal to each other or phase matched. Efficient power transfer occurs for $\delta = 0$, that is achieved when $\beta_b = \beta_a$ for the simple case $2\delta = \beta_b - \beta_a$. Then, the two modes have the same phase velocity and are synchronized (synchronous coupling). In case δ includes a contribution from an additional structure as a grating, phase matching of the two coupled modes can be accomplished by matching the difference $\Delta\beta = 2\delta = \beta_b - \beta_a$ with a grating phase factor to make $\delta = 0$. When considering phase matching between two modes, it is always important to include all contributions to the phase-mismatch parameter δ. Phase-matched coupling is always symmetric, meaning $\kappa_{ab} = \kappa^*_{ba}$ whenever $\delta = 0$. This statement is true even when $c_{ab} = 0$ and $\beta_b \neq \beta_a$. However, symmetric coupling does not necessary imply a phase-matched condition. Therefore it is also possible to have $\kappa_{ab} = \kappa^*_{ba}$ while $\delta \neq 0$.

4.2 FDTD METHOD

In 1966 *K.S. Yee* proposed a technique to solve Maxwell's curl equations using finite-difference time-domain (FDTD) technique [20]-[30]. FDTD numerically searches the solutions of Maxwell's equations directly in the time domain, without any restriction on geometry and physical properties of the propagation media, as well as on wave behaviour, i.e. without imposing exp($j\omega t$) time dependence as other methods operating in the frequency domain. Indeed, loss, anisotropic, time-varying and non-linear materials can be studied, within geometrical structures also very complicated. Yee's method has been used to solve numerous scattering and radiation problems on microwave circuits, dielectrics, and electromagnetic absorption in biological tissue at microwave frequencies. Examples of applications in computational electromagnetism are aircraft engineering (lightning strike, radar cross section, antenna placement, etc.), antennas, microwave ovens, electromagnetic compatibility (EMC) and so on. Initially, there was little interest in FDTD method, probably due to a lack of sufficient computing resources. However, with the advent of low cost, powerful computers and advances of the method itself, FDTD technique has become a very popular approach for a number of electromagnetism problems and simulation tools. In FDTD method, the structure is divided into cells to evaluate the discrete values of the fields at discrete time steps. FDTD solves the initial value problem and, therefore, a source must be appropriately chosen to propagate the field cell by cell, at each time step. Boundary conditions are very important in electromagnetism. FDTD supports a range of boundary conditions, such as perfectly matched layer (PML), periodic, and Bloch. Moreover, FDTD can support a number of different types of sources such as point dipoles, beams, plane waves, total-field scattered-field (TFSF) source, and guided-mode source for integrated optical components.

In this section the main rules of FDTD method are presented. FDTD method, also referred to as the Yee scheme [20] and further developed by Taflove in the 1970s [21], is mainly characterized on second-order central differences on staggered Cartesian grids. In Yee's algorithm, Maxwell's curl equations for an isotropic medium are:

$$\begin{aligned} \nabla \times \vec{E} &= -[\mu]\frac{d\vec{H}}{dt} - [\sigma^*]\vec{H} \\ \nabla \times \vec{H} &= [\sigma]\vec{E} + [\varepsilon]\frac{d\vec{E}}{dt} \end{aligned} \tag{36}$$

As it is well known, they can be written as six scalar equations in Cartesian coordinates (sixth-order hyperbolic system). The continuous form of these equations can be discretized in both space and time domains by the so-called Yee-cell technique, using a discrete mesh. Figure 10 illustrates a unit cell of a 3D mesh over a photonic structure cross section. Yee defines the grid coordinates (i, j, k) as $(i, j, k) = (i\Delta x, j\Delta y, k\Delta z)$, where Δx, Δy and Δz are the actual grid sizes.

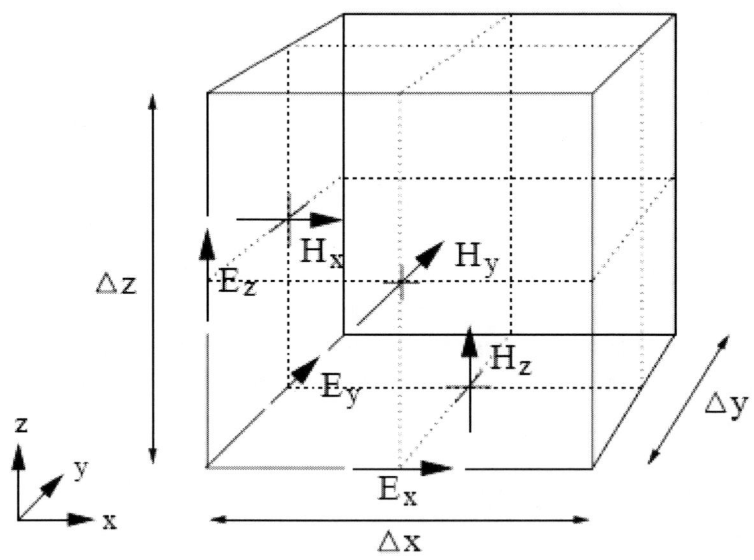

Figure 10. Scheme of Yee's 3D grid.

Any function of space and time is written as:

$$F^n(i,j,k) = F(i\Delta x, j\Delta y, k\Delta z, n\Delta t) = F(i\delta, j\delta, k\delta, n\Delta t) \tag{37}$$

where Δt is the time increment, n is the time index and $\delta = \Delta x = \Delta y = \Delta z$. The spatial and temporal derivatives of F are written using central finite difference approximations, as:

$$\frac{\partial F^n(i,j,k)}{\partial x}=\frac{F^n(i+1/2,j,k)-F^n(i-1/2,j,k)}{\delta}$$
$$\frac{\partial F^n(i,j,k)}{\partial t}=\frac{F^{n+1/2}(i,j,k)-F^{n-1/2}(i,j,k)}{\Delta t} \tag{38}$$

These two last equations are applied to the six scalar equations, resulting in six coupled explicit finite difference equations, in the form:

$$H_x\big|_{i,j+1/2,k+1/2}^{n+1/2}=\frac{M_x\big|_{i+1/2,j+1/2,k}^{-*}}{M_x\big|_{i+1/2,j+1/2,k}^{+*}}H_x\big|_{i,j+1/2,k+1/2}^{n-1/2}+ \\ +\frac{\Delta t}{\mu_{xx}\big|_{i,j+1/2,k+1/2}M_x\big|_{i,j+1/2,k+1/2}^{+*}}\left[\frac{E_y\big|_{i,j+1/2,k+1}^{n}-E_y\big|_{i,j+1/2,k}^{n}}{\Delta z}-\frac{E_z\big|_{i,j+1,k+1/2}^{n}-E_z\big|_{i,j,k+1/2}^{n}}{\Delta y}\right] \tag{39}$$

$$H_y\big|_{i+1/2,j,k+1/2}^{n+1/2}=\frac{M_y\big|_{i+1/2,j,k+1/2}^{-*}}{M_y\big|_{i+1/2,j,k+1/2}^{+*}}H_y\big|_{i+1/2,j,k+1/2}^{n-1/2}+ \\ +\frac{\Delta t}{\mu_{yy}\big|_{i+1/2,j,k+1/2}M_y\big|_{i+1/2,j,k+1/2}^{+*}}\left[\frac{E_z\big|_{i+1,j,k+1/2}^{n}-E_z\big|_{i,j,k+1/2}^{n}}{\Delta x}-\frac{E_x\big|_{i+1/2,j,k+1}^{n}-E_x\big|_{i+1/2,j,k}^{n}}{\Delta z}\right] \tag{40}$$

$$H_z\big|_{i+1/2,j+1/2,k}^{n+1/2}=\frac{M_z\big|_{i+1/2,j+1/2,k}^{-*}}{M_z\big|_{i+1/2,j+1/2,k}^{+*}}H_z\big|_{i+1/2,j+1/2,k}^{n-1/2}+ \\ +\frac{\Delta t}{\mu_{zz}\big|_{i+1/2,j+1/2,k}M_z\big|_{i+1/2,j+1/2,k}^{+*}}\left[\frac{E_x\big|_{i+1/2,j+1,k}^{n}-E_x\big|_{i+1/2,j,k}^{n}}{\Delta y}-\frac{E_y\big|_{i+1,j+1/2,k}^{n}-E_y\big|_{i,j+1/2,k}^{n}}{\Delta x}\right] \tag{41}$$

$$E_x\big|_{i+1/2,j,k}^{n+1}=\frac{M_x\big|_{i+1/2,j,k}^{-}}{M_x\big|_{i+1/2,j,k}^{+}}E_x\big|_{i+1/2,j,k}^{n} \\ +\frac{\Delta t}{\varepsilon_{xx}\big|_{i+1/2,j,k}M_x\big|_{i+1/2,j,k}^{+}}\left[\frac{H_z\big|_{i+1/2,j+1/2,k}^{n+1/2}-H_z\big|_{i+1/2,j-1/2,k}^{n+1/2}}{\Delta y}-\frac{H_y\big|_{i+1/2,j,k+1/2}^{n+1/2}-H_y\big|_{i+1/2,j,k-1/2}^{n+1/2}}{\Delta z}\right] \tag{42}$$

$$E_y|^{n+1}_{i,j+1/2,k} = \frac{M_y|^-_{i,j+1/2,k}}{M_y|^+_{i,j+1/2,k}} E_y|^n_{i,j+1/2,k} + \frac{\Delta t}{\varepsilon_{yy}|_{i,j+1/2,k} M_y|^+_{i,j+1/2,k}} \left[\frac{H_x|^{n+1/2}_{i,j+1/2,k+1/2} - H_x|^{n+1/2}_{i,j+1/2,k-1/2}}{\Delta z} - \frac{H_z|^{n+1/2}_{i+1/2,j+1/2,k} - H_z|^{n+1/2}_{i-1/2,j+1/2,k}}{\Delta x} \right] \tag{43}$$

$$E_z|^{n+1}_{i,j,k+1/2} = \frac{M_z|^-_{i,j,k+1/2}}{M_z|^+_{i,j,k+1/2}} E_z|^n_{i,j,k+1/2} + \frac{\Delta t}{\varepsilon_{zz}|_{i,j,k+1/2} M_z|^+_{i,j,k+1/2}} \left[\frac{H_y|^{n+1/2}_{i+1/2,j,k+1/2} - H_y|^{n+1/2}_{i-1/2,j,k+1/2}}{\Delta x} - \frac{H_x|^{n+1/2}_{i,j+1/2,k+1/2} - H_x|^{n+1/2}_{i,j-1/2,k+1/2}}{\Delta y} \right] \tag{44}$$

where $M_l|^{\pm *}_{i,j,k}$ and $M_l|^{\pm}_{i,j,k}$ are given by:

$$M_l|^{\pm}_{i,j,k} = 1 \pm \frac{\sigma_{ll}|_{i,j,k} \Delta t}{2\varepsilon_{ll}|_{i,j,k}} \quad \text{con} \quad l = x, y, z \tag{45}$$

$$M_l|^{\pm *}_{i,j,k} = 1 \pm \frac{\sigma_{ll}|_{i,j,k} \Delta t}{2\mu_{ll}|_{i,j,k}} \quad \text{con} \quad l = x, y, z \tag{46}$$

Equations (39)-(46) constitute the FDTD method [20]-[30], where the field distributions are evaluated in the computational region at time step t^{n+1}, depending from those at time step t^n. Because FDTD is an explicit scheme, there is a limit on the time step Δt to ensure stability, given by:

$$\Delta t < \frac{1}{c\sqrt{\frac{1}{(\Delta x)^2} + \frac{1}{(\Delta y)^2} + \frac{1}{(\Delta z)^2}}} \tag{47}$$

where c is the wave propagation speed. To yield accurate results, the grid spacing δ in the finite difference simulation must be less than the wavelength, usually less than $\lambda/10$. The stability condition relating the spatial and temporal step size is:

$$\upsilon_{\max} \Delta t = \frac{1}{\sqrt{\frac{1}{(\Delta x)^2} + \frac{1}{(\Delta y)^2} + \frac{1}{(\Delta z)^2}}} \tag{48}$$

where υ_{max} is the maximum wave velocity. When the step size δ is the same in all directions, the stability condition becomes:

$$\frac{\upsilon_{max}\Delta t}{\delta}=\frac{1}{\sqrt{N}} \tag{49}$$

where N is the number of spatial dimensions in the problem. A typical finite difference grid contains regions of different dielectric properties representing cell components. By specifying the dielectric properties of each region, $\varepsilon(i, j, k)$ and $\sigma(i, j, k)$, the fields within the enclosed region can be found. To model the scattering produced by cells at optical wavelengths, the grid size δ must be small enough. If a wavelength of 1 μm is used, the maximum step size will be $\delta = \lambda/10 = 100$ nm. A 2D model of a 20 μm diameter cell would require a grid with 200×200 points. The time step, based on the stability condition (49), would be 0.24 fs. Therefore, a large amount of memory and computation time is required. Other problems of stability can be given by boundary conditions, grid non-uniformity, loss, non-linear and dispersive media, if the number of iterations is too much increased. Another problem of this method consists of the numerical dispersion of waveguide modes propagating in the computational region, because the phase-velocity of FDTD numerical modes can vary by changing mode wavelength, propagation direction in the grid and grid resolution. This problem cannot be completely eliminated, being intrinsic to the method that operate in discrete domains, so giving spurious solutions which have not any physical meaning.

In the study of electromagnetic waves in optical waveguides, the analysis domain should be usually infinite, but in practical simulations it must be limited. For this reason, *Absorbing Boundary Conditions* (ABC) are applied at truncation planes of FDTD grid. By this way, an infinite extension of the grid can be simulated, indeed ABC assumption yields the absorption of the waves propagating out of the grid, without relevant back-reflections in the region of interest. Two important ABC models have been proposed by Mur and Berenger, respectively. The latter is more accurate and efficient because uses an absorptive layer all around the FDTD grid, absorbing waves of every polarization, frequency and incidence angle. This absorptive layer is called *Perfect Matched Layer* (PML) and introduces appropriately chosen loss parameters on both orthogonal field components.

The main advantages of FDTD are related to its easy implementation and efficiency for homogeneous domains. However, poor representation of curved boundaries (second-order accuracy is lost) and small details also occurs. FDTD is a powerful method to solve Maxwell's equations and find the field distributions in complex structures, but without any physical insight of optical phenomena and device behaviour like, for example, the coupled mode theory. For this reason, in the design of photonic devices is usually used only to simulate already known photonic structures, but not for design purposes, this task requiring other models more interpretative of physical processes.

4.3. Finite Element Method

The finite element method (FEM) is a procedure introduced to solve partial differential equations (PDE). It is one of the most famous techniques for solving eigenvalues problems, with important applications in scientific fields as mechanics, thermodynamics and electromagnetism. FEM is a completely general method [31], solving Maxwell's equations in a propagation medium of arbitrary characteristics. It is based on a discretization of the computational domain in subdomains of different geometries, usually triangles for 2D applications and tetrahedrons for 3D applications, inside which the field values are interpolated. The subdivision in triangular subdomains of variable shape and dimensions allows a detailed analysis of all the discontinuities of the structures being considered, with the possibility to thicken the computational grid where more accuracy is required.

Similarly to the Moment Method (MM), FEM approximates the field description by well analytically defined functions, called shape functions, with a not-known amplitude to be calculated. Both type and order of these functions are opportunely chosen. As it was seen in the previous section, the finite difference method (FDM) is an alternative approach for solving PDEs. Then, the differences between FEM and FDM can be so summarized:

- FDM is an approximation of the differential equation, while FEM is an approximation of its solution;
- The most attractive feature of FEM is its ability to handle complex geometries (and boundaries) with relative ease. While FDM in its basic form is restricted to handle rectangular shapes and simple alterations thereof, the handling of geometries in FEM is theoretically straightforward.
- There are reasons to consider the mathematical foundation of the finite element approximation more sound, for instance, because the quality of the approximation between grid points is poor in FDM.

FEM is often used to solve 2D problems with a planar domain, but it is valid also in three dimensions. When used in 2D domain, FEM can operate as a propagation method and an eigenvalue one, simply by choosing the longitudinal propagation direction as one of the two dimensions. In order to accommodate a wide domain of applications even for periodic structures, good numerical stability and moderate complexity in the numerical implementation, a formalism based on the use of FEM in the inhomogeneous region of the grating (where the Rayleigh expansion is not always valid) has been defined [32-33]. There are also many applications in which the grating has a simple structure and a small perturbation of the refractive index along the propagation direction, and, therefore, CMT is still valid. When the optical waveguide has a complex structure in the plane transversal to the propagation direction, FEM can be used as an eigenvalue method to the transversal plane, in order to evaluate both field distributions and propagation constants. By this mixed CMT-FEM approach, diffraction gratings can be designed and characterized. In this section we describe the FEM formulation used to describe a two-dimensional optical structure with a one-dimensional grating. This formulation can also be used to analyze every 2D photonic structure and not only periodic structures.

As an example, CMT-FEM analysis can be used to describe the spectral properties of a DFB quantum cascade laser (QCL). In this case, FEM is used to find the field distributions in the laser waveguide cross-section and the propagation constants. Then, these results are used through CMT relations to evaluate both coupling coefficients and reflection spectrum of DFB grating. Some interesting simulations on the effects of variations of structure geometrical parameters on device spectral properties are also presented. The structure being considered is a ridge-waveguide as in Figure 11, where DFB QCL cross section is represented. The effective index of such cross-section is evaluated by solving the following wave-equation using the 2D FEM analysis [34]:

$$\frac{\partial\Phi(x,y)}{\partial x}+\frac{\partial\Phi(x,y)}{\partial y}+k_0^2 n^2(x,y)\Phi(x,y)=\beta^2\Phi(x,y)=k_0^2 n_{eff}^2(x,y)\Phi(x,y) \tag{50}$$

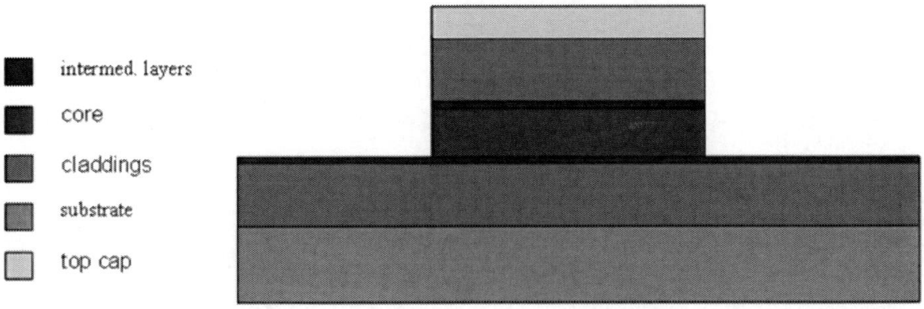

Figure 11. Cross section of ridge waveguide with InP epitaxial layers.

A non-uniform mesh has been set automatically by the FEM tool to have more accuracy in the domain regions of higher discontinuity. The following Figure 12 and 13 show the mesh and the calculated field distribution in the considered cross-section, respectively.

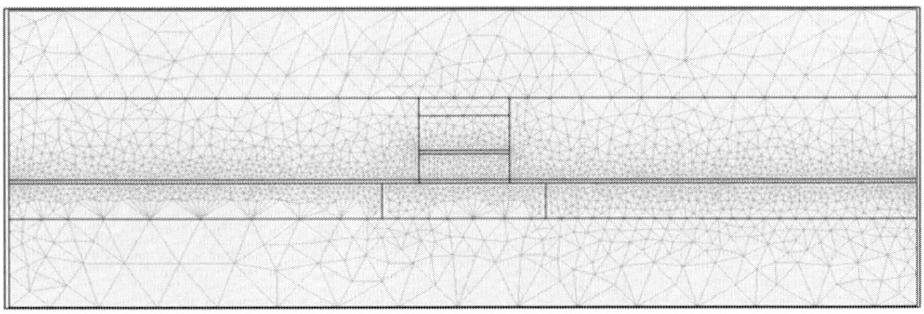

Figure 12. Mesh of QCL waveguide cross section, obtained by FEM.

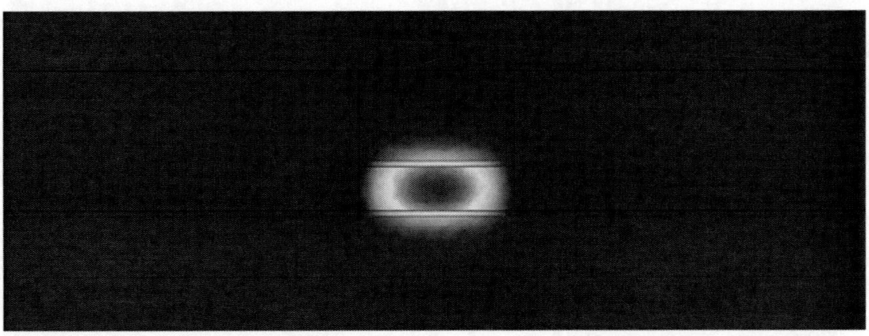

Figure 13. Plot of E_x for quasi-TM fundamental mode, obtained by FEM.

This procedure has been repeated for many different values of ridge width, in order to evaluate the optimal waveguide in terms of larger optical confinement and lower waveguide losses. To this aim, the structure has been designed and analyzed by FEM for each value of the ridge width, and results are reported in Figures 14 and 15.

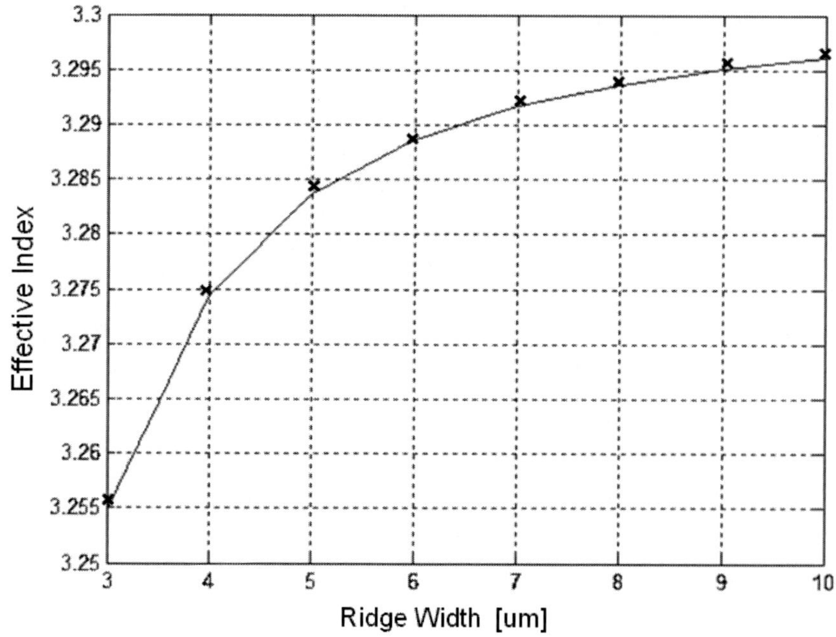

Figure 14. Effective index at wavelength 3.34 μm, for different ridge widths, calculated by FEM.

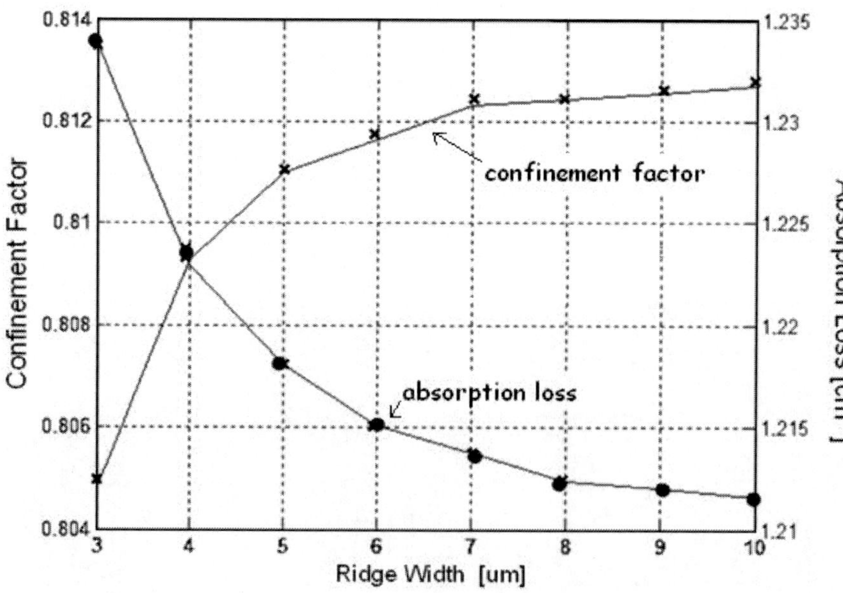

Figure 15. Confinement factor and modal losses at 3.34 μm, for different ridge widths, by FEM.

At the QCL laser edge, a partial reflection grating is assumed. Its structure along the propagation direction can be seen as a series of etched and unetched waveguide sections, as sketched in Figure 16. According to CMT, the coupling coefficient, that describes the power exchange between counter-propagating laser waves and gives a measure of grating strength, can be expressed by the following integral on the entire waveguide cross section:

$$k = \frac{k_0}{2n_{eff}} \int_{cs} [\chi_L(x,y,z,\omega_0) - \chi_L(x,y,z,\omega_0)] \Phi^2(x,y) \partial x \partial y \qquad (51)$$

where $\Phi(x,y)$ is the modal field distribution, n_{eff} the mode effective index, χ_S and χ_L the effective susceptibilities of etched and unetched waveguides, respectively.

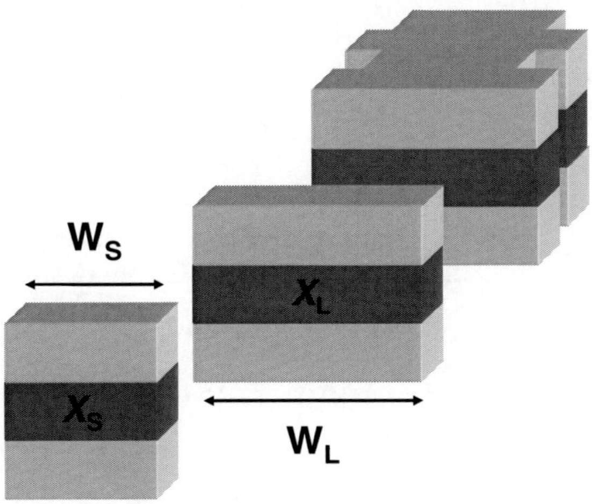

Figure 16. Grating longitudinal decomposition scheme in regions of width W_S and W_L, and susceptibility X_S and X_L, respectively.

The coupling coefficient previously defined, for our particular structure, can be expressed as:

$$k = \frac{2\varepsilon_0 \omega_B [n_{eff\_med}^2 - n_{air}^2] \Gamma_{ov}}{\pi} \qquad (52)$$

where $\omega_B = 2\pi c / \lambda_B$ is the Bragg frequency, Γ_{ov} is the overlap between field distributions in the grating region, $n_{air} = 1$ is the refractive index of air in the etched regions and n_{eff_med} is an average effective index, defined as:

$$n_{eff\_med} = \frac{(n_{effS} + n_{effL})}{\pi} \qquad (53)$$

where $n_{eff\ S}$ ed $n_{eff\ L}$ are the effective indices of etched and unetched regions, respectively. Thus, the grating period is given by $\Lambda = \lambda_B / 2n_{eff,med}$. FEM allows us to evaluate the grating overlap integral of the field distributions and, then, the coupling coefficient for different ridge widths. Results are given in the following Figures 17 and 18.

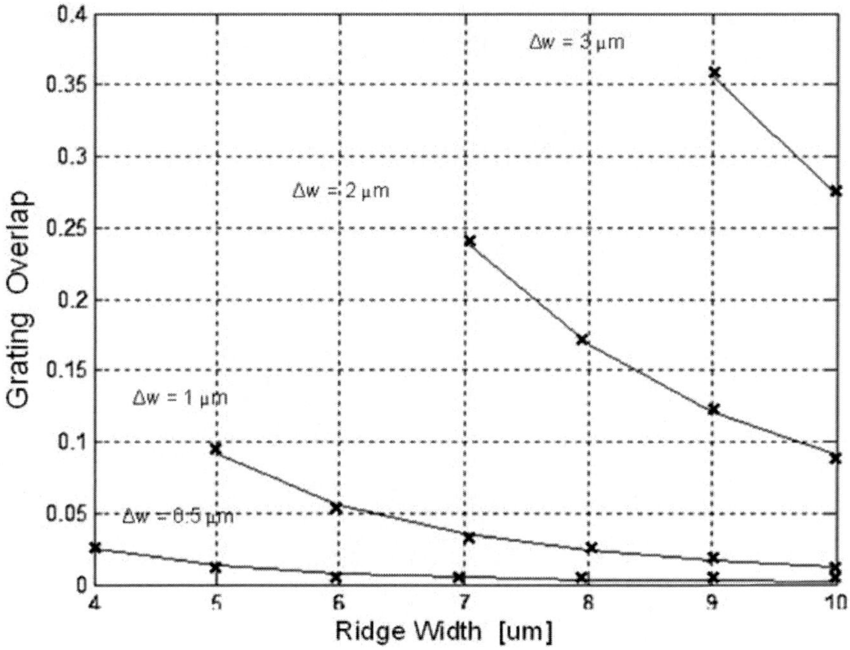

Figure 17. Grating overlap factor for various ridge widths, and $\Delta W = 0.5$, 1, 2 e 3 μm.

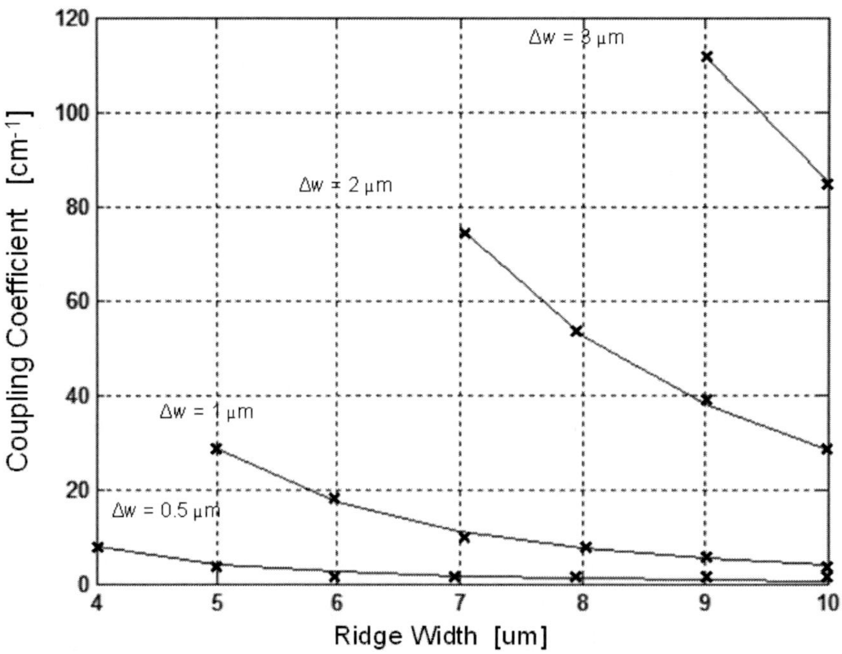

Figure 18. Coupling coefficient for various ridge widths, and $\Delta W = 0.5$, 1, 2 e 3 μm.

As it is clear in the plots, the coupling coefficient is larger with increasing ΔW and W_L, because in these cases the grating overlap increases as well as the effective index contrast between ridges of width W_L and W_S. Another parameter affecting the coupling coefficient is the grating duty cycle, defined as:

$$\gamma = \frac{\Lambda_L}{\Lambda_{TOT}} \tag{54}$$

where $\Lambda_{TOT} = \Lambda_S + \Lambda_L = \lambda_B / 2\, n_{eff\_med}$ and Λ_S and Λ_L are the longitudinal widths of etched and unetched grating regions, respectively.

The duty cycle effect can be taken into account modifying the expressions of average effective index and coupling coefficient, as follows:

$$n_{eff\_med} = [n_{effS}(1-\gamma) + n_{effL}(\gamma)] \tag{55}$$

$$k = \frac{2\varepsilon_0 \omega_B [n^2_{eff\_med} - n^2_{air}]\Gamma_{ov}}{\pi} \frac{\sin(m\pi\gamma)}{m} \tag{56}$$

where m is the grating order, i.e. 1 in this case. A significant limit of the laser fabrication process is the difficulty to realize lateral gratings with perfectly vertical etched sidewalls. This can affect the performance of both waveguide and grating, compared to the desired target. The tilting of etched sidewalls affects the laser behaviour in both its transversal and longitudinal section. This problem in transversal direction can be modelled with reference to the scheme of Figure 19.

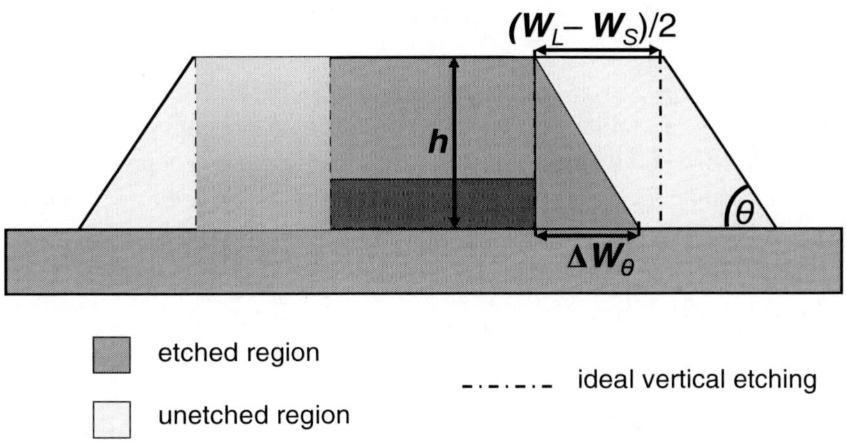

Figure 19. Laser transversal section, with vertical (ideal) and tilted (real) sidewalls of etched regions.

Ideally the grating is made of a periodic repetition of two perfectly vertical ridges of widths W_S and W_L, but, in real case, it is constituted by a periodic repetition of ridges with etched walls tilted of an angle $\pi/2 - \Theta$, compared to the vertical direction, having widths $W_S + 2\Delta W_\theta$ and $W_L + 2\Delta W_\theta$. ΔW_θ is given by:

$$\Delta W_\theta = h\tan(\pi/2-\theta) \tag{57}$$

We can assume that the real structure, with sidewalls tilted of angle Θ, has a behaviour in between that of ideal structure R_1, with ridge widths W_S and W_L, and that of ideal structure R_2, with ridge widths $W_S + 2\Delta W$ and $W_L + 2\Delta W$, where ΔW must be well appropriately chosen.

In fact, if we know the values P_1 and P_2 of a general design parameter P for the ideal structures R_1 and R_2 previously defined, we could approximate the value P_Θ of the real structure, with walls tilted of Θ, as:

$$P_\theta = P_1 + (P_2 - P_1)\left(\frac{A_t}{A_{tot}}\right)2\Gamma_{\Delta W} \tag{58}$$

where A_t / A_{tot} is the ratio between the area of the triangle, base ΔW_θ and height h, and that of rectangle, base ΔW and height h, and where $2\Gamma_{\Delta W}$ is the grating overlap, if ΔW is chosen to be $(W_L - W_S)/2$.

This relation has been used for calculation of effective index, optical confinement and losses changes as follows:

$$n_{eff\_\theta} = n_{neff\_1} + (n_{neff\_2} - n_{neff\_1})\left(\frac{h\tan(\pi/2\text{-}\theta)}{2\Delta W}\right)2\Gamma_{\Delta W} \tag{59}$$

$$\Gamma_{\_\theta} = \Gamma_{\_1} + (\Gamma_{\_2} - \Gamma_{\_1})\left(\frac{h\tan(\pi/2\text{-}\theta)}{2\Delta W}\right)2\Gamma_{\Delta W} \tag{60}$$

$$\alpha_{\_\theta} = \alpha_{f\_1} + (\alpha_{f\_2} - \alpha_{f\_1})\left(\frac{h\tan(\pi/2\text{-}\theta)}{2\Delta W}\right)2\Gamma_{\Delta W} \tag{61}$$

The total effect of angle variations on the coupling coefficient is reported in Figure 20. In the longitudinal section, a similar approach can be used. A scheme of grating longitudinal section with tilted etched sidewalls, compared to the vertical ones, is proposed in Figure 21. In this scheme we assume the grating period as unchanged, while a duty cycle change can be given by:

$$\gamma_\theta = \frac{(\Lambda_L + \Delta_{L\theta})}{\Lambda_{TOT}} \tag{62}$$

with $\Delta_{L\theta} = h\tan(\pi/2 - \Theta)$. Note that about 95% of modal power is confined in the active region and first higher layers, thus a reduced effective height h_{eff} can be considered, and not the total one h (in this case, 1.66 μm instead 4.66 μm). As it was already seen in the previous considerations, the duty cycle has a great effect on coupling coefficient. It can be also seen in Figure 22, referred to an angle Θ of only 3°.

The angle of the etched sidewalls also affects the average effective index and, then, the Bragg wavelength as:

$$n_{eff\_med\_\theta} = [n_{eff\_S}(1-\gamma_\theta) + n_{eff\_L}(\gamma_\theta)] \tag{63}$$

$$\lambda_{B\_\theta} = 2n_{eff\_med\_\theta}\Lambda_{TOT} \tag{64}$$

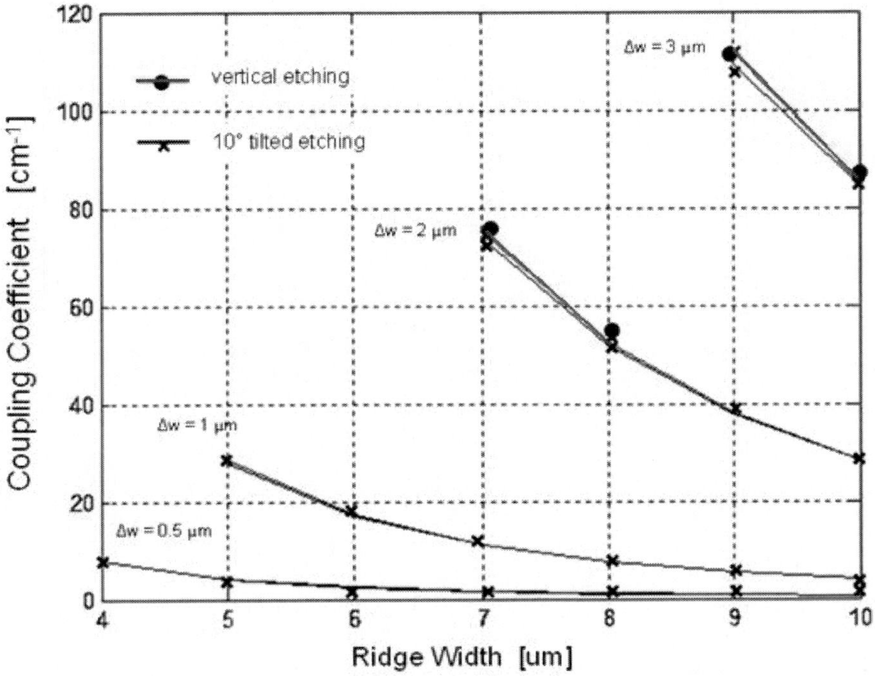

Figure 20. Coupling coefficient versus ridge width with ΔW = 0.5, 1, 2 e 3 μm, for vertical and 10° tilted etching.

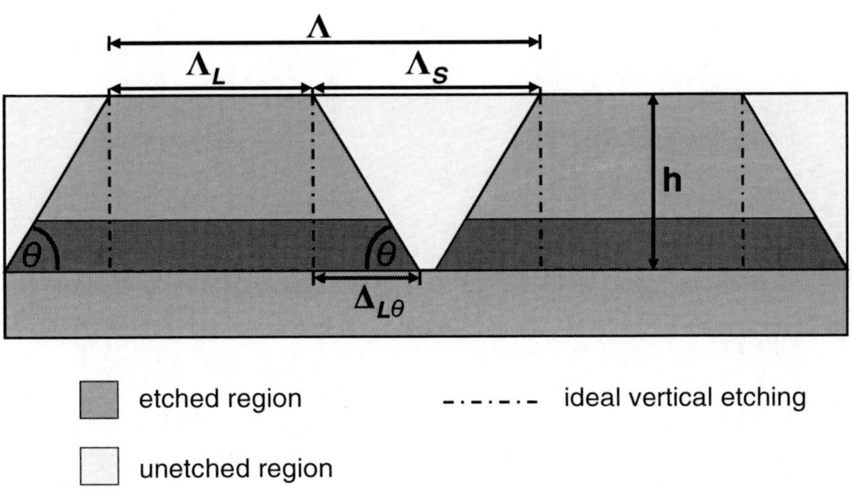

Figure 21. Laser longitudinal section scheme, with vertical (ideal) and tilted (real) sidewalls of etched regions.

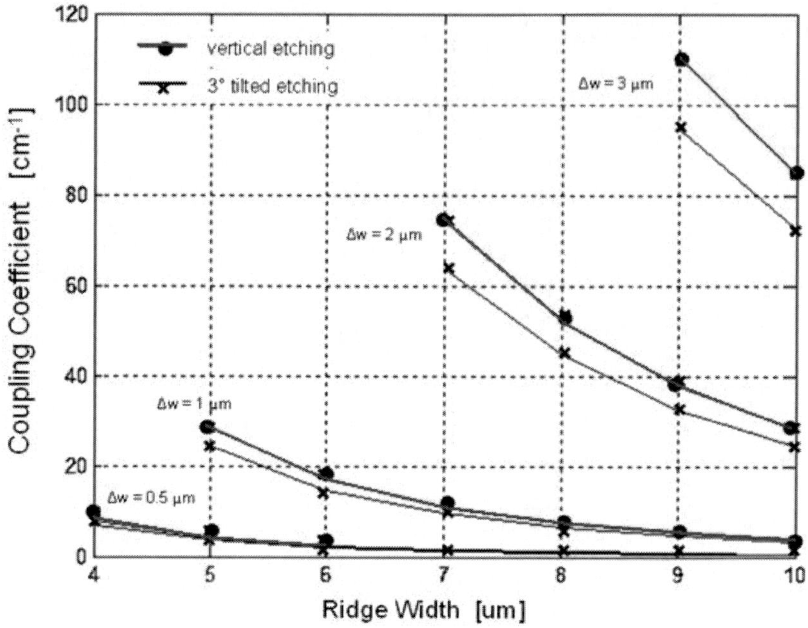

Figure 22. Grating coupling coefficient versus ridge width with ΔW = 0.5, 1, 2 e 3 μm, for vertical and 3° tilted etching.

According to CMT, the grating reflectivity is expressed as:

$$R(\delta) = \frac{\kappa^2 \tanh^2(\gamma L)}{\gamma^2 + \delta^2 \tanh^2(\gamma L)} \tag{65}$$

with

$$\delta = \frac{2\pi n_{eff}}{\lambda} - \frac{\pi}{\Lambda} \tag{66}$$

and

$$\gamma = \sqrt{\kappa^2 - \delta^2} \tag{67}$$

The maximum reflectivity, for $\lambda = \lambda_B$, that is for $\delta = 0$, is given by:

$$R_{\max} = R(0) = \tanh^2(\kappa^2 L) \tag{68}$$

The peak of reflectivity, calculated for different values of ΔW and W_L, has been estimated and results are shown in Figure 23. For each considered structure, the reflectivity peak can be increased by increasing the cavity length, but at the condition to have a stop band not exceeding the Fabry Perot mode distance.

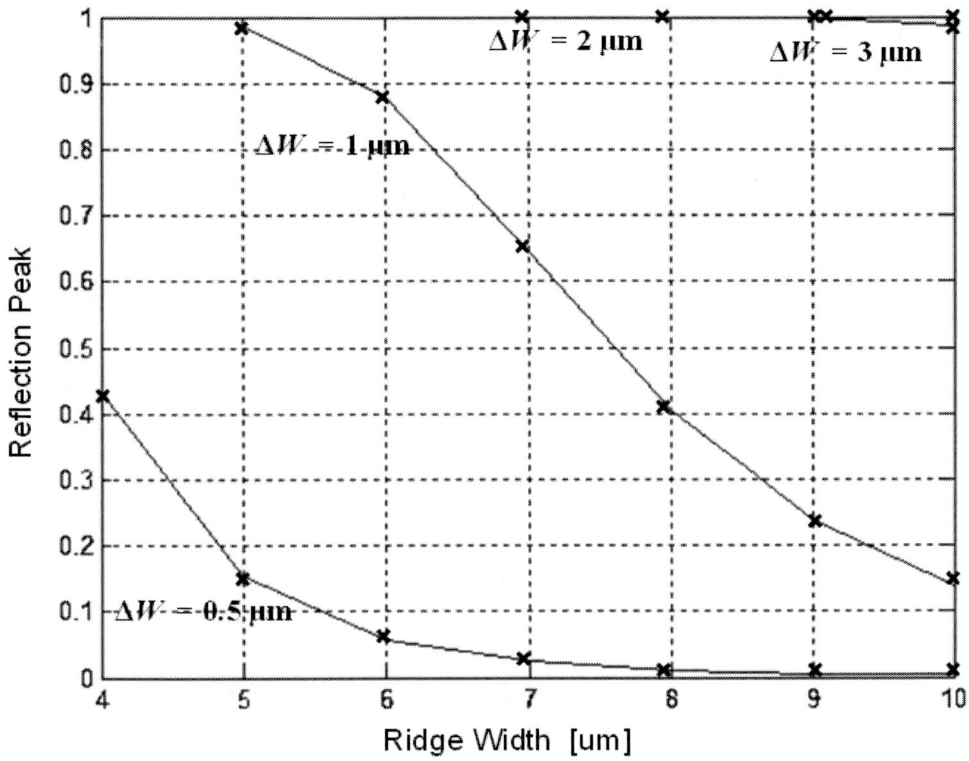

Figure 23. Grating reflection peak versus ridge width, for different values of ΔW and W_L and cavity length L = 1 mm.

Therefore, the maximum value of reflection peak for each structure, and maximum cavity length L_{max}, can be evaluated by imposing $\Delta\nu_0 = \Delta\nu_{FP}$:

$$\Delta\nu_0 = \frac{\kappa}{\pi n_g} = \Delta\nu_{FP}(\lambda_B) = \frac{1}{2L_{\max} n_g(\lambda_B)} \tag{69}$$

Thus, we obtain a reflection peak for $L = L_{max} = \pi / (2\,\kappa)$:

$$R(\delta = 0, L) < R(\delta = 0, L_{\max}) = \tanh^2(\kappa^2 L_{\max}) = \tanh^2(\pi/2) = 0.8412 \tag{70}$$

By using the above considerations, the best solution in terms of smallest stop band (0.3 nm) can be determined (ΔW and W_L). This solution has the highest cavity length, and so the largest laser threshold current. By the same way, one can determine the best solution, with the same reflection peak of 84%, in terms of highest coupling coefficient and shortest cavity length, and so the smallest threshold current and highest FP mode distance.

All the solutions between these last two will obviously show intermediate performance and can be chosen according to desired characteristics and applications. In Figure 24 the reflection spectra of two mentioned solutions are reported.

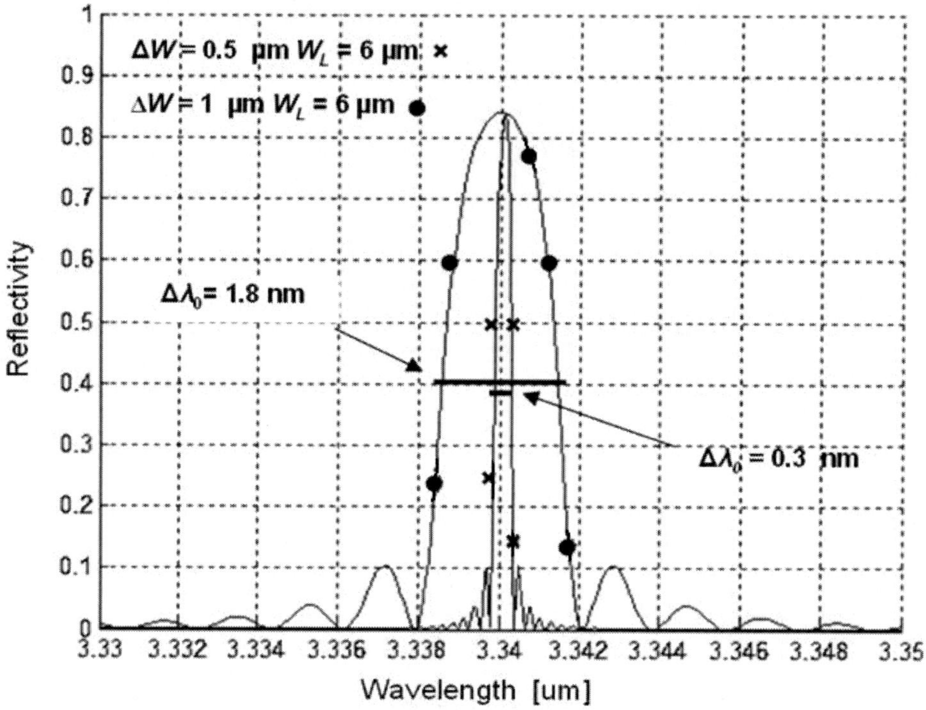

Figure 24. Grating reflection spectrum for two selected solutions and L = 1 mm.

4.4 Transfer Matrix Method

One possible approach to simplifying the grating analysis, whilst improving flexibility and robustness, is to employ matrix methods. Matrices have been used extensively in those engineering problems which are intrinsically highly numerical. In microwave engineering, matrices are used to find the electric and magnetic fields inside various microwave waveguides and devices. One major advantage of matrix methods is their flexibility. In fact, instead of repeatedly finding complicated analytical eigenvalue equations for each grating structure, a general matrix equation can be derived. However, because of the numerical nature of matrix methods, they cannot be used to verify the existence of analytical expressions in a particular problem.

In general, the transfer matrix approach uses a mode-matching technique to determine both power coupling and scattering. In all matrix methods, the structures involved are first divided into a number of smaller sections and all physical parameters are assumed to be homogeneous in each section. As a result, the total number of smaller sections can vary, mostly depending on the problem. From coupled wave equations, it is found that the field propagation inside a corrugated waveguide can be represented by a transfer matrix. Provided that the electric fields at the input plane are known, the matrix acts as a transfer function so that electric fields at the output plane can be determined. By joining these transfer matrices as a building block, a general N-sectioned grating structure model can be derived. Since the outputs from a transfer matrix automatically become the inputs of the next matrix, all boundary conditions inside the composite cavity are matched. The unsolved boundary conditions are those at the left and right facets, depending on particular application.

For an arbitrary one-dimensional grating structure as shown in Figure 25, the wave propagation is modelled by a 2×2 matrix A, such that any electric field leaving (i.e. $E_g(z_{i+1})$ and $E_S(z_i)$) and those entering (i.e. $E_g(z_i)$ and $E_S(z_{i+1})$) that section are related to each other by:

$$U = A\ V \tag{71}$$

where U and V are two column matrices each containing two electric wave components [35].

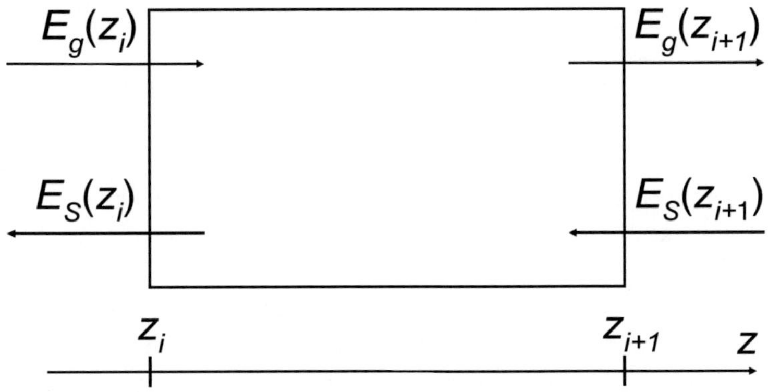

Figure 25. Wave propagation in a general 1-D grating structure.

Depending on the type of matrix method, the contents of U and V may vary. In scattering matrix method, matrix U includes all electric waves leaving the arbitrary section, whilst matrix V contains those entering the section. In both transmission line matrix (TLM) and transfer matrix methods (TMM), matrix U represents the electric wave components from one section side, whilst wave components from the other side are included in matrix V.

For the analysis of grating-based devices, such as semiconductor laser devices, both TLM and TMM are used. The difference between TLM and TMM lies in the domain of analysis, being TLM performed in time domain, whereas TMM works extremely well in frequency domain.

It has to be noted that both CMT and TMM are based on the assumption that the interacting modes are mutually orthogonal. These approaches give accurate results when the periodic perturbation is weak and the coupled sub-waveguides are almost identical (i.e., very similar index profiles and guiding layer thicknesses).

On the other hand, CMT and TMM approaches are less accurate if the grating perturbation is strong, and/or one or more layers of the structure being analyzed are lossy. Moreover, CMT and TMM approaches are inaccurate if one of the modes is leaky. However, a modified version of TMM allows the analysis of grating to be carried out, even for strongly perturbed structures or subject to significant non-linear effects.

In literature, a general formulation of TMM can be found [36] for Long Period Waveguide Grating (LPWG), but this formulation can be easily extended to each periodic structure. Non-uniform gratings can be also analyzed by transfer-matrix method [37]–[43].

4.5. Floquet-Bloch Theory

In this section, the rigorous method based on Floquet–Bloch theory (FBT) is briefly presented. This method has been developed in literature to investigate a number of devices, such as radiation loss in grating-assisted directional couplers [44], grating-assisted directional couplers (GADC) [45-46], fiber Bragg gratings [47-48], vertical cavity surface emitting laser (VCSEL) distributed Bragg reflectors (DBRs) [49]. The FBT general formulation is three dimensional, but in most applications it can be reduced to a two-dimensional one, in which one dimension is that of optical medium periodicity. In general, the periodic dielectric constant can be expressed by a Fourier series expansion. According to Floquet–Bloch theory, if a plane wave propagates in an infinite unbounded periodic medium, the complete field pattern should consist of an infinite number of space harmonics, and the power exchange between each pair of space harmonics is governed by Floquet's theorem. After the FBT field profile has been found inside the grating region, it is matched with the field distributions outside the grating (valuable by exact or numerical methods like TMM) on the interfaces of these regions, according to boundary conditions between dielectric layers. Then, the characteristic equation can be found, allowing to determine the propagation constants for the structure modes. FBT allows also to evaluate other optical properties of the periodic medium, giving a very useful and accurate help in the design of grating-based photonic devices. In the following, a simple derivation of the Floquet-Bloch theory is shown for designing DBR structures. A more general formulation of FBT can be found in literature [46, 50].

FBT has been used to evaluate the spectral properties of a DBR in terms of peak reflectivity and stop-band, allowing the calculation of the optimal period and index profile of the grating. The general scheme of a finite-length DBR is sketched in Figure 26.

Along the stratification and propagation direction z, three different sections can be identified: two regions with homogeneous refractive indices, n_t and n_a for $z < 0$ and $z > L$, respectively, and a multilayered mirror for $0 < z < L$, with periodic refractive index along z.

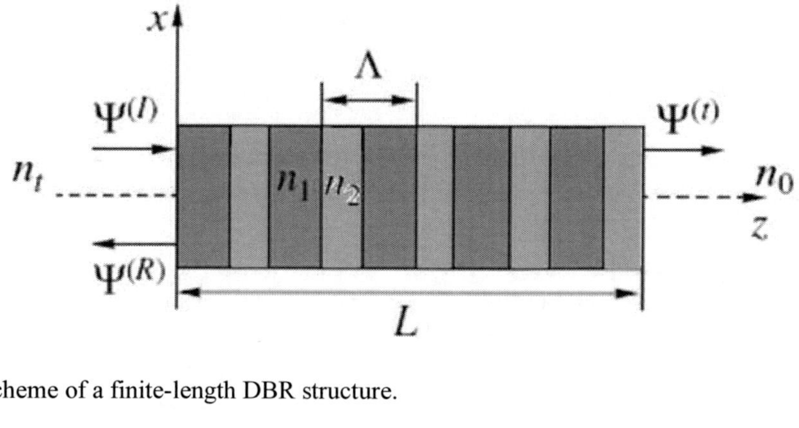

Figure 26. Scheme of a finite-length DBR structure.

If n_0 is the average refractive index of the structure and Λ is the grating period, the refractive index inside the periodic perturbation can be expressed as:

$$n(z) = n_0 + \Delta n\, f(Kz) \quad (72)$$

where $f(Kz)$ is an arbitrary periodic function describing the grating profile, Δn is the index modulation amplitude, and K is the amplitude of the grating vector, namely $K = 2\pi/\Lambda$. Thus, the grating shows two discontinuities along the z axis, at the sections $z = 0$ and L. Waves impinging on this finite-length DBR will be partially reflected and partially transmitted at the first interface within the perturbed medium up to the second interface. For DBRs, we consider waves with normal incidence, but the theory can be applied to non-orthogonal incidence too. To derive the equations of the model according to geometry of Figure 26, we assume the following expressions for incident and reflected field components:

$$\Psi_i^{(I)}(z) = \Psi_i \exp(j\beta_h z) \tag{73}$$

$$\Psi_i^{(R)}(z) = \rho_i \Psi_i \exp(-j\beta_h z) \qquad i = 1, 2, ... \tag{74}$$

respectively, where β_h is the propagation constant in the homogeneous region with refractive index n_t and ρ_i is the reflection coefficient. Eqs. (73) and (74) hold for each polarization (TE or TM) with appropriate field component Ψ_i (i.e., $\Psi_1 = E_y$ and $\Psi_2 = H_x$ for TE modes and $\Psi_1 = H_y$ and $\Psi_2 = E_x$ for TM modes). The permittivity function $\varepsilon(z)$ is periodic along z and can be expressed by a Fourier series expansion as:

$$\varepsilon(z) = \sum_n \varepsilon_n \mathrm{e}^{jnKz} \tag{75}$$

where

$$\varepsilon_n = \frac{1}{\Lambda} \int_{-\Lambda/2}^{\Lambda/2} \varepsilon(z) \mathrm{e}^{-jnKz} dz \tag{76}$$

are the expansion coefficients. According to Floquet's theorem, every field component (solution of the scalar Helmoltz equation) in the periodic region is expressed as the superposition of an arbitrary number of space harmonics, each characterized by the propagation constant k_{zn}:

$$\Psi_i(z) = \sum_n \Psi_{in} \exp(\pm jk_{zn} z) \qquad (i = 1, 2, ...) \tag{77}$$

which is related to the fundamental harmonic by:

$$k_{zn} = k_{z0} + nK = \beta + j\alpha + \frac{2\pi}{\Lambda} n \qquad n = 0, \pm 1, \pm 2, ... \tag{78}$$

where β and α are the leaky mode propagation constant and leakage factor in the perturbed structure, respectively. In Eq. (77), ψ_{in} is the n-th space harmonic amplitude of i-th electromagnetic field component. When the permittivity function is real and even, two independent solutions Ψ_1 and Ψ_2 can be discriminated in Eq. (77), having the same amplitude ψ_{in} but counter-propagating along z axis. Specifically, Ψ_1 is in positive (+) direction and Ψ_2 in negative (–) one. Hence, the general expression for the total field component can be written as a linear combination:

$$\Psi_i^{(T)}(z) = c_{i,1}\Psi_{i,1}(z) + c_{i,2}\Psi_{i,2}(z) \tag{79}$$

where $c_{i,1}$ and $c_{i,2}$ are two complex amplitude constants for the i-th field component. By substituting Eqs. (77) and (79) into Maxwell's equations, the following conditions are derived:

$$\begin{cases} c_{2,1} = c_{1,1} = c_1 \\ -c_{2,2} = c_{1,2} = c_2 \end{cases} \tag{80}$$

where c_1 and c_2 are two unknown independent complex amplitudes to be determined by means of the boundary conditions at the sections $z = 0$ and L. By means of further manipulations, the reflectivity and transmittivity are obtained (for TE or TM modes) as:

$$\rho = \left[\frac{\xi^{(+)}(L)\exp(-2\alpha L) - \xi^{(+)}(0)}{1 - \xi^{(+)}(0)\xi^{(+)}(L)\exp(-2\alpha L)} \right]^* \tag{81}$$

$$\tau = \left\{ \frac{\sum_{n^+} i_n e^{-jB_n L}}{\sum_{n^+} i_n e^{-j\beta_n L}} \frac{e^{-jKL} e^{-\alpha L}}{1 - \xi^{(+)}(0)\xi^{(+)}(L) e^{-2\alpha L}} \left[1 - \xi^{(-)}(0)\xi^{(+)}(L) \right] \right\}^* \tag{82}$$

with

$$\xi^{(\pm)}(z) = \frac{\sum_{n-} i_n e^{\pm jB_n L}}{\sum_{n^+} i_n e^{\pm jB_n L}}, \tag{83}$$

where $B_n = |\beta + nK| - K$, the asterisk denotes the conjugate operator, and indices n^+ and n^- indicate the space harmonics along the positive and negative direction of z axis, respectively. Hence, it is evident that Floquet–Bloch formalism does not introduce any approximation, apart from the unavoidable truncation of Fourier series expansion of the periodic permittivity. Moreover, the scattering losses due to the presence of the periodic structure are taken into account by the leakage factor α, the same for each space harmonic.

As depicted before, the FBT approach has been applied to many photonic devices based on grating structures. Among them, narrow grating-assisted filters for telecommunications systems [51], silicon oxynitride grating-assisted directional couplers [52], distributed Bragg reflectors for vertical cavity surface emitting lasers (VCSEL) [53], and dual-grating directional couplers for efficient coupling between an optical fiber and a thin SOI waveguide [54-58], have been investigated. In these studies, FBT has been used as a rigorous and very powerful CAD tool for accurate design of these grating-assisted photonic devices, some of them being fabricated and experimentally characterized in order to verify the good agreement with FBT predictions. Recently, theoretical and experimental investigations have been devoted to higher order diffraction gratings in SOI rib waveguides, in order to achieve very good filtering and sensing behaviour [59-60].

CONCLUSION

In this chapter, a brief overview of the most popular and known numerical techniques for design and simulation of grating-assisted photonic devices is presented. Both properties and characteristics of Coupled Mode theory and Finite Difference Time Domain, as well as Finite Element Method and Floquet-Bloch Theory, have been summarized.

REFERENCES

[1] Loewen E. (Ed.), *Diffraction Gratings Handbook*, Bausch and Lomb, Rochester, 1970.
[2] Liu, J.-M., *Photonic Devices*, Cambridge University Press, 2005.
[3] Faist, J., Gmachl, C., Capasso, F., Sirtori, C., Sivco, D. L., Baillargeon, J. N., Hutchinson, A. L., Cho, A. Y., *Applied Physics Letters*, 1997, 70, 2670–2672.
[4] Baillargeon, J. N., Cho, A. Y., Thiel, F. A., Fischer, R. J., Pearah, P. J., Cheng, K. Y., *Applied Physics Letters*, 1994, 65, 207–209.
[5] Kennedy, K., Krysa, A. B., Roberts, J. S., Groom, K. M., Hogg, R. A., Revin, D. G., Wilson, L. R., Cockburn, J. W., *Applied Physics Letters*, 2006, 89, 201117-9.
[6] Littman, M. G., Metcalf, H. J., *Applied Optics*, 1978, 17, 2224-2227.
[7] Liu, K., Littman, M. G., *Optics Letters*, 1981, 6, 117-119.
[8] Hawthorn, C. J., *et al.*, *Review of Scientific Instruments*, 2001, 72, 4477-4479.
[9] Litchinitser, N. M., *et al.*, *Optical and Fiber Communications Reports*, 2007, 4, 41-85.
[10] Vengsarkar, A. M., *et al.*, *J. Lightwave Technology*, 1996, 14, 58-64.
[11] Albert, J., *et al.*, *Applied Physics Letters*, 2006, 89, 101127-9.
[12] Hill, K. O., *et al.*, *Applied Physics Letters*, 1978, 32, 647-649.
[13] Meltz, G., *et al.*, *Optics Letters*, 1989, 14, 823-825.
[14] Hill, K. O., *Applied Physics Letters*, 1993, 62, 1035-1037.
[15] Ouellette, F., *Optics Letters*, 1987, 12, 847-849.
[16] Mihailov, S. J., *et al.*, *Optics Letters*, 2003, 28, 995-997.
[17] Martinez, A., *et al.*, *Optics Letters*, 2006, 31, 1603-1605.
[18] Archambault, J. L., *et al.*, *Electronics Letters*, 1993, 29, 453-455.
[19] Yariv, A., *IEEE J. Quantum Electronics*, 1973, QE-9, 919-933.
[20] Yee, K., *IEEE Trans. on Antennas and Propagation*, 1966, 14, 302–307.
[21] Taflove, A., Brodwin, M. E., *IEEE Trans. on Microwave Theory and Techniques*, 1975, 23, 623–630.
[22] Taflove, A., *IEEE Trans. on Electromagnetic Compatibility*, 1980, 22, 191–202.

[23] Mur, G., *IEEE Trans. on Electromagnetic Compatibility*, 1981, 23, 377–382.
[24] Umashankar, K. R., Taflove, A., *IEEE Trans. on Electromagnetic Compatibility*, 1982, 24, 397–405.
[25] Liao, Z. P., Wong, H. L., Yang, B. P., Yuan, Y. F., *Scientia Sinica A*, 1984, 27, 1063–1076.
[26] Berenger, J., *J. Computational Physics*, 1994, 114, 185–200.
[27] Gedney, S. D., *IEEE Trans. on Antennas and Propagation*, 1996, 44, 1630–1639.
[28] Kunz, K. S., Luebbers, R. J., *The Finite Difference Time Domain Method for Electromagnetics*, CRC Press, 1993.
[29] Taflove, A., Hagness, S. C., *Computational Electrodynamics: The Finite-Difference Time-Domain Method,* Artech House Publ., 2005.
[30] Yu, W., Mittra, R., Su, T., Liu, Y., Yang, X., *Parallel Finite-Difference Time-Domain Method*, Artech House Publ., 2006.
[31] Clough, R. W., Wilson. E. L., Early Finite Element Research at Berkeley, see www.edwilson.org/History/fe-history.pdf.
[32] Delort, T., Maystre, D., *J. Optical Society of America A*, 1993, 10, 2592-2601.
[33] Petit, R., *Electromagnetic Theory of Gratings*, Springer-Verlag, Berlin, 1980.
[34] Comsol Multiphysics by COMSOL AB, ver. 3.2, *single license*, 2005.
[35] Ghafouri–Shiraz, H., *Distributed Feedback Laser Diodes and Optical Tunable Filters*, Wiley, 2003.
[36] Liu, Q., Chiang, K. S., *J. Lightwave Technology*, 2006, 24, 3540-3546.
[37] Ke, H., Chiang, K. S., Peng, J. H., *IEEE Photon. Technology Letters*, 1998, 10, 1596–1598.
[38] Yamada, M., Sakuda, K., *Applied Optics*, 1987, 26, 3474–3478.
[39] Erdogan, T., *J. Lightwave Technology*, 1997, 15, 1277–1294.
[40] Kwon, M. S., Shin, S. Y., *Optics Communications*, 2005, 250, 41–47.
[41] Liu, Q., Chiang, K. S., Lor, K. P., *IEEE Photon. Technology Letters*, 2005, 17, 2340–2342.
[42] Chan, F. Y. M., Chiang, K. S., *Optics Communications*, 2005, 244, 233–243.
[43] Lim, K. Y., Low, A. L. Y., Chien, S. F., You, A. H., Foo, Y. C., *IEICE Electronics*, 2004, 1, 103-104.
[44] Passaro, V. M. N., Armenise, M. N., *IEEE J. Quantum Electronics*, 1995, 31, 1691-1697.
[45] Butler, J. K., Sun, N.-H., Evans, G. A., Pang, L., Congdon, P., *J. Lightwave Technology*, 1998, 16, 1038–1048.
[46] Passaro, V. M. N., *J. Lightwave Technology*, 2000, 18, 973-984.
[47] Passaro, V. M. N., Diana, R., Armenise, M. N., *J. Optical Society of America A*, 2002, 19, 1844-1854.
[48] Passaro, V. M. N., Diana, R., Armenise, M. N., *J. Optical Society of America A*, 2002, 19, 1855-1866.
[49] Passaro, V. M. N., Magno, F., Rizzato, A., De Leonardis. F., *Laser Physics*, 2006, 16, 1029–1038.
[50] Sun, N.-H., Butler, J. K., Evans, G. A., Pang, L., Congdon, P., *J. Lightwave Technology*, 1997, 15, 2301-2315.
[51] Passaro, V. M. N., *Optics Express*, 2002, 10, 1354-1360.
[52] Passaro, V. M. N., Masanovic, G. Z., *Optics Express*, 2006, 14, 1055-1063.

[53] De Leonardis, F., Passaro, V. M. N., Magno, F., *J. Computational Electronics*, 2007, 6, 289-292.

[54] Masanovic, G. Z., Passaro, V. M. N., Reed, G. T., *IEEE Photon. Technology Letters*, 2003, 15, 1395-1397.

[55] Passaro, V. M. N., Masanovic, G., Reed, G. T., *Dual-Grating Assisted Optical Coupler*, patent n. P51262C WO, 2003.

[56] Masanovic, G. Z., Passaro, V. M. N., Reed, G. T., *IEE Proc. Pt J – Optoelectronics*, 2005, 152, 41-48.

[57] Masanovic, G., Reed, G. T., Headley, W., Timotijevic, B., Passaro, V. M. N., Atta, R., Ensell, G., Evans, A. G. R., *Optics Express*, 2005, 13, 7374-7379.

[58] Reed, G. T., Masanovic, G. Z., Headley, W. R., Timotijevic, B., Gardes, F. Y., Chan, S. P., Waugh, P., Emerson, N. G., Png, C. E., Paniccia, M. J., Liu, A., Hak, D., Passaro, V. M. N., *IEEE J. Selected Topics in Quantum Electronics*, 2006, 12, 1335-1344.

[59] Chan, S. P., Passaro, V. M. N., Mashanovich, G. Z., Ensell, G., Reed, G. T., 2007, *Journal of European Optical Society: Rapid Publications*, 2007, 2, 07029.

[60] Passaro, V. M. N., Loiacono, R., D'Amico, G., De Leonardis, F., *IEEE Sensors Journal*, 2008, 8.

In: Modeling of Photonic Devices
Editor: Vittorio M. N. Passaro
ISBN 978-1-60456-980-3

Chapter 7

NONLINEAR AND DISPERSIVE EFFECTS IN SOI OPTICAL WAVEGUIDES

Francesco De Leonardis*[*,1] and Vittorio M. N. Passaro[2]
[1]Photonics Research Group,
Ingegneria dell'Ambiente e per lo Sviluppo Sostenibile,
Politecnico di Bari, viale del Turismo n. 8, 74100 Taranto, Italy
[2]Photonics Research Group, Dipartimento di Elettrotecnica ed Elettronica,
Politecnico di Bari, via Edoardo Orabona n. 4, 70125 Bari, Italy

ABSTRACT

In this chapter, nonlinear optical propagation of ultrafast pulses in silicon-on-insulator rib waveguides is theoretically investigated. Two photon absorption, free carrier dispersion, self- and cross-phase modulation induced by Kerr effect, walk-off, group velocity dispersion, third-order dispersion, self-steeping and polarization coupling are taken into account by a very general modeling under sub-picosecond regime. Pulse compression as induced by the soliton generation are presented and discussed.

Keywords: Integrated Optics, Nonlinear Optics, Silicon-On-Insulator Technology, Modeling, Raman Scattering, Optical Solitons

1. INTRODUCTION

Silicon is the ideal platform for Integrated Optics and Optoelectronics. The quality of commercial silicon wafers driven by the Microelectronics industry still continues to improve while the cost continues to decrease. Moreover, the compatibility with silicon integrated circuits manufacturing and silicon Micro-Electro-Mechanical Systems (MEMS) technology is

* E-mail address: f.deleonardis@poliba.it

another important reason for interest in Silicon Photonics [1-2]. As a transmission medium, silicon has much higher nonlinear effects than the commonly used silicon dioxide, in particular Kerr and Raman effects. Additionally, silicon-on-insulator (SOI) waveguides can confine the optical field to an area that is approximately 100 times smaller than the modal area in a standard single-mode optical fiber. The combination of these properties enables efficient nonlinear interaction of optical pulses at relatively low power levels inside SOI waveguides, over the millimetre-scale interaction length. For this reason, in recent years considerable effort has been directed towards investigation of nonlinear phenomena such as Raman amplification [3-8], Stokes and anti-Stokes Raman conversion [9-11], cross phase modulation [12], lossless modulation [13].

Although the third-order nonlinear effects have been extensively studied for silica fibers [14] and these studies can be used as a guidance for SOI waveguides, it is important to remember that silicon is a semiconductor crystal exhibiting unique features such as Two-Photon Absorption (TPA), plasma effect and Free Carrier Absorption (FCA) [15-16]. These features can provide new functionalities [17-22], but on the other hand they may become obstacles in some cases, such as light generation in silicon based on Raman effect [23-30]. Moreover, TPA and FCA imposes non trivial difficulties for mathematical modelling in SOI waveguiding structures with respect to silica fibers [31-32].

More recently, with the aim to avoid the detrimental influence of the TPA effect, a large interest for optical devices operating in the mid-IR has driven the research to exploit the Raman effect in a large wavelength range, 2.2 ÷ 5.3 μm [33-34]. Simultaneously, another research topic has involved the propagation of ultrafast optical pulses in SOI waveguides [35-38], in which FCA influence can be neglected due to the use of sub-picosecond pulses.

In section 2 the theoretical model is presented in a very general formalism, taking into account TPA, FCA, where the free carriers are generated mainly by TPA of the input pulses, plasma dispersion, Self-Phase-Modulation (SPM) and Cross-Phase-Modulation (XPM) as induced by the Kerr nonlinearity, Group-Velocity Dispersion (GVD), Third-Order Dispersion (TOD), self steeping, and higher order nonlinear effects as induced by the delayed Raman response. The model includes the interaction between two input pulses under the condition of neglecting the Stimulated Raman Scattering (SRS). The mathematical formalism to treat these nonlinear processes involves a system of partial differential equations in order to calculate the space-time evolution of both polarization components (quasi-TE and quasi-TM) for each wave, and the rate equation that governs the free carrier dynamics. In section 3 a number of numerical results are shown, including comparisons between our solutions and some results proposed in literature, and a parametric study based on finite element method (FEM) to find the optimal SOI rib waveguide cross-section to reduce the detrimental TOD effects on the soliton propagation.

2. MODELING

In this section, the complete mathematical model to investigate the propagation of optical pulses in sub-picosecond regime is developed. When such optical pulses propagate inside a waveguide, both dispersive and nonlinear effects largely influence their shapes and spectra. Then, the equation system must take into account a number of nonlinear effects, such as TPA,

SPM and XPM induced by Kerr effect, GVD effect, third-order dispersion (TOD), self-steeping, nonlinear birefringence effect, and higher order nonlinear effects induced by the delayed Raman response. The mathematical model described in this section can be seen as a generalization of the theoretical study proposed in literature [36, 38], considering the energy exchange between two different pulses and polarization coupling induced by the waveguide birefringence. It is worth noting that, in the coupling between the two pulses, Raman effect is not involved because the frequency difference is assumed smaller than Raman frequency shift in silicon, i.e. $\Omega_R = 15.6$ THz. However, the Raman effect is indirectly taken into account since the delayed response of the Raman scattering influences the nonlinear response function and, then, both SPM and XPM effects. In addition, the introduction of the delayed response for the Raman scattering leads to extend the investigation for pulse time widths under 1 ps, ranging from 10 to 1000 fs. The starting point is the nonlinear response function $R(t)$, written as [14]:

$$R(t) = (1 - f_R)\delta(t) + f_R h_R(t) \tag{1}$$

where $\delta(t)$ indicates the Dirac delta function and f_R represents the fractional contribution of the delayed Raman response to nonlinear polarization P_{NL} [14]. Thus, the response function $R(t)$ includes both electronic and vibrational (Raman) contributions. In addition, the Raman response function $h_R(t)$ is responsible for the Raman gain spectrum. In particular, the functional form of $h_R(t)$ can be deduced from the Raman gain spectrum [14], $g_R(\Delta\omega)$, known to exhibit a Lorentzian shape [34].

Then, the complete determination of the nonlinear response function $R(t)$ requires the knowledge of the fraction f_R. It can be estimated from the numerical value of the Raman gain peak and normalization condition $\int h_R(t)dt = 1$ [14]. Assuming a gain peak value of 20 cm/GW (typical for silicon at λ = 1550 nm), it results f_R = 0.043 [37], smaller than in silica fibers, i.e. 0.18 [14]. In the following analysis, we assume a typical SOI waveguide as sketched in Figure 1, having rib total height H, slab height H_s and rib width W.

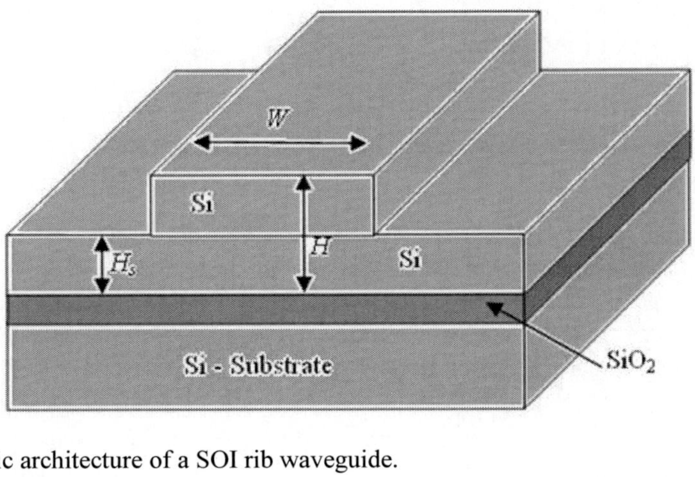

Figure 1. Schematic architecture of a SOI rib waveguide.

In a single-mode SOI waveguide, two propagating modes are typically confined, one quasi-TE (dominant horizontal x-component of electric field) and one quasi-TM (dominant vertical y-component). The coupled-mode approach describes the power exchange among two optical pulse (p_1, p_2) considering both polarizations for each wave.

Since the assumption of translational invariance along the propagation direction (z) can be made, the electric field in the single mode SOI waveguide is written in terms of variable separation as $E(x,y,z,t) = C \cdot F(x,y)A(z,t)e^{j\beta_0 z}$, where β_0 is the propagation constant and $F(x,y)$ is the optical mode distribution in the waveguide cross section $x-y$ (solutions of Helmoltz wave equation), and C is a normalization constant, $C = 1 \Big/ \sqrt{\int\int_{-\infty}^{+\infty} \left|F(x,y)\right|^2 dxdy}$.

Thus, the total electric field inside the SOI waveguide can be written without any leak of generality, as:

$$\mathbf{E}(x,y,z,t) = \hat{\mathbf{x}}\begin{bmatrix} C_1 A_1(z,t)F_1(x,y)e^{j\left(\beta_{0,1}z-\omega_{p1}t\right)} + \\ +C_3 A_3(z,t)F_3(x,y)e^{j\left(\beta_{0,3}z-\omega_{p2}t\right)} \end{bmatrix} + \hat{\mathbf{y}}\begin{bmatrix} C_2 A_2(z,t)F_2(x,y)e^{j\left(\beta_{0,2}z-\omega_{p1}t\right)} + \\ +C_4 A_4(z,t)F_4(x,y)e^{j\left(\beta_{0,4}z-\omega_{p2}t\right)} \end{bmatrix} \tag{2}$$

with the following meaning for superscripts and subscripts: $1 = p_1^{(TE)}$, $2 = p_1^{(TM)}$, $3 = p_2^{(TE)}$, $4 = p_2^{(TM)}$. In Eq. (2), ω_{p1}, ω_{p2}, are the angular frequencies of pulses p_1 and p_2, respectively. Then, by following the procedure outlined in [31] and assuming the nonlinear contributions to $\mathbf{P}_{NL}$ as a small perturbation of refractive index, we obtain the following coupled equations for input pulses:

$$\begin{aligned}
&\frac{\partial A_\xi}{\partial z} + \beta_{1\xi}\frac{\partial A_\xi}{\partial t} + j\frac{1}{2}\beta_{2\xi}\frac{\partial^2 A_\xi}{\partial t^2} - \frac{1}{6}\beta_{3\xi}\frac{\partial^3 A(z,t)}{\partial t^3} = -\frac{\left(\alpha_\xi^{(prop)} + \alpha_\xi^{(FCA)}\right)}{2}A_\xi + \\
&-0.5\beta^{(TPA)} f_{\xi,\xi}\left|A_\xi\right|^2 A_\xi - 0.5\beta^{(TPA)} f_{\xi,\upsilon}\left|A_\upsilon\right|^2 A_\xi - 0.5\beta^{(TPA)} f_{\xi,\kappa}\left|A_\kappa\right|^2 A_\xi + \\
&+j\gamma_\xi(1-f_R)A_\xi(\left|A_\xi\right|^2 + 2\left|A_\upsilon\right|^2) + j2\gamma_\xi(1-f_R)\left|A_3\right|^2 A_\xi + j2\gamma_\xi(1-f_R)\left|A_4\right|^2 A_\xi + \qquad \xi = 1,2 \\
&j\frac{2\pi}{\lambda_{p1}}\Delta n_\xi A_\xi + j2k_{\xi,3,\upsilon,4}A_\upsilon A_3^* A_4 e^{j(\beta_{0,\upsilon}+\beta_{0,4}-\beta_{0,\xi}-\beta_{0,3})z} + \\
&j2k_{\xi,4,\upsilon,3}A_\upsilon A_4^* A_3 e^{j(\beta_{0,\upsilon}+\beta_{0,3}-\beta_{0,\xi}-\beta_{0,4})z} + jk_{\xi,\xi,\upsilon,\upsilon}A_\upsilon A_\xi^* A_\upsilon e^{j(2\beta_{0,\upsilon}-2\beta_{0,\xi})z} + \\
&+R_{p1}(z,t) - \frac{\gamma_{\xi,\xi}}{\omega_{p1}}\frac{\partial(\left|A_\xi\right|^2 A_\xi)}{\partial t}
\end{aligned} \tag{3}$$

$$\frac{\partial A_\kappa}{\partial z}+\beta_{1\kappa}\frac{\partial A_\kappa}{\partial t}+j\frac{1}{2}\beta_{2\kappa}\frac{\partial^2 A_\kappa}{\partial t^2}-\frac{1}{6}\beta_{3k}\frac{\partial^3 A_\kappa(z,t)}{\partial t^3}=-\frac{\alpha_\kappa^{(prop)}}{2}A_\kappa-\frac{1}{2}\alpha_\kappa^{(FCA)}A_\kappa+$$
$$-0.5\beta^{(TPA)}f_{\kappa,2}\left|A_2\right|^2A_\kappa-0.5\beta^{(TPA)}f_{\kappa,1}\left|A_1\right|^2A_\kappa-0.5\beta^{(TPA)}f_{\kappa,\xi}\left|A_\xi\right|^2A_\kappa+$$
$$+j2(1-f_R)\gamma_\kappa\left|A_1\right|^2A_\kappa+j2(1-f_R)\gamma_\kappa\left|A_2\right|^2A_\kappa+j(1-f_R)\gamma_\kappa\left|A_\kappa\right|^2A_\kappa+\qquad \kappa=3,4$$
$$+j2(1-f_R)\gamma_\kappa\left|A_\nu\right|^2A_\kappa+j\frac{2\pi}{\lambda_2}\Delta n_\kappa A_\kappa+j2k_{\kappa,1,2,\nu}A_{\mu+1}A_\mu^*A_\nu e^{j(\beta_{0,2}+\beta_{0,\nu}-\beta_{0,1}-\beta_{0,\kappa})z}+$$
$$+j2k_{\kappa,2,1,\nu}A_1A_2^*A_\nu e^{j(\beta_{0,1}+\beta_{0,\nu}-\beta_{0,2}-\beta_{0,\kappa})z}+jk_{\kappa,\kappa,\nu,\nu}A_\nu A_\kappa^*A_\nu e^{j(2\beta_{0,\nu}-2\beta_{0,\kappa})z}+$$
$$+R_{p2}(z,t)-\frac{\gamma_{\kappa,\kappa}}{\omega_2}\frac{\partial(\left|A_\kappa\right|^2A_\kappa)}{\partial t} \tag{4}$$

In Eq. (3), subscript $\upsilon=2,1$ holds for $\xi=1,2$, respectively. In Eq. (4), if $\kappa=3,4$, it results $\nu=4,3$. In conclusion, Eqs. (3)-(4) describe the time-space evolution of both quasi-TE and quasi-TM pump pulses, while the asterisk denotes the complex conjugate. The last terms in Eqs. (3)-(4) take into account the self-steeping effect as induced by including the first time derivative of P_{NL} in the wave equation. Moreover, in the previous equations we have indicated with:

$$R_{p1}(z,t)=j\gamma_\xi f_R A_\xi(z,t')\int h_R(t-t')\left[\left|A_\xi(z,t')\right|^2+\left|A_\kappa(z,t')\right|^2\right]dt' \text{ with } \xi=1,2\ \ \kappa=3,4 \tag{5}$$

$$R_{p2}(z,t)=j\gamma_\kappa f_R A_\kappa(z,t')\int h_R(t-t')\left[\left|A_\kappa(z,t')\right|^2+\left|A_\xi(z,t')\right|^2\right]dt' \text{ with } \kappa=3,4\ \ \xi=1,2 \tag{6}$$

The relationships (7)-(8) take into account the SRS effect and higher order nonlinear effects induced by the delayed Raman response, such as the intrapulse Raman scattering [14]. Rigorously, the relationships (5)-(6) should include additional terms proportional to $e^{-j\Omega_R(t-t')}$ (see [14]), that we have neglected since the small frequency difference between the two input pulses does not induce SRS effect.

Moreover, the coefficients $\gamma_{i,i}=n_2\omega_i f_{i,i}/c$, being c the light velocity and n_2 the nonlinear refractive index, depend on SPM and XPM effects, as induced by Kerr nonlinearity. The terms $k_{i,j,k,l}=n_2\omega_i f_{i,j,k,l}/c$ with $i,j=1,...,8$, lead to consider the birefringence effects. The overlap integrals $f_{i,j}$ and $f_{i,j,k,l}$ are given by:

$$f_{i,j}=\frac{\iint\left|F_i(x,y)\right|^2\left|F_j(x,y)\right|^2dxdy}{\iint\left|F_i(x,y)\right|^2dxdy\iint\left|F_j(x,y)\right|^2dxdy}\qquad i,j=1,...,4$$

$$f_{i,j,k,l} = \frac{\int\int_{-\infty}^{+\infty} F_i^* F_j^* F_k F_l dxdy}{\sqrt{\left(\int\int_{-\infty}^{+\infty}|F_i|^2 dxdy\right)\left(\int\int_{-\infty}^{+\infty}|F_j|^2 dxdy\right)\left(\int\int_{-\infty}^{+\infty}|F_k|^2 dxdy\right)\left(\int\int_{-\infty}^{+\infty}|F_l|^2 dxdy\right)}} \qquad i,j,k,l = 1,...,4$$

In particular, $f_{i,i}^{-1} = A_{eff,i}$ represents the effective area of the optical mode relevant to i-th pulse, $i = 1,...,4$, influencing the efficiency of any nonlinear device, in particular in SOI technology. The terms $k_{i,i,j,j}$ are dominant ($k_{i,i,j,j} > k_{i,j,k,l}$ for $i \neq j$, $k \neq l$), and represent the coherent coupling between the two polarization components for waves at the same frequency, giving degenerate four-wave mixing (FWM) [14]. In general, it is possible to define a characteristic length for the modal birefringence as $L_B = 2\pi/(\beta_{0,k} + \beta_{0,l} - \beta_{0,j} - \beta_{0,i})$, whose meaning is as follows: for long waveguides with large birefringence it occurs $L \gg L_B$, then the terms containing $k_{i,j,k,l}$ often change sign and the total phase contribution due to birefringence averages out to zero. However, since $k_{i,i,j,j}$ is usually dominant, the phase contribution is surely negligible even if $L > L_B(k_{i,i,j,j})$. In contrast, this contribution cannot be neglected if $L \leq L_B(k_{i,i,j,j})$, as occurs in short waveguides showing moderate birefringence.

As indicated by the terms proportional to $\beta_{1,i}$, $\beta_{2,i}$, $\beta_{3,i}$, and $\beta^{(TPA)}$, the model includes walk-off effect, group-velocity dispersion, TOD effect, and TPA effect as induced by the input pulses propagation, respectively. Moreover, the total loss coefficient in system (4)-(6) has been written as the summation of two contributions, $\alpha_i^{(tot)} = \alpha_i^{(prop)} + \alpha_i^{(FCA)}$, where $\alpha_i^{(prop)}$ is the propagation loss coefficient in the rib waveguide, depending on the material absorption and fabrication process, and $\alpha_i^{(FCA)}$ is the contribution due to FCA, as induced by the change of the free carriers generated mainly by TPA of the pump pulse. According to [39], we have evaluated $\alpha_i^{(FCA)}$ by means of the following:

$$\alpha_i^{(FCA)} = 8.5 \cdot 10^{-18} \cdot \left(\frac{\lambda_i}{1.55}\right)^2 \Delta N_e + 6.0 \cdot 10^{-18} \cdot \left(\frac{\lambda_i}{1.55}\right)^2 \Delta N_h = \sigma_i \cdot N_c = \sigma_0 \cdot \left(\frac{\lambda_i}{1.55}\right)^2 N_c$$

where $N_c = \Delta N_e = \Delta N_h$ is the density of electron-hole pairs generated by the TPA process. The coefficient $\sigma_0 = 1.45 \times 10^{-17}$ cm^{-2} [15] is the FCA cross section measured at $\lambda = 1.55$ μm, and λ_i is the wavelength of relevant mode. In the system (4)-(5), $\Delta n = -8.8 \cdot 10^{-22} \cdot \left(\frac{\lambda_i}{1.55}\right)^2 \Delta N_e - 8.5 \cdot 10^{-18} \cdot \left(\frac{\lambda_i}{1.55}\right)^2 (\Delta N_h)^{0.8} \approx -1.66 \cdot \delta_i \cdot N_c$ is the change of effective index due to the plasma dispersion effect [39] as induced by the free carriers, where $\delta_i = 8.8 \cdot 10^{-22} \cdot (\lambda_i / 1.55)^2$ [15]. In order to obtain a consistent mathematical model, we have coupled the system (3)-(4) to the rate equation governing the free carrier dynamics into the waveguide core, given by [5]:

$$\frac{dN_c}{dt} = -\frac{N_c}{\tau_{eff,\xi}} + \frac{\beta^{(TPA)}}{2\hbar\omega_p}\left(\left|A_{\xi,\xi}(z,t)\right|^2 f_{\xi,\xi}\right)^2 \tag{7}$$

where τ_{eff} is the relevant effective recombination lifetime for free carriers.

The mathematical complexity of our model can be reduced without any leak of accuracy if we assume input pulse FWHM widths T_{FWHM} □ 10 fs, i.e. above 80 fs. In this case, the following Taylor-series expansion is applicable [14]:

$$\left|A_i(z,t-t')\right|^2 \cong \left|A_i(z,t)\right|^2 - t'\frac{\partial\left|A_i(z,t)\right|^2}{\partial t} \quad i = 1,\ldots,4 \tag{8}$$

The previous relationship leads to avoid the integral terms in Eqs.(5)-(6). Then, by substituting, for example, Eq. (8) in (5), it becomes:

$$R_{p1}(z,t) = j\gamma_\xi f_R A_\xi \left[\left|A_\xi\right|^2 - \frac{\partial\left|A_\xi\right|^2}{\partial t}\frac{d\tilde{h}_R}{d\Delta\omega}\bigg|_{\Delta\omega=0} + \left|A_\kappa\right|^2 - \frac{\partial\left|A_\kappa\right|^2}{\partial t}\frac{d\tilde{h}_R}{d\Delta\omega}\bigg|_{\Delta\omega=0}\right] \tag{9}$$

where $\tilde{h}_R(\Delta\omega)$ represents the Fourier transform of Raman response $h_R(t)$, which is directly related to the complex Raman gain by means of the Raman susceptibility. The great utility of assumption (8) is due to conversion of the equation system (3)-(4) in a partial differential equation system in the space-time domain, without using any integral term. This leads to exploit the potential of the collocation method as a fast and accurate solution method [31]. Hereinafter, normalized variables are used for numerical purposes, $U_i(z,t) = A_i(z,t)/\sqrt{P_0}$ and $\tau = (t - z\cdot\beta_{1,1})/T_0$, where P_0 is the Gaussian pump peak power and T_0 is related to pulse FWHM width by $T_0 = T_{FWHM}/1.665$. Moreover, we will use normalized parameters to simplify the physical discussion, i.e. dispersion, walk-off and nonlinear lengths, as:

$$L_D^{'} = \frac{T_0^2}{|\beta_3|}; \qquad L_D = \frac{T_0^2}{|\beta_2|};\; L_w = \frac{T_0}{|d|};\; L_{NL} = \frac{1}{\gamma P_0}$$

being $d = \beta_{1p} - \beta_{1s}$ the group velocity mismatch. To numerically solve the coupled equation system (3)-(4) and (7), the collocation method has been used by developing the unknown functions (i.e. A_i, N_c) as a linear combination of M orthogonal functions $\phi_m(\tau)$ through M weight functions $c_m(z)$, as

$$X(z,\tau) = \sum_{m=1}^{M} c_m(z)\phi_m(\tau) \tag{10}$$

where $X(z,\tau)$ states for each unknown function. Hermite-Gauss functions have been chosen as basis functions, i.e. $\phi_m(\tau)=H_{m-1}(\tau)\exp(-0.5\tau^2)$. The unknown coefficients $c_m(z)$ are determined by satisfying the differential equation system at M collocation points τ_m, such that $H_M(\tau_m)=0$, $m=1,\dots,M$. The collocation method has been already demonstrated to improve the accuracy of about three orders of magnitude [40] for the same computation time if compared to the well known split-step Fourier method [31]. Moreover, improvements in the calculation time of about three orders of magnitude are expected for the same accuracy. Then, the system of partial differential equations (3)-(4) in z,t becomes a system of ordinary differential matrix equations in z, which has been solved by a fourth-order Runge-Kutta algorithm (for algebraic details, see [31]). The overlap integrals have been evaluated with high accuracy using the full-vectorial finite element method (FEM), with not less than 60,000 elements mesh [41].

3. Numerical Results

3.1 Single Pulse Propagation

In this section, the case in which only one optical pulse propagates inside the SOI waveguide experiencing together the GVD-induced pulse broadening, SPM-induced chirp and TOD effect, is analyzed. Therefore, this operation regime can be investigated by considering Eq. (7) coupled with the following nonlinear equation:

$$\frac{\partial A}{\partial z}+j\frac{1}{2}\beta_2\frac{\partial^2 A}{\partial \tau^2}-\frac{1}{6}\beta_3\frac{\partial^3 A}{\partial \tau^3}=-\frac{\left(\alpha_2^{(prop)}+\alpha_2^{(FCA)}\right)}{2}A-0.5\beta^{(TPA)}f|A|^2A$$
$$+j\gamma(1-f_R)A|A|^2+j\frac{2\pi}{\lambda}\Delta nA-\frac{\gamma}{\omega}\frac{\partial(|A|^2A)}{\partial\tau} \qquad (11)$$

Thus, two limit cases could be considered depending on the relative magnitudes of the dispersion lengths L_D and L'_D. If $L_D \gg L'_D$ throughout the SOI waveguide, GVD effects are negligible. In that case, TOD effect acts with the SPM-induced spectral broadening. On the contrary, if $L_D \ll L'_D$, GVD effect dominates on TOD, and SPM-induced spectral broadening could be compensated by the effects of group-velocity dispersion. These two cases can be physically obtained by appropriately selecting the wavelength of the input optical pulse and the waveguide cross-section to meet the condition $\beta_2=0$ and $\beta_3=0$, respectively.

In many applications of Silicon Photonics, it is very important to find the design rules for any SOI rib waveguide, i.e. to simultaneously meet single-mode and birefringence free conditions. Although the high refractive index contrast between waveguide cladding and core facilitates the light confinement in submicron-scale structures, it also makes the control of waveguide birefringence extremely challenging. The research in birefringence control is driven primarily by the requirement of polarization insensitivity in interferometer devices such as arrayed waveguide gratings, ring resonators and all-optical devices, such as ultrafast switches and all-optical gates. Waveguide birefringence is defined in this chapter as the difference between the effective indices of two orthogonally polarized waveguide modes, quasi-TM (dominant component along vertical direction) and quasi-TE (dominant component along horizontal direction) polarized mode, i.e. $\Delta n_{eff} = n_{eff}^{TM} - n_{eff}^{TE}$. In general, a different birefringence coefficient is involved in the analysis for each wave. Thus, the SOI waveguide cross-section can be chosen such to meet the birefringence free condition ($\Delta n_{eff} = 0$) only at a specific wavelength of a particular wave (e.g., 1550 nm), and not for each wave.

In our discussion, the design criteria to find the birefringence free condition are not critical, in fact the goal of the following analysis is to determine step by step the optimal waveguide cross-section to inhibit a particular dispersion effect, then analyzing the influence of remaining nonlinear and dispersion effects. Rigorously speaking, the waveguide core geometry influences not only the modal birefringence and dispersion coefficients, but also other critical parameters such as the effective carrier recombination lifetime τ_{eff}. As it is known, free carrier diffusion usually needs to be considered in addition to recombination lifetime [15]. In fact, if diffusion carriers move out of the modal area, this results in an effective lifetime shorter than the recombination lifetime in SOI structures, namely τ_r. If τ_t is the transit time, then $1/\tau_{eff,\xi} = 1/\tau_r + 1/\tau_t$. In [15] τ_r has been demonstrated to be dependent only on the surface-recombination velocity ($S = 10^3 cm/s$) at the interface between top silicon and buried oxide, with a typical value of about 100 ns. In a different way, contribution τ_t also depends on the optical mode size, i.e. $\tau_t = 0.5w\sqrt{H/(SD)}$, where w is the optical mode width and D is the ambipolar diffusion coefficient [15]. Thus, a small cross section is usually necessary to obtain a low value for τ_{eff}. However, since the dispersion effects (GVD, TOD) take mainly place in sub-picosecond regime (< 1 ps), and FCA induced by TPA produces negligible effects for short pulses, then the design criteria to manage the dispersive effects indirectly induce small free carrier recombination lifetimes too.

To investigate the influence of rib waveguide sizes on GVD and TOD coefficients, we have evaluated β_2 and β_3 coefficients as a function of rib total height, $r = H_s/H$ parameter, and wavelength of quasi-TE and quasi-TM modes. Figure 2 shows the locii of points where $\beta_2 = 0$ for both polarizations in the plane r-wavelength, for different values of H, assuming W = 500 nm. It is worth noting that the curves have been evaluated with high accuracy using the full-vectorial finite element method (FEM), with not less than 60,000 elements mesh [41].

Several features of Figure 2 are noteworthy. Quasi-TE modes show for wavelengths larger than a particular value, namely $\bar{\lambda}$, that it is possible to find only one value of r parameter where condition $\beta_2 = 0$ holds. Moreover, $\bar{\lambda}$ increases with increasing H. A completely different behaviour is induced by TM polarization. In fact, the plot indicates that the locus of the points is split in two arms, one placed at low wavelengths and overlaps all values of r parameter in the range 0.15 ÷ 0.5. On the contrary, the second arm is placed for large values of wavelengths and r. Thus, for quasi-TM modes condition $\beta_2 = 0$ could not be obtained. In addition, this range shifts towards large wavelengths with increasing H.

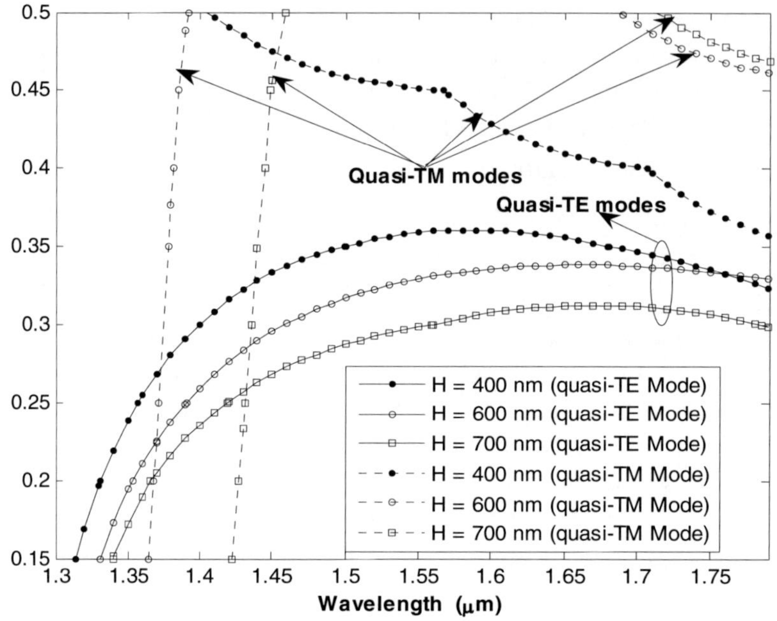

Figure 2. Zero group-velocity dispersion spectrum versus rib parameter r and rib height.

With the aim to isolate the influence of TOD effect, it is important to evaluate β_3 coefficient spectrum as a function of rib parameter *r*. Figures 3 and 4 show the level curves of GVD coefficient in the plane wavelength-r for quasi-TE modes and different values of H. We have focused the simulations on quasi-TE modes since, as discussed before, this polarization can more easily meet the condition $\beta_2 = 0$.

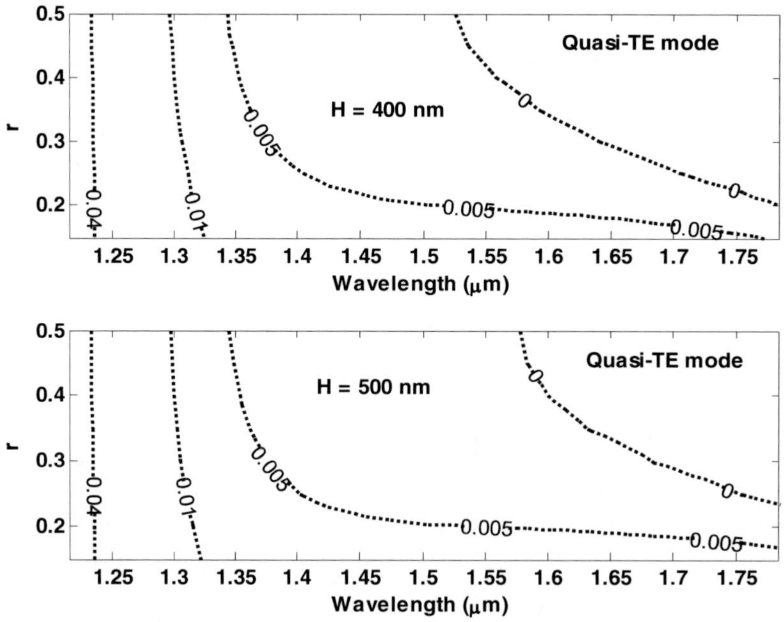

Figure 3. TOD level curves for quasi-TE modes, with $H = 400$ nm and $H = 500$ nm.

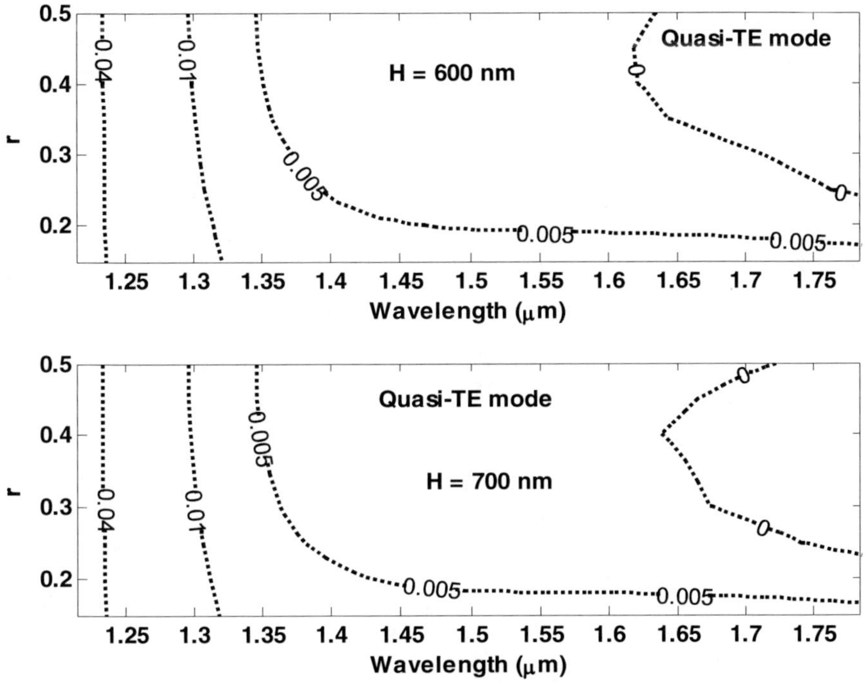

Figure 4. TOD level curves for quasi-TE modes with $H = 600$ nm and 700 nm.

The plots show that zero-level curves involve mainly wavelengths larger than 1.6 μm. Moreover, Figs.3-4 show that condition $\beta_3 = 0$ occurs at wavelengths around 1.5 μm with reducing H. Thus, a SOI rib waveguide with $H = 600$ nm, $W = 500$ nm, and $r = 0.15$ can be chosen in order to isolate the TOD effect. Figure 5 shows the electric field spatial distribution of such waveguide for quasi-TM and quasi-TE modes, respectively.

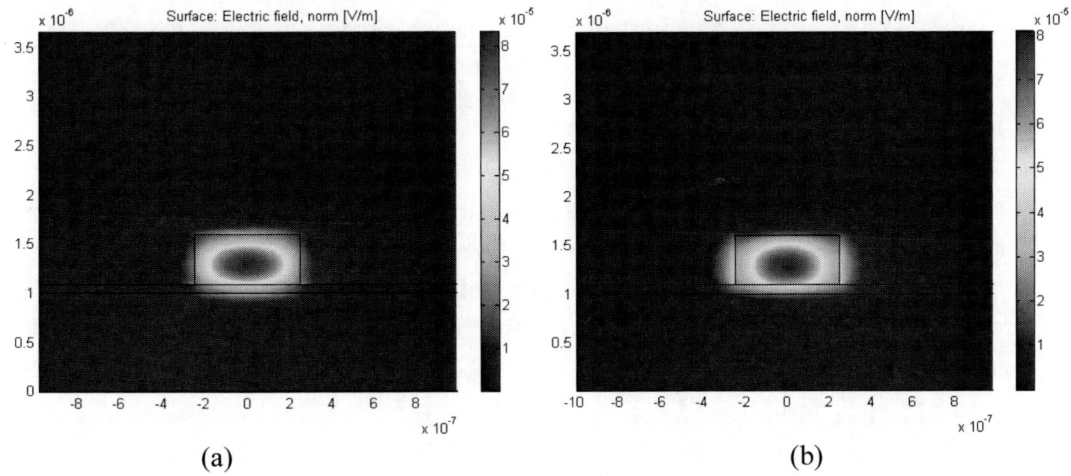

(a) (b)

Figure 5. Electric field distribution: (a) quasi-TM mode; (b) quasi-TE mode.

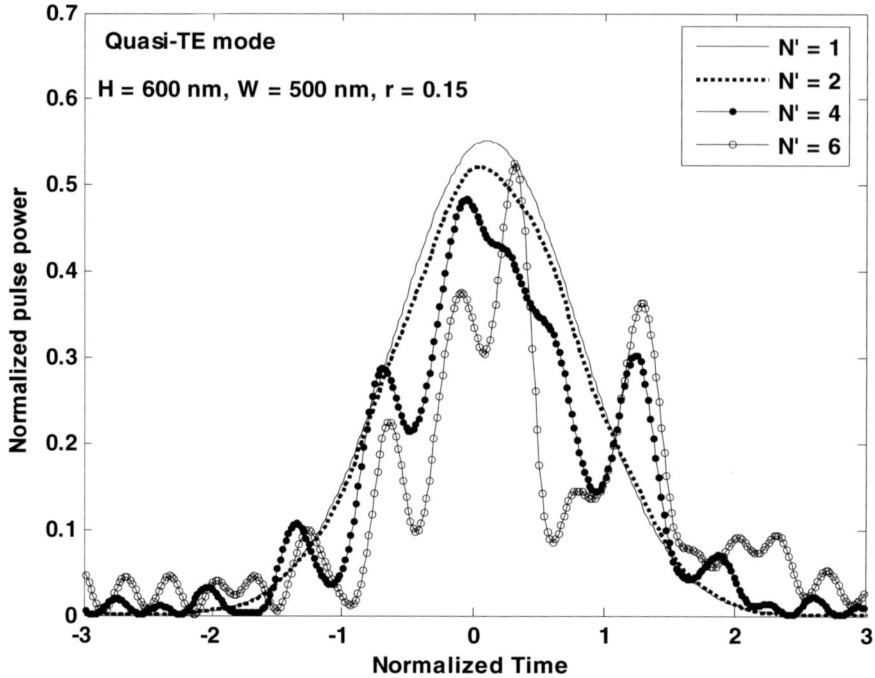

Figure 6. Temporal evolution of Gaussian pulse after 5 mm propagation, for different values of N'.

In case of $\beta_2 = 0$, it is convenient to introduce the number $N' = \sqrt{\gamma P_0 T_0^2 / |\beta_3|}$, that governs the relative importance of TOD and SPM effects during pulse evolution [14]. In particular, TOD dominates for N'□ 1, while SPM dominates for N'□ 1. In Figure 6, the temporal profiles of pulses at the waveguide output (after 5 mm propagation), calculated for several values of N' with T_{FWHM} = 100 fs, are sketched.

The dominant nonlinear effect seen in Figure 6 is self-phase modulation (SPM), its signature being the temporal compression. When acting alone, SPM broadens the pulse spectrum and leaves the temporal pulse profile unchanged [14]. In addition, the optical pulses preserve their symmetric profiles both in time and frequency domains. However, with increasing the values of N', TOD breaks the temporal symmetry of pulse profiles, as it is clearly illustrated in Figure 6. In fact, as the interplay between SPM and TOD increases, a strong pulse asymmetry and pulse reshaping occur. Thus, an oscillation in the pulse shape can be observed as N' increases from 1 (P_0 = 2 mW) to 6 (P_0 = 50 mW).

The next step is to isolate the GVD effect, by setting to zero the TOD coefficient. Using the parametric simulations of β_2 and β_3 as a function of wavelength, we have found for quasi-TE modes the condition $\beta_3 = 0$ and $\beta_2 < 0$ for H = 400 nm, λ = 1641.6 nm, W = 500 nm and r = 0.3. Thus, if this SOI waveguide is used to generate optical solitons by means of compensation between SPM and GVD effects at 1641.6 nm, a birefringence value $\Delta n_{eff} \approx -0.1785$ and a characteristic birefringence length $L_B \cong 9.19$ μm occur, too. Since waveguides with L□ L_B are needed to generate optical solitons, then the terms containing $k_{i,j,k,l}$ often change sign and the total phase contribution due to birefringence averages out to zero. In the following simulations, we have considered the SOI waveguide as excited by only one Gaussian optical pulse with $T_{FWHM} = 80$ fs, and peak power selected to satisfy the condition $N = \sqrt{L_D / L_{NL}} = 2.2$ ($P_0 = 8.38$ W). Figure 7 shows the temporal shape of optical pulse aligned as a quasi-TE mode at different propagation lengths, assuming $\beta^{(TPA)} = 0.45$ cm/GW, $n_2 = 6 \times 10^{-5}$ cm^2/GW, $\alpha^{(prop)} = 5$ dB/cm. The other parameters are 1.25ns, $n_{eff} = 2.8025$, $A_{eff} = 0.204$ μm^2, $\beta_2 = $ -0.449 ps^2/m.

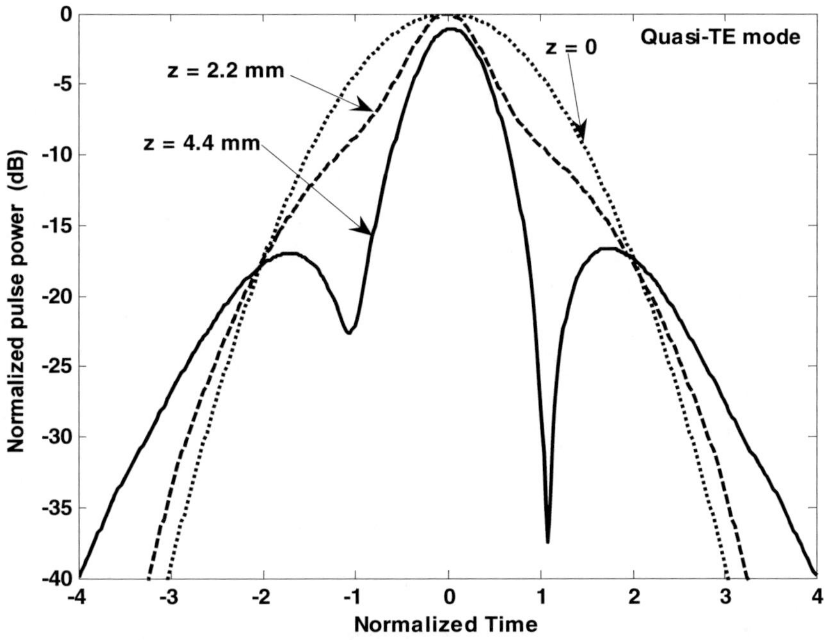

Figure 7. Temporal evolution of Gaussian pulse at different propagation lengths for TE polarization.

The input Gaussian pulse propagates inside the SOI waveguide as a soliton, having a *sech*-like profile starting from a propagation length of $z = 2.2$ mm and not influenced by TOD effect. In addition, the compression factor shows a reduction from the value 2.22 to 2.44 with changing the propagation length from z =2.2 mm to z =4.4 mm. The physical comments induced by Figure 7 are similar to the conclusions presented in a experimental work in literature [38], where a SOI waveguide with $H = 400$ nm, $W = 860$ nm and $H_s = 100$ nm is used to demonstrate, for the first time, the formation of solitons. According with [38], a quasi-TM ultrafast pulse is launched at λ = 1.484 μm with a Gaussian shape and FWHM width $T_{FWHM} = 116$ fs. The waveguide length is L=5 mm, larger than both nonlinear length $L_{NL} = (\gamma P_0)^{-1}$ and dispersion length $L_D = T_0^2 / |\beta_2|$. As a result, the interplay between SPM and GVD causes the pulses to evolve into a soliton. Figure 8 shows the simulated shape of output pulse by assuming $\beta^{(TPA)} = 0.45$ cm/GW, $n_2 = 6 \times 10^{-5}$ cm^2/GW and $\alpha_i^{(prop)} =$ 5 dB/cm [38]. The waveguide proposed has also an anomalous GVD of -2.15 ps^2/m for quasi-TM modes. The input peak power has been selected to realise the condition $\sqrt{L_D / L_{NL}} \approx 1.35$. In the plot of Figure 8, the dotted line designates the input Gaussian pulse, while the curve with triangle markers indicates the output pulse calculated with M =45. Our numerical investigations show a very good agreement with results presented in [38], by choosing a number M of Hermite-Gauss functions larger than 35. The accuracy improves by increasing M, but this improvement is not significant for $M > 45$.

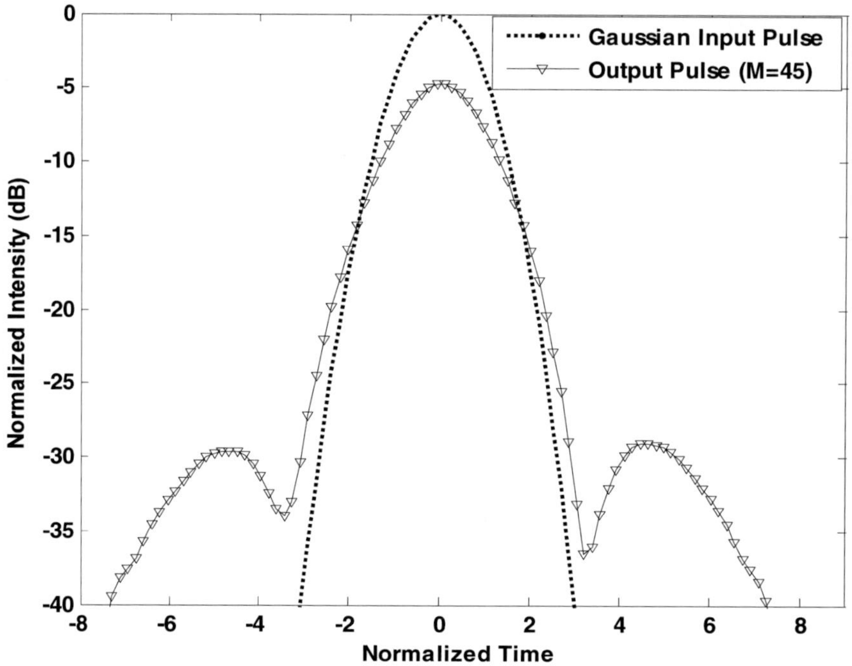

Figure 8. Normalized intensity versus normalized time for input and output pulses.

Thus, $35 \le M \le 45$ represents the best trade-off between accuracy and calculation time, demonstrating that the collocation method can improve the calculation time (as in our case) of about three orders of magnitude for the same accuracy [40], if compared with well known split-step Fourier method [14]. The curve asymmetry in the temporal shape is demonstrated by our results to be induced by self-steeping effect, included in Eq. (11) as first order time derivative. Moreover, by comparing the main lobes of Gaussian input and output pulses (see Figure 8), the output pulse can be appropriately fitted by a *sech*-like pulse, confirming that the interplay between the SPM and GVD effects has caused the formation of the fundamental soliton [38]. Because solitons result from a balance between the nonlinear and dispersive effects, the pulse must maintain its peak power in order to preserve its soliton character. Waveguide losses represent detrimental effects simply because they reduce the peak power of the optical pulse along the waveguide length. Differently from optical fibers where the linear propagation losses represent the main contribute, FCA induced by TPA effect adds its detrimental effect to propagation losses in SOI waveguides. Thus, it is interesting to compare the results proposed in Figure 7 with the virtual case in which both linear losses, FCA and TPA are neglected. Figure 9 shows the time-space evolution of a Gaussian pulse polarized as a quasi-TE mode with $T_{FWHM} = 80$ fs, and $N = \sqrt{L_D / L_{NL}} = 2.2$, as in Figure 7.

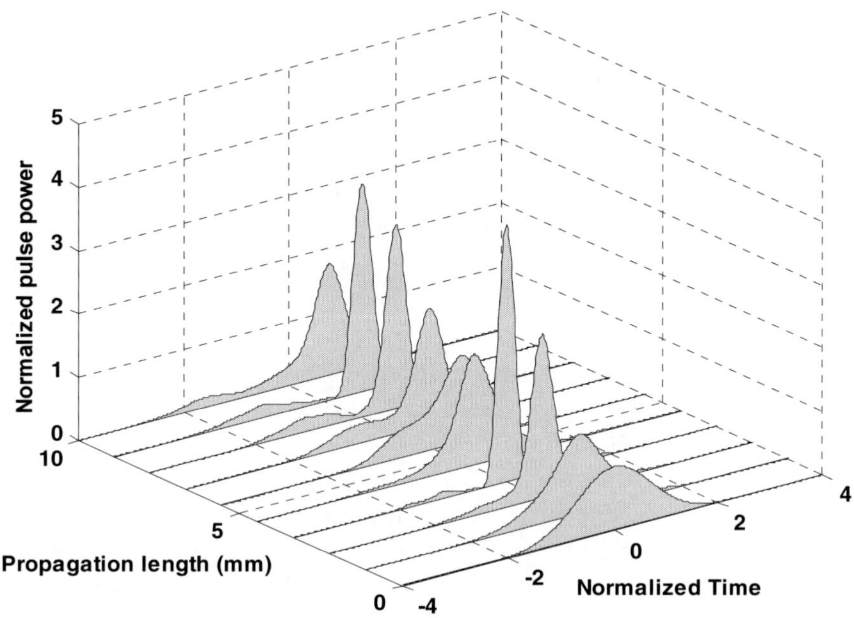

Figure 9. Space-time evolution of optical soliton.

Periodic evolution of a second-order soliton over one period is shown in Figure 9. As the pulse propagates along the SOI waveguide, it first converts its shape into a hyperbolic-secant pulse starting from the Gaussian profile. Moreover, after a propagation length of 5 mm, the optical pulse is affected by a temporal broadening and distortion, and then merges again to recover the compressed hyperbolic-secant shape. This pattern is rigorously periodic. The origin of this periodic evolution results from an interplay between SPM and GVD effects. It is worth to remind that the SOI waveguide considered in the simulation has $\beta_3 = 0$ for quasi-TE modes, and both linear propagation losses and TPA effect have been set to zero to avoid the instability effects on soliton propagation. Thus, in this condition SPM generates spectral broadening and dominates initially but GVD soon catches up and leads to pulse contraction. In conclusion, we can observe as the presence of optical losses, TPA and FCA enable the periodic evolution of this higher order soliton, but not allowing its sech-shape.

In conclusion, in case when GVD, TOD, and SPM act together, the input Gaussian pulse tends to become asymmetric and distort with an oscillatory behaviour near one of pulse trailing edge. In addition, the generation of a soliton peak can appear if the pulse wavelength drops into the anomalous region. In fact, when optical pulses propagate relatively far from the zero-dispersion wavelength of SOI waveguide, TOD effects on solitons are small and can be considered like a perturbation. Physically speaking, TOD slows down the soliton and, as a result, the soliton peak is delayed by an amount that increases linearly with distance.

3.2. Two Pulse Propagation

Due to relatively large physical dimensions of wavelength-division-multiplexing (WDM) components on a chip, it has recently been proposed to use ultrashort pulses in conjunction with optical time-division multiplexing (OTDM) for achieving high transmission rates [36]. Thus, it is important to understand the dynamics of ultrashort pulse propagation in SOI waveguides, both in linear and nonlinear regimes. In particular, as discussed before, the dispersive properties of these waveguides can be tailored through the most appropriate choice of waveguide dimensions. While in the previous section only a single optical pulse is assumed to propagate inside the SOI waveguide with optimized cross section (for TE polarization), in this section we investigate the pulse compression effect considering two input Gaussian pulses. This condition could correspond to a situation good for WDM schemes, in which wavelength difference between two neighbouring pulses can be very small, around 0.4 ÷ 1 nm.

When two or more optical waves having different wavelengths propagate simultaneously inside the SOI waveguide, they can interact to each other through the waveguide nonlinearity. Generally speaking, such an interaction can generate new waves under appropriate conditions through a variety of nonlinear phenomena such as SRS and four-wave mixing (FWM). If the input pulses do not meet these appropriate conditions, the two waves can also couple by means of XPM, without inducing any energy transfer. As it is known, XPM effect is always accompanied by SPM effect and it occurs because the effective index seen by the optical wave in a nonlinear medium, such as silicon, depends not only on the intensity of the wave itself, but also on the intensity of other co-propagating pulses [14]. The presence of XPM and SPM effects can induce interesting phenomena, depending on normal or anomalous dispersion regime in the SOI waveguide. Before showing the numerical solutions of the system equations (3)-(7), it is interesting to evaluate how β_2 and β_3 coefficients are changed as a function of wavelength. Figure 10 shows both GVD and TOD coefficients spectra for both polarizations, for a SOI waveguide having $H = 700$ nm, $W = 500$ nm, and $r = 0.15$.

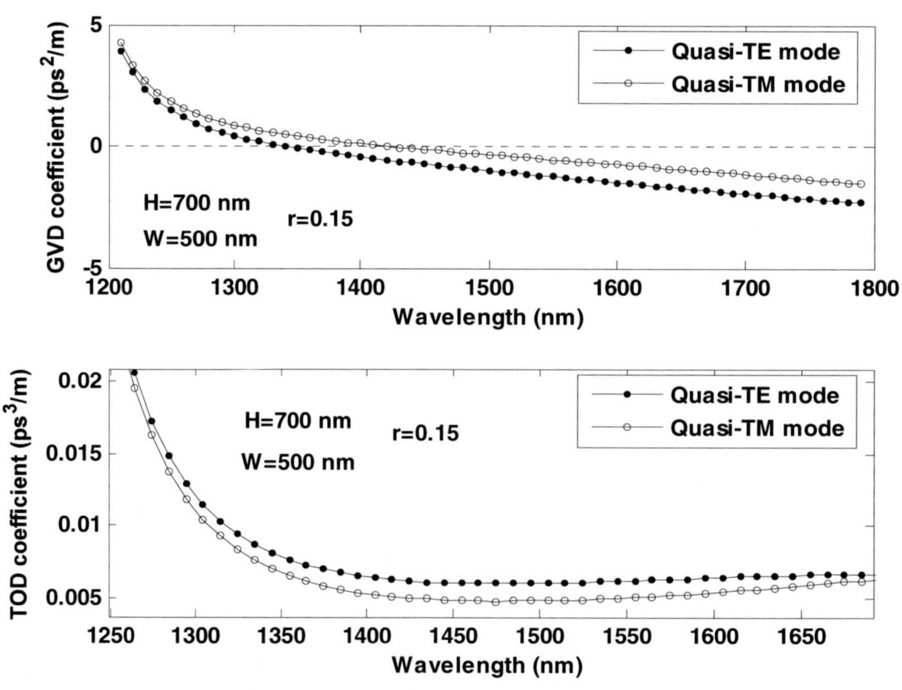

Figure 10. GVD and TOD coefficient spectra for both polarizations.

The plot indicates for the assumed SOI waveguide a GVD coefficient into anomalous region for both polarizations. In particular, the zero-GVD wavelength is larger for quasi-TM than for quasi-TE modes. The second subplot shows as TOD coefficient presents a flat shape in the wavelength range 1450 ÷ 1550 nm, for both polarizations. Thus, we can conclude that the designed SOI waveguide allows two optical pulses to be coupled through XPM effect, without any appreciable change of GVD and TOD coefficients, even though quasi-TE modes are used.

Figure 11 shows the space-time evolution of two Gaussian pulses aligned as a quasi-TM mode, and having $\lambda_1 = 1500$ nm, and $\lambda_3 = 1500.8$ nm, respectively. The parameters used include $T_{FWHM} = 100$ fs, $\beta^{(TPA)} = 0.45$ cm/GW, $n_2 = 6\times10^{-5}$ cm^2/GW, $\alpha^{(prop)} = 5$ dB/cm, $\tau_{eff} = 1.44$ ns $n_{eff,1} = 3.0856$, $n_{eff,3} = 3.0855$, $A_{eff,1} \cong A_{eff,3} = 0.283$ μm^2, $\beta_{2,1} = -0.3337$ ps^2/m, $\beta_{2,3} = -0.334$ ps^2/m, $\beta_{3,1} = 4.817\times10^{-3}$, and $\beta_{3,3} = 4.819\times10^{-3}$. The peak powers of input pulses are selected to satisfy the condition $N = \sqrt{L_{D,1}/L_{NL,1}} = \sqrt{L_{D,3}/L_{NL,3}} = 3$ ($P_{01} \cong P_{03} =$ 9.38 W).

It is worth noting that the pulse walk-off and the Raman amplification are negligible because of the small difference between the optical pulse wavelengths. This can correspond to some WDM schemes, where the wavelength difference between two pulses can be 0.4 ÷ 1 nm (in the simulation a separation of 100 GHz was assumed). In case of Figure 11, the pulses are coupled mainly due to XPM and TPA effects. The XPM-induced coupling among optical fields give rise to a number of interesting nonlinear effects in optical SOI waveguides. In Figure 11, the numerical solutions indicate the compression of two pulses at $z \cong 2$ mm.

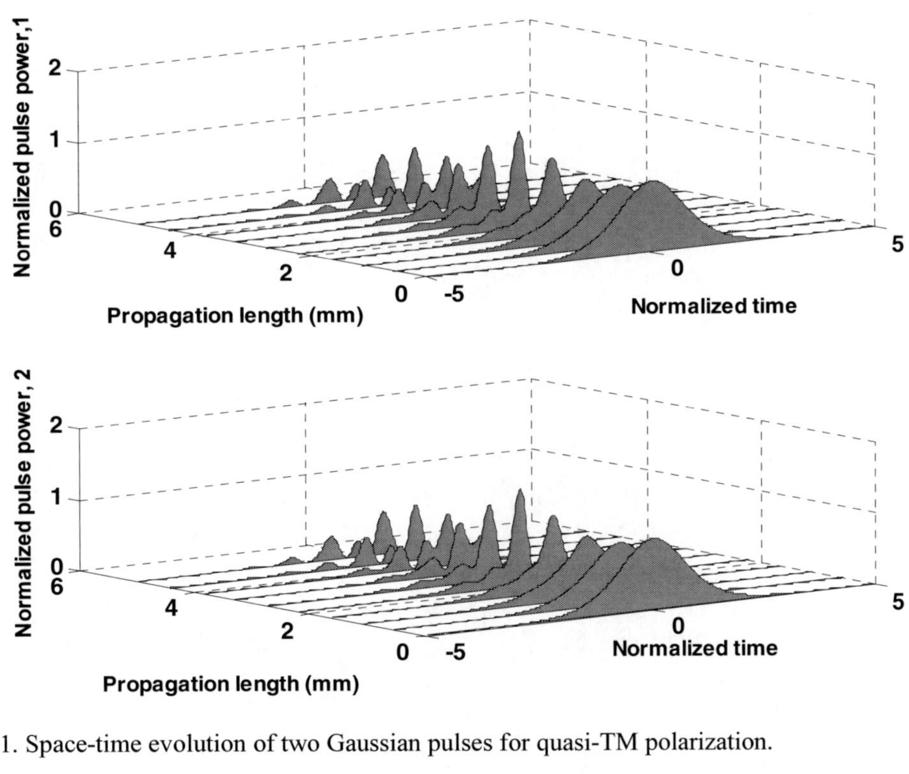

Figure 11. Space-time evolution of two Gaussian pulses for quasi-TM polarization.

This compression is induced by the conversion of input Gaussian pulses into a *sech*-like profile. Rigorously speaking, the sech-like profiles do not designate solitons in a strict mathematical sense, but they should be more accurately referred to solitary waves, providing to realize a compression factor of 4.67, stating from two input Gaussian pulses with $T_{FWHM} = 100$ fs. In addition, 3D simulations indicate a reshaping with an oscillatory behavior for a propagation length $z \cong 5$ mm. Our investigations show that, using a value $N = \sqrt{L_{D,1}/L_{NL,1}} = \sqrt{L_{D,3}/L_{NL,3}} = 2$ ($P_{01} \cong P_{03} =$ 4.17 W), *sech*-like profile until $z \cong 5$ mm with a compression factor of 2.68 can be achieved.

Finally, Table 1 shows the compression factor values calculated by solving the equations system (3)-(10) for different values of T_{FWHM}, assuming $N = \sqrt{L_{D,1}/L_{NL,1}} = \sqrt{L_{D,3}/L_{NL,3}} = 2$, and $z \cong 5$ mm.

Table 1. Compression factor for soliton pulses

FWHM time width	Compression factor
80 fs	~ 3.58
90 fs	~ 3.58
100 fs	~ 2.68
110 fs	~ 1.79
120 fs	~ 1.38
150fs	~ 1.2
180 fs	~ 1

CONCLUSION

In this chapter a general model has been presented and applied to simulate ultrafast nonlinear pulses in submicrometer-scale SOI optical rib waveguides. Soliton formation and pulse compression, considering one or two input pulses, have been investigated by considering optimal rib structures.

REFERENCES

[1] Reed, G. T., *Nature*, 2004, 427, 595-596.

[2] Reed, G. T., Knights, A. P., *Silicon Photonics: An Introduction*, John Wiley, 2004.

[3] Claps, R., Dimitropoulos, D., Han, Y., Jalali, B., *Optics Express*, 2002, 10, 1305-1313.

[4] [4] Claps, R., Dimitropoulos, D., Raghunathan, V., Han, Y., Jalali, B., *Optics Express*, 2003, 11, 1731-1739.

[5] Liu, A., Rong, H., Paniccia, M., Cohen, O., Hak, D., *Optics Express*, 2004, 12, 4261-4268.

[6] Xu, Q., Almeida, V. R., Lipson, M., *Optics Express*, 2004, 12, 4437-4442.

[7] Xu, Q., Almeida, V. R., Lipson, M., *Optics Letters*, 2005, 30, 35-37.

[8] Espinola, R. L., Dadap, J. I., Osgood, Jr., R. M., McNab, S. J., Vlasov, Y. A., *Optics Express*, 2004, 12, 3713-3718.

[9] Dimitropoulos, D., Raghunathan, V., Claps, R., Jalali, B., *Optics Express*, 2004, 12, 149-160.

[10] Claps, R., Raghunathan, V., Dimitropoulos, D., Jalali, B., *Optics Express*, 2003, 11, 2862-2872.

[11] Raghunathan, V., Claps, R., Dimitropoulos, D., Jalali, B., *J. Lightwave Technology,* 2005, 23, 2094-2102.

[12] Boyraz, O., Koonath, P., Raghunathan, V., Jalali, B., *Optics Express*, 2004, 12, 4094-4102.

[13] Jones, R., Liu, A., Rong, H., Paniccia, M., Cohen, O., Hak, D., *Optics Express*, 2005, 13, 1716-1723.

[14] Agrawal, G. P., *Nonlinear fiber optics*, Academic Press, London, 2001.

[15] Claps, R., Raghunathan, V., Dimitropoulos, D., Jalali, B., *Optics Express*, 2004, 12, 2774-2780.

[16] Liang, T. K., Tsang, H. Ki, *IEEE J. Selected Topics in Quantum Electronics,* 2004, 10, 1149-1153.

[17] Almeida, V. R., Barrios, C. A., Panepucci, R. R., Lipson, M., *Nature*, 2004, 431, 1081-1084.

[18] Preble, S. F., Xu, Q., Schmidt, B. S., Lipson, M., *Optics Letters*, 2005, 30, 2891-2893.

[19] Manolatou, C., Lipson, M., *J. Lightwave Technology,* 2006, 24, 1433-1439.

[20] Liang, T. K., Nunes, L. R., Sakamoto, T., Sasagawa, K., Kawanishi, T., Tsuchiya, M., Priem, G. R., Van Thourhout, D., Dumon, P., Beats, R., Tsang, H. K., *Optics Express*, 2005, 13, 7298-7303.

[21] Moss, D. J., Fu, L., Eggleton, B. J., *Electron. Letters*, 2005, 41, 320-321.

[22] Liang, T. K., Nunes, L. R., Tsuchiya, M., Abedin, K. S., Mayazaki, T., Van Thourhout, D., Bogaerts, W., Dumon, P., Beats, R., *Optics Communications*, 2006, 265, 171-174.

[23] Claps, R., Raghunathan, V., Boyraz, O., Koonath, P., Dimitropoulos, D., Jalali, B., *Optics Express*, 2005, 13, 2459-2466.

[24] Boyraz, O., Jalali, B., *Optics Express*, 2004, 12, 5269-5273.

[25] Krause, M., Renner, H., Brinkmeyer, E., *Optics Express*, 2004, 12, 5703-5710.

[26] Rong, H., Jones, R., Liu, A., Cohen, O., Hak, D., Fang, A., Paniccia, M., *Nature*, 2005, 433, 725-728.

[27] Rong, H., Liu, A., Jones, R., Cohen, O., Hak, D., Nicolaescu, R., Fang, A., Paniccia, M., *Nature*, 2005, 433, 292-294.

[28] Boyraz, O., Jalali, B., *Optics Express*, 2005, 13, 796-800.

[29] Rong, H., Kuo, Y-H., Xu, S., Liu, A., Jones, R., Paniccia, M., Cohen, O., Raday, O., *Optics Express,* 2006, 14, 6705-6712.

[30] De Leonardis, F., Passaro, V. M. N., *New Journal of Physics*, 2007, 9, art. 25.

[31] Passaro, V. M. N., De Leonardis, F., *J. Lightwave Technology,* 2006, 24, 2920-2931.

[32] De Leonardis, F., Passaro, V. M. N., *J. Lightwave Technology,* 2007, 25, 2352-2366.

[33] Jalali, B., Raghunathan, V., Shori, R., Fathpour, S., Dimitropoulos, D., Stafsudd, O., *IEEE J. Selected Topics in Quantum Electronics*, 2006, 12, 1618-1627.

[34] Raghunathan, V., Borlaug, D., Rice, R., Jalali, B., *Optics Express,* 2007, 15, 14355-14362.

[35] Dekker, R., Niehusmann, J., Forst, M., Driessen, A., *Proc. Symp. IEEE/LEOS,* Benelux Chapter, Eindhoven, 2006.
[36] Chen, X., Panoiu, N. C., Hsieh, I., Dadap, J. I., Osgood, Jr., R. M.*, IEEE Photonics Technology Letters*, 2006, 18, 2617-2619.
[37] Yin, L., Lin, Q., Agrawal, G. P., *Optics Letters*, 2007, 32, 391-393.
[38] Zhang, J., Lin, Q., Piredda, G., Boyd, R. W., Agrawal, G. P., Fauchet, P. M.*, Optics Express*, 2007, 15, 7682-7688.
[39] Soref, R. A., Bennett, B. R., *IEEE J. Quantum Electronics*, 1987, QE-23, 123-129.
[40] Deb, S., Sharma, A., *Optical Engineering*, 1993, 32, 695-699.
[41] COMSOL Multiphysics by COMSOL AB, Stockholm, ver. 3.2, *single license*, 2005.

In: Modeling of Photonic Devices
Editor: Vittorio M. N. Passaro

ISBN 978-1-60456-980-3

Chapter 8

RAMAN EFFECT IN SOI OPTICAL STRUCTURES

Francesco De Leonardis*[*1] and Vittorio M. N. Passaro[2]

[1]Photonics Research Group, Ingegneria dell'Ambiente e per lo Sviluppo Sostenibile, Politecnico di Bari, viale del Turismo n. 8, 74100 Taranto, Italy

[2]Photonics Research Group, Dipartimento di Elettrotecnica ed Elettronica, Politecnico di Bari, via Edoardo Orabona n. 4, 70125 Bari, Italy

ABSTRACT

In this chapter, stimulated Raman scattering, two photon absorption, free carrier dispersion, self- and cross-phase modulation induced by Kerr effect, walk-off, pump depletion and polarization coupling are investigated theoretically in silicon-on-insulator (SOI) optical waveguides. The propagation of both pump and Stokes pulses in picosecond and sub-picosecond regimes is presented. Comparisons between Raman amplification in straight waveguides and in optical resonators iscarried out. Finally, the guidelines for the design of Raman ring lasers based on SOI technology are proposed.

Keywords: Integrated Optics, Nonlinear Optics, Silicon-On-Insulator Technology, Modeling, Stimulated Raman Scattering, Kerr Effect, Picosecond and Femtosecond Pulses

1. INTRODUCTION

Over the next 5 to 10 years, the computing and telecommunications industries will face increasing challenges to deliver more data faster. Consumers will also require faster access to these huge amounts of data. While microprocessors (see Intel® platform) are projected to meet these future demands, the bandwidth of the interconnects needs to be increased to meet the speed of microprocessors [1]. As an example related to the limitations induced by the electrical interconnects, it is interesting to consider the total interconnection length. In fact,

---

* E-mail address: f.deleonardis@poliba.it

nowadays chips have total interconnection lengths per unit area of the chip of the order of 5 km/cm^2, with a chip area of 450 mm^2 and an expectation in the next ten years of 20 km/cm^2 for a chip area of 800 mm^2. But the problem is not only related to the length of interconnects, but also to the complexity of their architecture. Nowadays, there are six layers of metal levels, but in ten years from now there will be more than 12. All these facts introduce problems related to signal propagation delay causing *RC* coupling, signal latency, signal cross-talk and *RL* delays due to reduction in dimension and increase in density of the metal line [2]. A possible solution to these problems is the use of optical interconnects. Even if the optical interconnects through optical fibres and III–V laser sources are already used to connect different computers, the trend for the next 10-15 years is to realise optical interconnects for computer boards, and within the same chip. This trend could be achieved if the photonics industry will overcome some actual limitations. In fact, if one compares the Photonics industry with Microelectronics today, one can see the following differences: 1) a variety of different materials is used: InP as substrate for source development, silica as material for fibres, lithium niobate for modulators, other materials for DWDM and EDFA; 2) no single material or single technology is leading the market; 3) the industry is characterized by many different small companies which are specialized in specific devices; 4) the production technology is still very primitive. Chip scale integration of optical components, which enables low cost and high reproducibility, is not yet achieved.

It is commonly accepted that the industrial model of Microelectronics, if applied to Photonics, should be a booster to the development and implementation of photonics. In this sense, Silicon is the ideal platform for Integrated Optics and Optoelectronics. The quality of commercial silicon wafers driven by the Microelectronics industry continues to improve while the cost continues to decrease. Moreover, in addition to optical interconnects, the compatibility with silicon integrated circuits manufacturing and silicon Micro-Electro-Mechanical Systems (MEMS) technology is another important reason for the interest in Silicon Photonics [3-4]. In order to develop Silicon Photonics, there are six main areas or building blocks for investigation, including selectively guiding and transporting of the light within the silicon, encoding light, detecting light, light amplification and generation, packaging the devices and intelligent control of these photonic functions.

While a wide variety of passive devices were developed in the 1990's, recent activities have focused on achieving active functionality, mostly light amplification and generation, in silicon on insulator (SOI) waveguides. To this aim, various approaches have been investigated based on silicon engineered materials, including nanocrystals [5-6], Er-doped silicon oxides [7], Si/SiGe structures [8] and porous silicon [9]. However, each of the previous approaches can produce non trivial limitations. Recently a different approach for light generation has been proposed in literature, based on Stimulated Raman Scattering (SRS) effect. In fact, as a transmission medium, silicon has much higher nonlinear effects than the commonly used silicon dioxide, in particular the Raman effect. Raman gain has been already successfully exploited in fiber amplifiers and lasers, but usually several kilometres of fibre are required to create a useful device. Fortunately, the gain coefficient for SRS is approximately four orders of magnitude higher in silicon than in silica. Additionally, SOI waveguides can confine the optical field to an area that is approximately 100 times smaller than the modal area in a standard single-mode optical fiber. The combination of these properties is making SRS observable over the millimetre-scale interaction length, usually encountered in integrated optical devices. Recently, several experimental and theoretical studies have been proposed in literature exploiting this effect, such as Raman amplification in SOI waveguides [10-16], Stokes and anti-Stokes Raman conversion [17-19], cross phase modulation-based interferometer switch [20], two-photon absorption [21-22] and lossless modulation [23]. Moreover, the need for active devices in a SOI integrated platform has stimulated an increasing amount of experimental and theoretical research in light generation in silicon based on Raman effect [24-32]. More recently, the large interest for optical devices operating in the mid-IR has driven the research to exploit the Raman effect in the wavelength range 2.2 ÷ 5.3 μm. The works proposed by *Jalali et al.* [33-34] represent the first studies to investigate the prospects for Raman silicon laser and amplification in the mid-IR wavelength range.

This chapter summarizes a very generalized modeling to investigate all nonlinear effects involved in the propagation of optical pulses in SOI rib waveguides in picosecond and subpicosecod regimes. In the modeling section, the theoretical model is presented in a very general formalism, considering Stimulated Raman Scattering, Two Photon Absorption (TPA), Free Carrier Absorption (FCA) where the free carriers are generated mainly by TPA of the pump pulse, plasma dispersion, polarization coupling induced by the waveguide birefringence, Self-Phase-Modulation (SPM) and Cross-Phase-Modulation (XPM) as induced by Kerr nonlinearity, Group-Velocity Dispersion (GVD), Third Order Dispersion (TOD), self steeping, and higher order nonlinear effects as induced by the delayed Raman response. The model includes the interaction between the pump pulse and the fundamental Stokes wave, as well as the interactions with first and second order Stokes waves. The mathematical formalism to treat these nonlinear processes involves a system of partial differential equations in order to calculate the space and time evolution of both polarization components (quasi-TE and quasi-TM) for each wave in SOI structures. In the Numerical Results section, a number of theoretical results are shown, including comparisons between our solutions and some results proposed in literature. The numerical results proposed are classified in picosecond regime and sub-picosecond regime. In the former, we have mainly focused on the evaluation of net Raman gain in straight waveguides and optical resonators. In the latter, a parametric study based on finite element method (FEM) is developed to individuate the optimal waveguide cross-section with low GVD effects, the threshold conditions for generating higher order Stokes waves and the space-time pulse evolution in femtosecond regime.

2. MODELING

Despite the large number of experimental works, only a few models have been used to investigate the Raman effect in waveguide structures, usually neglecting the time evolution of propagating pulses. In addition, some important nonlinear effects (pump depletion, Kerr effect, plasma dispersion, walk-off, delayed Raman response) are usually not considered. For example, in the theory proposed in [12, 21, 26], the mathematical models involve only the differential equations for the intensities of pump and Stokes waves under continuous wave (CW) operation. Further, other models [17-20] include the equations for studying both the intensity and phase of the electric field of pump and Stokes beams only in CW regime. Raman gain dynamics in SOI waveguides is investigated by two coupled rate equations where an effective pump power for time resolution is used [15]. Finally, the coupled-mode theory as proposed in [35] does not include the rate equation for the pump pulse, considers a partial polarization coupling and neglects both the pump depletion induced by Raman and TPA effects, as well as the carrier density time dynamics. The quasi-CW regime model of SRS in SOI waveguides, as considered in literature, applies for experimental investigation where the pump pulses used has a time width $T_0 \geq 1$ ns, since the walk-off length L_w, generally exceeds the waveguide length L. However, for ultrashort pulses with $T_0 < 100$ ps, typically $L_w \leq L$. SRS is limited by the group-velocity mismatch and occurs only over distances $z \,\square\, L_w$, although the actual waveguide length L is considerable larger than L_w. At the same time, the non linear effects such as SPM and XPM can become important and affect the evolution of both pump and Raman pulses. Thus, in this situation the time derivative cannot be neglected and appropriate numerical techniques are necessary to solve the partial equations system that govern the mutual interaction between Raman and pump pulses.

Thus, in the case of ultrashort pulses, the starting point of the model is the wave equation written in terms of linear and nonlinear polarization vector [36]:

$$\nabla^2 \mathbf{E} - \frac{1}{c^2}\frac{\partial^2 \mathbf{E}}{\partial t^2} = \mu_0 \frac{\partial^2 \mathbf{P}_L}{\partial t^2} + \mu_0 \frac{\partial^2 \mathbf{P}_{NL}}{\partial t^2} \qquad (1)$$

where c is light velocity, μ_0 is the vacuum permeability, and the linear ($\mathbf{P}_L$) and nonlinear ($\mathbf{P}_{NL}$) parts of induced polarization vector are related to the electric field $\mathbf{E}(\mathbf{r},t)$ through the relationship [36]:

$$\mathbf{P}_L(\mathbf{r},t) = \varepsilon_0 \int_{-\infty}^{+\infty} \chi^{(1)}(t-t') \cdot \mathbf{E}(\mathbf{r},t)dt' \qquad (2)$$

$$\mathbf{P}_{NL}(\mathbf{r},t) = \varepsilon_0 \iiint_{-\infty}^{+\infty} \chi^{(3)}(t-t_1, t-t_2, t-t_3) \vdots \mathbf{E}(\mathbf{r},t_1)\mathbf{E}(\mathbf{r},t_3)\mathbf{E}(\mathbf{r},t_3)dt_1 dt_2 dt_3 \qquad (3)$$

being ε_0 the vacuum permittivity, $\chi^{(1)}(t)$, and $\chi^{(3)}(t)$ the linear and third order susceptibility respectively . It is worth to note that the Fourier transform of $\chi^{(1)}(t)$ can be related to both refractive index and absorption coefficient of silicon [36]. In addition, the third-order effects such as Kerr effect, TPA, and SRS are included by means of the susceptibility $\chi^{(3)}(t)$.

Considerable simplification occurs if the nonlinear response is assumed to be instantaneous, so that the time dependence of $\chi^{(3)}(t)$ in Eq. (3) is given by the product of three Dirac delta functions of the form $\delta(t-t_i)$ ($i = 1,2,3$) [36]. This assumption can be accepted for Raman effect if the optical pulses involved in the process have pulse time widths larger that 1 ps [16]. Rigorously speaking, both electrons and nuclei respond to the optical field in a nonlinear way. The nuclei response is inherently slower if compared with the electronic response. Thus, to remove the hypothesis of an instantaneously responding medium as assumed in [16] and unify the theory for both picosecond and subpicosecond regimes, the functional form for the third-order susceptibility has to be assumed as [36]:

$$\chi^{(3)}(t-t_1,t-t_2,t-t_3)=\chi^{(3)}R(t-t_1)\delta(t-t_2)\delta(t-t_3) \tag{4}$$

where $R(t)$ is the nonlinear response function normalized in a manner similar to delta function, and given by [36]:

$$R(t)=(1-f_R)\delta(t)+f_R h_R(t) \tag{5}$$

where $\delta(t)$ indicates the Dirac delta function and f_R represents the fractional contribution of the delayed Raman response to nonlinear polarization P_{NL}. Thus, the response function $R(t)$ includes both electronic and vibrational (Raman) contributions. In addition, the Raman response function $h_R(t)$ is responsible for the Raman gain spectrum. In particular, the functional form of $h_R(t)$ can be deduced from the Raman gain spectrum, $g_R(\Delta\omega)$, known to exhibit a Lorentzian shape as induced by $g_R=4\omega_{s0}\mu_0\chi^R(\Omega_R)/(n_{eff,p}\cdot n_{eff,s0})$ [34], being μ_0 the magnetic susceptibility, $n_{eff,p}$ and $n_{eff,s0}$ the effective indices of pump and fundamental Stokes waves, respectively, and ω_{s0} the angular frequency of the Stokes pulse. Finally, the term $\chi^R(\Omega_R)$ represents the Raman–resonant susceptibility as [17]:

$$\chi^R(\omega_{s0};\omega_p-\omega_{s0},\omega_{s0})=\frac{2\Omega_R\Gamma_R\xi_R}{2j\Gamma_R\Delta\omega+\Omega_R^2-\Delta\omega^2} \tag{6}$$

where $\xi_R=11.2\times10^{-18}$ (m^2V^{-2}) [17] is the value of Raman susceptibility under resonance condition ($\Delta\omega=\Omega_R$), $\Gamma_R=2\pi\times53$ GHz is the resonance half-width, $\Omega_R=15.6$ THz is the Raman frequency shift between pump and fundamental Stokes wave in silicon, and $\Delta\omega=\omega_p-\omega_{s0}$. Now, the complete determination of nonlinear response function $R(t)$ still requires the knowledge of the fraction f_R. It can be estimated from the numerical value of the Raman gain peak through the normalization condition $\int h_R(t)dt=1$, as in [37, 38]. Assuming a gain peak value of 20 cm/GW (typical for silicon at $\lambda=1550$nm), it results $f_R=0.043$ [37], smaller than in silica fibers, i.e. 0.18 [36].

Eqs. (1)-(5) are the basic relationships used in this chapter to derive the mathematical modelling of the stimulated Raman scattering in the presence of any other nonlinear effect. In our analysis, we assume a SOI waveguide geometry as sketched in Figure 1, having rib total height H, slab height H_s and rib width W. In a single-mode SOI waveguide, two propagating modes are typically confined, one quasi-TE (dominant horizontal x-component of electric field) and one quasi-TM (dominant vertical y-component). Further, the assumption of translational invariance along the propagation direction (z) can be made in Eq. (1), thus the variable separation is applicable in the form $E(x,y,z,t)=C\cdot F(x,y)A(z,t)e^{j\beta_0 z}$, where β_0 is the propagation constant and $F(x,y)$ is the optical mode distribution in the waveguide cross section $x-y$ (solutions of Helmoltz wave equation), and C is a normalization constant, $C=1\Big/\sqrt{\int\int_{-\infty}^{+\infty}\left|F(x,y)\right|^2 dxdy}$.

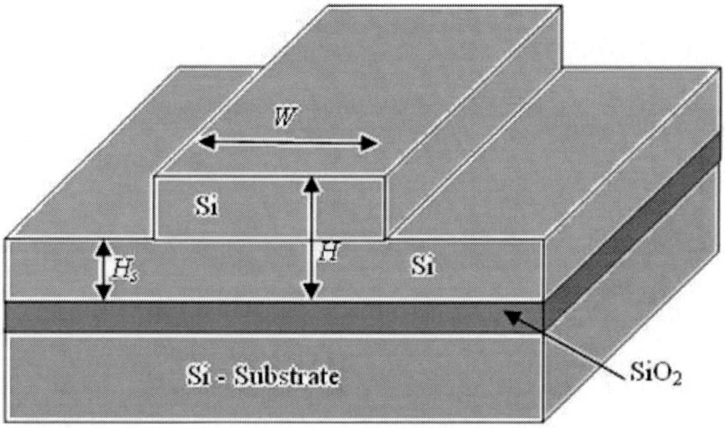

Figure 1. Schematic geometry of a SOI rib waveguide.

Thus, with the aim to obtain a set of coupled equations describing the power transfer among pump wave (p), fundamental Stokes wave (s_0), and higher-order Stokes (s_1, s_2) in presence of both polarized modes, the total electric field inside the SOI waveguide can be written without any leak of generality as [38]:

$$\mathbf{E}(x,y,z,t)=\hat{\mathbf{x}}\begin{bmatrix} C_1A_1(z,t)F_1(x,y)e^{j(\beta_{0,1}z-\omega_p t)}+ \\ +C_3A_3(z,t)F_3(x,y)e^{j(\beta_{0,3}z-\omega_{s0} t)}+ \\ +C_5A_5(z,t)F_5(x,y)e^{j(\beta_{0,5}z-\omega_{s1} t)}+ \\ +C_7A_7(z,t)F_7(x,y)e^{j(\beta_{0,7}z-\omega_{s2} t)} \end{bmatrix} +\hat{\mathbf{y}}\begin{bmatrix} C_2A_2(z,t)F_2(x,y)e^{j(\beta_{0,2}z-\omega_p t)}+ \\ +C_4A_4(z,t)F_4(x,y)e^{j(\beta_{0,4}z-\omega_{s0} t)}+ \\ +C_6A_6(z,t)F_6(x,y)e^{j(\beta_{0,6}z-\omega_{s1} t)}+ \\ +C_8A_8(z,t)F_8(x,y)e^{j(\beta_{0,8}z-\omega_{s2} t)} \end{bmatrix} \quad (7)$$

with the following meaning for superscripts and subscripts: $1=p^{(TE)}$, $2=p^{(TM)}$, $3=s_0^{(TE)}$, $4=s_0^{(TM)}$, $5=s_1^{(TE)}$, $6=s_1^{(TM)}$, $7=s_2^{(TE)}$, $8=s_2^{(TM)}$. In Eq. (7), ω_p, ω_{s0}, ω_{s1}, ω_{s2} are the angular frequencies of pump, fundamental Stokes, first and second order Stokes wave, respectively. Then, by following the procedure outlined in [16, 38] and assuming the nonlinear contributions to $\mathbf{P}_{NL}$ as a small perturbation of refractive index, we obtain the following coupled equations for pump and Stokes waves:

$$\frac{\partial A_\xi}{\partial z}+\beta_{1\xi}\frac{\partial A_\xi}{\partial t}+j\frac{1}{2}\beta_{2\xi}\frac{\partial^2 A_\xi}{\partial t^2}-\frac{1}{6}\beta_{3\xi}\frac{\partial^3 A_\xi}{\partial t^3}=-\frac{\left(\alpha_\xi^{(prop)}+\alpha_\xi^{(FCA)}\right)}{2}A_\xi-0.5\beta^{(TPA)}f_{\xi,\xi}\left|A_\xi\right|^2A_\xi+$$
$$-0.5\beta^{(TPA)}f_{\xi,\upsilon}\left|A_\upsilon\right|^2A_\xi+j\gamma_\xi(1-f_R)A_\xi\left(\left|A_\xi\right|^2+2\left|A_\upsilon\right|^2\right)+j2\gamma_\xi(1-f_R)\left|A_3\right|^2A_\xi+$$
$$+j2\gamma_\xi(1-f_R)\left|A_4\right|^2A_\xi+j\frac{2\pi}{\lambda_p}\Delta n_\xi A_\xi+j2k_{\xi,3,\upsilon,4}A_\upsilon A_3^*A_4e^{j(\beta_{0,\upsilon}+\beta_{0,4}-\beta_{0,\xi}-\beta_{0,3})z}+\qquad \xi=1,2$$
$$+j2k_{\xi,4,\upsilon,3}A_\upsilon A_4^*A_3e^{j(\beta_{0,\upsilon}+\beta_{0,3}-\beta_{0,\xi}-\beta_{0,4})z}+jk_{\xi,\xi,\upsilon,\upsilon}A_\upsilon A_\xi^*A_\upsilon e^{j(2\beta_{0,\upsilon}-2\beta_{0,\xi})z}+$$
$$+R_{p0}(z,t)-\frac{\gamma_{\xi,\xi}}{\omega_p}\frac{\partial\left(\left|A_\xi\right|^2A_\xi\right)}{\partial t}\qquad (8)$$

the following for quasi-TE and quasi-TM fundamental ($\kappa=3,4$) and first order Stokes ($\kappa=5,6$) pulses:

$$\frac{\partial A_\kappa}{\partial z}+\beta_{1\kappa}\frac{\partial A_\kappa}{\partial t}+j\frac{1}{2}\beta_{2\kappa}\frac{\partial^2 A_\kappa}{\partial t^2}-\frac{1}{6}\beta_{3\kappa}\frac{\partial^3 A_\kappa}{\partial t^3}=-\frac{\alpha_\kappa^{(prop)}}{2}A_\kappa-\frac{1}{2}\alpha_\kappa^{(FCA)}A_\kappa-\beta^{(TPA)}f_{\kappa,2}\left|A_2\right|^2A_\kappa+$$
$$+j2(1-f_R)\gamma_\kappa\left|A_\mu\right|^2A_\kappa+j2(1-f_R)\gamma_\kappa\left|A_{\mu+1}\right|^2A_\kappa+j(1-f_R)\gamma_\kappa\left|A_\kappa\right|^2A_\kappa+j2(1-f_R)\gamma_\kappa\left|A_\nu\right|^2A_\kappa+\qquad \kappa=3,4,5,6$$
$$+j\frac{2\pi}{\lambda_\kappa}\Delta n_\kappa A_\kappa+j2k_{\kappa,\mu,\mu+1,\nu}A_{\mu+1}A_\mu^*A_\nu e^{j(\beta_{0,\mu+1}+\beta_{0,\nu}-\beta_{0,\mu}-\beta_{0,\kappa})z}+j2k_{\kappa,\mu+1,\mu,\nu}A_\mu A_{\mu+1}^*A_\nu e^{j(\beta_{0,\mu}+\beta_{0,\nu}-\beta_{0,\mu+1}-\beta_{0,\kappa})z}+$$
$$+jk_{\kappa,\kappa,\nu,\nu}A_\nu A_\kappa^*A_\nu e^{j(2\beta_{0,\nu}-2\beta_{0,\kappa})z}+j2k_{\kappa,\mu+4,\nu,\mu+5}A_\nu A_{\mu+4}^*A_{\mu+5}e^{j(\beta_{0,\nu}+\beta_{0,\mu+5}-\beta_{0,\kappa}-\beta_{0,\mu+4})z}-\beta^{(TPA)}f_{\kappa,1}\left|A_1\right|^2A_\kappa+$$
$$+j2k_{\kappa,\mu+5,\nu,\mu+4}A_\nu A_{\mu+5}^*A_{\mu+4}e^{j(\beta_{0,\nu}+\beta_{0,\mu+4}-\beta_{0,\kappa}-\beta_{0,\mu+5})z}+R_p(z,t)+R_s(z,t)-\frac{\gamma_{\kappa,\kappa}}{\omega_\kappa}\frac{\partial\left(\left|A_\kappa\right|^2A_\kappa\right)}{\partial t}\qquad (9)$$

and, finally, for quasi-TE and quasi-TM second order Stokes (q = 7,8) pulses:

$$\frac{\partial A_q}{\partial z}+\beta_{1q}\frac{\partial A_q}{\partial t}+j\frac{1}{2}\beta_{2q}\frac{\partial^2 A_q}{\partial t^2}-\frac{1}{6}\beta_{3q}\frac{\partial^3 A_q}{\partial t^3}=-\frac{\left(\alpha_q^{(pt)}+\alpha_q^{(FCA)}\right)}{2}A_q-\beta^{(TPA)}f_{q,2}\left|A_2\right|^2A_q+$$
$$-\beta^{(TPA)}f_{q,1}\left|A_1\right|^2A_q+j2\gamma_q(1-f_R)\left|A_5\right|^2A_q+j\gamma_q(1-f_R)\left|A_6\right|^2A_q+$$
$$+j\gamma_q(1-f_R)\left|A_q\right|^2A_q+j2\gamma_q(1-f_R)\left|A_\zeta\right|^2A_q+j\frac{2\pi}{\lambda_{s2}}\Delta n_qA_q+\qquad q=7,8$$
$$+j2k_{q,5,6,\zeta}A_6A_5^*A_\zeta e^{j(\beta_{0,6}+\beta_{0,\zeta}-\beta_{0,q}-\beta_{0,5})z}+j2k_{q,6,5,\zeta}A_5A_6^*A_\zeta e^{j(\beta_{0,5}+\beta_{0,\zeta}-\beta_{0,6}-\beta_{0,q})z}+$$
$$+jk_{q,q,\zeta,\zeta}A_\zeta A_q^*A_\zeta e^{j(2\beta_{0,\zeta}-2\beta_{0,q})z}+R_{s2}(z,t)-\frac{\gamma_{q,q}}{\omega_{s2}}\frac{\partial\left(\left|A_q\right|^2A_q\right)}{\partial t}\qquad (10)$$

In Eq. (8), subscript $\upsilon = 2,1$ holds for $\xi = 1,2$, respectively. In Eq. (9), if $\kappa = 3,4$, it results $\omega_\kappa = \omega_{s0}$, $\mu = 1$ and $\nu = 4,3$, while if $\kappa = 5,6$, it results $\omega_\kappa = \omega_{s1}$, $\mu = 3$ and $\nu = 6,5$. In conclusion, Eq. (8) describes the time-space evolution of both quasi-TE and quasi-TM pump pulses, Eq. (9) is written for fundamental and first order Stokes pulses of both polarizations, and Eq. (10) for the second order Stokes pulses of both polarizations. In equation system (8-(10), the asterisk denotes the complex conjugate. The last terms in Eqs. (8)-(10) take into account the self-steeping effect, as induced by including the first time derivative of P_{NL} in the wave equation. Moreover, in the previous equations we have indicated with:

$$\begin{aligned} R_{p0}(z,t) &= j\gamma_\xi f_R A_\xi(z,t') \int h_R(t-t') \left[\left| A_\xi(z,t') \right|^2 + \left| A_\kappa(z,t') \right|^2 \right] dt' + \\ &+ j\gamma_\xi f_R A_\kappa(z,t') \int h_R(t-t') A_\xi(z,t') A_\kappa^*(z,t') e^{-j\Omega_R(t-t')} dt' \end{aligned} \tag{11}$$

with $\xi = 1,2$ κ=3,4,

$$\begin{aligned} R_p(z,t) &= j\gamma_\kappa f_R A_\kappa(z,t') \int h_R(t-t') \left[\left| A_\kappa(z,t') \right|^2 + \left| A_{\bar{\kappa}}(z,t') \right|^2 \right] dt' + \\ & j\gamma_\kappa f_R A_{\bar{\kappa}}(z,t') \int h_R(t-t') A_\kappa(z,t') A_{\bar{\kappa}}^*(z,t') e^{-j\Omega_R(t-t')} dt' \end{aligned} \tag{12}$$

with $\kappa = 3,4;$ $\bar{\kappa} = 5,6$,

$$\begin{aligned} R_s(z,t) &= j\gamma_\kappa f_R A_\kappa(z,t') \int h_R(t-t') \left[\left| A_\kappa(z,t') \right|^2 + \left| A_\eta(z,t') \right|^2 \right] dt' + \\ & j\gamma_\kappa f_R A_\eta(z,t') \int h_R(t-t') A_\kappa(z,t') A_\eta^*(z,t') e^{+j\Omega_R(t-t')} dt' \end{aligned} \tag{13}$$

with $\eta = 1,2$ if $\kappa = 3,4$ or $\eta = 3,4$ if $\kappa = 5,6$. Finally, it results:

$$\begin{aligned} R_{s2}(z,t) &= j\gamma_q f_R A_q(z,t') \int h_R(t-t') \left[\left| A_\kappa(z,t') \right|^2 + \left| A_q(z,t') \right|^2 \right] dt' + \\ & j\gamma_q f_R A_k(z,t') \int h_R(t-t') A_q(z,t') A_\kappa^*(z,t') e^{+j\Omega_R(t-t')} dt' \end{aligned} \tag{14}$$

with $\kappa = 5,6;$ q=7,8.

The relationships (11)-(14) take into account both SRS effect and higher order nonlinear effects induced by the delayed Raman response, such as the intrapulse Raman scattering [36]. In particular, the terms proportional to $e^{-j\Omega_R(t-t')}$ produce the pulse depletion induced by SRS effect, while the terms including $e^{+j\Omega_R(t-t')}$ are responsible for the Raman gain. Moreover, the coefficients $\gamma_{i,i} = n_2 \omega_i f_{i,i} / c$, being c the light velocity and n_2 the nonlinear refractive index, depend on SPM and XPM effects as induced by Kerr nonlinearity. The terms $k_{i,j,k,l} = n_2 \omega_i f_{i,j,k,l} / c$ with $i, j = 1,...,8$, lead to consider the birefringence effects, as explained in previous chapter 7. The overlap integrals $f_{i,j}$ and $f_{i,j,k,l}$, $i, j, k, l = 1,...,8$, are also given in chapter 7. Of course, the system of equations degenerates for a polarization insensitive structure, where the coupling between polarizations vanishes. Whenever the pump is aligned with the quasi-TE polarized mode (as usually occurs in the most experimental set-ups), the second equation for the quasi-TM pump in (8) is not applicable and all coefficients $k_{i,j,k,l}$ are zero. However, Eqs. (9)-(10) hold their validity, because a quasi-TE pump pulse could in principle generate Stokes waves of both polarizations. As indicated by terms proportional to $\beta_{1,i}$, $\beta_{2,i}$, $\beta_{3,i}$ and $\beta^{(TPA)}$, the model includes walk-off effect, group-velocity dispersion, third order dispersion, and TPA effect as induced by the pump pulse propagation, respectively. Moreover, the total loss coefficient in system (8)-(10) has been written as summation of two contributions, $\alpha_i^{(tot)} = \alpha_i^{(prop)} + \alpha_i^{(FCA)}$, where $\alpha_i^{(prop)}$ is the propagation loss coefficient in the rib waveguide (depending on the material absorption and fabrication process), and $\alpha_i^{(FCA)}$ is the FCA contribution, as induced by the change of the free carriers generated mainly by TPA of the pump pulse (see chapter 7).

In order to obtain a consistent mathematical model, we have coupled the system (8)-(10) to the rate equation governing the free carrier dynamics into the waveguide core as [12]:

$$\frac{dN_c}{dt} = -\frac{N_c}{\tau_{eff,\xi}} + \frac{\beta^{(TPA)}}{2\hbar\omega_p}\left(\left|A_{\xi,\xi}(z,t)\right|^2 f_{\xi,\xi}\right)^2 \tag{15}$$

where $\tau_{eff,\xi}$ is the relevant effective recombination lifetime for free carriers, $\xi = 1,2$.

Our model can be simplified without any leak of accuracy if we assume input pulse FWHM widths T_{FWHM} □ 10 fs (i.e., above 80 fs). In this case, the following Taylor-series expansion is applicable [38]:

$$A_i(z,t-t') \cong A_i(z,t) - t'\frac{\partial A_i(z,t)}{\partial t} \qquad i = 1,\ldots,8 \tag{16}$$

The previous relationship leads to avoid the integral terms in Eqs. (11)-(14). Then, by substituting for example Eq. (16) in (14), it becomes:

$$\begin{aligned} R_{s2}(z,t) = {} & j\gamma_q f_R A_q \left\{ \begin{bmatrix} \left|A_q\right|^2 - A_q \dfrac{\partial A_q^*}{\partial t}\dfrac{d\tilde{h}_R}{d\Delta\omega}\Big|_{\Delta\omega=0} + \\ -A_q^* \dfrac{\partial A_q}{\partial t}\dfrac{d\tilde{h}_R}{d\Delta\omega}\Big|_{\Delta\omega=0} + \\ + \dfrac{\partial A_q^*}{\partial t}\dfrac{\partial A_q}{\partial t}\dfrac{d^2\tilde{h}_R}{d\Delta\omega^2}\Big|_{\Delta\omega=0} \end{bmatrix} + \begin{bmatrix} \left|A_\kappa\right|^2 - A_\kappa \dfrac{\partial A_\kappa^*}{\partial t}\dfrac{d\tilde{h}_R}{d\Delta\omega}\Big|_{\Delta\omega=0} + \\ -A_\kappa^* \dfrac{\partial A_\kappa}{\partial t}\dfrac{d\tilde{h}_R}{d\Delta\omega}\Big|_{\Delta\omega=0} + \\ + \dfrac{\partial A_\kappa^*}{\partial t}\dfrac{\partial A_\kappa}{\partial t}\dfrac{d^2\tilde{h}_R}{d\Delta\omega^2}\Big|_{\Delta\omega=0} \end{bmatrix} \right\} + \\ & + j\gamma_q f_R A_\kappa \left\{ \left[A_q A_\kappa \tilde{h}_R(\Omega_R) - A_q \frac{\partial A_\kappa^*}{\partial t}\frac{d\tilde{h}_R}{d\Delta\omega}\Big|_{\Delta\omega=\Omega_R} - A_\kappa \frac{\partial A_q^*}{\partial t}\frac{d\tilde{h}_R}{d\Delta\omega}\Big|_{\Delta\omega=\Omega_R} + \frac{\partial A_\kappa^*}{\partial t}\frac{\partial A_q}{\partial t}\frac{d^2\tilde{h}_R}{d\Delta\omega^2}\Big|_{\Delta\omega=\Omega_R} \right] \right\} \end{aligned} \tag{17}$$

where $\tilde{h}_R(\Delta\omega)$ represents the Fourier transform of the Raman response $h_R(t)$, which is directly related to the complex Raman gain by means of the Raman susceptibility, as in Eq.(6).

Several features of Eq. (17) are noteworthy. The term proportional to $\tilde{h}_R(\Omega_R)$ considers SRS effect, while all terms proportional to first and second order derivative of $\tilde{h}_R(\Delta\omega)$ are related to higher order nonlinear effects. The great utility of assumption (17) is due to conversion of equation system (8)-(10) in a partial differential equation system in the space-time domain, without using any integral term. This leads to exploit the potential of collocation method as a fast and accurate solution approach [40]. Hereinafter, normalized variables are used for numerical purposes, $U_i(z,t) = A_i(z,t)/\sqrt{P_0}$ and $\tau = (t - z\beta_{1,\xi})/T_0$, where P_0 is the Gaussian pump peak power and T_0 is related to pulse FWHM width by $T_0 = T_{FWHM}/1.665$. Moreover, we will use normalized parameters as in chapter 7, i.e. dispersion, third order dispersion, walk-off and nonlinear lengths, as:

$$L_D = \frac{T_0^2}{|\beta_2|}; \qquad L_D^{'} = \frac{T_0^3}{|\beta_3|}; \; L_w = \frac{T_0}{|d|}; \; L_{NL} = \frac{1}{\gamma P_0}$$

being $d = \beta_{1p} - \beta_{1s}$ the group velocity mismatch [36]. To numerically solve the coupled equation system (8)-(10) and (15) with assumption (16), the collocation method [40] has been used by developing the unknown functions (i.e. A_i, N_c) as a linear combination of M orthogonal functions $\phi_m(\tau)$ through M weight functions $c_m(z)$, as:

$$X(z,\tau) = \sum_{m=1}^{M} c_m(z)\phi_m(\tau) \tag{18}$$

where $X(z,\tau)$ states for each unknown function. Hermite-Gauss functions have been chosen as basis functions, i.e. $\phi_m(\tau) = H_{m-1}(\tau)\exp(-0.5\tau^2)$. The unknown coefficients $c_m(z)$ are determined by satisfying the differential equation system at M collocation points τ_m, such that $H_M(\tau_m) = 0$, $m = 1,\ldots,M$. Advantages of collocation method are summarized in chapter 7. Once again, the system of partial differential equations (8)-(10) in z,t becomes a system of ordinary differential matrix equations in z under assumption (16), whose solutions have been found by a fourth-order Runge-Kutta algorithm. In total, up to seven matrix equations have been simultaneously solved, with at least $M = 35$ collocation points, so giving a system of 35x7 ordinary differential equations. In some cases, better accuracy is achieved with $M = 45$. The overlap integrals have been evaluated with high accuracy using the full-vectorial finite element method (FEM), with at least 60,000 elements mesh [41].

3. Numerical Results

In this section all numerical results are presented, both for picosecond and sub-picosecond pulse regimes.

3.1.Picosecond Regime

In this sub-section, a comparison between the numerical solution of equation system (8)-(10) and (15), and some numerical and experimental results proposed in literature for Raman effect in SOI waveguides with large cross section [12] is presented and analysed.

The SOI waveguide proposed in [12] has rib total height $H = 1.45$ µm, rib width $W = 1.52$ µm and slab height $H_s = 0.82$ µm. The pump pulse is launched at wavelength 1.545µm with FWHM width $T_{FWHM} = 17$ ns, the Stokes probe signal has a peak power of 2mW at 1.680 µm (according to Raman shift). The pump and Stokes signals interact for a length L = 4.8 cm and are aligned with quasi-TE and quasi-TM waveguide modes, respectively. In Figure 2 the carrier density versus time is plotted at different interaction lengths, L = 0, 2.4 and 4.8 cm, assuming g_R = 10.5 cm/GW and $\alpha_i^{(prop)}$ = 0.22 dB/cm [12]. According to parameters in [12], the effective carrier recombination lifetime has been assumed $\tau_{eff,1} = 25$ ns. Figure 2 shows a very good agreement with the results presented in [12], by choosing a number of Hermite-Gauss functions M ranging from 25 to 35.

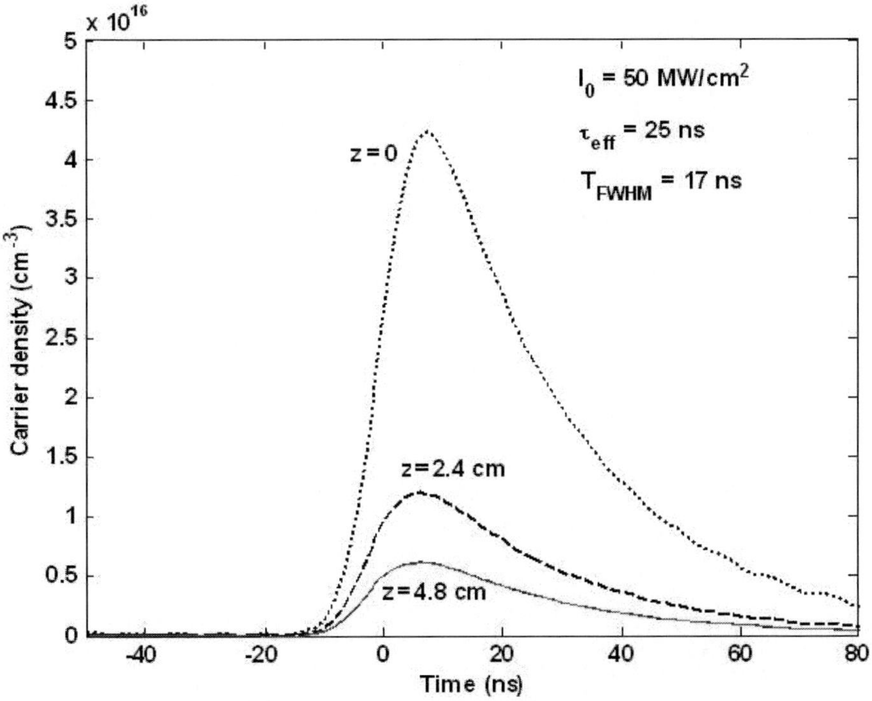

Figure 2. Free carrier density profile generated by TPA versus time for different z positions.

The accuracy improves by increasing M up to a saturation value, being the improvement not significant for $M > 35$. Thus, $M = 35$ represents the best trade-off between accuracy and calculation time. Some comments can be made on results of Figure 2. When the pump pulse FWHM width T_{FWHM} is of the same order of magnitude of carrier recombination lifetime $\tau_{eff,1}$ (as in the experiment [12]), the linear combination given by Eq. (18) leads to a very good fitting of the carrier density profile ($M = 35$). In contrast, the situation is much different for $T_{FWHM} << \tau_{eff,1}$. In fact, in this case the carrier density profile is not similar to the curves plotted in Figure 2, but tends to assume an *erf* function shape. However, this profile can be still fitted by means of Eq. (12), using $M > 35$.

Our rigorous prediction has been also compared with approximated results obtained by decoupling the partial differential equations (8)-(10) from the time equation of the carrier density (15). For instance, this approximation is very often used in both experimental and theoretical papers. Thus, our rigorous numerical results are compared in Figure 3 with the approximated solution and experimental points in literature [12], as a function of the pump peak intensity.

Experimental data are referred to the net Raman gain ($10\log I_{z=L}^{Stokes} / I_{z=0}^{Stokes}$) [12], solid line designates our rigorous simulation for quasi-TM fundamental Stokes mode, dotted line represents the approximated numerical result for quasi-TM Stokes mode, and finally dashed line with markers is sketched for the rigorous simulation in case of quasi-TE Stokes mode. The curves put in evidence that the quasi-TM case matches much better the experimental points that the quasi-TE case, as confirmed by the experiment. It is clear that using the approximation, a relative error of about 3.3% is introduced in comparison with the exact numerical result (see solid and dotted lines).

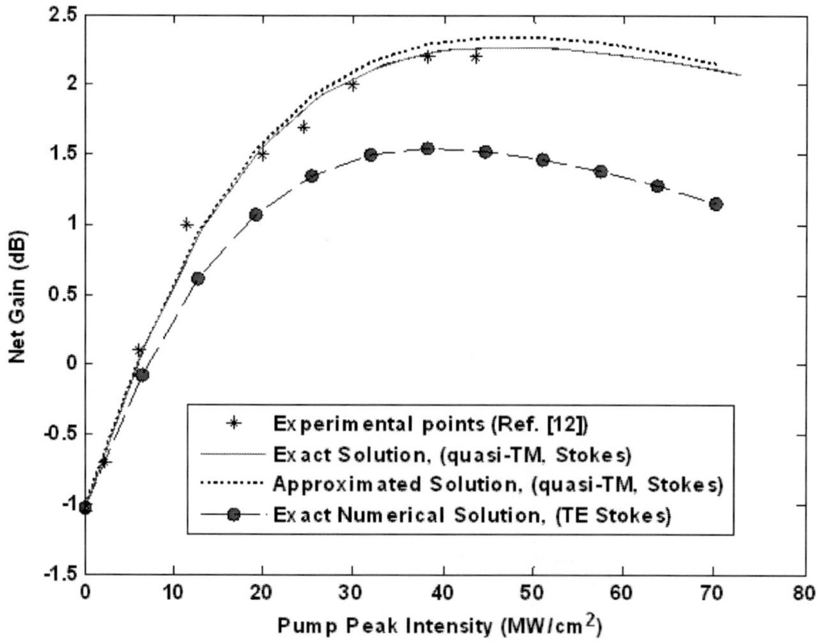

Figure 3. Net Raman gain versus pump peak intensity for a 4.8 cm long SOI waveguide.

This error tends to reduce by decreasing the pump pulse FWHM with respect to the effective carrier recombination lifetime. Anyway, the exact solution shows a very good agreement with experimental measurements, also in the region where the gain saturation effect is observable. In this region the model used in [12] produces large errors.

In addition, it is important to note that, in principle, two probe signals aligned in orthogonal polarizations could be simultaneously introduced in SOI waveguides and then amplified by the same pump pulse without any reciprocal interference (polarization multiplexing). Their different behavior depends on the effective area, smaller in quasi-TM than in quasi-TE case, while the Raman gain is almost the same in both cases. Thus, to obtain the same net gain for quasi-TE and quasi-TM Stokes waves, it is necessary to design the SOI waveguide to obtain the same mode area (free birefringence). It is known that the insensitive polarization condition can be achieved using large waveguide cross sections [42]. Anyway, the waveguide core geometry influences not only the modal birefringence, but also other critical parameters such as the effective carrier recombination lifetime $\tau_{eff,1}$. Usually, free carrier diffusion needs to be considered in addition to the recombination lifetime, as shown in [21] and chapter 7 of this book. Therefore, assuming a submicron structure with $H = 700$ nm, $r = H_s / H = 0.15$, and $W = 500$ nm, it is possible to obtain a low value of effective lifetime as $\tau_{eff,1} = 1.65$ ns.

In the following, we investigate the time-space evolutions of very fast pulses propagating along the SOI waveguide, $L = 1$ cm long. In the first simulation, we have assumed $P_0 = 1.5$ W, $T_{FWHM} = 1$ ps and $\lambda_p = 1.4332$μm, keeping the probe power 20 dB below the pump power. The probe (Stokes) wavelength is $\lambda_{s0} = 1.5487$μm, and the Raman gain is $g_R = 10.5$ cm/GW. The other calculated parameters include $n_{eff,p}(TE) = 3.0887$, $A_{eff,p}(TE) = 0.2305$ μm^2, $n_{eff,s0}(TE) = 3.0134$, $n_{eff,s0}(TM) = 3.0863$, $A_{eff,s0}(TE) = 0.2464$ μm^2 and $A_{eff,s0}(TM) = 0.2282$ μm^2. Figure 4 shows both pump and fundamental Stokes pulse evolutions, as conditioned by all nonlinear effects. Under the assumed conditions, a number of significant nonlinear effects are excited and strongly influence the pulse propagations. Moreover, these conditions are not usually encountered in ordinary experimental set-ups.

Several features of Figure 4 are noteworthy. The Raman pulse builds up after $z \approx 3$ mm. The energy transfer from the pump to Stokes pulse induces two peaks in the pump pulse shape to be formed, as a result of pump depletion. Of course, the relevant hole corresponds exactly to time location of Stokes wave. The time locations of pump and Stokes pulses are separated because of the walk-off effect (evaluated by considering the material index dispersion). Moreover, the Raman pulse is narrower than the input pulse, as a consequence of the spectral broadening induced by SPM, XPM, and plasma dispersion effects. Thus, the main nonlinear effects play an important role in changing the pulse shape even over a very small interaction length (1 cm in this case), since the assumed pump pulse is very fast. It is worth to note that, due to the large pulse width (1 ps), GVD and TOD effects do not influence the dynamics of the process. On the other side, as it is possible to note from equation system (8)-(10), SPM, XPM and plasma dispersion coefficients heavily influence the time dynamics of wave complex amplitudes.

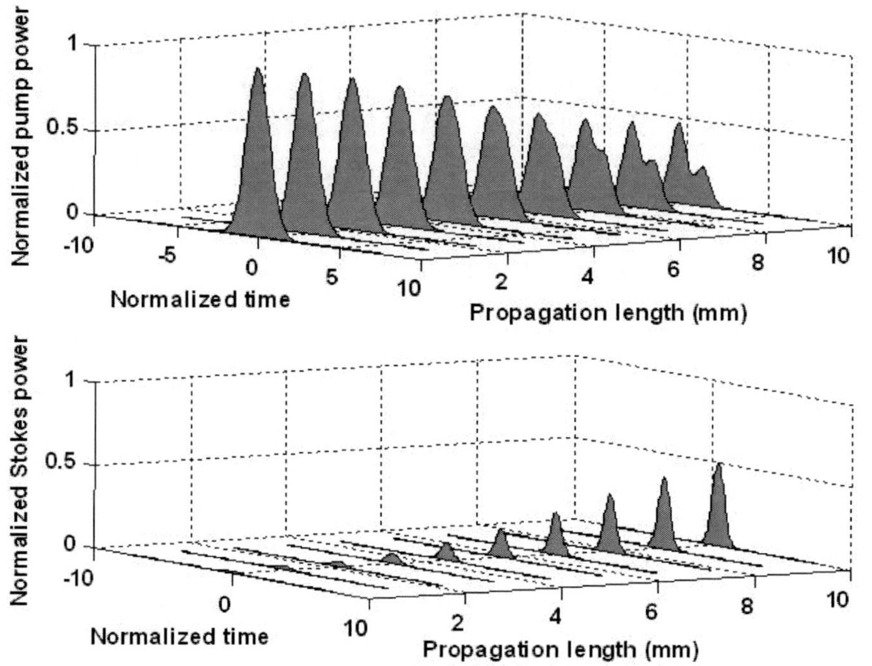

Figure 4. Time-space evolution of both pump and fundamental Stokes pulses.

In Figure 5 the nonlinear phase shift time evolution of fundamental Stokes pulse as evaluated at $L = 10$mm is sketched for different values of T_{FWHM}. In the simulation, the probe power is still kept 20 dB below the pump power ($P_0 = 1.5$ W).

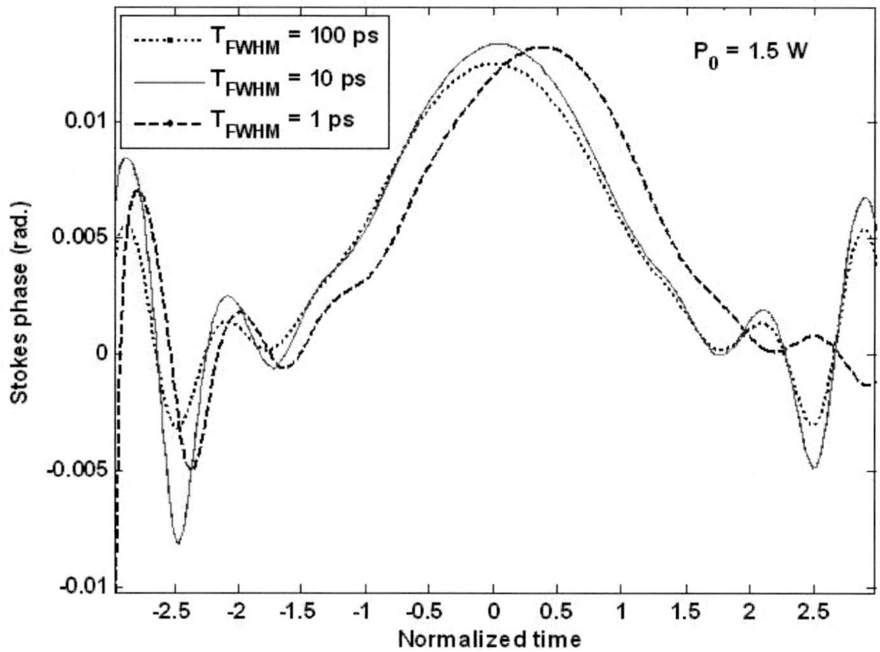

Figure 5. Time evolution of Stokes phase at L = 10mm, for various values of T_{FWHM}.

We can see as the total phase profile for $T_{FWHM} = 1$ ps show a main lobe centred at $\tau \neq 0$, as induced by walk-off effect between pump and Stokes pulses. In addition, a detailed analysis show that the phase contribution due to plasma dispersion effect is dominant over SPM and XPM effects for $T_{FWHM} = 100$ ps (large pulses). On the contrary, for short pulses $T_{FWHM} = 1$ ps, SPM and XPM effects give similar contribution to the total phase as the plasma dispersion effect [16].

A very important application of SRS effect in picosecond regime can be represented by the amplification of a Stokes pulse inside an optical resonator. In fact, the utilization of a resonant microcavity to induce the Raman effect can represent a strong stimulus towards the development and implementation of Raman lasers with micro-scale sizes and low threshold, as demonstrated in some pioneering works [43]-[44]. In general, the high quality factors allowed by these microcavities are of wide interest in Optics for a variety of studies ranging from fundamental Physics such as Quantum-Electrodynamics to applied areas such as low-threshold and narrow-linewidth lasers, nonlinear all optical gates, as well as high sensitivity transducers for optical sensing. Therefore, differently from other Raman architectures [24]-[29], where the gain medium is represented by a simple SOI straight waveguide, in this sub-section we theoretically analyze the possibility to induce the SRS effect in integrated micro-cavities based on SOI technology, where the resonant microcavity is coupled with an external bus straight waveguide, as sketched in Figure 6.

Our approach is motivated by the need to design high performance active devices in SOI integrated platform, i.e. exploiting the possibility that a micro-cavity offers to reduce the threshold of nonlinear optical effects, due to its intrinsic high enhancement factor. The mathematical model and physical features involved in this sub-section can be obtained by particularizing Eqs. (8)-(10) under the assumptions (16). The details of this procedure have been already presented [31].

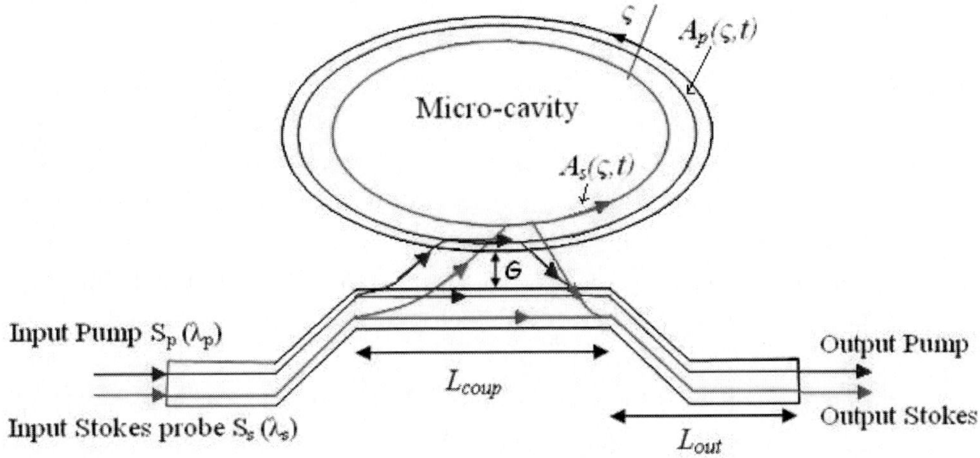

Figure 6. Schematic architecture of SOI resonant micro-cavity coupled to external waveguide.

For a reason of easiness, we refer here only to pump (p) and fundamental Stokes (s) pulses. Basically, all losses induced by the resonant cavity (such as propagation, bending and radiation loss induced by sidewall roughness) on the optical mode are to be related to the loss quality factor Q_i^l ($i = p,s$) of cold cavity. This loss quality factor is found to be $1/Q_i^l = 1/Q_{i,prop} + 1/Q_{i,bend} + 1/Q_{i,scatt} + \ldots$, where $Q_{i,prop}$, $Q_{i,bend}$ and $Q_{i,scatt}$ designate the contributions related only to propagation, bending or scattering loss, respectively. Further, Q_i^l can be also given as a function of overall linear loss coefficient (α_{loss}) as $Q_i^l = \omega_i \tau_{l,i}$, being $\tau_{l,i} = 1/(\alpha_{loss,i} v_{g,i})$ the decay time related to losses and $v_{g,i}$ the wave group velocity. Moreover, from a phenomenological point of view, it is necessary to consider the optical power leaving the cavity, due to coupling with the external waveguide, as a loss term, while the optical power injected into the cavity from the waveguide bus as a source term. Thus, indicating with Q_i^c the quality factor relevant only to waveguide-cavity coupling process, it can be evaluated as $Q_i^c = \omega_{p,s} \tau_{c,(p,s)}$, where $\tau_{c,(p,s)}$ is the time constant of the decay process for the pump and Stokes photons as induced by this coupling. Finally, the overall quality factor of the resonant cavity is $1/Q_i = 1/Q_i^l + 1/Q_i^c$. In conclusions, the energy exchange between pump and fundamental Stokes pulses inside the optical cavity can be analysed by means of the following equations [31]:

$$\begin{aligned} v_{g,p}\frac{\partial A_p}{\partial \varsigma} + \frac{\partial A_p}{\partial t} &= j\left(\omega_{p,\bar{m}} - \omega_p\right)A_p - \frac{1}{2}\frac{1}{\tau_p}A_p - \frac{1}{2}v_{g,p}\alpha_p^{(FCA)}A_p - 0.5 v_{g,p}\beta^{(TPA)} f_{p,p}\left|A_p\right|^2 A_p \\ &+ j v_{g,p}\gamma_{p,p}\left|A_p\right|^2 A_p + j2 v_{g,p}\gamma_{p,s}\left|A_s\right|^2 A_p + j v_{g,p}\frac{2\pi}{\lambda_p}\Delta n_p A_p - \frac{1}{2}v_{g,p} g_R f_{p,s}\frac{\omega_p}{\omega_s}\left|A_s\right|^2 A_p + \xi_p S_p \end{aligned} \quad (19)$$

$$\begin{aligned} v_{g,s}\frac{\partial A_s}{\partial \varsigma} + \frac{\partial A_s}{\partial t} &= j\left(\omega_{s,\bar{q}} - \omega_s\right)A_s - \frac{1}{2}\frac{1}{\tau_s}A_s - \frac{1}{2}v_{g,s}\alpha_s^{(FCA)}A_s - v_{g,s}\beta^{(TPA)} f_{s,s}\left|A_p\right|^2 A_s \\ &+ j v_{g,s}\gamma_{s,s}\left|A_s\right|^2 A_s + j2 v_{g,s}\gamma_{s,p}\left|A_p\right|^2 A_s + j v_{g,s}\frac{2\pi}{\lambda_s}\Delta n_s A_s + \frac{1}{2}v_{g,s} g_R f_{s,p}\left|A_p\right|^2 A_s + \xi_s S_s \end{aligned} \quad (20)$$

In Eq. (19)-(20), GVD and TOD effects are not included as second and third order time derivative, since this effect is negligible in SOI micro-scale structures under picosecond regime [16]. Moreover, the power transfer to higher order Stokes waves is also negligible for short interaction lengths and relatively large values of T_{FWHM}. The partial differential equation system (19)-(20) clearly represents a model generalization over a simple SOI straight waveguide [16]. This aspect is mainly evident by observing the first, second and last terms on the right hand of Eq. (19)-(20). In fact, ($\omega_{p,\bar{m}} - \omega_p$) and ($\omega_{s,\bar{q}} - \omega_s$) indicate the mismatch from the resonance condition of input pump and Stokes probe frequency, respectively. The term τ_p (τ_s) represents the overall photon decay time of the pump (Stokes) pulse inside the cavity. It is related to the overall resonator quality factor by $Q_{\bar{m}} = \omega_p \tau_p$, where τ_p is given by:

$$\frac{1}{\tau_p} = \frac{1}{\tau_{l,p}} + \frac{1}{\tau_{c,p}} \tag{21}$$

being the two contributions related to loss ($\tau_{l,p}$) and coupling ($\tau_{c,p}$) time constants, respectively. Similar definition holds for τ_s. Moreover, the coefficients ξ_p and ξ_s are related to the power fraction transferred into the resonator from the input pump (S_p) and input Stokes probe (S_s) as $\xi_p = \sqrt{v_{gp} / \left(L_{cavity} \tau_{c,p}\right)}$ and $\xi_s = \sqrt{v_{gs} / \left(L_{cavity} \tau_{c,s}\right)}$, being v_{gp} and v_{gs} the pump and Stokes wave group velocities, respectively. For convenience, it is useful to write these coefficients in terms of the coupling factor κ_p^2 (κ_s^2), defined as the power fraction of the input pump (Stokes) pulse injected into the resonator as coming from the bus waveguide. It is possible to demonstrate [45] that the coupling factor κ_p^2 (κ_s^2) is related to the coupling time constant $\tau_{c,p}$ ($\tau_{c,s}$) through the following relations:

$$\kappa_p^2 = \frac{1}{\tau_{c,p}} \frac{L_{cavity}}{v_{gp}}; \quad \kappa_s^2 = \frac{1}{\tau_{c,s}} \frac{L_{cavity}}{v_{gs}} \tag{22}$$

Some physical comments can be derived from Eqs. (19)-(20). Through the general formalism, the walk-off effect (first order time derivative) between pump and Stokes waves is considered inside the resonator. As demonstrated in [16], this effect is dominant in SOI waveguides for pulses with short full-wave-half-maximum (FWHM) time widths $T_{FWHM} \approx 1\text{ps}$, where the walk-off length $L_w = T_0 / \left(\left|v_{gp}^{-1} - v_{gs}^{-1}\right|\right)$ is typically shorter than the waveguide length. Then, SRS is limited by the group velocity mismatch and occurs only over distances $z \ll L_w$, even when the propagation length is much larger than L_w. On the contrary, in case of SRS effect induced in the resonant microcavity, the limiting factor for the process should not be the group velocity mismatch, but the increase of pump energy inside the resonator, i.e. the enhancement factor of the pump pulse. It can be defined by $\Gamma = \left|A_p / S_p\right|$, where A_p is the maximum of the pump amplitude inside the cavity and S_p is the input pump peak, outside the cavity. Therefore, the design of the architecture of Figure 6 requires a condition $\Gamma \gg 1$, in order to achieve better performance over a simple straight waveguide, in terms of lower threshold and smaller sizes. To better understand the influence of enhancement factor on the performance of architecture in Figure 6, we can intuitively consider the propagation of the pump under pulsed excitation and neglect all nonlinear effects in Eq. (19), obtaining:

$$\frac{dA_p}{dt} = j\left(\omega_{p,\bar{m}} - \omega_p\right) A_p - \frac{1}{2\tau_p} A_p + \xi_p S_p \tag{23}$$

Thus, under this approximation numerical investigations show that, for given κ_p^2, the enhancement factor Γ assumes a monotonically decreasing shape as a function of cavity length for each value of T_{FWHM}. Anywhere, a very small optical microcavity requires large pulse widths to realise a high enhancement factor. In fact, as the input pulse is narrower than the time constant of evanescent coupling process ($\tau_{c,p}$), the field inside the ring cannot be enhanced. Moreover, additional improvements on the enhancement factor can be achieved by reducing κ_p^2. In fact, numerical simulations show that Γ shape versus κ_p^2 presents an absolute maximum, and this peak shifts towards smaller values of κ_p^2 with increasing T_{FWHM}.

Thus, the analysis of pump pulse propagation in linear regime leads to conclude that the pulsed excitation of micro-resonators should require $T_{FWHM} >> \tau_{c,p}$, in order to increase the enhancement factor and then to reduce the threshold due to nonlinear effects. However, as it will be clear, the previous discussions can be considered with good approximation as still valid even in the presence of nonlinear interaction between pump and Stokes pulses, if FCA is negligible and $T_{FWHM} \square \tau_{eff}$. In particular, the condition $T_{FWHM} \geq$ 100 ps allows an increasing of Γ even in the presence of FCA effect. Moreover, using $T_{FWHM} \geq$ 100 ps and λ_p around 1.55 μm, the walk-off length is of the order of $L_w \square$ 3.3 m. This means that the group velocity mismatch does not influence again the SRS process for microcavity resonators.

Then, we can assume pump and Stokes waves as travelling with the same velocity inside the resonator, and thus the partial differential equations (19)-(20) can be transformed in ordinary differential equations by writing [31]:

$$\frac{dA_{p,s}}{dt} = \frac{\partial A_{p,s}}{\partial t} + \frac{1}{\beta_1}\frac{\partial A_{p,s}}{\partial \varsigma} \tag{24}$$

where $\beta_{1,p} \cong \beta_{1,s} = \beta_1$.

To point out the positive influence of a large pump enhancement factor, we compare the cavity resonator with respect to the simple straight waveguide. To this aim, the results, solution of equation system (19)-(20)-(15) under condition (24), have been compared with both numerical results proposed in Figure 3 and experimental results proposed in literature for SOI waveguides with large cross section [12]. In order to achieve a consistent comparison using the architecture shown in Figure 6, we have assumed the same values as in [12] for T_{FWHM}, input pump and input probe wavelengths, peak power of input probe and, in particular, effective recombination lifetime of free carriers, τ_{eff}. Since it is directly proportional to the optical mode width [21], the estimated value $\tau_{eff} \square$ 25 ns is obtained using a cavity resonator with the same cross-section as in the straight waveguide [12]. This induces to consider a racetrack resonator in the architecture of Figure 6 as the best choice to realise our comparison. In addition, we have assumed the same cross section for the external rib waveguide to achieve technological compatibility and improve the coupling with the resonator. Thus, the structure used in the simulation consists of a racetrack resonator with a total length $L_{cavity} = 0.3770$ cm and bend radius of 50.28 μm, gap $G = 0.7$ μm, and coupling length $L_{coup} = 191.5$ μm. Simulations performed by means of beam propagation method (BPM) [46] have given the following coupling factors: $\kappa_p^2 = 17.55\%$ and $\kappa_s^2 = 10\%$.

Figure 7 shows the net Raman gain as a function of the input pump peak intensity. In particular, markers represents the experimental data measured in [12] for the straight SOI waveguide, dotted line designates the numerical results obtained by Eqs. (8)-(10)-(15) for SOI waveguides, and solid line represents the numerical solution given by Eqs. (15)-(19)-(20). In each case, we have assumed $\beta^{(TPA)} = 0.5$ cm/GW, g_R= 10.5 cm/GW and α_{loss} = 0.22 dB/cm [12]. Before any comment on the physical features derived from Figure 7, some considerations must be emphasized about the definition of net Raman gain used when SRS effect is excited inside the microcavity. As usually reported in literature [12-16], the net Raman gain in a straight waveguide is defined as:

$$G_{net} = 10\log P_{z=L}^{Stokes} / P_{z=0}^{Stokes} \qquad (25)$$

where $P_{z=L}^{Stokes}$ represents the Stokes wave power after an interaction length L, depending on SRS effect. The term $P_{z=0}^{Stokes}$ indicates the Stokes wave power at $z = 0$ and, then, it represents the input probe. The same definition is not directly applicable to micro resonators because, in this case, the input probe (signal S_s) experiences the enhancement effect induced by the presence of resonator, too. However, by adapting the definition used for straight waveguide in Eq. (25), the net Raman gain in the presence of the micro-cavity should become:

$$G_{net} = 10\log\left|A_s / A_s^L\right|^2 \qquad (26)$$

where A_s represents the Stokes field amplitude inside the micro resonator (solution of equation system (15)-(19)-(20) with assumption (24)), while A_s^L represents the solution of Eq. (20) when all non linear and loss contributions are neglected. Thus, Eq. (26) allows to calculate the Raman amplification separated from the enhancement effect experienced from the input probe (signal S_s). Moreover, definition (26) allows to compare in a consistent way the SRS effect as induced in micro resonators and straight waveguides, as shown in Figure 7.

The plot shows that the racetrack resonator with $L_{cavity} = 0.377$ cm and peak pump intensity larger than 45 MW/cm^2, induces a saturation net gain similar to waveguide case, but with an interaction length approximately 13 times smaller. In addition, Figure 7 emphasizes that the SRS effect in micro-cavity produces a linear net gain of about 3.2 dB, three times larger than that measured in the SOI waveguide proposed in [12]. In turn, this gives a threshold for the input pump of 0.46 MW/cm^2 in the racetrack resonator, considerably lower than 6.74 MW/cm^2 as measured in [12]. Thus, the advantages obtainable with the micro resonator in terms of reduced threshold and sizes are evident, as due to a pump enhancement factor Γ>>1. However, our calculations show a Γ value smaller than that estimated by assuming the pump as propagating in linear regime. In fact, in this case the pulse width and free carrier recombination lifetime are both of the same order of magnitude, so inducing FCA effect to become dominant, and increasing the energy depletion inside the resonator.

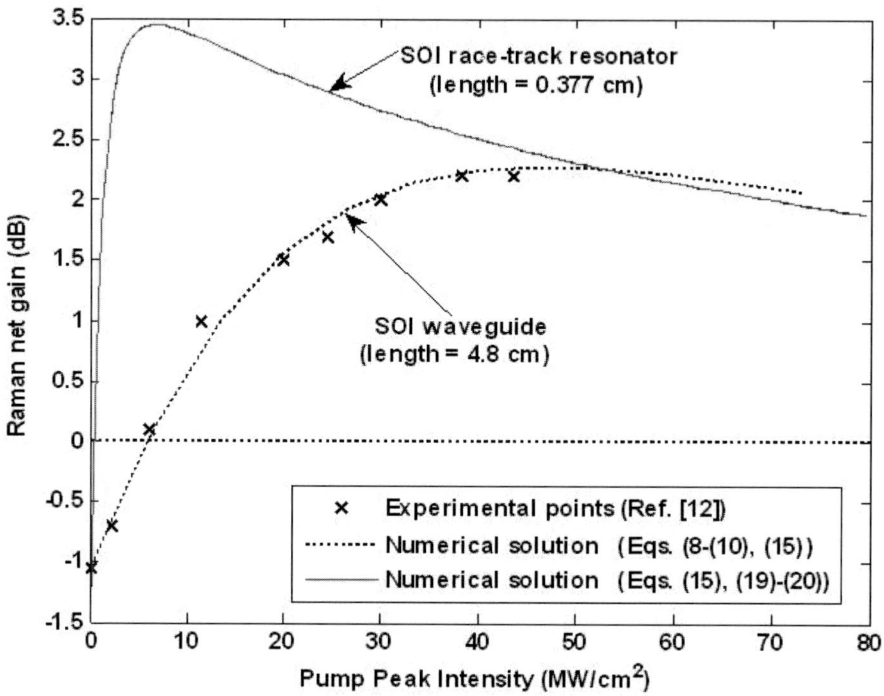

Figure 7. Comparison between waveguide and micro resonator in terms of net Raman gain versus pump peak intensity (experimental points and numerical results).

The influence of interaction length and FCA on Γ can be better understood by considering the simulations presented in Figure 8 and 9. Figure 8 shows the ratio between the Raman net gain for an optical resonator and a straight waveguide, respectively, as a function of T_{FWHM} for various interaction lengths L. In particular, we have considered three racetrack cavities with $L = L_{cavity}$ = 0.0628 cm, 0.1885 cm and 0.3770 cm and bend radii 8.4 μm, 25.14 μm and 50.28 μm, respectively.

For both racetrack resonators and straight waveguides, we have assumed H = 1.45 μm, W = 1.52 μm, $H_s = 0.82$ μm, λ_p = 1.545 μm and λ_s = 1.680 μm. We have set Stokes probe signal and pump pulse with a peak power of 2 mW and 0.4 W, respectively, and τ_{eff} = 25 ns. Coupling factors, coupling length and gap hold the same values as before, i.e. $\kappa_p^2 = 17.55\%$, $\kappa_s^2 = 10\%$, $L_{coup} = 191.5$ μm and $G = 0.7$ μm. Finally, it is important to outline that the relatively large cavity perimeter means to have a very small FSR, i.e. λ_p and λ_s matched to the microcavity resonant wavelengths. Curves in Figure 8 do not have a monotonically decreasing shape for each value of L_{cavity}, but they present an absolute maximum for low values of T_{FWHM}, depending on peaks in the enhancement factor shape (see Figure 9).

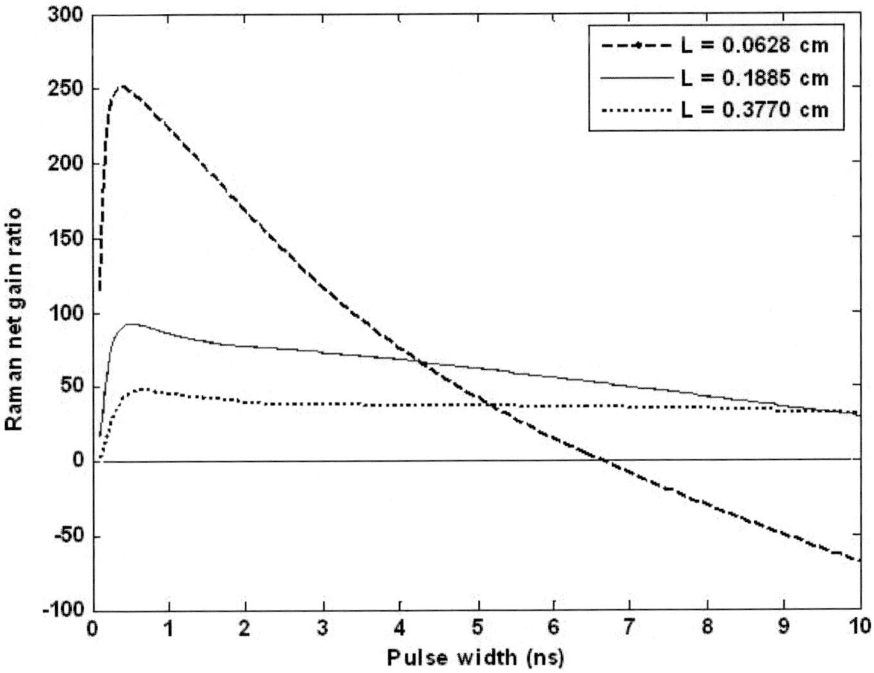

Figure 8. Raman net gain ratio versus pump pulse width for various microcavity lengths.

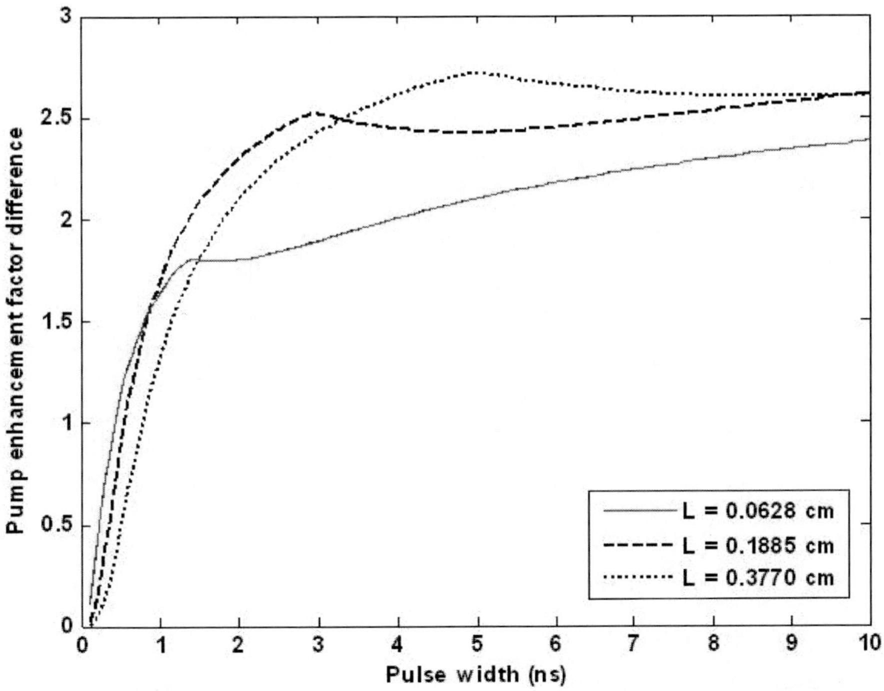

Figure 9. Pump enhancement factor difference versus pump pulse width for various interaction lengths (approximated and rigorous solutions).

Moreover, these peaks shift toward higher values of T_{FWHM} with increasing L_{cavity}. In fact, larger values of L_{cavity} lead to extend the range of T_{FWHM} where the detrimental effect induced by FCA can be partially compensated. The net Raman gain is monotonically decreasing by increasing T_{FWHM}, with an approximately exponential shape as a result of the strong FCA influence. Moreover, we have found that for L_{cavity}= 0.0628 cm, a value of T_{FWHM} exists, where the Raman net gain ratio becomes negative (SRS effect under threshold). This occurs because the energy depletion by FCA inside the cavity induces a pump enhancement factor, which is too low to enforce the pulse along such short interaction length.

Anyway, Figure 8 well emphasizes that it is always possible to find a range of T_{FWHM} where the pump enhancement factor is improved and, thus, the Raman net gain is considerably larger than in straight waveguides, even for short interaction lengths. Figure 9 shows the difference between the pump enhancement factor (Γ) evaluated by solving the system (19)-(20)-(15) with assumption (24), as well as the approximated value for Γ obtained using Eq. (23), as a function of T_{FWHM} for various interaction lengths *L*. From Figure 9 we derive that, in any case, this approximated value is well larger than the rigorous solution. However, the approximated solutions are in good agreement with rigorous predictions only in a short range, namely 100 ps ÷ 150 ps for L_{cavity} = 0.0628 cm, 100ps ÷ 200ps for L_{cavity} = 0.1885 cm and 100ps ÷ 299ps for L_{cavity} = 0.3770 cm. This means that FCA induced by TPA effect is negligible in these ranges, i.e. energy depletion is not induced inside the micro-cavity. In fact, the influence of FCA becomes dominant only when T_{FWHM} is of the same order of magnitude as τ_{eff}, in this case equal to 25 ns (see Figure 9). This behaviour explains the presence of two different slopes in the curves of Figure 9.

The previous discussion outlines that the potential of the architecture proposed is enhanced if microcavity resonators with high quality factor Q and large Γ are used. Higher Q factors can be obtained by reducing the total resonator losses (propagation, bending, scattering, and so on), decreasing the coupling factor and increasing the ring radius. It is evident that the reduction of bending losses requires to use larger ring radii, a conflicting requirement with monolithic integration needs. Fortunately, SOI technology generally leads to overcome this problem. In fact, the high index contrast facilitates light guiding in micro-scale structures, thus making either negligible or very small the bending loss in SOI microcavities. In that case, microcavity losses are mainly limited by sidewall roughness scattering. At the same time, low values of both coupling factor and ring radius induce large pump enhancement factors, providing to realise low values of τ_{eff} and use large pulse widths, $T_{FWHM} > 10$ ps . In fact, the last two conditions lead to minimize FCA effect inside the resonator. The reduction of τ_{eff} can be carried out by an appropriate design of resonator cross section. Thus, a small cross section is needed to obtain a low value for τ_{eff}. We have considered a resonator with small cross section, a few hundreds of squared nanometres. We have chosen a rib total height H = 500 nm (~ λ/3), $r = H_s / H = 0.3$ and W = 300 nm, in order to achieve very high confinement inside the rib and negligible bending losses. For these values, τ_{eff} is estimated as about 1 ns.

Figure 10 shows the net Raman gain versus pump coupling factor for different micro cavities. In particular, three ring resonators with ring radius R = 10, 15, and 20 μm, are considered, respectively. In the simulations, we assume T_{FWHM} = 100ps, $\beta^{(TPA)} = 0.5$ cm/GW, g_R = 10.5 cm/GW , $\alpha_{loss} = 0.6$ dB/cm, $|S_p|^2$ = 0.4 W and $|S_s|^2$ = 2 mW. The input pump wavelength is tuned at the same resonance wavelength for the ring microcavities ($\lambda_{p,\bar{m}} = 1432.5358$ nm). The input Stokes probe is launched at the wavelength corresponding to Raman shift with respect to the input pump. In this condition, the shift between input probe and resonance wavelengths is estimated as 1.99 nm for R = 10 or 20 μm, and 2.49 nm for R = 15 μm, i.e. the Stokes pulse is not at resonance for the cavity (off-resonance condition). Thus, Figure 10 shows the realistic case of off-resonance condition, and the "ideal" case where the Stokes wave is assumed to be at the resonance for the ring resonator, as the pump. The coupling factors κ_s^2 and κ_p^2 have been obtained by Finite Difference Time Domain (FDTD) simulations, and interpolating functions have been extracted on a number of these simulation points, a few of which are reported in Table 1.

The plots in Figure 10 clearly show the detrimental effect induced by mismatch between Stokes and resonant wavelengths, in terms of reduced Raman gain for each κ_p^2. This is due to the partial destructive interference that the Stokes wave experiences as travelling inside the microring. In addition, off-resonance condition induces a decreasing shape of the net gain, with a weak ripple as a function of κ_p^2, where the ideal case assumes a flat shape.

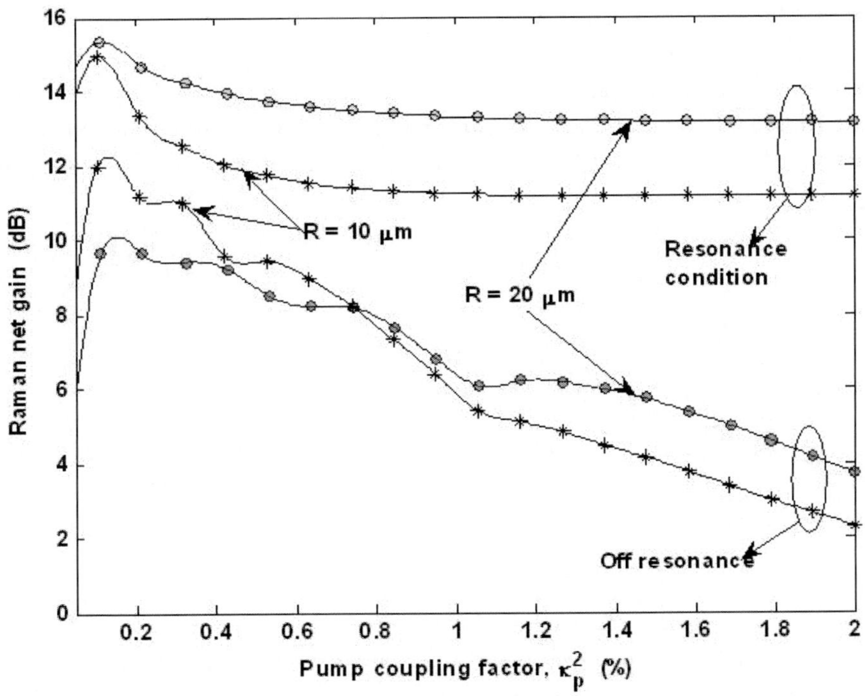

Figure 10. Raman net gain versus pump coupling factor for various ring radii, under resonance or off-resonance conditions for Stokes pulse.

Table 1. FDTD results for pump (1433.2 nm) and Stokes (1586.1 nm) coupling factors

Gap, G (μm)	R = 10 μm		R = 15 μm		R = 20 μm	
	κ_p^2 (%)	κ_s^2 (%)	κ_p^2 (%)	κ_s^2 (%)	κ_p^2 (%)	κ_s^2 (%)
0.1	19.32	29.54	28.05	41.97	46.57	62.67
0.2	3.41	4.55	3.66	7.41	6.94	10.67
0.25	1.43	1.78	1.32	3.11	2.67	4.40
0.3	0.6	0.7	0.47	1.30	1.03	1.81
0.35	0.25	0.27	0.17	0.55	0.4	0.75

Anyway, some significant physical aspects can be noted by examining the ideal case. The curve shape shows an absolute maximum whose coupling factor corresponds to critical coupling. As it is well known [45], the coupling condition can be described in terms of ratio $K = \tau_0/\tau_{c,p}$ between total loss decay time $\tau_0 = 1/\left[v_{gp}\left(\alpha_{loss,p} + \alpha_p^{FCA}\right)\right]$ and pump coupling decay time, $\tau_{c,p}$. Following the standard conventions, under-coupling is denoted by $K < 1$, over-coupling by $K > 1$, and critical coupling by $K = 1$. This last condition corresponds to a vanishing transmitted pump at the bus waveguide end, and thus it represents the optimal condition to simultaneously maximise the Raman net gain and minimise the Raman threshold. In any case, it is clear that, in our general approach, τ_0 is related to both linear propagation loss and FCA, differently from definition used in [45]. Thus, an estimation of the maximum position from the relationship $\tau_{l,p} = \tau_{c,p}$ is only a first approximation prediction, while the rigorous numerical solution of equation system (15)-(19)-(20) can determine the exact value.

In addition, it is evident that an accurate design of micro-scale resonator is required to avoid the detrimental effects due to off-resonance over net Raman gain and Stokes, as much as possible. To this aim, a very efficient method to design the microcavity resonator consists of an appropriate distribution control of resonant angular frequencies, to achieve $\omega_{p,\bar{m}} - \omega_{s,\bar{q}} = \Omega_R$. It is evident that this condition gives a discrete set of cavity lengths, as:

$$L_{cavity} = \frac{2\pi l c}{n_{eff,p}\omega_{p,\bar{m}} - n_{eff,s}\omega_{p,\bar{m}} + n_{eff,s}\Omega_R} \tag{27}$$

where l is an integer number, counting the difference between pump and Stokes wave longitudinal orders in the cavity. Then, each value of l determines the cavity length needed to minimize the off-resonance effects. However, the evaluation of $\omega_{p,\bar{m}}$ in Eq. (27) also requires the knowledge of L_{cavity}. Thus, solutions are to be found using a recursive procedure. For example, by setting $l = 26$, we obtain a radius $R = 19.78$ μm in case of microring resonator. This value determines a quasi ideal resonance condition (the difference being 15 pm), considerably smaller than the mismatch of 1.99 nm previously obtained in case of $R = 20$ μm. Therefore, under this quasi ideal resonance, the detrimental effects induced by off-resonance are found to be negligible.

In several applications, it is important to use the stimulated Raman scattering for CW laser operation. From the system of equations (15)-(19)-(20), we can observe that a quasi-CW regime occurs if the input pump time width T_{FWHM} in the bus waveguide satisfies the condition: $T_{FWHM} >> \max(\tau_p, \tau_{eff})$. Since τ_p is of the order of a few tens of picoseconds in the most cases, it must be $T_{FWHM} \geq 100$ ps and, thus, τ_{eff} becomes the dominant parameter for quasi-CW regime. Moreover, a large T_{FWHM} means that the group-velocity mismatch does not influence the SRS process for microcavities.

In the following part, we derive the basic formulas to determine the threshold condition of SRS effect in the resonator for lasing under quasi-CW regime. Under this hypothesis, a number of considerations can be made on the equation system (15)-(19)-(20) with assumption (24). In fact, quasi-CW regime means a steady-state analysis, where the threshold condition for SRS effect leads to neglect the terms depending on Kerr and plasma dispersion effects. In particular, SPM and XPM effects can be considered under threshold when the Raman effect is assumed at the threshold. In fact, Raman dominates over Kerr effect in SOI technology. Finally, in order to minimize the input pump power at the threshold, we assume that the angular frequencies of both pump and Stokes pulses are very close to resonance ($\omega_{p,s} \cong \omega_{\overline{m},\overline{q}}$). Therefore, the equation system (15)-(19)-(20) is simplified as follows:

$$-\frac{1}{2\tau_p}A_p - v_{gp}\frac{g_P}{2}|A_s|^2 A_p - \frac{1}{2}v_{gp}\frac{\beta^{TPA}}{A_{eff,p}}|A_p|^2 A_p - \frac{1}{2}v_{gp}\alpha_p^{FCA}A_p + \xi_p S_p = 0 \tag{28}$$

$$-\frac{1}{2\tau_s}A_s + v_{gs}\frac{g_s}{2}|A_p|^2 A_s - v_{gs}\frac{\beta^{TPA}}{A_{eff,p}}|A_p|^2 A_s - \frac{1}{2}v_{gs}\alpha_s^{FCA}A_s = 0 \tag{29}$$

$$\frac{\beta^{TPA}}{2\hbar\omega_p}\left(\frac{|A_p|^2}{A_{eff,p}}\right)^2 A_p - \frac{N_c}{\tau_{eff}} = 0 \tag{30}$$

Of course, in case of CW Raman laser operation, $S_s = 0$ (absence of input probe) is assumed in Eq. (20). Using the definition of α_p^{FCA} and α_s^{FCA}, where the free carrier density is evaluated by Eq. (30), two solutions of Eq. (28) are found as:

$$A_p = \begin{cases} A_{p,1} = \sqrt{\dfrac{v_{gs}\left(\dfrac{g_s}{2} - \dfrac{\beta^{TPA}}{A_{eff,p}}\right) + \sqrt{v_{gs}^2\left(\dfrac{g_s}{2} - \dfrac{\beta^{TPA}}{A_{eff,p}}\right)^2 - \dfrac{v_{gs}\sigma_s\eta}{\tau_s}}}{v_{gs}\sigma_s\eta}} \\ A_{p,2} = \sqrt{\dfrac{v_{gs}\left(\dfrac{g_s}{2} - \dfrac{\beta^{TPA}}{A_{eff,p}}\right) - \sqrt{v_{gs}^2\left(\dfrac{g_s}{2} - \dfrac{\beta^{TPA}}{A_{eff,p}}\right)^2 - \dfrac{v_{gs}\sigma_s\eta}{\tau_s}}}{v_{gs}\sigma_s\eta}} \end{cases} \tag{31}$$

with $\eta = \dfrac{\tau_{eff}\beta^{TPA}}{2\hbar\omega_p A_{eff,p}^2}$. In turn, the solution for the pump amplitude A_p includes two possible threshold levels for the input pump (S_p), in order to excite the Stokes wave by SRS in the microcavity resonator. In fact, by substituting Eq. (31) into (29) and considering $A_s = 0$ (threshold condition), we derive:

$$P_{th} = \left|S_p^{th}\right|^2 = \begin{cases} \left\| \dfrac{1}{\xi_p}\left(\dfrac{A_{p,1}}{2\tau_p} + \dfrac{v_{gp}\beta^{TPA}}{2A_{eff,p}} \left|A_{p,1}\right|^2 A_{p,1} + \dfrac{v_{gp}\sigma_p\eta}{2}\left|A_{p,1}\right|^4 A_{p,1} \right) \right\|^2 = P_{th,1} \\ \left\| \dfrac{1}{\xi_p}\left(\dfrac{A_{p,2}}{2\tau_p} + \dfrac{v_{gp}\beta^{TPA}}{2A_{eff,p}} \left|A_{p,2}\right|^2 A_{p,2} + \dfrac{v_{gp}\sigma_p\eta}{2}\left|A_{p,2}\right|^4 A_{p,2} \right) \right\|^2 = P_{th,2} \end{cases} \qquad (32)$$

Several considerations can be derived from Eqs. (31)-(32). First, the two values of SRS threshold depend on both TPA and FCA effects. In addition, for a given coupling factor κ_p^2 for the input pump, the upper threshold ($P_{th,1}$) increases while the smaller level ($P_{th,2}$) decreases by decreasing κ_s^2. In contrast, for a given value of κ_s^2, both threshold levels increase by decreasing κ_p^2. Finally, considering a real field amplitude of input pump S_p, i.e. both real solutions A_{p1} and A_{p2}, SRS effect in the resonator can have a finite threshold only if it occurs:

$$v_{gs}^2\left(\frac{g_s}{2} - \frac{\beta^{TPA}}{A_{eff,p}}\right)^2 - \frac{v_{gs}\sigma_s\eta}{\tau_s} \geq 0 \qquad (33)$$

Therefore, Eq. (33) imposes a limitation on the total cavity decay time τ_s for Stokes wave. In particular, it leads to individuate the range for κ_s^2 needed to find the SRS effect threshold into the microcavity resonator. In fact, using Eq.(22b), Eq. (33) can be re-written as:

$$\kappa_s^2 < \frac{L_{cavity}}{v_{gs}} \frac{\left(\tau_{l,s} - \overline{\tau}\right)}{\tau_{l,s}\overline{\tau}} \text{ if } \tau_{l,s} > \overline{\tau} \qquad (34)$$

where $\overline{\tau} = \eta\sigma_s v_{gs}^{-1}\left(\dfrac{g_s}{2} - \dfrac{\beta^{TPA}}{A_{eff,p}}\right)^{-2}$.

The time constant $\overline{\tau}$, related to nonlinear effects, includes SRS, TPA and FCA. Thus, Eq. (33) or similarly Eq. (34), states that the Stokes photon decay rate related to coupling mechanism has to assume an appropriate value to guarantee that the photon generation rate induced by SRS effect can compensate the overall photon decay rate (due to linear losses, FCA and coupling process).

Then, the Raman emission power can be evaluated by Eqs. (28) and (31) inside a microcavity having $L_{cavity} < L_{NL}$, where $L_{NL} = 1/(|A_p|^2 \gamma_{p,p(s)})$ represents the (nonlinear) length over which SPM and XPM effects become important. After straightforward algebra, we obtain:

$$P_s = |A_s|^2 = \xi_s\left(\sqrt{P_{inc}} - \sqrt{P_{th,i}}\right)\frac{2}{v_{gp} g_p A_{p,i}} \qquad i = 1,2 \qquad (35)$$

where $P_{inc} = |S_p|^2$ is the pump power injected into the bus waveguide. It is evident that two levels of emitted Raman power can be evaluated, depending on the considered threshold.

According to a worst case scenario (higher pump power), assuming $|A_p|^2$ of the order of Watts, Eq. (35) holds being L_{NL} □ 9 m, i.e. it is applicable to a very large number of experimental cases. Moreover, the laser Stokes power calculated at the bus waveguide end is given by:

$$P_{out} = \kappa_s^2 P_s e^{-\alpha_s L_{out}} \qquad (36)$$

where α_s is the linear propagation loss inside the bus waveguide for Stokes wave, and L_{out} represents the waveguide length from the coupling region to the end (see Figure 1). Finally, the laser external efficiency η_{ex} is evaluated by linearizing the previous expression for the input pump power $P_{inc} = |S_p|^2$ close to the threshold condition, as:

$$\eta_{ex} = \frac{dP_{out}}{d(P_{inc})} = e^{-\alpha_s L_{out}} \kappa_s^2 \xi_p \frac{2}{g_p v_{gp} A_{p,i}} \frac{1}{2\sqrt{P_{th,i}}} \qquad i = 1,2 \qquad (37)$$

Differently from Raman lasers based on a Fabry-Perot cavity, Eq. (35) demonstrates that the output Stokes power presents a square root dependence from the input pump power, as a consequence of the pump-to-Raman conversion influenced by the coupling between resonator and waveguide (see ξ_p in Eq. (28)). In addition, the previous equations show that, by increasing κ_s^2, both output power and external efficiency increase at the expenses of a larger threshold level, i.e. $P_{th,2}$. Thus, Eq. (32)-(36)-(37) are very useful relationships to find the best trade-off among different physical parameters, with the aim to design efficient CW Raman lasers based on SOI microcavity resonators.

To test the analytical formulas and the physical assumptions for the Raman effect induced in microcavities under CW operation, we have compared our numerical results with some experiments proposed in literature. The architecture used in the experimental setup proposed in [30] is equal to that in Figure 6, where the microcavity is constituted by a SOI racetrack resonator. The device proposed has a rib width $W = 1.5$ μm, height $H = 1.55$ μm, and etch depth equal to 0.7 μm. The total length of racetrack cavity is 3 cm and the bend radius is 400 μm. Finally, the measured optical parameters include $\alpha_{loss,p} = \alpha_{loss,s} = \alpha_{loss} = 0.6 \pm 0.1$ dB/cm, $\tau_{eff} = 1$ns, $g_R = 9.5$ cm/GW, and $\beta^{TPA} = 0.5$cm/GW. In experiments [30], Raman laser devices with different coupling factors for pump and Stokes waves were considered. Thus, Figure (11) shows the laser output power P_{out} versus P_{inc} (pump input power) for three devices characterized by different κ_p^2 and κ_s^2. The markers represent the experimental data, the solid lines with markers designate numerical results as evaluated by means of Eq. (36). The numbers in the parenthesis indicate the experimental coupling factors for pump and Stokes waves at 1550 nm and 1686 nm, respectively. We have to focus on two fundamental aspects. First, the upper threshold value ($P_{th,1}$) assumes a level too high to be practical because of the large racetrack length and, second, Eq. (34) does not impose any limitation on the coupling factor κ_s^2 to be used.

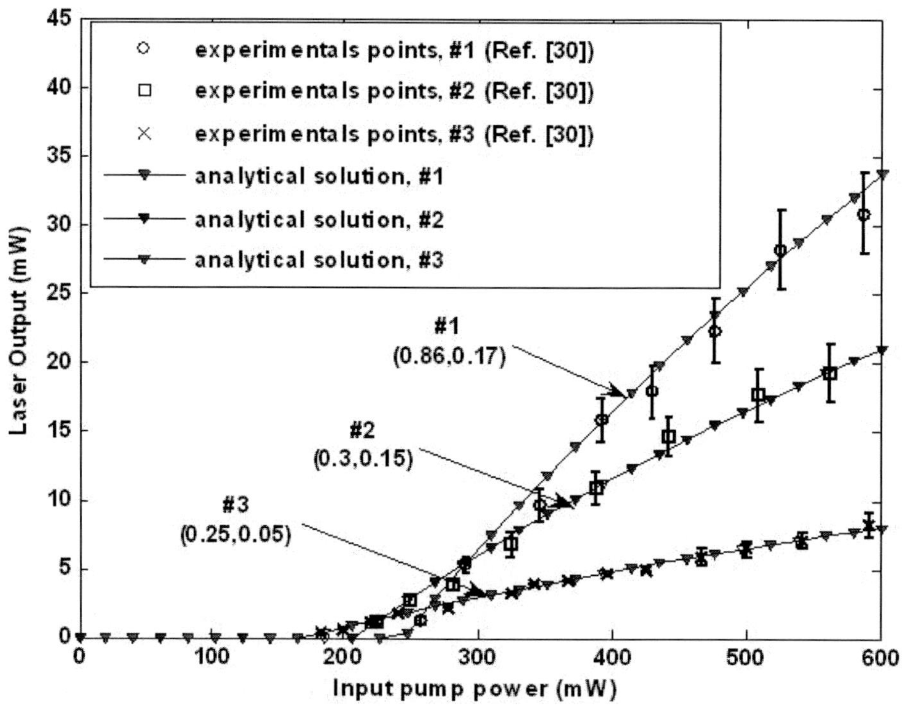

Figure 11. Comparison between experimental data and rigorous modeling in terms of CW Raman laser output power versus input pump power for three racetrack resonators.

The same coupling conditions as in [30] have been assumed for these comparisons, so arbitrary fitting parameters were not used. The plots show a very good agreement with the experimental data in terms of threshold values, output powers above threshold, and external efficiencies. This comparison has been summarized in Table 2 in terms of measured and simulated threshold values for three laser devices (#1, #2, #3).

Table 2. Comparison between measured [30] and calculated Raman laser thresholds

Laser device	Measured $P_{th,2}$ [30] (mW)	Calculated $P_{th,2}$ (mW)	Δ (%)	Calculated $P_{th,2}$ [30] (mW)	Δ (%)
#1	231	245.45	6.25	256	10.82
#2	208	210.17	1.04	211	1.44
#3	170	174.52	2.66	162	4.70

As shown in this table, our formula leads to evaluate the threshold power ($P_{th,2}$) with high accuracy, the percentage error being well smaller than in the model proposed in [30]. In addition, differently from mathematical approach followed in [30], our model allows to estimate the output power and the external efficiency by means of Eq. (36)-(37), in case of CW Raman lasers based on a microcavity with $L_{cavity} < L_{NL}$. This it is possible essentially because Eq. (24) allows the nonlinear interaction between pump and Stokes waves inside the microcavity to be investigated in the time domain, instead of space domain.

It is important to outline that these experimental devices [30] cannot show the double threshold levels, since the SOI racetrack resonators have too large cavity perimeters to induce a double solution in Eq. (32).

Moreover, we have investigated the CW Raman threshold for micro-scale SOI cavity resonators. Hereinafter, microring resonator will be considered as laser resonant microcavity. In particular, three ring resonators have been assumed with ring radius $R = 10$, 15, and 20 μm, respectively. In order to apply Eq. (32), it is essential to estimate both pump and Stokes coupling factors. To this aim, a number of simulations were carried out by means of FDTD method [46], and results are summarized in Table 3. The pump and Stokes wavelengths have been assumed as 1433.2 nm and 1586.1 nm, respectively. As it is evident in Table 3, for each value of ring radius R and gap G, it always results $\kappa_s^2 > \kappa_p^2$.

The physical reason for this behavior is related to the different refractive index contrast between cover and guiding structure (larger for pump than for Stokes wave). Moreover, coupling to micro-size ring resonators is not negligible only in the region of smallest separation. This means that, for such small structures, the coupling length cannot be a design parameter, as it usually occurs in large-scale racetrack resonators [30].

Table 3. FDTD results for pump (1433.2 nm) and Stokes (1586.1 nm) coupling factors

Gap, G (μm)	$R = 10$ μm		$R = 15$ μm		$R = 20$ μm	
	κ_p^2 (%)	κ_s^2 (%)	κ_p^2 (%)	κ_s^2 (%)	κ_p^2 (%)	κ_s^2 (%)
0.1	19.32	29.54	28.05	41.97	46.57	62.67
0.2	3.41	4.55	3.66	7.41	6.94	10.67
0.25	1.43	1.78	1.32	3.11	2.67	4.40
0.3	0.6	0.7	0.47	1.30	1.03	1.81
0.35	0.25	0.27	0.17	0.55	0.4	0.75

Using the simulation results in Table 3, the threshold conditions have been found. In Figure 12 the laser external efficiency is sketched as a function of Stokes coupling factor κ_s^2, for different ring radii and $\beta^{(TPA)} = 0.5$ cm/GW, $g_R = 10.5$ cm/GW, $\alpha_{loss,p} = \alpha_{loss,s} = \alpha_{loss} = 0.6$ dB/cm, $\tau_{eff} = 1$ ns, $L_{out} = 2$ cm. Hereinafter, calculations of field distributions, effective indices and modal areas were carried out by full-vectorial finite element method (FEM) [41].

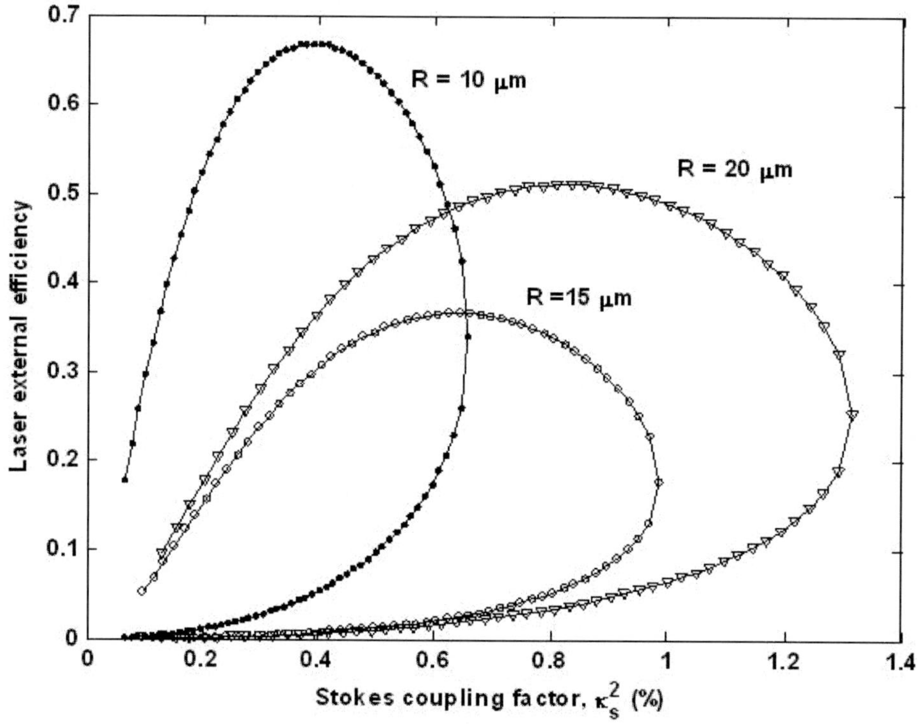

Figure 12. Laser external efficiency versus Stokes coupling factor for various ring radii.

The plots clearly demonstrate that the double values of laser threshold induces two arms in the shape of external efficiency for each ring radius. This is true only if κ_s^2 satisfies Eq. (34), i.e. when both TPA and FCA effects induce two possible solutions (A_{p1}, A_{p2}) for the field amplitude A_p, as in Eq. (31). The two arms of curves in Figure 12 merge at one limit point, different for each ring radius. Thus, SRS effect can be excited in CW regime into the microring cavity only if the coupling factor κ_s^2 is not larger than this critical value, depending on the ring radius. Finally, the upper arm of the curves in figure decreases by decreasing R, as a result of the improved pump enhancement factor.

Finally, the laser external efficiency is sketched in Figure 13 as a function of Stokes coupling factor κ_s^2, for various propagation loss coefficients and $\beta^{(TPA)} = 0.5$ cm/GW, $g_R =$ 10.5 cm/GW, $R = 20$ μm, $\tau_{eff} = 1$ ns. It is possible to observe that, with increasing α_{loss}, the upper arm of curves shifts towards smaller values, while the other arm is weakly influenced. In addition, the limit point shifts to lower values of κ_s^2.

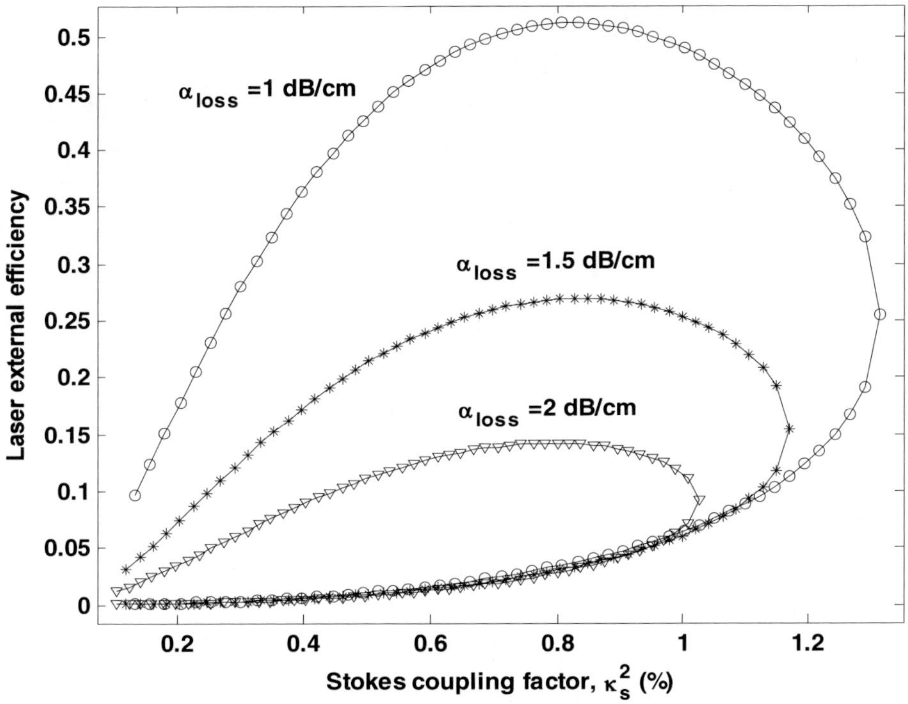

Figure 13. Laser external efficiency versus Stokes coupling factor for various propagation loss coefficients.

This latter effect indicates that exist a particular value of propagation loss coefficient where κ_s^2 cannot induce the SRS threshold. Thus, for the case considered the SRS effect cannot be obtained for $\alpha_{loss} \geq 5.8$ dB/cm.

3.2. Sub-Picosecond Regime

In this sub-section, the space-time evolution of sub-picosecond Raman pulses propagating along the SOI waveguide is investigated. First, a set of simulations have to be carried out in order to evaluate the dispersion coefficient of the SOI waveguides. Figures 14 and 15 show the GVD coefficient spectra for different values of r and quasi-TE and quasi-TM modes, respectively. The simulations have been obtained by means of FEM method assuming $H = 700$ nm and $W = 500$ nm. Figure 14 shows β_2 coefficient in both normal and anomalous regimes with $r < 0.3$.

On the contrary, quasi-TM modes (see Figure 15) leads to individuate the zero dispersion condition for each considered value of r. Moreover, for both polarizations these zero dispersion points move towards smaller values of wavelength with decreasing r. In addition, it is worth to note that quasi-TM modes present a reduced sensitivity to r change with respect to quasi-TE modes.

For the considered waveguide, GVD and TOD coefficients are listed in Table 4, as obtained from FEM calculations. Using the results of previous simulations, assuming

$H = 700$ nm, $W = 500$ nm and $r = 0.15$, we obtain the condition $\Delta n_{eff} \approx 0.073$. Thus, if this SOI waveguide is used to realise a SOI Raman amplifier at 1.55 μm, this birefringence induces a characteristic birefringence length $L_B \cong 21.23$ μm. Since waveguides with $L \square L_B$ are needed to realise the Raman amplifier, then terms containing $k_{i,j,k,l}$ often change sign and the total phase contribution due to birefringence averages out to zero.

Table 4. GVD coefficients for W = 500 nm, H = 700 nm, r = 0.15

Parameter	Quasi-TE	Quasi-TM	Quasi-TE	Quasi-TM
	β_2 (ps^2/m)	β_2 (ps^2/m)	β_3 (ps^3/m)	β_3 (ps^3/m)
Pump wavelength (λ = 1434.4 nm)	-0.64	-0.056	+0.0061	+0.0049
0th order Stokes wavelength (λ = 1550 nm)	-1.23	-0.53	+0.0062	+0.0050
1st order Stokes wavelength (λ = 1685.9 nm)	-1.87	-1.07	+0.0067	+0.0062
2nd order Stokes wavelength (λ = 1847.9 nm)	-2.5	-1.8	+0.0056	+0.0085

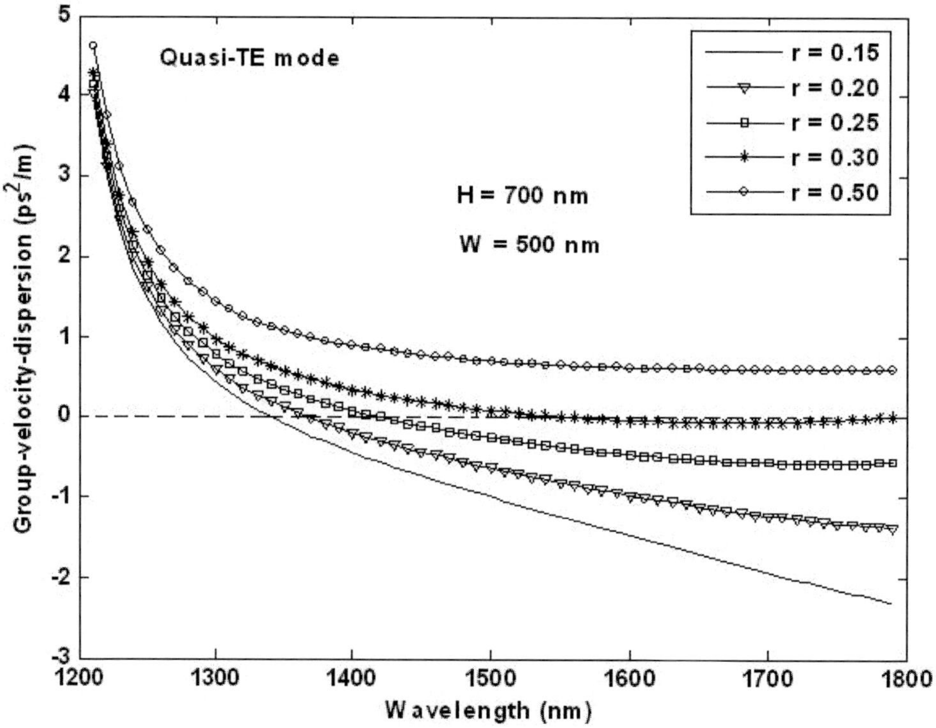

Figure 14. Group-velocity dispersion spectra for various rib parameters and quasi-TE mode.

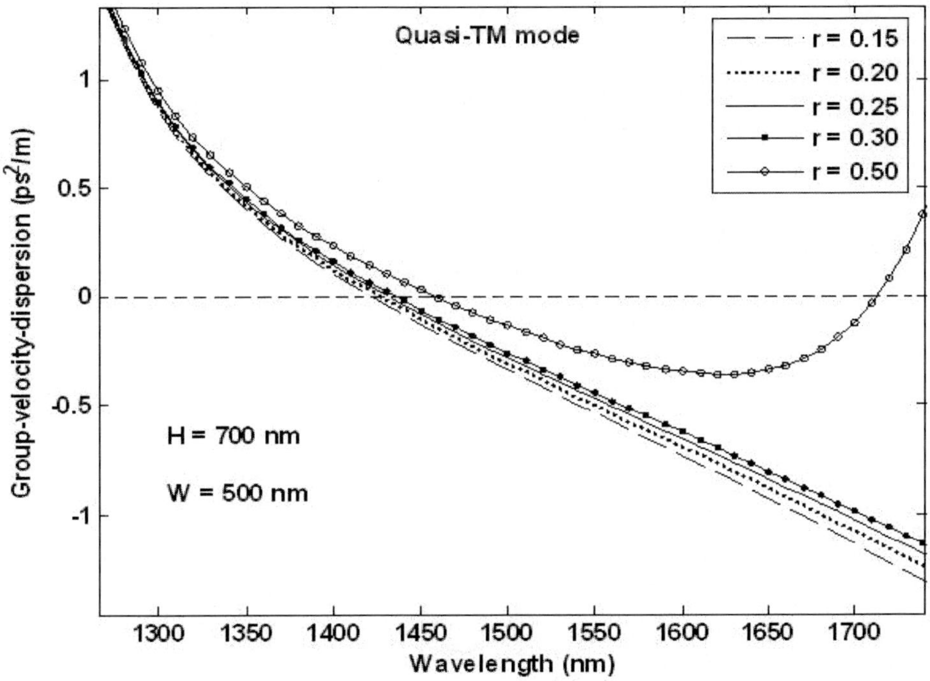

Figure 15. Group-velocity dispersion spectra for various rib parameters and quasi-TE mode.

In the first sub-picosecond simulation, we have assumed $P_0 = 1\,\text{W}$, $T_{FWHM} = 80\,\text{fs}$ and λ_p= 1.434.4μm, keeping the fundamental Stokes probe power 10 dB below the pump power, while the higher order Stokes waves have a peak power comparable with the spontaneous emission. To put into evidence the influence of GVD effects on the amplification of an optical signal matched with the fundamental Stokes wave, we have considered a relatively low pump power to avoid the excitation of higher order Stokes waves. The probe (Stokes) wavelength is λ_{s0}= 1.55 μm and Raman gain is $g_R = 10.5$ cm/GW [16]. Figure 16 shows the temporal shape of fundamental Stokes pulse at different propagation lengths assuming $\beta^{(TPA)} = 0.45$ cm/GW, $n_2 = 6\times10^{-5}$ cm^2/GW, $\alpha_i^{(prop)} = 5$ dB/cm and both input probe polarizations. The other calculated parameters are $A_{eff,p} = 0.230$ μm^2, $A_{eff,s0} = 0.2467$ μm^2 (quasi-TE) and 0.2283 μm^2 (quasi-TM), $\tau_{eff,1} = 1.65$ ns.

We can observe from Figure 16 as quasi-TE and quasi-TM modes present a very different behaviour, due to their different dispersion lengths, namely $L_D(TE)$= 2.9 mm for quasi-TE and $L_D(TM)$= 6.77 mm for quasi-TM, having the same walk-off and nonlinear lengths L_w=1.81 mm and L_{NL}=1.06 mm, respectively, for both polarizations. Since each characteristic length represents the length scale over which the relevant physical effect becomes important, it seems clear as the Raman amplification of quasi-TM signal probe is less influenced than quasi-TE mode from the temporal broadening induced by GVD effect, as it can be seen by the pulse shape at different propagation lengths.

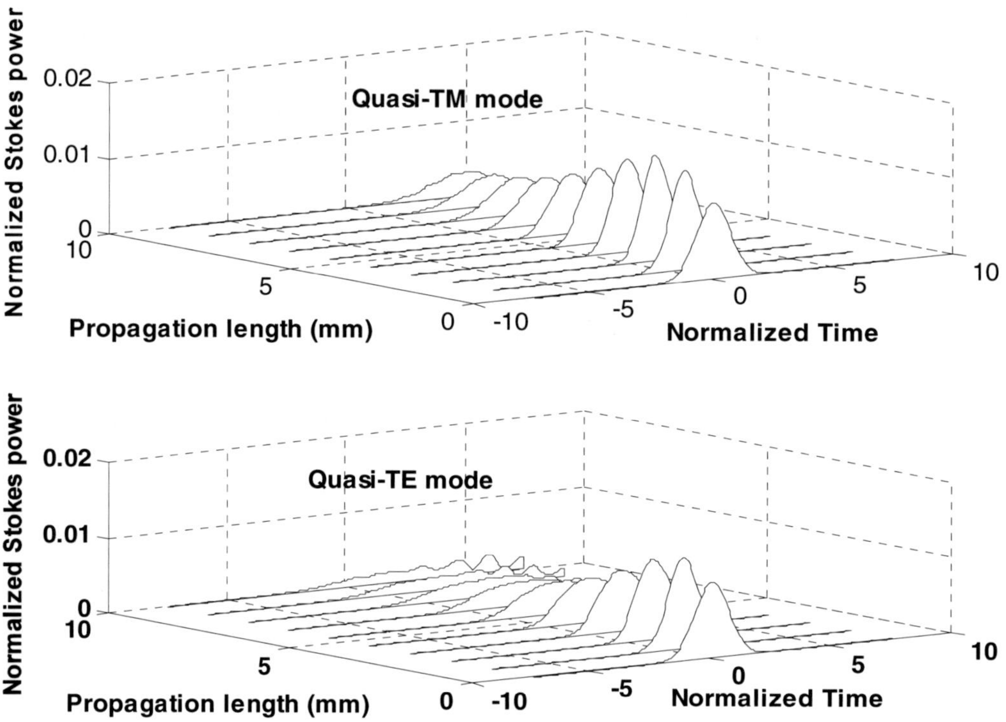

Figure 16. Time evolution of fundamental Stokes pulse at different propagation lengths for both polarizations.

Moreover, the TOD effect does not influence the dynamics of the process, since the characteristic length $L_D^{'}$ exceeds the waveguide length for both polarizations. Moreover, at the end (z = 10 mm), quasi-TE Stokes signal is completely destroyed, while the quasi-TM probe presents a depletion and a temporal broadening preserving the Gaussian-like shape.

As it is possible to observe from Figure 16, the quasi-TM probe signal is favourite over quasi-TE probe to experience SRS effect. The previous considerations can be scaled by changing the pulse width in the same SOI waveguide. Figure 17 shows the net gain versus FWHM width for various waveguide lengths and both mode polarizations. The plot shows that, for each waveguide length, quasi-TM modes induce a net gain larger than quasi-TE ones. This essentially depends on smaller optical mode area and larger dispersion length, allowed by quasi-TM modes for considered waveguides, than quasi-TE case.

It is worth to note that the trend previously described is not general, since it depends critically on the values assumed by dispersion length for quasi-TE and quasi-TM modes, and thus it is influenced on the waveguide cross-section. In fact, as demonstrated in [38] using $H = 400$ nm, $r = 0.2$, and $W = 368$ nm, it is possible to realise $L_D(TE) = 32.9$ mm for quasi-TE and $L_D(TM) = 0.73$ mm for quasi-TM, being walk-off and nonlinear lengths, $L_w = 1.45$ mm and $L_{NL} = 3.49$ mm, respectively, the same for both polarizations.

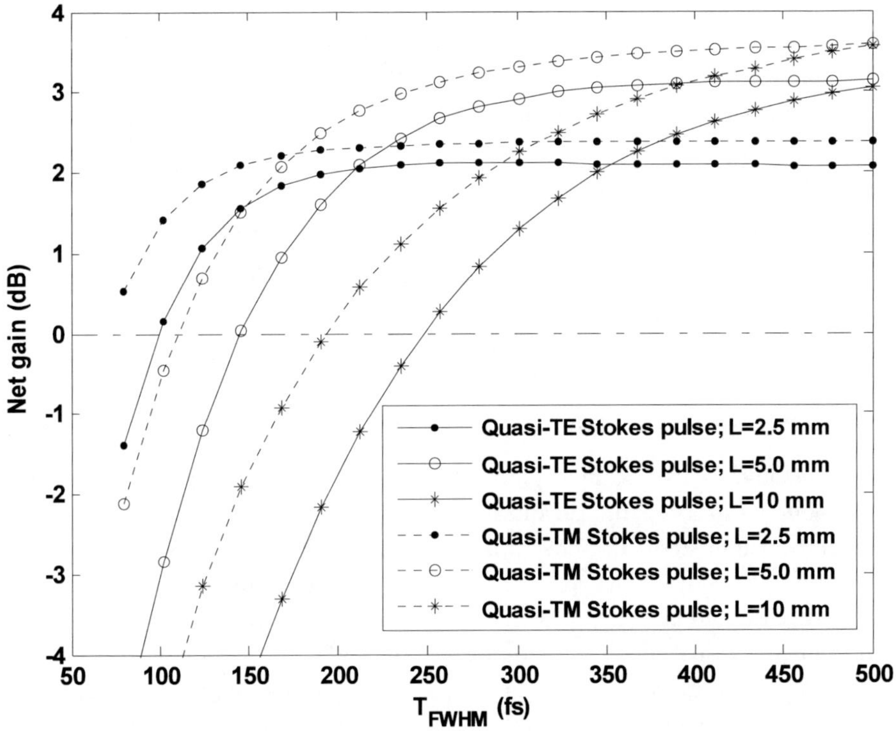

Figure 17. Raman net gain versus FWHM width for various waveguide lengths (W=500nm, H=700nm, r = 0.15).

Thus, Raman amplification of quasi-TE signal probe is not influenced by temporal broadening induced by GVD effect. On the contrary, quasi-TM probe signal can experience the SRS effect only over a short distance, less than 0.73 mm. In fact, by increasing the propagation distance, the anomalous GVD effect becomes dominant, so inducing a temporal broadening and destroying the pulse shape. Thus, for this last SOI waveguide, simulations similar to those in Figure 17 demonstrate that, for each waveguide length, the curves related to quasi-TE and quasi-TM modes have crossing points for a particular value of FWHM width, namely $\overline{T}_{FWHM}$. For values $T_{FWHM} \leq \overline{T}_{FWHM}$, quasi-TE modes experience a larger net gain with respect to quasi-TM ones, having a larger anomalous GVD effect for the designed waveguide (see discussion above).

However, the opposite behaviour is revealed for $T_{FWHM} > \overline{T}_{FWHM}$. In fact, larger values of FWHM width induces the TM dispersion length $L_D(TM)$ to be larger than the waveguide length. In this situation, the anomalous GVD effect is not dominant even for quasi-TM modes and, then, the net gain for TM polarization can increase, becoming larger than for quasi-TE because of its smaller optical mode area.

With the goal to improve the performance of Raman amplifiers within a sub-micrometer-scale cross section waveguide, it can be investigated the possibility to control the pulse dynamics by changing the excitation condition. In this sense, an important parameter useful for the optimisation is the soliton number, defined as $N = \sqrt{L_D / L_{NL}}$. In the following, we consider both pump pulse and fundamental Stokes probe signal as Gaussian pulses with

$T_{FWHM} = 80$ fs and peak power selected to satisfy the condition $N = \sqrt{L_{D,p}/L_{NL,p}} = \sqrt{L_{D,s0}/L_{NL,s0}} = 1$. Soliton numbers $N > 1$ are not considered here to avoid the generation of higher order Stokes waves.

Figures 18 and 19 show the time evolution of quasi-TE pump and Stokes probe signal with quasi-TM and quasi-TE polarization, respectively. The simulations have been performed assuming $H = 400$ nm, $r = 0.2$, $W = 368$ nm, $g_R = 20$ cm/GW [34], $\beta^{(TPA)} = 0.45$ cm/GW, $n_2 = 6\times10^{-5}$ cm^2/GW, $\alpha_i^{(prop)} = 5$ dB/cm and both input probe polarizations. The other calculated parameters are $A_{eff,p} = 0.158\ \mu\text{m}^2$, $A_{eff,s0} = 0.185\ \mu\text{m}^2$ (quasi-TE) and 0.156 μm^2 (quasi-TM), $\tau_{eff,1} = 0.9$ ns and optical mode width $w = 0.3\ \mu$m. Figure 18 shows that, for a quasi-TM signal probe, the soliton condition $N = \sqrt{L_{D,s0}/L_{NL,s0}} = 1$ does not allow an effective improvement of net Raman gain to be achieved. On the contrary, the net gain is estimated as about 3.4 at z = 5 mm for quasi-TE polarization (see Figure 19).

These results demonstrate the possibility to achieve Raman amplification with ultrafast pulses over relatively long SOI waveguides, without any pulse distortion and with predictable time delay. A slight compression 1.7 is also obtained. Pulse compression is an important task in many nonlinear optical devices. Compressed ultrafast pulses can be excited in SOI technology by exploiting the Raman effect. This excitation can be obtained with high performance if the fundamental Stokes wave is amplified by the pump, starting as a Gaussian signal probe. In fact, by setting its parameters as a higher order soliton, the Stokes wave evolves in the SOI waveguide as a compressed pulse in form of a soliton.

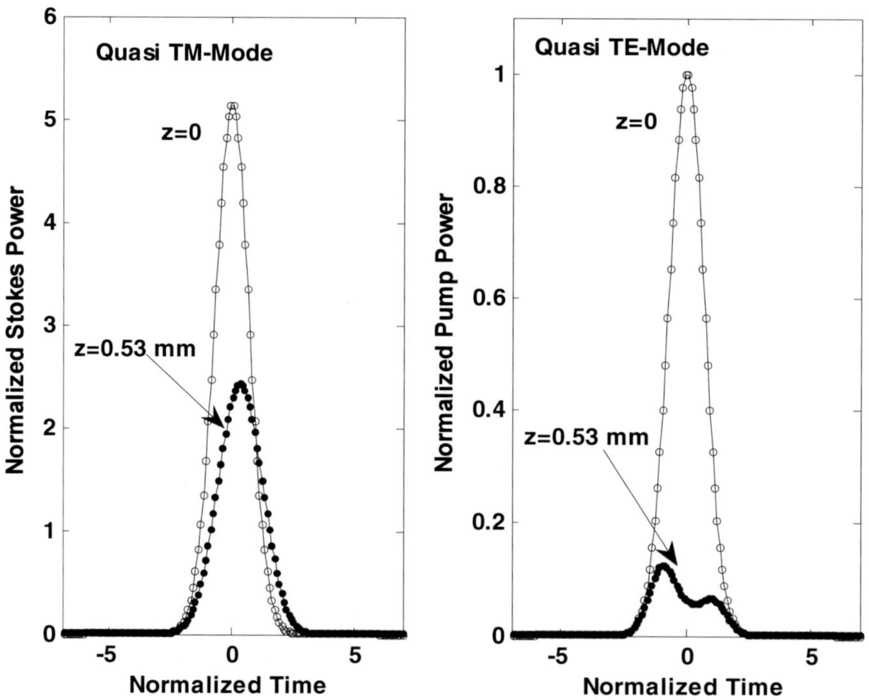

Figure 18. Time evolution of pump and Stokes pulses (TM mode) at different propagation lengths.

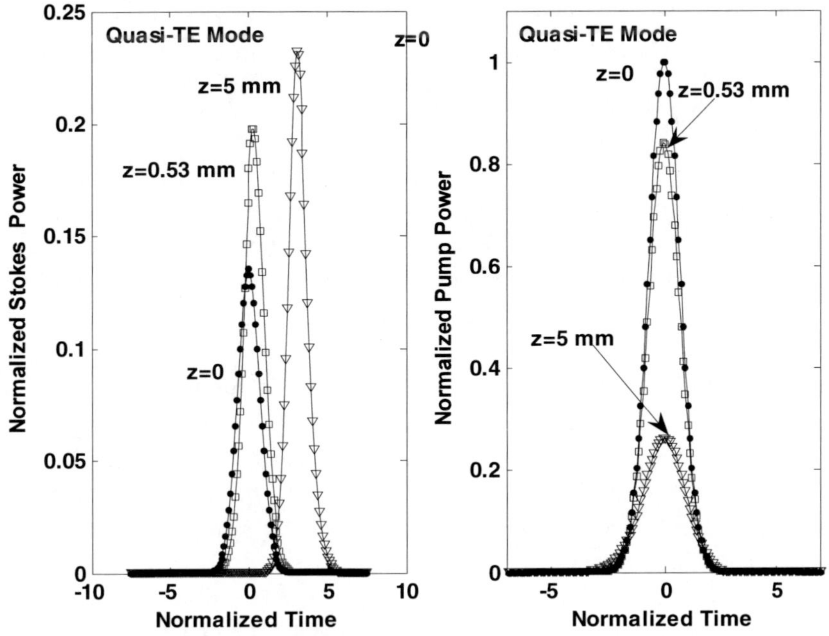

Figure 19. Time evolution of pump and quasi-TE Stokes pulses at different propagation lengths.

To demonstrate this behavior, we have performed some simulations and results are shown in Figure 20 in terms of normalized intensity of pump and fundamental Stokes versus normalized time, at the waveguide output z = 10 mm. A 3dB compression of 2.67, i.e. from 80 to 30 fs, is so proved.

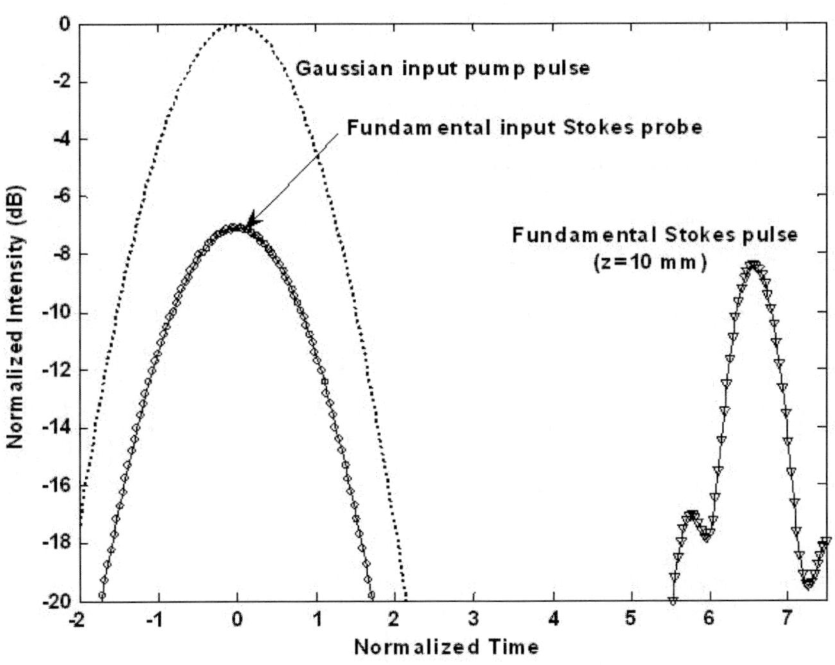

Figure 20. Normalized intensity of pump and fundamental Stokes wave versus normalized time.

This results has been obtained by exciting the Stokes wave as a third order soliton with $N=\sqrt{L_{D,s0}/L_{NL,s0}}=3$ (2.06 W) and Gaussian signal probe, and by using a Gaussian pump with T_{FWHM} = 80 fs and $N=\sqrt{L_{D,p}/L_{NL,p}}=2$ (10.6 W). Moreover, the first order Stokes wave is set as a Gaussian pulse with $N=\sqrt{L_{D,s1}/L_{NL,s1}}=1$ (6.54 W) in order to avoid any power coupling with the fundamental Stokes pulse. As a result, fundamental Stokes propagates as a soliton, having a *sech*-like profile at z = 10 mm. This offers the possibility to excite compressed Raman pulses at relatively low powers, because Raman effect allows to partially compensate optical losses. To achieve the same aim, much larger power has been used in literature [28].

In some applications, including Raman amplifiers, it is important to know the threshold condition for higher order Stokes pulses. It is known that the fundamental Stokes wave can itself acts as a secondary pump pulse and generate further higher order Stokes modes by Raman effect. This process of cascaded Raman scattering has been described by including other coupled equations in the partial differential equation system and new coupling terms into the coupled mode equations.

Then, not only the fundamental Stokes wave but also the first and second order Stokes waves are included, as well as both polarizations (quasi-TE, quasi-TM) for each wave and relative coupling effects. The analysis of cascaded Raman scattering is very important because it could allow the simultaneous generation of higher order Stokes pulses at different wavelengths (continuum generation). The curves in Figure 21 and 22 represent design criteria to select the quasi-TE pump requirements to generate (or avoid) higher order Stokes pulses in case of quasi-TE or quasi-TM polarization, respectively.

Although similar trends could be revealed in optical fibres, the physical effects involved in SOI waveguides are substantially different. In fact, while the threshold conditions in optical fibres mainly depend on Raman scattering and walk-off effect, the threshold values in SOI waveguides significantly depend also on TPA and FCA effects in picosecond regime [30], and on GVD effect in sub-picosecond (femtosecond) regime.

Both Figures 21 and 22 show the pump peak power required to generate the net Raman gain for the first (s_1) and second order (s_2) Stokes waves as a function of interaction length and different pump pulse widths, namely $T_{FWHM}=100$, and 500 fs. We have assumed g_R = 20 cm/GW and $\alpha_i^{(prop)}$ = 5 dB/cm and pump wavelength λ_p = 1434.4 nm. The Stokes waves are generated at wavelengths λ_{s0} = 1550 nm, λ_{s1} = 1685.9 nm, and λ_{s2} = 1847.9 nm, respectively. The effective carrier recombination lifetime is $\tau_{eff,1}=0.9$ ns, as estimated using the relationship given in previous section. In some cases, the pump peak power at Stokes threshold decreases by reducing T_{FWHM}. Similarly to picosecond regime [16], numerical data in sub-picosecond regime can be interpolated with good accuracy by using an exponential formulation in the form $a/\left(1-e^{-b\cdot L}\right)$ [36]. Therefore, fitting parameters are extracted to predict the higher order Stokes thresholds by using this simple function, as shown from the solid lines in Figures 21 and 22.

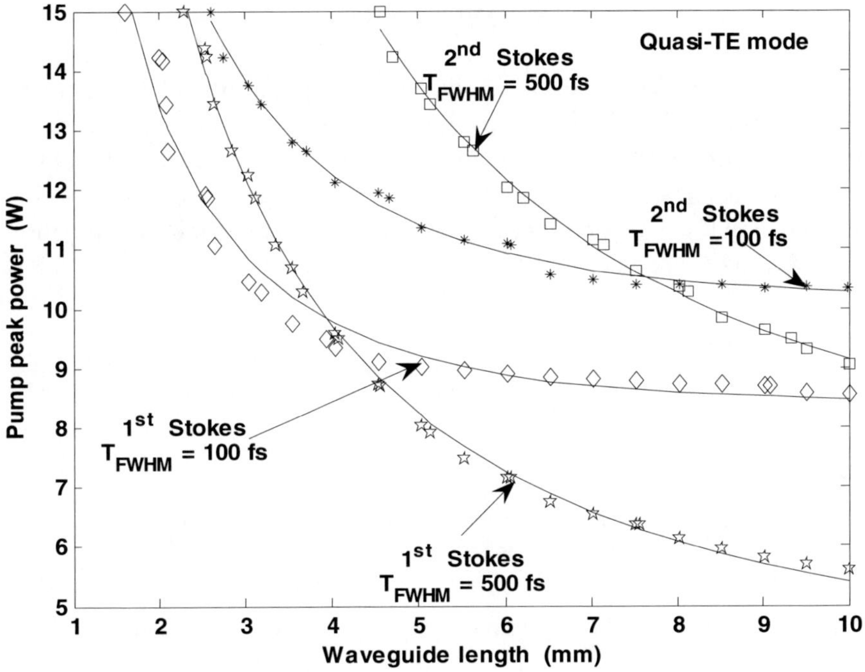

Figure 21. Peak pump power generating quasi-TE higher Stokes waves versus interaction length for various T_{FWHM} .

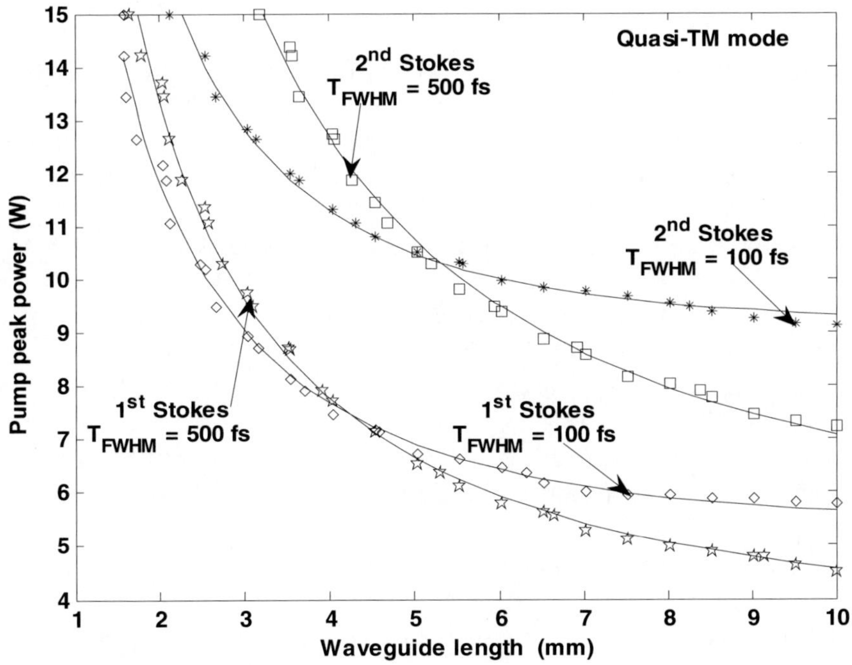

Figure 22. Pump peak power generating quasi-TM higher Stokes waves versus interaction length for various T_{FWHM} .

Conclusion

In this chapter, a general model is presented and applied to simulate ultrafast nonlinear pulses in sub-micrometer-scale SOI optical waveguides. Comparisons with experimental measurements and theoretical predictions in literature are presented for straight waveguides and optical resonators under picosecond regime. Some applications for nonlinear signal processing, including Raman amplification, wavelength conversion, soliton formation and pulse compression, are discussed by considering optimal SOI structures in sub-picosecond regime.

References

[1] Intel Corporation WEB site.

[2] Pavesi, L., *J. Physics: Condensed Matter,* 2003, 15, 1169-1196.

[3] Reed, G. T., *Nature*, 2004, 427, 595-596.

[4] Reed, G. T., Knights, A. P., *Silicon Photonics: An Introduction*, John Wiley, 2004.

[5] Pavesi, L., Negro, L. D., Mazzoleni, G., Franzo, G., Priolo, S., *Nature*, 2000, 408, 440-444.

[6] Ruan, J., Fauchet, P. M., Negro, L. D., Cazzanelli, M., Pavesi, L., *Applied Physics Letters*, 2003, 83, 5479-5481.

[7] Polman, A., Van Veggel, F. C. J. M., *J. Optical Society of America B*, 2004, 21, 871-892.

[8] Claps, R., Raghunathan, V., Boyraz, O., Koonath, P., Dimitropoulos, D., Jalali, B., *Optics Express*, 2005, 13, 2459-2466.

[9] Sirleto, L., Raghunathan, V., Rossi, A., Jalali, B., *Electronics Letters*, 2004, 40, 1221-1222.

[10] Claps, R., Dimitropoulos, D., Han, Y., Jalali, B., *Optics Express*, 2002, 10, 1305-1313.

[11] Claps, R., Dimitropoulos, D., Raghunathan, V., Han, Y., Jalali, B., *Optics Express*, 2003, 11, 1731-1739.

[12] Liu, A., Rong, H., Paniccia, M., Cohen, O., Hak, D., *Optics Express*, 2004, 12, 4261-4268.

[13] Xu, Q., Almeida, V. R., Lipson, M., *Optics Express*, 2004, 12, 4437-4442.

[14] Xu, Q., Almeida, V. R., Lipson, M., *Optics Letters*, 2005, 30, 35-37.

[15] Espinola, R. L., Dadap, J. I., Osgood, Jr., R. M., McNab, S. J., Vlasov, Y. A., *Optics Express*, 2004, 12, 3713-3718.

[16] Passaro, V. M. N., De Leonardis, F., *J. Lightwave Technology,* 2006, 24, 2920-2931.

[17] Dimitropoulos, D., Raghunathan, V., Claps, R., Jalali, B., *Optics Express*, 2004, 12, 149-160.

[18] Claps, R., Raghunathan, V., Dimitropoulos, D., Jalali, B., *Optics Express*, 2003, 11, 2862-2872.

[19] Raghunathan, V., Claps, R., Dimitropoulos, D., Jalali, B., *J. Lightwave Technology,* 2005, 23, 2094-2102.

[20] Boyraz, O., Koonath, P., Raghunathan, V., Jalali, B., *Optics Express*, 2004, 12, 4094-4102.

[21] Claps, R., Raghunathan, V., Dimitropoulos, D., Jalali, B., *Optics Express*, 2004, 12, 2774-2780.
[22] Liang, T. K, Ki Tsang, H., *IEEE J. Selected Topics in Quantum Electronics,* 2004, 10, 1149-1153.
[23] Jones, R., Liu, A., Rong, H., Paniccia, M., Cohen, O., Hak, D., *Optics Express*, 2005, 13, 1716-1723.
[24] Claps, R., Raghunathan, V., Boyraz, O., Koonath, P., Dimitropoulos, D., Jalali, B., *Optics Express*, 2005, 13, 2459-2466.
[25] Boyraz, O., Jalali, B., *Optics Express*, 2004, 12, 5269-5273.
[26] Krause, M., Renner, H., Brinkmeyer, E., *Optics Express*, 2004, 12, 5703-5710.
[27] Rong, H., Jones, R., Liu, A., Cohen, O., Hak, D., Fang, A., Paniccia, M., *Nature*, 2005, 433, 725-728.
[28] Rong, H., Liu, A., Jones, R., Cohen, O., Hak, D., Nicolaescu, R., Fang, A., Paniccia, M., *Nature*, 2005, 433, 292-294.
[29] Boyraz, O., Jalali, B., *Optics Express*, 2005, 13, 796-800.
[30] Rong, H., Kuo, Y-H., Xu, S., Liu, A., Jones, R., Paniccia, M., Cohen, O., Raday, O., *Optics Express,* 2006, 14, 6705-6712.
[31] De Leonardis, F., Passaro, V. M. N., *New Journal of Physics*, 2007, 9, 25.
[32] De Leonardis, F., Dimastrodonato, V., Passaro, V. M. N., *Semiconductor Science and Technology*, 2008, 23, 064008.
[33] Jalali, B., Raghunathan, V., Shori, R., Fathpour, S., Dimitropoulos, D., Stafsudd, O., *IEEE J. Selected Topics in Quantum Electronics*, 2006, 12, 1618-1627.
[34] Raghunathan, V., Borlaug, D., Rice, R., Jalali, B., *Optics Express,* 2007, 15, 14355-14362.
[35] Dimitropoulos, D., Houshmand, B., Claps, R., Jalali, B., *Optics Letters,* 2003, 28, 1954-1956.
[36] Agrawal, G. P., *Nonlinear fiber optics*, Academic Press, 2001.
[37] Yin, L., Lin, Q., Agrawal, G. P., *Optics Letters*, 2007, 32, 391-393.
[38] De Leonardis, F., Passaro, V. M. N., *IEEE J. Selected Topics in Quantum Electronics*, 2008, 14.
[39] Soref, R. A., Bennett, B. R., *IEEE J. Quantum Electronics*, 1987, QE-23, 123-129.
[40] Deb, S., Sharma, A., *Optical Engineering*, 1993, 32, 695-699.
[41] COMSOL Multiphysics by COMSOL AB, Stockholm, ver. 3.2, single license, 2005.
[42] Soref, R. A., Schmidtchen, J., Petermann, K., *IEEE J. Quantum Electronics,* 1991, 27, 1971-1974.
[43] Qian, S. X., Snow, J. B., Tzeng, H. M., Chang, R. K., *Science,* 1986, 231, 486-488.
[44] Zhang, J. Z., Chang, R. K., *J. Optical Society of America B,* 1989, 6, 151-153.
[45] Kippenberg, T. J., Spillane, S. M., Min B., Vahala, K. J., *IEEE J. Selected Topics in Quantum Electronics,* 2004, 10, 1219-1228.
[46] OptiBPM by Optiwave Systems Inc., Ottawa, Canada, ver. 7.0, *single license*, 2005.

In: Modeling of Photonic Devices
Editor: Vittorio M. N. Passaro
ISBN 978-1-60456-980-3

Chapter 9

THERMO-OPTIC EFFECT IN SILICON STRUCTURES

Francesca Magno* and Vittorio M. N. Passaro
Photonics Research Group, Dipartimento di Elettrotecnica ed Elettronica,
Politecnico di Bari, via Edoardo Orabona n. 4, 70125 Bari, Italy

ABSTRACT

In this chapter, an extensive analysis of silicon guided-wave structures modulated by thermo-optic effect is carried out by a multiphysics integrated approach. Results are obtained and presented for silicon-on-insulator rib waveguides, as well as for silicon waveguide arrays and silicon slot waveguides. Conditions to achieve best performance in terms of high switching frequency and low power consumption are derived.

Keywords: Thermo-optic Modulation, silicon waveguides, thermal effects, modeling

1. INTRODUCTION

Recently, there has been a great interest in the development of optical waveguide modulators, switches, ring resonators, filters, and couplers using silicon-on-insulator (SOI) technology [1-2]. SOI-based optical devices are characterized by small optical losses over telecommunications wavelengths and have the potential for integration with SOI-based CMOS electronics, reduced parasitics, absence of latch-up and high-speed operation at low power.

One significant technological issue is associated with the ability to perform optical modulation. We have already analyzed in chapter 5 optical modulation in silicon through plasma dispersion effect. A different and attractive way to modulate the refractive index in SOI rib waveguides is the thermo-optic effect [3], allowing low transmission loss, low cost, high stability, low power consumption and very large scale of integration. Apart from the well

* E-mail address: francesca.magno@libero.it

known optical characteristics, silicon also possesses good thermal features, with its high thermo-optical coefficient and thermal conductivity. Silicon possesses a very high thermo-optic coefficient ($1.86 \cdot 10^{-4} K^{-1}$ at 1550 nm), so silicon refractive index change produced by thermo-optic effect is a widely used approach for switching, modulation and tuning in a several silicon photonic device. In the last few years, a significant number of experimental and theoretical papers on thermo-optic modulation in silicon technology, polymers and optical fibers has been published [4-22], demonstrating the great interest for thermo-optic modulation in long-period gratings, matrix switches, compensation devices and sensors.

Static and dynamic thermo-optical analysis of small rib waveguides and arrays can be carried out by a multiphysics approach using full-vectorial finite element method (FV-FEM) [23]. In this chapter, we perform a fully integrated thermo-optic simulation. In fact, in the same simulation environment, thermal and optical investigations are performed, without any approximation on refractive index distribution and, consequently, on structure effective indices. Typically, the two analyses are developed in literature in two different simulation environments, so require some approximations to export the refractive index distribution.

2. Thermo-Optical Analysis in SOI Waveguides

In this section, we have investigated the thermo-optic effect in SOI rib waveguides, considering an aluminum heater laying on the top of the structure. The finite element method [24] has been used to solve both thermal and wave propagation problems (at the wavelength of 1.55μm) in rib optical waveguides. The thermal problem is solved and the obtained temperature distribution is used to evaluate the refractive index change, caused by the thermo-optic effect, in the heated regions. This new distribution of refractive index is then used to solve the optical problem. The analysis of waveguides is carried out in the *x-y* plane (2D), perpendicular to the light propagation direction *z*.

2.1. Thermal Model

The mathematical equation for heat transfer by conduction is the general heat equation:

$$\rho C \frac{\partial T}{\partial t} + \nabla \cdot \left(-k \nabla T\right) = Q \qquad (1)$$

where T is the temperature, Q is a heat source or heat sink, ρ is the density, C is the heat capacity and k is the thermal conductivity of the medium. These three thermo-physic properties determine the thermal diffusivity α, which gives information on how rapidly a temperature variation at the medium surface propagates through the medium itself. If the thermal conductivity is isotropic, Eq. (1) becomes:

$$\rho C_p \frac{\partial T}{\partial t} - k \nabla^2 T = Q \qquad (2)$$

In a stationary problem, the temperature does not change with time and the first term of the equation disappears. If there is no production or consumption of heat in the medium, Q sets to zero. To solve a particular heat conduction problem and find the temperature distribution inside a medium, it is necessary to specify the medium geometry and thermo-physic properties, the distribution of possible sources and the initial and boundary conditions. These conditions are essentially of two types: Dirichlet or Neumann. In the first type, a temperature is imposed on the medium surface, as:

$$T = T_0 \tag{3}$$

The second type of condition sets the heat flux q on a boundary as:

$$-\mathbf{n} \cdot \mathbf{q} = q_0 \tag{4}$$

where n stands for the surface normal. The heat flux is defined by the Fourier law as:

$$\mathbf{q} = -k\nabla T \tag{5}$$

where the negative sign is justified by the opposite directions of heat flux and temperature gradient.

SOI waveguides normally have a silicon core (refractive index n ~3.5) surrounded by cladding layers of air or silica, with typical refractive indices between 1 and 2. Hence there is a very high index contrast between waveguide core and cladding, allowing good confinement of optical modes and reduction of device dimensions. A typical SOI rib waveguide is sketched in Figure 1, with an aluminum heater on the top having a width W. The buried oxide layer is required to achieve thermal isolation of the waveguide from the substrate, because the thermal conductivity in SiO_2 is about 100 times smaller than in silicon. The distance H between the heater and the top of rib must be thick enough to insulate the waveguide core from the metallic electrode. For all the analyzed structures, we have set this distance to 100 nm, being the buried oxide thickness of 1 μm.

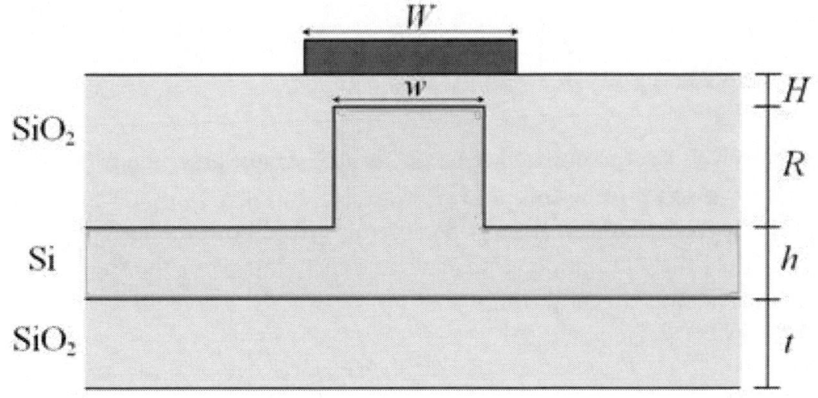

Figure 1. General view of SOI rib waveguide with the heater on the top.

We have considered the absence of generation terms and a scalar conductivity for both static and dynamic analysis. Table 1 summarizes the thermo-physical values used in simulations for each material in SOI waveguide. Constant values for the conductivity, density and heat capacity of the air are used. However, since air conductivity and density are almost linear in the temperature range 30-100°C (our range of investigation), we can substitute their constant values with linear relationships depending on the temperature. As initial condition, a constant temperature T_0 has been set at each point of the structure. The used boundary conditions are:

$\mathrm{T} = \mathrm{T}_0$ on the top and bottom boundary

$\mathrm{T} = \mathrm{T}_{\mathrm{heater}}$ on heater surfaces

$\mathbf{n} \cdot (k\nabla T) = 0$ on lateral surfaces

$\mathbf{n} \cdot (k_1\nabla T_1 - k_2\nabla T_2) = 0$ on internal boundaries

where T_{heater} becomes a temperature step in the dynamic analysis, and a constant value in the static analysis. In both cases, the maximum heater temperature change is assumed as 70°C.

Table 1. Thermo-optic constants in thermal simulations

Material	k [W/m·K]	ρ [kg/m^3]	C [J/kg·K]
Silicon	163	2330	703
Silica	1.38	2203	703
Aluminum	160	2700	900
Air (30°C)	0.026	1.166	1005

2.2. Optical Model

The partial differential equation that describes the light propagation is the wave equation:

$$\nabla^2 \mathrm{E}_t + \nabla\left(\frac{\nabla n^2(x,y) \times \mathrm{E}_t}{n^2(x,y)}\right) + \left(\frac{2\pi}{\lambda}\right)^2 n^2(x,y)\mathrm{E}_t = \left(\frac{2\pi}{\lambda}\right)^2 n_{eff}\mathrm{E}_t \tag{6}$$

derived from Maxwell's equations, where E_t is the transverse electric field, λ is the wavelength, $n(x,y)$ is the refractive index distribution and n_{eff} is the effective index (eigenvalue of the wave equation). It must be solved to find both mode field distribution and effective index. Thus, the 2D temperature distribution obtained in the thermal investigation has been used to evaluate the new refractive indices of the waveguide, as a result of the thermo-optic effect in silicon and silica:

$$\mathrm{n}'_{\mathrm{Si}} = \mathrm{n}_{\mathrm{Si}} + \Delta\mathrm{n}_{\mathrm{Si}} = 3.477 + 1.86 \cdot 10^{-4}\Delta\mathrm{T}$$

$$\mathrm{n}'_{\mathrm{SiO_2}} = \mathrm{n}_{\mathrm{SiO_2}} + \Delta\mathrm{n}_{\mathrm{SiO_2}} = 1.477 + 10^{-5}\Delta\mathrm{T}$$

at the wavelength 1.55μm. ΔT is the temperature change caused by heating and the *bulk* thermo-optic coefficients of silicon and silica are used. The refractive index distribution is used to solve the eigenvalue optical problem and find the field distributions and the effective indices of the structure modes. Particularly, we have focused our attention on the fundamental quasi-TE and quasi-TM modes of the rib waveguides. By means of the static analysis, performed at various temperatures of the heater, we have derived the dependence of the effective indices on temperature as an *effective* thermo-optic coefficient, $\partial n_{eff}/\partial T$. This procedure is significantly different from the one commonly followed in literature, where only the bulk thermo-optic coefficient is taken into account. The effective thermo-optic coefficient depends on the waveguide materials (refractive index, density, heat capacity, thermal conductivity), geometry, temperature change in each layer, heater material and width. If a guided wave optical device employs the thermo-optic effect for switching [25], filtering or tuning [26] operations, its characteristics depend on this effective thermo-optic coefficient, that can be seen as a property of the whole waveguide structure. For example, in a Mach-Zehnder thermo-optic switch, the phase shift is written as:

$$\Delta\varphi = \frac{2\pi}{\lambda}\Delta n_{eff} L_h = \frac{2\pi}{\lambda} L_h \frac{\partial n_{eff}}{\partial T}\Delta T \tag{7}$$

where L_h is the heater length. By means of the dynamic analysis, the time dependence of the effective index can be evaluated, so estimating the thermo-optic response time of the analyzed structures.

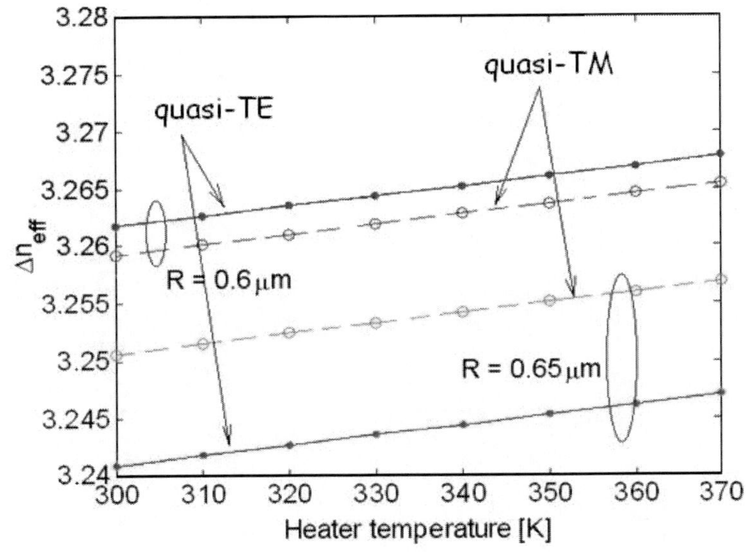

Figure 2. Effective index change versus heater temperature.

Our investigations have been focused on waveguides with the following dimensions: $h+R=1\mu m$, $w=0.6\mu m$, $R=0.6\mu m$ and $W=0.6\mu m$. Then, we have varied the etch depth R and h (keeping the sum $h+R$ as constant), the waveguide width w and the heater width W, and derived their influence on the effective index. Figure 2 shows the linear dependence of the effective index change over heater temperature, for quasi-TE and quasi-TM modes and waveguides with different etch depth *R*.

In Figure 3(a), the 2D static distribution of refractive index change Δn_{Si} and Δn_{SiO2} in silicon core and silica cladding is sketched, and the relevant dynamic behavior of the refractive index change in a point placed in the region of waveguide where the field intensity is maximum, is also shown. By means of an exponential fit as in Figure 3(b), the time constant τ of the thermo-optic effect for each considered structure can be found.

The time constant gets better in structures where the effective thermo-optic coefficient is improved, too. Namely, the characteristics improve with increasing the etch depth, as it can be noted in Figure 4 where quasi-TE case exhibits a slightly larger time constant than quasi-TM (0.93μs instead of 0.91μs for $R = 0.6$μm, and 0.86μs instead of 0.845μs for $R = 0.7$μm). Furthermore, the time constant and the thermo-optic coefficient are always improved when considering the quasi TM mode of the rib instead of quasi-TE, because the field maximum is placed in a more heated region [27].

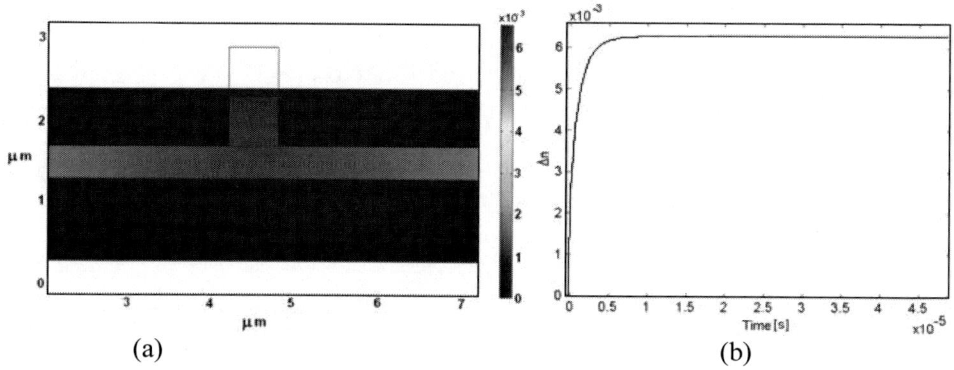

Figure 3(a). Static 2D distribution of refractive index change in silicon core and silica cladding and (b) dynamic behavior in the rib.

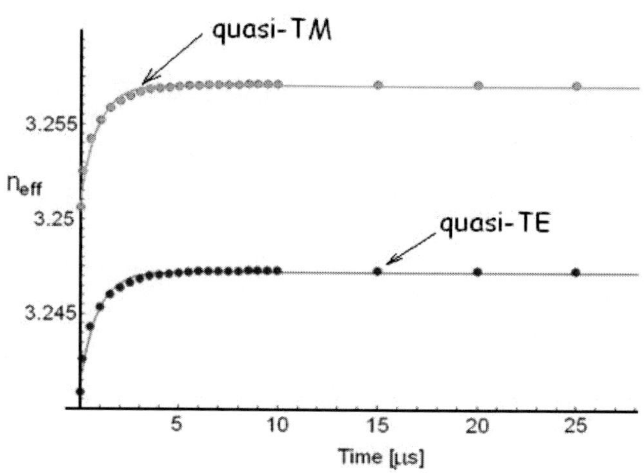

Figure 4. Time response for quasi-TE and quasi-TM modes.

3. Analysis of SOI Rib Waveguide Arrays

Waveguide arrays can be also important in some guided-wave optical applications [28-29]. In this section different arrays have been considered, namely types #1, #2 and #3. Figures 5-7 show the refractive index change in these different types of waveguide array for type #1, #2 and #3, respectively, if the distance between waveguides is large enough to obtain optical decoupling. In type #1, two rib waveguides are clearly sketched (see Figure 5), having a common slab layer. It is evident as the presence of another waveguide influences the temperature distribution in the waveguide under the heater. Stronger heating is revealed for waveguide under heater, differently from the other one. Types #2 and #3 (Figures 6-7) are two possible solutions for obtaining thermal isolation between adjacent waveguides of the array. The former includes two rib waveguides with a separation of the slab layer in between. The latter (*double-rib*) is characterized by full isolation of each rib waveguide with silicon oxide all around.

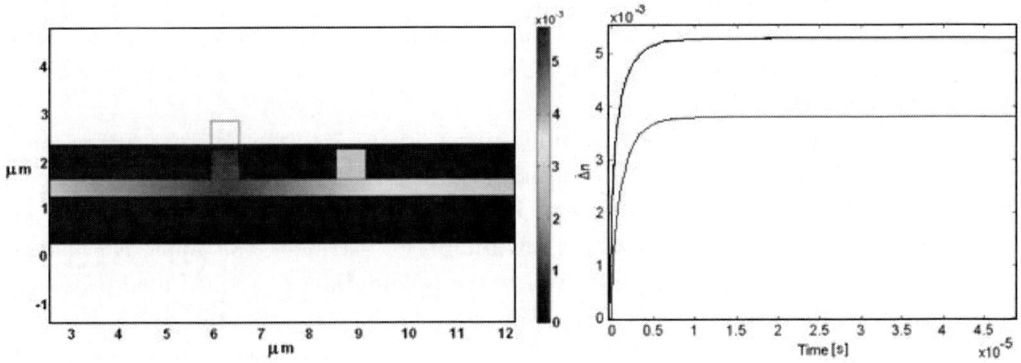

(a) (b)

Figure 5. (a) Static 2D distribution of refractive index change in silicon core and silica cladding and (b) dynamic behavior in the heated (top curve) and adjacent (bottom curve) rib waveguides (type #1 array).

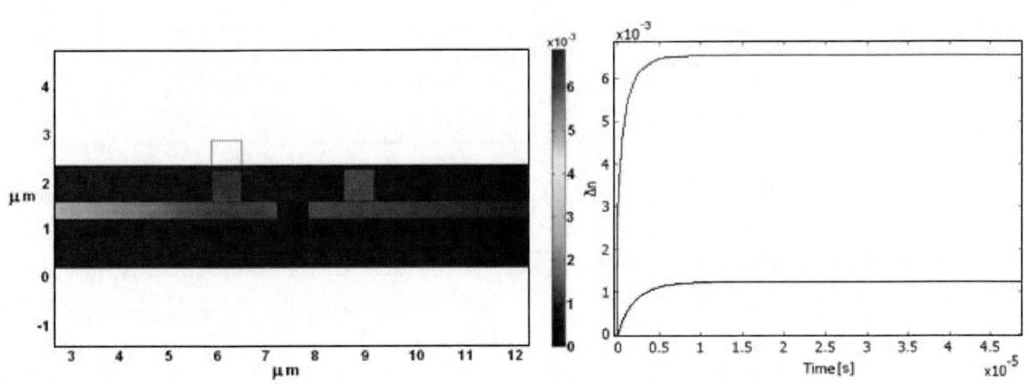

(a) (b)

Figure 6. (a) Static 2D distribution of refractive index change in silicon core and silica cladding and (b) dynamic behavior in heated (top curve) and adjacent (bottom curve) rib waveguides (type #2 array).

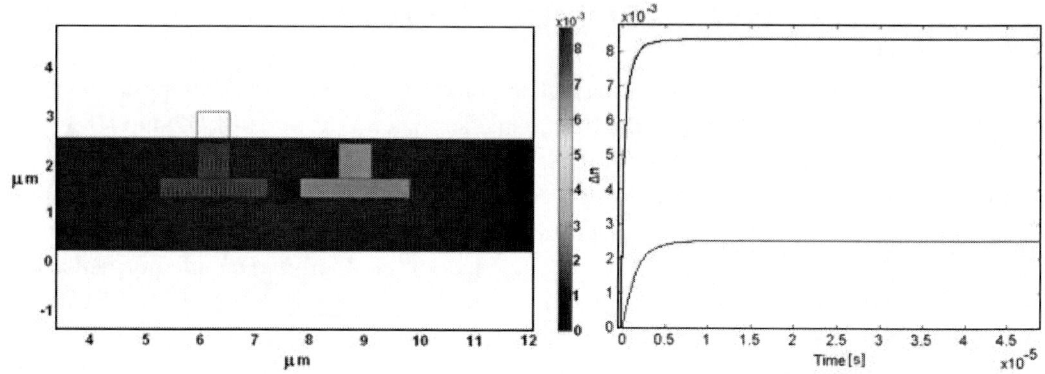

(a) (b)

Figure 7. (a) Static 2D distribution of refractive index change in silicon core and silica cladding and (b) dynamic behavior in the heated (top curve) and adjacent (bottom curve) rib waveguides (type #3 array).

Table 2 collects both time constant values and effective thermo-optic coefficients of two adjacent waveguides in the three array configurations. Type #3 array exhibits the best performance, showing improved results for both thermo-optic coefficient and time constant because of its high thermo-optic isolation.

By means of this fully integrated multiphysics analysis, the approximation on the refractive index distribution in the waveguides, usually assumed in literature, can be overcome to obtain a much more rigorous analysis. This is especially important in the core of the waveguide, where the field intensity is maximum. Furthermore, this approach is crucial, especially for operating with waveguides where the guiding layer is constituted of materials with low thermal diffusivity.

Table 2. Static and dynamic characteristics of SOI rib waveguide arrays (first row: waveguide under heater; second row: adjacent waveguide)

Parameter	Type #1	Type #2	Type #3
$\partial n_{eff}/\partial T$	7.8410^{-5} $5.68\ 10^{-5}$	9.7310^{-5} $1.84\ 10^{-5}$	1.2410^{-4} $3.76\ 10^{-5}$
τ (s)	$7.87\ 10^{-7}$ $1.25\ 10^{-6}$	$6.10\ 10^{-7}$ $2.12\ 10^{-6}$	$4.08\ 10^{-7}$ $1.67\ 10^{-6}$

The importance of the diffusivity can be highlighted in the study of a waveguide where silica is placed over the heater instead air. Both static and dynamic behaviors get better, because the diffusivity of silica is approximately two orders of magnitude lower than for air. Thus, the thermal power can be better confined in the guiding layer. Some numerical examples are shown below. Firstly, we have calculated effective thermo-optic coefficients and other parameters for an array structure of type #1 (shown in Figure 8), using parameters h = 0.6 µm, R = 0.9 µm, w = 1.1 µm, H = 0.1 µm and array period d = 8.4 µm. Results are summarized in Table 3. The heater is placed on the top of second waveguide (WG 1), and we have investigated effective index and thermal parameters for WG 1 and two adjacent waveguides, WG 2 and WG 3.

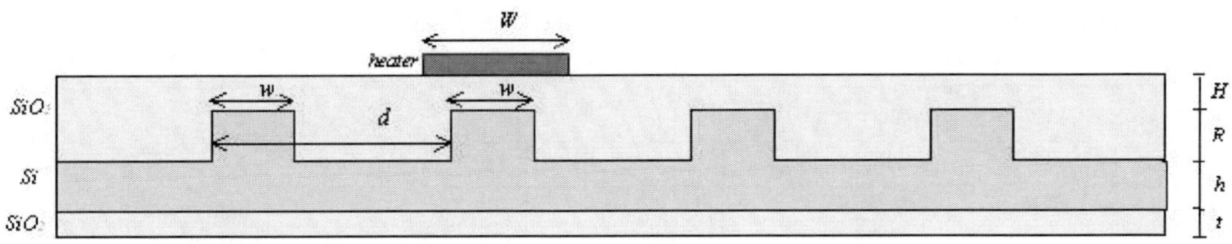

Figure 8. Array architecture (type #1, with array period *d*).

For each waveguide two numbers are indicated, the first for TE polarization and the second for TM polarization. Better thermo-optic coefficients are obtained when the heater width is well larger than the rib waveguide being heated. A moderate effect is also revealed on adjacent waveguides.

In following Tables 4-6, other results of our thermo-optical analysis are given in terms of switching power to achieve a phase shift (in equivalent interferometer geometry), response time to a temperature step and maximum frequency of thermo-optic modulation, respectively. In Tables 5 and 6, for each heater width W_H only one number is exhibited: actually, time constant and maximum modulation frequency formulations are independent from thermo-optic coefficient.

Table 3. Effective thermo-optic coefficients for $d = 8.4$ μm (type #1 array)

WG	Effective thermo-optic coefficient: $dn_{eff}/dT \cdot 10^{-4}$		
	W_H= 0.5μm	W_H= 1.1μm	W_H= 2μm
1	0.64671289 0.64672714	0.93077012 0.93068285	1.03165114 1.03140177
2	0.28420429 0.28410575	0.40939874 0.40931834	0.45669716 0.45659985
3	0.15543625 0.15545579	0.22397044 0.22397144	0.24991765 0.24978815
WG	$dn_{eff}/dT \cdot 10^{-4}$		
	W_H= 3μm	W_H= 4μm	
1	1.07051815 1.07025574	1.10037412 1.10016384	
2	0.47818549 0.47808166	0.49694379 0.49689513	
3	0.26162720 0.26161827	0.27185296 0.27184127	

In another example for the same structure (type #1 array), using h = 0.6 μm, R = 0.9 μm, w = 1.1 μm, H = 0.1 μm and d = 29.5 μm, we note from Table 7 as the thermo-optic coefficient in the two adjacent waveguides is negligible, because a complete optical and thermal decoupling occurs due to well larger array period *d*. Tables 8-10 show results in terms of switching power, response time and maximum frequency of thermo-optic modulation, respectively.

Table 4. Switching power for π phase shift, for d = 8.4 μm (type #1 array)

WG	Switching power: P_π (mW)				
	W_H= 0.5μm	W_H= 1.1μm	W_H= 2μm	W_H= 3μm	W_H= 4μm
1	6.6780 6.6779	9.9433 9.9442	16.1481 16.1520	23.2468 23.2525	30.0925 30.0982
2	15.1960 15.2013	22.6061 22.6106	36.4775 36.4852	52.0427 52.0540	66.6333 66.6398
3	27.7848 27.7813	41.3221 41.3219	66.6586 66.6931	95.1204 95.1236	121.8048 121.8101

Table 5. Response time of thermo-optic modulation for d = 8.4 μm (type #1 array)

τ (μs)				
W_H= 0.5μm	W_H= 1.1μm	W_H= 2μm	W_H= 3μm	W_H= 4μm
202.446	97.348	56.480	39.592	31.096

Table 6. Maximum modulation frequency for d = 8.4 μm (type #1 array)

f (kHz)				
W_H= 0.5μm	W_H= 1.1μm	W_H= 2μm	W_H= 3μm	W_H= 4μm
1.817	3.779	6.513	9.292	11.830

Table 7. Effective thermo-optic coefficients for d = 29.5 μm (type #1 array)

WG	Effective thermo-optic coefficient: $dn_{eff}/dT \cdot 10^{-4}$		
	W_H= 0.5μm	W_H= 1.1μm	W_H= 2μm
1	0.52319787 0.52303813	0.78945098 0.78927169	0.88898136 0.88874814
2	~ 10^{-6}	~ 10^{-6}	~ 10^{-6}
3	~ 10^{-7}	~ 10^{-7}	~ 10^{-7}
WG	$dn_{eff}/dT \cdot 10^{-4}$		
	W_H= 3μm	W_H= 4μm	
1	0.92771534 0.92746111	0.94742077 0.96316167	
2	~ 10^{-6}	~ 10^{-6}	
3	~ 10^{-7}	~ 10^{-7}	

Table 8. Switching power for d = 29.5 μm (type #1 array)

WG	P_π (mW)				
	W_H= 0.5μm	W_H= 1.1μm	W_H= 2μm	W_H= 3μm	W_H= 4μm
1	8.2546 8.2571	11.7232 11.7259	18.7396 18.7445	26.8251 26.8325	34.9507 34.3795
2	-	-	-	-	-
3	-	-	-	-	-

Table 9. Response time of thermo-optic modulation for $d = 29.5$ µm (type #1 array)

τ (µs)				
W_H= 0.5µm	W_H= 1.1µm	W_H= 2µm	W_H= 3µm	W_H= 4µm
202.447	97.349	56.480	39.592	31.096

Table 10. Maximum modulation frequency for $d = 29.5$ µm (type #1 array)

f (kHz)				
W_H= 0.5µm	W_H= 1.1µm	W_H= 2µm	W_H= 3µm	W_H= 4µm
1.817	3.779	6.513	9.292	11.830

Some plots evaluated by FEM for the array with period $d = 8.4$ µm are also shown. Heat transfer simulation for the examined waveguide array is given in Figure 9 for a heater width W_H = 2 µm and heating at 100 °C. Moreover, quasi-TM fundamental mode in the heated waveguide (after 100°C heating), quasi-TM fundamental mode in one period-distant waveguide (indicated with WG 2 in tables) after 100 °C heating, and quasi-TM fundamental mode in the two period-distant waveguide (WG 3 in the tables) after 100 °C heating, are sketched in Figures 10-12, respectively.

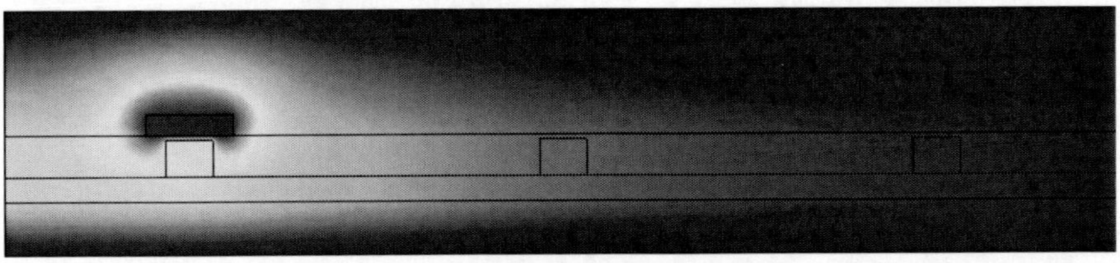

Figure 9. Heating of array waveguide (WG1) at 100 °C (W_H= 2 µm).

Figure 10. Quasi-TM fundamental mode of array waveguide (WG 1), after heating at 100 °C.

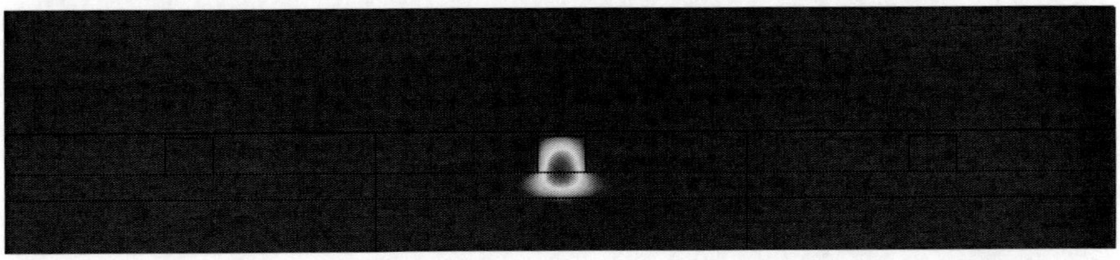

Figure 11. Quasi-TM fundamental mode of array waveguide (WG 2), after heating at 100 °C.

Figure 12. Quasi-TM fundamental mode of array waveguide (WG 3), after heating at 100 °C.

Finally, a detailed investigation has been carried out on the double-rib (type #3) array having h=0.6μm, R = 0.9μm, w = 1.1μm, w_2 = 1.5μm, H = 0.1μm, d > 8.5 μm, as in Figure 13.

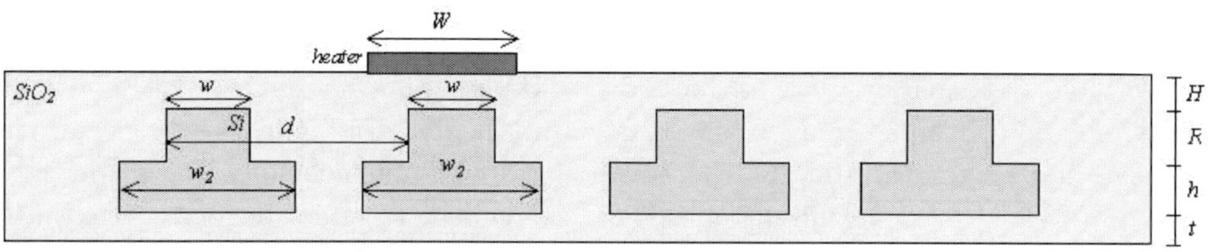

Figure 13. Geometric scheme of double-rib waveguide array.

As it is evident from Table 11, the thermo-optic coefficient in the far waveguides is once again negligible. In the following figures, some plots by FEM are shown. In Figure 14, the heat transfer simulation for the examined waveguide array with a heater width W_H = 2 μm is performed again for heating at 100°C.

We can see how the heat spread is limited to the heated waveguide, without expanding up to the neighbouring waveguides of the array, as on the contrary occurs in the previous structures. The fundamental quasi-TM mode is sketched in Figure 15 in the heated waveguide (again after 100°C heating).

Table 11. Effective thermo-optic coefficients for various waveguides of array #3

WG	Effective thermo-optic coefficient: $dn_{eff}/dT \cdot 10^{-4}$		
	W_H = 0.5μm	W_H = 1.1μm	W_H = 2μm
1	1.14631576 1.15057175	1.40534639 1.40510625	1.48570747 1.48543778
2	0.01469329 0.01530466	0.01703001 0.01708761	0.01948981 0.01954685
3	~ 10^{-8}	~ 10^{-8}	~ 10^{-8}
WG	$dn_{eff}/dT \cdot 10^{-4}$		
	W_H = 3μm	W_H = 4μm	
1	1.51950454 1.51476970	1.5250160 1.5388769	
2	0.02219612 0.02163795	0.0299454 0.0309225	
3	~ 10^{-8}	~ 10^{-7}	

Figure 14. Heating of waveguide WG 1 at 100 °C (type #3 array).

The fundamental mode in one period distant waveguide, indicated as WG 2 in Table 3, is sketched in Figure 16, again after 100 °C heating. Fundamental mode change in the two period distant waveguide (WG 3 in table), is negligible after 100 °C heating.

Figure 15. Quasi-TM fundamental mode of waveguide WG 1, after heating at 100 °C (type #3 array).

An estimation of power consumption has been also evaluated by using an interferometer scheme to estimate a π phase shift. Results are summarized in Table 12. The power budget is moderate if the heater width is small enough. An estimation of response time and dynamic frequency is also given in next Tables 13 and 14. Other details can be found in literature [23, 28].

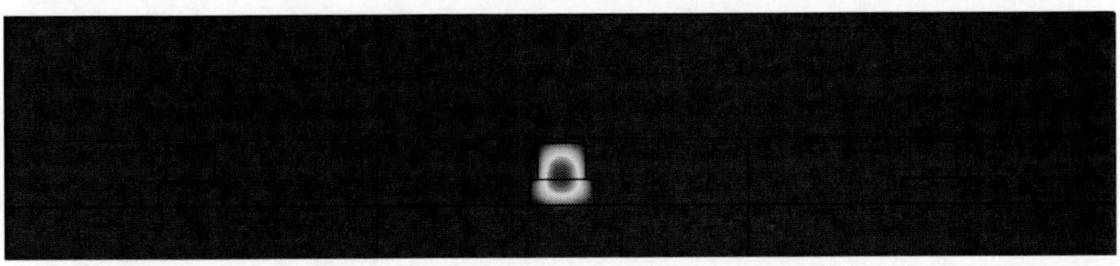

Figure 16. Quasi-TM fundamental mode of waveguide WG2, after heating at 100 °C (type #3 array).

Table 12. Power consumption to achieve π phase shift for various waveguides (type #3 array)

WG	P_π (mW)				
	W_H= 0.5μm	W_H= 1.1μm	W_H= 2μm	W_H= 3μm	W_H= 4μm
1	3.7675 3.7536	6.5855 6.5866	11.2129 11.2150	16.3778 16.4290	21.7132 21.5176
2	293.9281 282.1865	543.4478 541.6160	854.7622 852.2678	1121.1903 1150.1124	1105.7770 1070.8382

Table 13. Response time for waveguide array (type #3 array)

τ (μs)				
W_H = 0.5μm	W_H = 1.1μm	W_H = 2μm	W_H = 3μm	W_H = 4μm
202.447	97.349	56.480	39.592	31.096

Table 14. Maximum modulation frequency (type #3 array)

f (kHz)				
W_H = 0.5μm	W_H = 1.1μm	W_H = 2μm	W_H = 3μm	W_H = 4μm
1.817	3.779	6.513	9.292	11.830

4. Analysis of SOI Slot Waveguides

Nanometer slot optical waveguides realized by Silicon-on-Insulator (SOI) technological platform enable high optical confinement in a very small low-index region, as described in chapter 2 of this book. This kind of guiding structures is emerging as very attractive for fabrication of both active and passive photonic devices. In the last two years, a great variety of optical devices has been realized based on SOI slot waveguides, including micro-ring resonators, optical modulators, electrically pumped light emitting devices, directional couplers, all-optical logic gates and beam splitters. Very recently, slot guiding structures have been demonstrated as very useful for chemical and biochemical optical sensing because of their high sensitivity to either refractive index change of cover medium or thickness increase of deposited receptor layer.

In this section the thermo-optic effect in silicon slot waveguides (see Figure 17) is theoretically investigated. We have again carried out a full-vectorial FEM-based static and dynamic thermo-optical analysis of slot waveguides under heating.

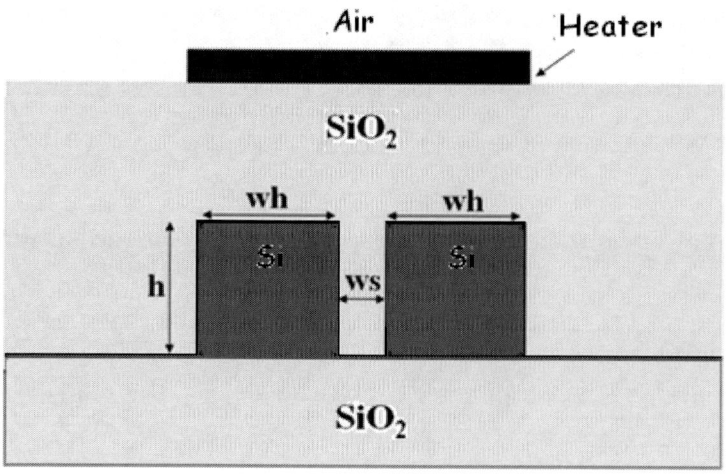

Figure 17. Geometric scheme of silicon slot waveguide with heater on the top.

Waveguide effective index sensitivity (S) to temperature changes produced by an heater laying on the top of the guiding slot structure has been investigated. Assuming to produce an instantaneous temperature change within the metallic heater, dynamic behaviour of waveguide effective index has been also calculated. Influence of slot waveguide geometrical features on thermal sensitivity and response time has been studied, taking also into account the effect of cover medium optical and thermal properties.

The typical 2D distribution of electric field in the slot waveguide is sketched in Figure 18 for quasi-TE polarization. 2D temperature distribution after 50K heating is shown in Figure 19, assuming w_s = 50nm, w_h =250 nm and h =200nm. The temperature is nearly constant in the whole waveguide (oxide slot and silicon regions), with a point to point relative change of 10^{-3} K, and outside rapidly decreases.

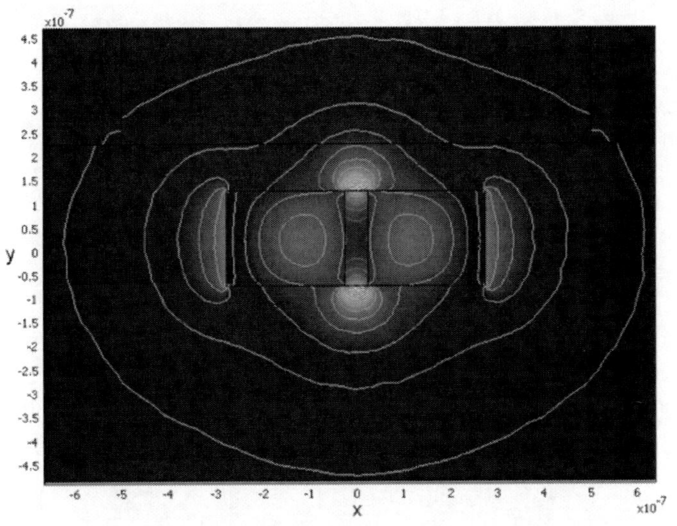

Figure 18. 2D electric field distribution in SOI slot waveguide by FEM.

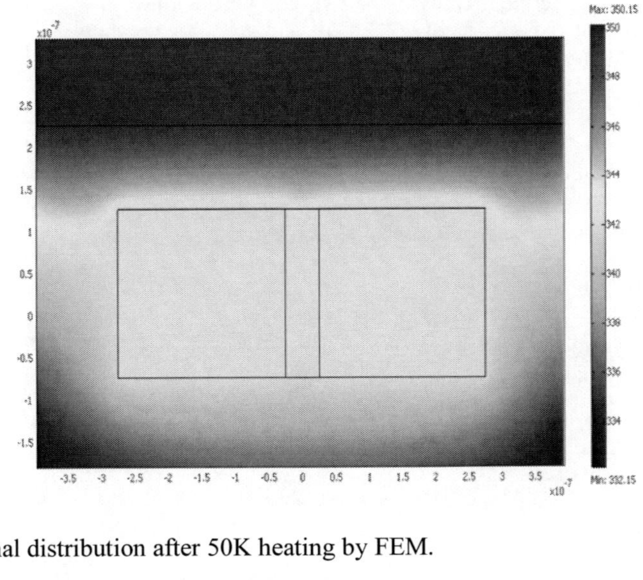

Figure 19. 2D thermal distribution after 50K heating by FEM.

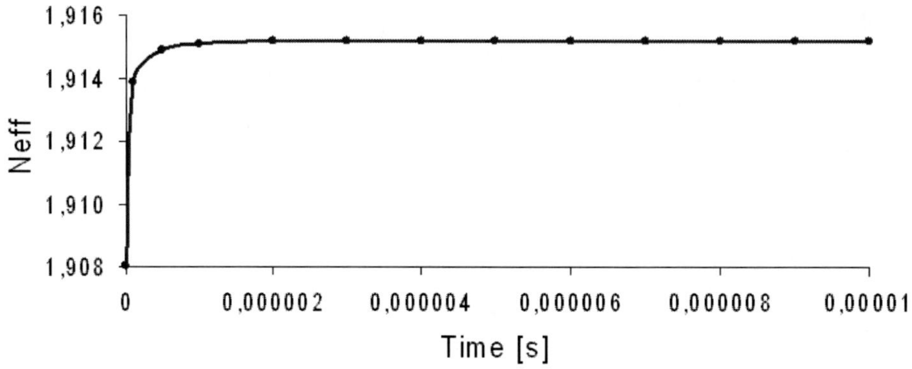

Figure 20. Effective index change versus time after 50K heating.

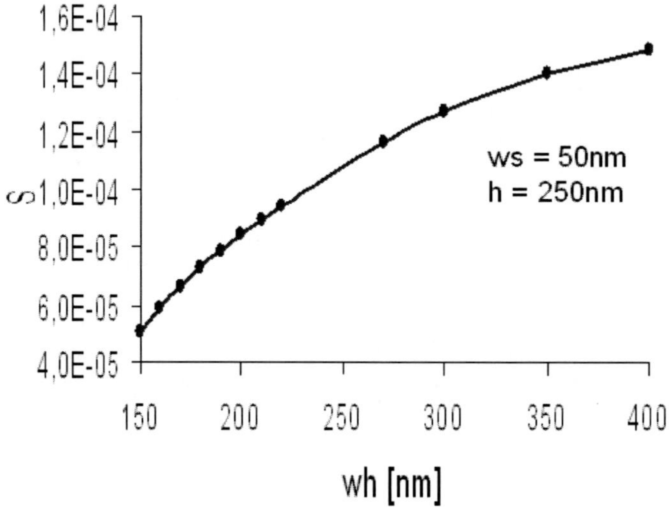

Figure 21. Slot sensitivity to thermal heating versus silicon wire width.

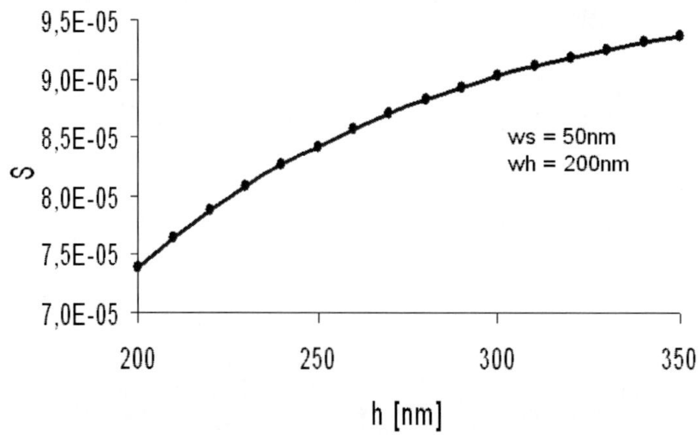

Figure 22. Slot sensitivity to thermal heating versus slot height.

The effective index time response for quasi-TE polarization is given in Figure 20. Waveguide effective index sensitivity (S) to temperature changes has been investigated as a function of slot height h and silicon regions dimension w_h, and results are shown in Figures 21 and 22 in case of $w_s = 50$ nm and $h = 250$ nm, or $w_s = 50$ nm and $w_h = 200$nm, respectively. The sensitivity generally improves by increasing the silicon region dimensions, because of silicons large thermo-optic coefficient.

Conclusion

A theoretical investigation and simulation of thermo-optic effect in SOI waveguides and waveguide arrays has been carried out in this chapter by means of the finite element method. The central key of this analysis is the calculation and use of the effective thermo-optic coefficient, that is a property of the whole waveguide structure, taking into account all geometrical and material parameters, and the multiphysics integration of both optical and thermal investigations. This integration is fundamental to remove the approximation on the refractive index distribution. The same study could be performed with other materials, including polymers, and applied to the design of devices employing the thermo-optic effect, such as optical switches and ring resonators.

References

[1] Reed, G. T., *Nature*, 2004, 427, 595-596.

[2] Reed, G. T., Knights, A. P., *Silicon Photonics: An Introduction*, John Wiley and Sons, 2004.

[3] Agrawal, G. P., *Lightwave Technology: Components and Devices*, Wiley-Interscience, 2004.

[4] Kwon, M.-S., Shin, S.-Y., *J. Lightwave Technology*, 2004, 22, 1968-1975.

[5] Green, W., Lee, R., DeRose, G., Scherer, A., Yariv, A., *Optics Express*, 2005, 13, 1651-1659.

[6] Wang, X., Howley, B., Chen, M. Y., Chen, R. T., *J. Lightwave Technology*, 2006, 24, 1558-1565.

[7] Jang, C.-H., Chen, R. T., *J. Lightwave Technology*, 2003, 21, 1053-1058.

[8] Rasras, M. S., Gill, D. M., Patel, S. S., Tu, K.-Y., Chen, Y.-K., White, A. E., Pomerene, A. T. S., Carothers, D. N., Grove, M. J., Sparacin, D. K., Michel, J., Beals, M. A., Kimerling, L. C., *J. Lightwave Technology*, 2007, 25, 87-92.

[9] Goh, T., Yasu, Mitsuho, H., Kuninori; H. A., Okuno, M., Ohmori, Y., *J. Lightwave Technology*, 2001, 19, 371-379.

[10] Maru, K., Tanaka, K., Chiba, T., Nonen, H., Uetsuka, H., *J. Lightwave Technology*, 2004, 22, 1523-153.

[11] Kasahara, R., Yanagisawa, M., Goh, T., Sugita, A., Himeno, A., Yasu, M., Matsui, S., *J. Lightwave Technology*, 2002, 20, 993-1000.

[12] Yan, Q., Yu, J., Xia, J., Liu, Z., *Chinese Optics Letters*, 2003, 1, 217-219.

[13] Hurvitz, T., Ruschin, S., Brooks, D., *et al.*, *J. Lightwave Technology*, 2005, 23, 1918-1922.

[14] Takiguchi, K., Okamoto, K., *et al.*, *J. Lightwave Technology*, 2003, 21, 2463-2469.

[15] Schlipf, T. R., Street, M. W., Pandavenes, J., McBride, R., Cumming, D. R. S., *J. Lightwave Technology*, 2003, 21, 1944-1952.

[16] Asakawa, S., Kobayashi, M., Nagase, R., Kominato, T., *J. Lightwave Technology*, 2003, 21, 821-830.

[17] Howley, B., Wang, X., Chen, M., Chen, R. T., *J. Lightwave Technology*, 2007, 25, 883-890.

[18] Suzuki, K., Takiguchi, K., Hotate, K., *J. Lightwave Technology*, 2000, 18, 66-72.

[19] Madsen, C. K., Cappuzzo, M., Laskowski, E. J., Chen, E., Gomez, L., Griffin, A., Wong-Foy, A., Chandrasekhar, S., Stulz, L., Buhl, L., *J. Lightwave Technology*, 2004, 22, 1041-1050.

[20] Dabarsyah, B., Goh, C. S., Khijwania, S. K., Set, S. Y., Katoh, K., Kikuchi, K., *J. Lightwave Technology*, 2007, 25, 2711-2718.

[21] Emelett, S. J., Soref, R., *Optics Express*, 2005, 13, 7840-7853.

[22] Emelett, S. J., Soref, R., *J. Lightwave Technology*, 2005, 23, 1800-1807.

[23] Passaro, V. M. N. (Ed.), *Silicon Photonics*. Research Signpost Publ., Kerala, India, 2006.

[24] Comsol Multiphysics by COMSOL AB, ver. 3.2, *single license*, 2005.

[25] Nawrocka, M. S., et al., *Applied Physics Letters*, 2006, 89, 071110.

[26] Della Corte, F., et al., *J. Applied Physics*, 2000, 88, 7115-7119.

[27] Iodice, M., et al., *Opt. Express*, 2006, 14, 5266-5278.

[28] Passaro, V. M. N., Magno, F., Tsarev, A., *Optics Express*, 2005, 13, 3429-3437.

[29] Tsarev, A.V., De Leonardis, F., Passaro, V. M. N., *Optics Express*, 2008, 16, 3101-3113.

In: Modeling of Photonic Devices
Editor: Vittorio M. N. Passaro

ISBN 978-1-60456-980-3

Chapter 10

GAIN IN OPTICALLY OR ELECTRICALLY PUMPED ERBIUM-DOPED SILICON NANOCLUSTERS

***Roberto Pagano**[*1] **and Vittorio M. N. Passaro**[2]*
[1]Now at Tyndall National Institute, Cork, Ireland,
but work made while at Politecnico di Bari, Italy
[2]Photonics Research Group, Dipartimento di Elettrotecnica ed Elettronica,
Politecnico di Bari, via Edoardo Orabona n. 4, 70125 Bari, Italy

ABSTRACT

In this chapter, the rate equation system describing the interaction between silicon nanoclusters and erbium ions is theoretically investigated. The system solution is found both for optical and electrical pumping. Then, a possible DBR laser cavity design is proposed. Moreover, a simplified model of rate equation is presented, that permits one to obtain a very fast gain calculation, and the evaluation of the saturation power of the system. The main purpose of this chapter is to investigate the theoretical conditions to reach a net gain inside this material system, i.e. to find the critical physical parameters for both pumping conditions and to make a comparison between them.

Keywords: Nanocrystals and nanoparticles, optical pumping, Optical Waveguides, Integrated Optical Devices and Systems, modeling

1. INTRODUCTION

Recently, there have been many research efforts [1] on the sensitization of rare-earth ions (especially erbium) by silicon nanoclusters (Si:nc). In fact, Si:nc system typically exhibits a strong absorption cross-section for pump light in the visible region, and can efficiently transfer this energy to nearby Er ions [2]. This can help in developing Er doped waveguide

[*] E-mail address: roberto.pagano@tyndall.ie

amplifiers and lasers in a silicon-compatible material, transversely pumped by broadband sources such as LED arrays [3]. Therefore, an active medium based on silicon nanoclusters (Si-nc) doped with erbium (in order to obtain infrared emission), and embedded in a sub-stechiometric silicon dioxide matrix, seems to be one of the most promising techniques for the fundamental goal of achieving a silicon-based laser source. Moreover, a considerable reduction of detrimental phenomena can be observed in this material system, such as Auger de-excitation of erbium ions, free carrier absorption (FCA) and energy back-transfer, typical of any silicon-based source. In the future, these properties could be brought to the first silicon-based light source.

However, the modeling of this complex material system has not yet become well established in literature because of the uncertainty on some values of important parameters (i.e., excitation and emission cross sections of erbium inside a sub-stechiometric matrix of silicon dioxide) [1, 3, 6, 7], and some aspects are still open to achieve the best fitting between experiments and theory. From this point of view, it is important to improve the accuracy allowed by the modeling to find the best design criteria and obtain a net active gain of the system.

This chapter is structured in two main sections. The first part concerns the simulation of system behaviour in the case of optical pumping, while the second one is focused on electrical pumping. In each Section, we derive all conditions needed to reach the gain by appropriate rigorous modeling. Finally, in the last part of the chapter we propose a possible optical cavity design in order to obtain laser action of the system. Moreover, we will show a simplified model of rate equations that enables the fast calculation of gain and other parameters, and we will make a comparison between this method and the exact solution of rate equations to verify its accuracy.

2. MODELING OF OPTICAL PUMPING

For a few years, a number of researchers have tried to describe the complex physical mechanisms involved in the optical excitation of silicon nanoclusters with erbium doping ($Er:Si\text{-}nc:SiO_2$ material system). The largely popular model applied to study the behaviour of any erbium-doped silicon nanocluster $Er:Si\text{-}nc:SiO_2$ system is based on seven coupled rate equations [1,4], describing the interactions between the Er^{3+} ions, modeled as a five-level system, and Si-nc as a two-level system (see Figure 1). However, there are some common assumptions that need to be discussed. The first concerns the erbium's absorption and emission cross-sections. In fact, many authors assume [1] an effective enhancement of $Er:Si\text{-}nc:SiO_2$ absorption and emission cross sections of about two orders of magnitude ($\square\ 10^{-19}\ \chi\mu^2$) with respect to the simple $Er:SiO_2$ system ($\square\ 10^{-21}\ cm^2$), on the basis of work by Han et al. [5]. This could explain the large gain achieved in their experiments [6]. However, Daldosso et al. [7] have shown that an effective enhancement of emission and absorption cross sections was not really present, and that the behaviour of their samples, prepared with different annealing times, was generally variable, in particular in terms of refractive index. In fact, they observed that a refractive index decrease was related to shorter annealing times, with a resulting increase of effective values of emission and absorption cross sections (related through the McCumber's law). On the basis of these experimental evidences, a semi-empirical law has been extrapolated by exploiting the data reported in [7], describing the variation of effective absorption cross section versus refraction index. The best interpolation law, compatible with that system physics, is an exponential function as:

$$\sigma_{abs}(n^2) = 4\times10^{-18}\exp\left(-2.6897n^2\right)\ (\text{cm}^2) \quad (1)$$

The second assumption concerns the absence of stimulated emission terms in literature [1]. On the contrary, they have been considered here, because we are definitely interested in our model to the lasing action of the system.

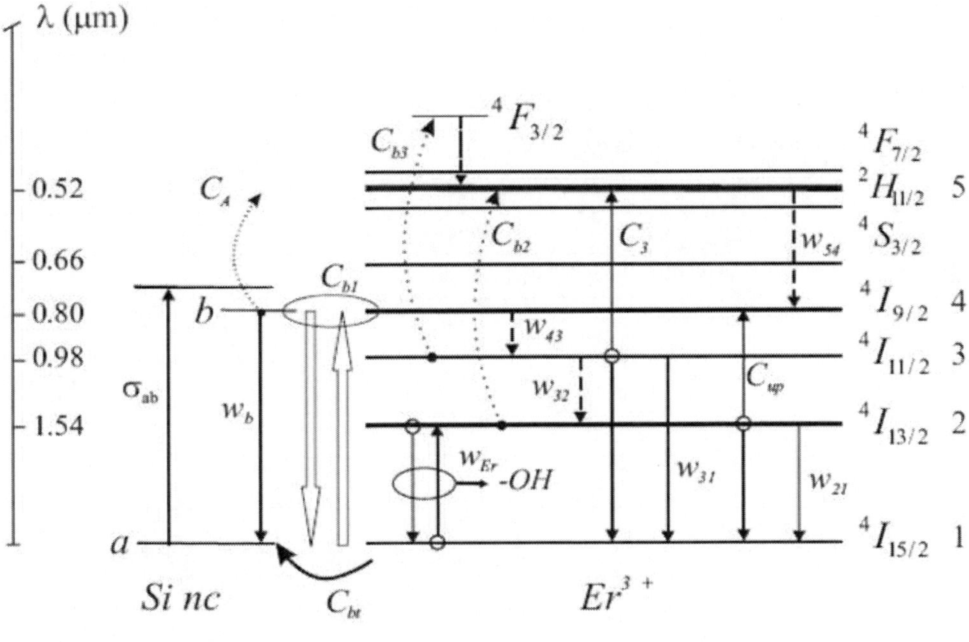

Figure 1. General scheme of transitions in seven-level Er:Si-nc:SiO_2 material system [4].

2.1. Simulations of Optical Pumping

The parameters used in the rate equations system for optical pumping are listed in Table 1. The simulations have been carried out as a function of erbium concentration and optical pumping power. In particular, the erbium concentration has been varied in the range $10^{19}\div10^{22}$ cm^{-3}, while the pump photon flux has been changed in the range $10^{20}\div10^{22}$ cm^{-2}s^{-1}. The gain, calculated from the rate equation system under stationary conditions, is shown in Figure 2. As we can see, if the photon flux has a fixed value, the net optical gain curve (obtained by considering the FCA losses) reaches a maximum value and, then, decays with increasing the erbium concentration, because of the concentration quenching and up-conversion mechanisms. The maximum optical gain, 1.5266 cm^{-1}, corresponds to an erbium concentration of 7.1073×10^{20} cm^{-3} and, obviously, to the maximum pumping optical power.

Table 1. Parameters used for optical pumping simulations [1,5,7]

Symbol	Parameter	Value (units)
N_c	Nanocrystal density	3×10^{19} cm^{-3}
σ_d	Direct absorption cross section	1×10^{-20} cm^2
σ_{nc}	Absorption cross section of nanoclusters	1×10^{-16} cm^2
λ_{exc}	Pumping wavelength	488 nm
w_b	Inverse lifetime of nanoclusters	4×10^4 s^{-1}
w_{21}	Erbium's inverse lifetime between 2nd and 1st energetic levels	526.3 s^{-1}
w_{32}	Erbium's inverse lifetime between 3rd and 2nd energetic levels	4.2×10^5 s^{-1}
w_{43}	Erbium's inverse lifetime between 4th and 3rd energetic levels	1×10^7 s^{-1}
w_{54}	Erbium's inverse lifetime between 5th and 4th energetic levels	1×10^7 s^{-1}
w_{31}	Erbium's inverse lifetime between 3rd and 1st energetic levels	1.4×10^3 s^{-1}
C_{b1}	Erbium-nanocluster coupling coefficient for 1st erbium energetic level	3×10^{-15} cm^3/s
C_{b2}	Erbium-nanocluster coupling coefficient for 2nd erbium energetic level	3×10^{-19} cm^3/s
C_{b3}	Erbium-nanocluster coupling coefficient for 3rd erbium energetic level	3×10^{-19} cm^3/s
C_{up}	Cooperative up-conversion coefficient	5×10^{-17} cm^3/s
C_A	Auger recombination coefficient	3×10^{-19} cm^3/s
C_3	Cooperative up-conversion coefficient	1.8×10^{-17} cm^3/s
σ_{fca}	FCA effective cross section	1×10^{-18} cm^2
n	Refractive index	1.46

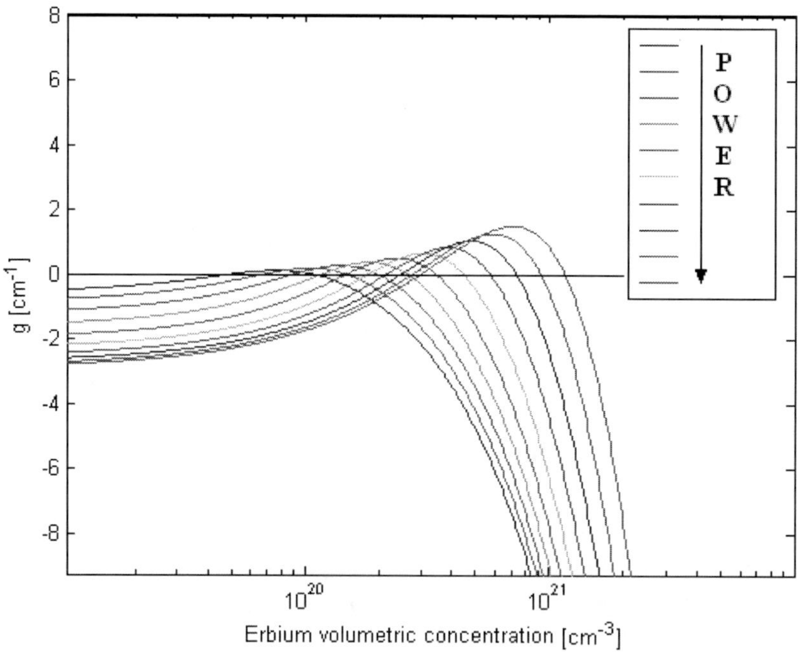

Figure 2. Net optical gain versus erbium concentration and pumping photon flux (ten values in the range $10^{20} \div 10^{22}$ cm^{-2}/s).

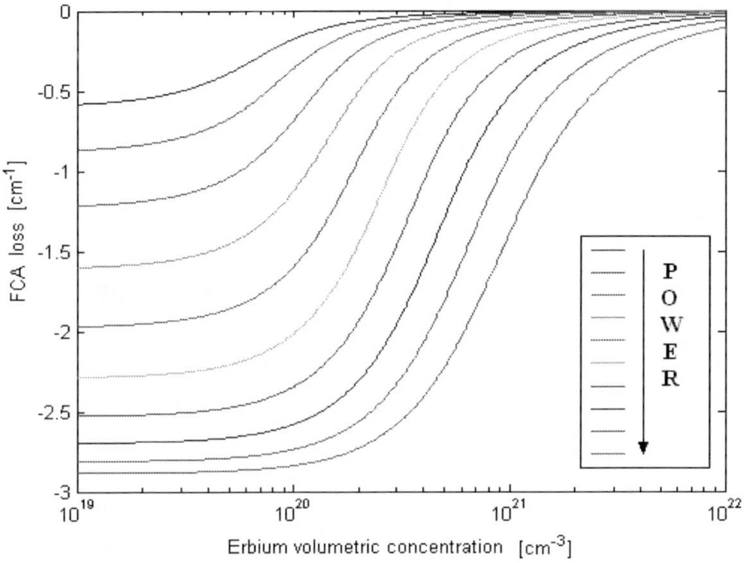

Figure 3. Free carrier absorption versus erbium concentration and pump photon flux (ten values in the range $10^{20} \div 10^{22}$ cm^{-2}/s).

Figure 3 shows the FCA losses versus erbium concentration and pump photon flux (the variation ranges are the same as before). The theory behind FCA losses in erbium-doped silicon nanoclusters can be found in literature [8]. The free carrier absorption loss increases with the optical power, because of the increase of carrier generation rate inside the silicon nanoclusters, while it decreases with increasing the erbium concentration, because the larger erbium concentration the bigger the probability of an exciton to decay inside a nanocluster, transferring its energy to the nearby erbium ion.

Therefore, the possibility to obtain a net optical gain is strictly related to FCA losses. In fact, our results show that a positive gain is achievable only for FCA effective cross sections below 1×10^{-18} cm^2. However, it is generally shown in literature [8] a higher theoretical critical value, around 1×10^{-17} cm^2. Thus, on the basis of our rigorous evaluation, this last value cannot allow any optical gain for this material system to be achieved.

It is important to remark that FCA effect is the most important source of loss inside this active medium, but not the only one. In fact, losses related to light scattering caused by nanoclusters are still present, but their influence is not critical and they will not be considered in our model. In Figure 4 net optical gain, total optical gain of the system (without FCA loss) and FCA loss are sketched as a function of erbium concentration and optical pump power. The net optical gain is the difference between total optical gain and FCA losses.

The maximum value of system total gain is found to be 1.5266 cm^{-1} for an Er concentration of 7.1073×10^{20} cm^{-3}. This value is larger than that found by *Pacifici* [4], that was 0.24 cm^{-1}, even with the higher values of emission and absorption cross sections that he used in the model because of the effective value of FCA absorption cross section. In fact he used the same value as in our model (1×10^{-18} cm^2), but in the model proposed here, the effective FCA absorption cross section is ten times lower, considering the relationship of the cross section with the transfer time, as made by *Polman et al.* [8].

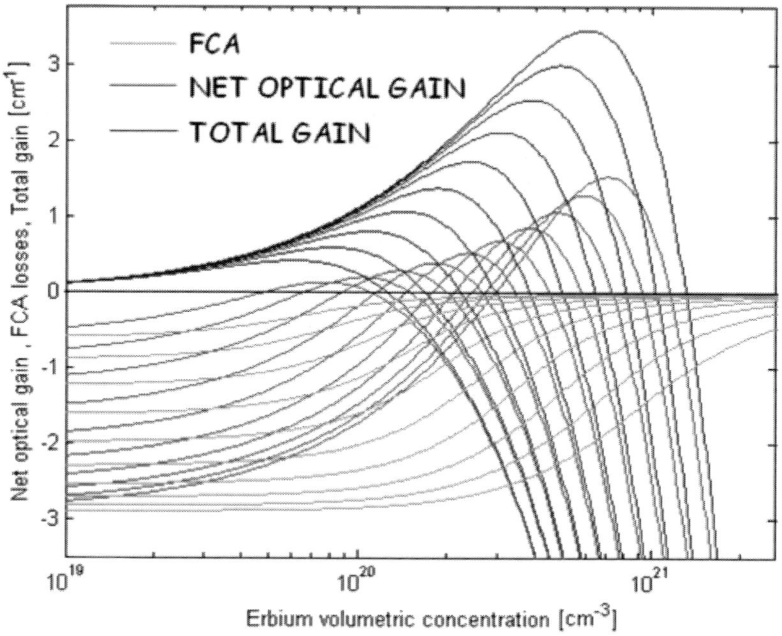

Figure 4. Comparison between total optical gain (blue curves), net optical gain (red curves) and FCA losses (green curves) versus erbium volumetric concentration and optical pumping power (ten values in the range $10^{20} \div 10^{22}$ cm^{-2}/s).

There are a lot of parameters that could give the possibility to obtain a net optical gain at 1540nm and, among them, the number of Er^{3+} ions excitable per nanocluster is one of the most important. In fact, large concentrations of erbium are necessary to obtain a positive gain. However, if only small fractions of these concentrations are coupled to the nanoclusters, all the uncoupled erbium ions will strongly absorb the light emission, due to phenomena such as excited-state absorption (ESA), concentration quenching, and so on.

In the following Figure 5, the curves that describe the number of erbium ions coupled to nanoclusters are sketched as a function of erbium concentration and optical pumping power. As it is shown, the number of erbium ions excited per nanocluster increases with Er concentration up to a maximum value (this is clear for higher pumping powers). After that, non-radiative phenomena related to the concentration of erbium will reduce the number of ions excited per nanocluster.

The erbium excitation is a sequential process, then only one erbium ion per time could be excited by a nanocluster. This means that one nanocluster could excite more nearby erbium ions. The transfer time of the energy of an excited nanocluster to an erbium ion at ground level is about 1μs, while the de-excitation time (decay time) of an excited erbium ion is of the order of milliseconds. For low erbium concentrations (up to□ 1×10^{19} cm^{-3}), the fraction of excited nanoclusters may be considered as independent from the erbium concentration, but it increases with photon flux by increasing the erbium concentration with a lower rate, because of the more probable de-excitation by erbium ions. All these considerations mean that the nanoclusters excitation cross section is not a fixed value, but it is rather an effective excitation cross section, depending on both erbium concentration and photon flux, as shown in Figure 6.

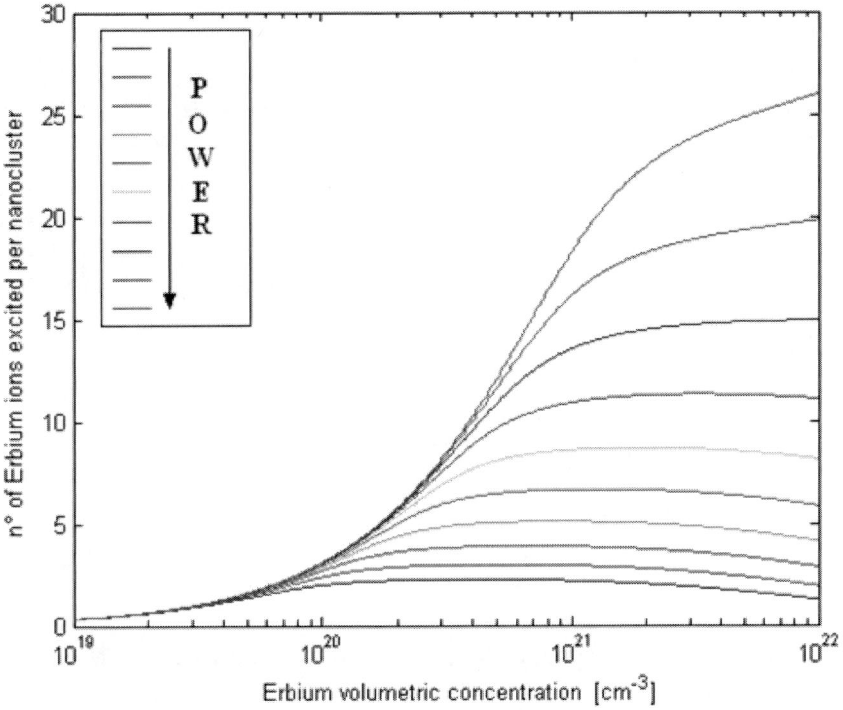

Figure 5. Number of erbium ions excited per nanocluster at level 2 as a function of erbium volumetric concentration and optical pumping power (ten values in the range $10^{20} \div 10^{22}$ cm^{-2}/s).

The effective cross section has been calculated by using the following relation:

$$\sigma_{eff} = \frac{w_{32} N_3}{\phi_p \left(N_{Er} - N_2 \right)} \tag{2}$$

where w_{32} is transition rate (the inverse of decay time) between 3rd and 2nd erbium level, N_3 is the erbium population at level 3, ϕ_p is the photons flux and N_{Er} is the total erbium concentration inside the sample. The considered nanocluster concentration corresponds to a value of 3×10^{19} cm^{-3}.

As we can see, the various curves have maxima corresponding to specific values of photon flux. After those peaks, the effective cross section tends to diminish because of parasitic effects as up-conversion, approaching values related to direct excitation of erbium ions. In fact, the minimum effective excitation cross section is that for direct optical Er excitation at 488nm, namely 1×10^{-20} cm^2. We can note that an effective excitation cross section can be larger than the excitation cross section of nanoclusters used in the simulation, i.e. $\sigma_{nc} = 1\times10^{-16}$ cm^2. The explanation of this behavior is that a single nanocluster can in principle excite more erbium ions.

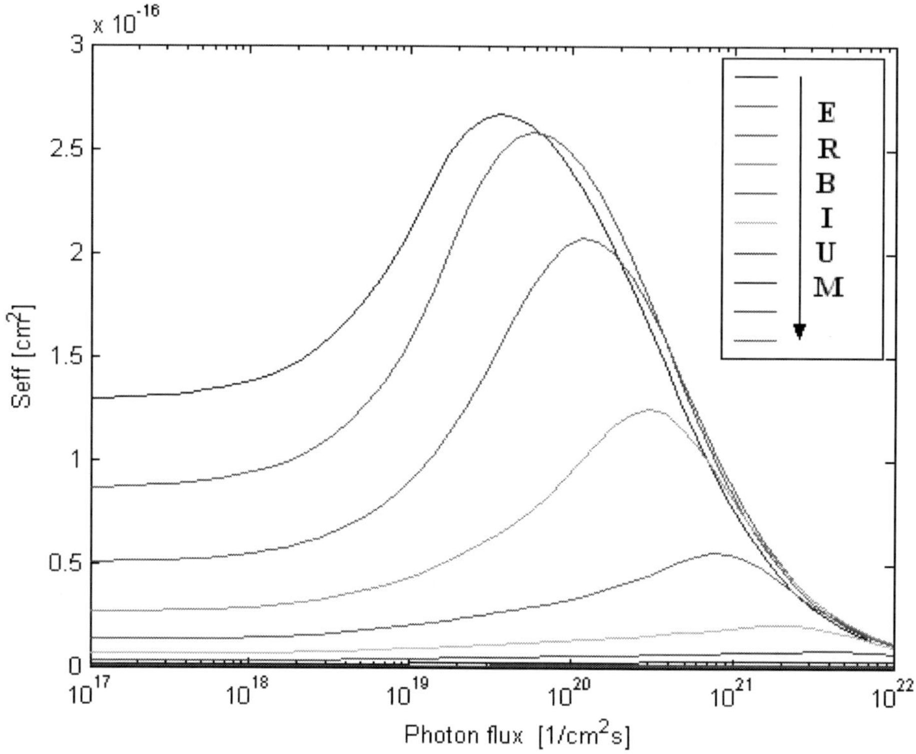

Figure 6. Effective excitation cross-section of the system versus photon flux density and erbium concentration (ten values in the range $10^{19} \div 10^{22}$ cm^{-3}). Nanocluster concentration is 3×10^{19} cm^{-3}.

In Figure 7 the percentage rates of erbium excitable population as a function of erbium concentration are shown. The simulations confirm that a larger excitation rate of erbium ions is possible if the concentrations of erbium and nanocluster are similar. If the concentration of erbium ions is larger than that of nanoclusters, the fraction of excited erbium ions increases only with optical power.

Anyway, the excitation rate decreases with increasing the Er concentration inside the sample. As one can see, the up-conversion is one of the most detrimental effects for high volumetric concentrations of erbium.

To evaluate its impact on the system, simulations as a function of up-conversion coefficient have been carried out. Curves of net optical gain related to different values of up-conversion coefficient in the range $1\times10^{-17} \div 1\times10^{-16}$ cm^3/s as a function of Er concentration are sketched in Figure 8. The photon flux was fixed to value 1×10^{22} cm^{-2}/s.

Maximum value of achievable gain (without FCA) is 10.4985 cm^{-1}, while the gain is 9.0379 cm^{-1} by considering FCA losses. This means that losses due to up-conversion are larger than those related to FCA. To the best of our knowledge, similar simulations as a function of up-conversion coefficient are not yet available in literature.

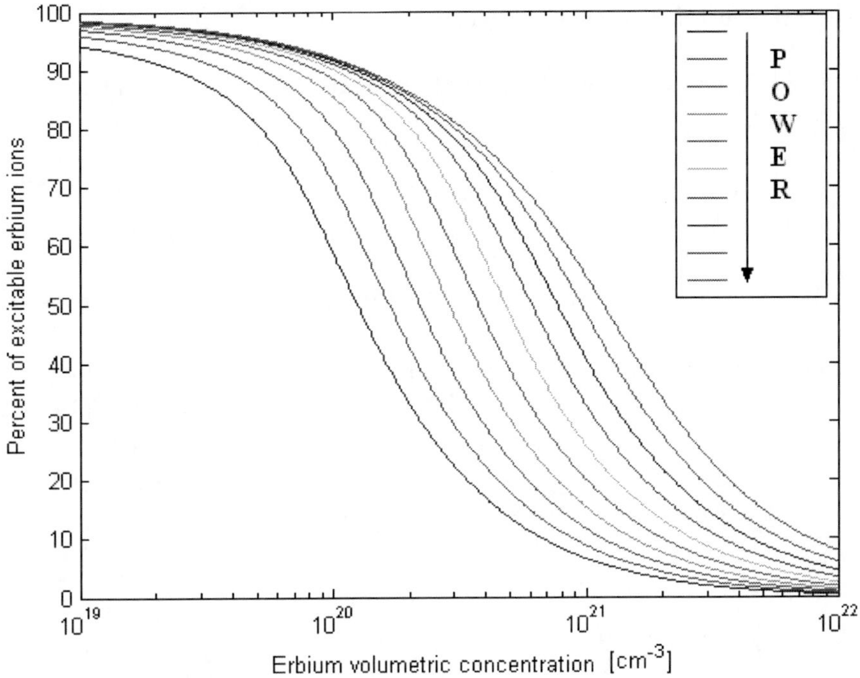

Figure 7. Percentage rates of excitable erbium ions versus erbium volumetric concentration and photon flux density (ten values in the range $10^{20} \div 10^{22}$ cm^{-2}/s).

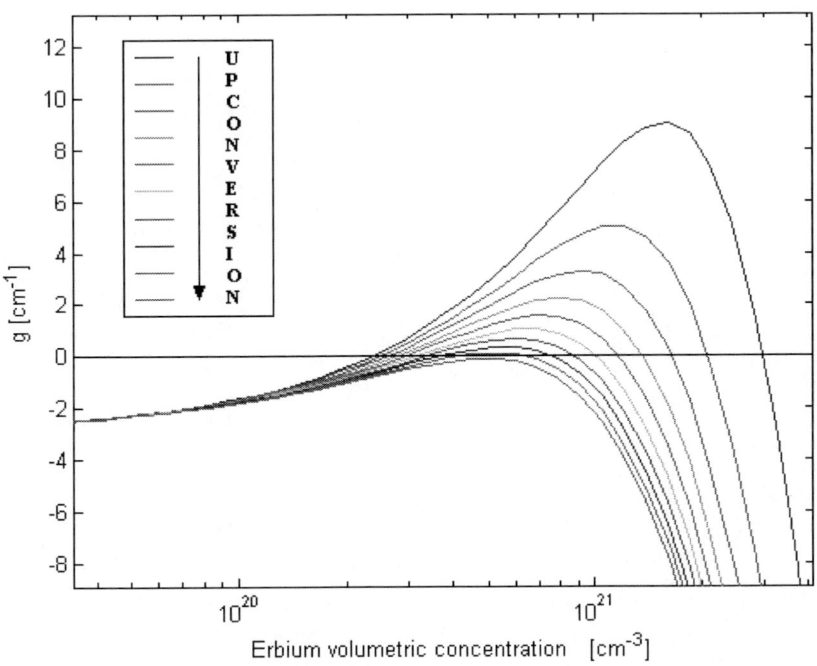

Figure 8. Net optical gain versus erbium volumetric concentration for various up-conversion coefficients, in the range $1\times10^{-17} \div 1\times10^{-16}$ cm^3/s.

2.2. Simplified Model of Optical Pumping

In this sub-section we derive a simplified model of rate equations to achieve the following advantages:

1. a simple evaluation of the system saturation power;
2. moderate memory consumption and fast simulations, in comparison with the complete model;
3. global understanding of variables involved in the physical process.

There properties have reached by finding only the stationary solutions, then the transient solutions of differential equations are lost. Moreover, the simplified model requires some coefficients that could be only found by using the complete approach model, but the advantage now is that the exact model is to be applied only one time. With reference to Table 1 for model coefficients, the simplified system can be written as:

$$\sigma_{nc}\phi_p n_a - w_b n_b - \sum_{i=1}^{3} C_{bi} n_b N_i = 0 \quad (3)$$

$$-\sigma_{nc}\phi_p n_a - w_b n_b + \sum_{i=1}^{3} C_{bi} n_b N_i = 0 \quad (4)$$

$$\sigma_d \phi_p N_1 + \sum_{i=1}^{3} C_{bi} n_b N_i + C_3 N_3^2 - (w_{51} + w_{54}) N_5 = 0 \quad (5)$$

$$C_{b1} n_b N_1 + C_{up} N_2^2 + w_{54} N_5 - w_{43} N_4 = 0 \quad (6)$$

$$w_{43} N_4 - (w_{32} + w_{31}) N_3 + C_{b3} n_b N_3 - 2C_3 N_3^2 = 0 \quad (7)$$

$$w_{32} N_3 - (w_{21} + w_{Er}) N_2 - \phi_s (\sigma_e N_2 - \sigma_a N_1) - 2C_{up} N_2^2 - C_{b2} n_b N_2 - C_A n_b N_2 = 0 \quad (8)$$

$$(w_{21} + w_{Er}) N_2 + \phi_s (\sigma_e N_2 - \sigma_a N_1) + C_{up} N_2^2 + C_A n_b N_2 + w_{31} N_3 + C_3 N_3^2 + w_{51} N_5 - \sigma_d \phi_p N_1 - C_{b1} n_b N_1 = 0 \quad (9)$$

The unknown quantities of the system are the erbium populations at every energetic level and the population of excited nanocrystals (N_1, N_2, N_3, N_4, N_5, n_b). The known parameters are the sum of all erbium ions allocated at various levels, equal to the total bulk Er concentration inside the sample N_T, while the sum of excited nanoclusters n_b and non-excited nanoclusters n_a is equal to the total concentration of nanoclusters inside the sample. This means that:

$$N_1 + N_2 + N_3 + N_4 + N_5 = N_T \quad (10)$$

$$n_a + n_b = n_c \qquad (11)$$

As we can see, the equation system (3)–(9) represents a homogeneous system of non linear equations. The exact solution of this system requires long calculation times. Then, the system has been linearized in order to simplify the system resolution, with the aim to extract some global parameters for description of the physical process. The assumption made is that the term $C_{up}N_2^2$ in equations (6), (8) and (9) can be replaced with the term $C_{up}N_{t2}N_2$. Similarly, the term $C_3N_3^2$ is substituted by $C_3N_{t3}N_3$ (the meaning of constants N_{t2} and N_{t3} will be explained in the following). Another unknown quantity of the system is the population of excited states of nanoclusters, n_b. In the equation system, this variable is usually multiplied by the Er population, from level 1 to level 3. To dissociate this quantity from Er population at a certain level, we have substituted the variables N_1, N_2, N_3, with constants N_{t1}, N_{t2}, N_{t3}. After these assumptions, we have got the following linearized system:

$$\sigma_{ab}\phi_p n_a - w_b n_b - \left(C_{b1}n_bN_{t1} + C_{b2}n_bN_{t2} + C_{b3}n_bN_{t2}\right) = 0 \qquad (12)$$

$$-\sigma_{ab}\phi_p n_a - w_b n_b + C_{b1}n_bN_{t1} + C_{b2}n_bN_{t2} + C_{b3}n_bN_{t2} = 0 \qquad (13)$$

$$\sigma_d\phi_p N_1 + \sum_{i=2}^{3} C_{bi}n_bN_i + C_3N_{t3}N_3 - \left(w_{51} + w_{54}\right)N_5 = 0 \qquad (14)$$

$$C_{b1}n_bN_1 + C_{up}N_{t2}N_2 + w_{54}N_5 - w_{43}N_4 = 0 \qquad (15)$$

$$w_{43}N_4 - \left(w_{32} + w_{31}\right)N_3 + C_{b3}n_bN_3 - 2C_3N_{t3}N_3 = 0 \qquad (16)$$

$$w_{32}N_3 - \left(w_{21} + w_{Er}\right)N_2 - \phi_s\left(\sigma_eN_2 - \sigma_aN_1\right) - 2C_{up}N_{t2}N_2 - C_{b2}n_bN_2 - C_An_bN_2 = 0 \qquad (17)$$

$$\begin{aligned}&\left(w_{21} + w_{Er}\right)N_2 + \phi_s\left(\sigma_eN_2 - \sigma_aN_1\right) + C_{up}N_{t2}N_2 + C_An_bN_2 + \\ &w_{31}N_3 + C_3N_tN_3 + w_{51}N_5 - \sigma_d\phi_pN_1 - C_{b1}n_bN_1 = 0\end{aligned} \qquad (18)$$

Here the term $w_{51}N_5$ is practically negligible. Moreover, starting from Eq. (17), we obtain:

$$N_2 = \frac{\left(w_{32} + w_{31}\right)N_3 + \phi_s\sigma_aN_1}{\left(w_{21} + w_{Er}\right) + 2C_{up}N_{t2} + C_{b2}n_b + C_An_b + \phi_s\sigma_e} = aN_3 + bN_1 \qquad (19)$$

By the same way, Eq. (18) can be rewritten as:

$$N_1 = \frac{\left(w_{21} + w_{Er} + C_{up}N_{t2} + C_A n_b + \sigma_e \phi_s\right)N_2 + C_3 N_{t3} N_3}{\sigma\phi_p + C_{b1}n_b + \sigma_a \phi_s} = cN_2 + dN_3 \quad (20)$$

Finally, the coefficients N_{t1}, N_{t2}, N_{t3} are defined as follows:

$$\begin{cases} N_{t2} = \overline{r}_{21} N_{t1} \\ N_{t3} = \overline{r}_{32} N_{t2} \\ N_T = N_{t1} + N_{t2} + N_{t3} \end{cases} \quad (21)$$

where $\overline{r}_{21}$ and $\overline{r}_{32}$ are given by the standard deviation of population ratios between level 2 and 1, and between level 3 and 2, respectively. With data exposed in Table 1, we have found $\overline{r}_{21} = 6.6271$ and $\overline{r}_{32} = 0.0226$. Then, Er populations at higher levels are:

$$N_3 = \frac{w_{43} N_4}{w_{32} + w_{31} + C_{b3}n_b + 2C_3 N_{t3}} = eN_4 \quad (22)$$

$$N_5 = \frac{\sigma\phi N_1 + C_{b2}n_b N_2 + (C_{b3}n_b + C_3 N_{t3})N_3}{w_{54}} = fN_1 + gN_2 + hN_3 \quad (23)$$

Finally, we find the gain expression as:

$$g = \left(\sigma_e N_2 - \sigma_a N_1\right) = \frac{N_T}{\alpha \frac{\delta}{\gamma} + \beta}\left(\sigma_e - \sigma_a \frac{\delta}{\gamma}\right) \quad (24)$$

where

$$\alpha = \left(1 - b/a - b/ae + f - bh/a\right) \quad (25)$$

$$\beta = \left(1 + 1/a + 1/ae + g + h/a\right) \quad (26)$$

$$\delta = c + d/a \quad (27)$$

$$\gamma = 1 + bd/a \quad (28)$$

2.3. Comparison between Simplified and Complete Models

Some simulations are needed to compare the simplified approach with the exact model predictions. In Figure 9 the gain curves (without FCA) are shown as calculated by exact (blue curves) and approximated method (red curves), respectively. It is clear from the figure as the difference between the two methods is low in the central region of the graph, while it tends to increase in the lateral sides. Then, the approximated approach is very useful to have a fast qualitative insight of the system behaviour.

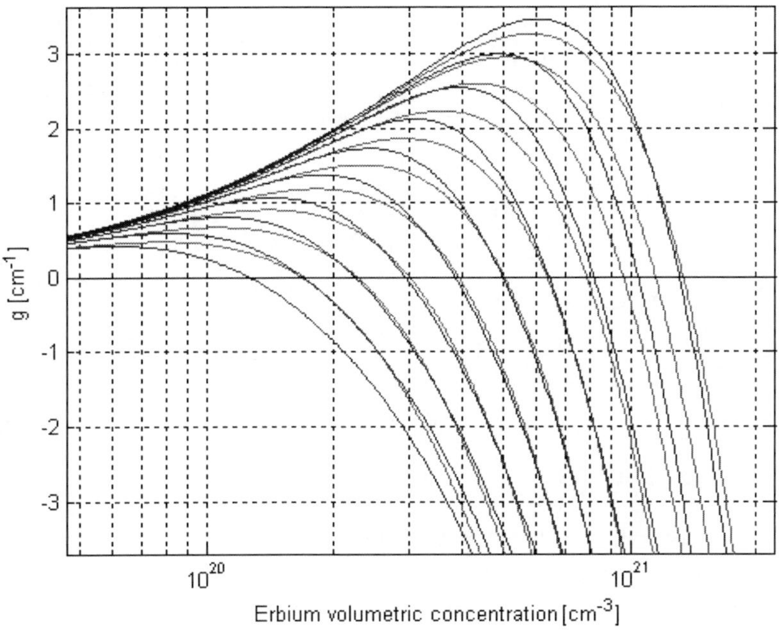

Figure 9. Comparison between exact (blue curves) and simplified (red curves) model predictions (photon flux in the range $10^{20} \div 10^{22}$ cm^{-2}/s).

The significant advantage of the approximated model is the fast estimation of system saturation power. To calculate this value, it is necessary to rewrite the modal gain as [9]:

$$g = \frac{g_0}{1 + I / I_{SAT}} \tag{29}$$

where I_{SAT} is the input signal intensity needed to saturate the gain, while g_0 is the non-saturated gain coefficient. It is simple to verify that at saturation it holds $g = g_0 / 2$. In Figure 10 the curves of saturation intensity versus pump power, evaluated for ten values of Er volumetric concentration, are shown. If the erbium doping increases, then the saturation intensity increases too, because a larger excitable Er population at level 1 is available.

However, it is important to note that an increase of saturation intensity with pump power tends to “saturate” itself, because an increase of pump power brings to a boost of parasitic phenomena such as Auger recombination and up-conversion. In Figure 11 the saturation curves versus erbium concentration are sketched for ten values of photons flux. Similar behavior as before can be observed. Now the “saturation” can be seen as a closer distance between curves with increasing the pump power.

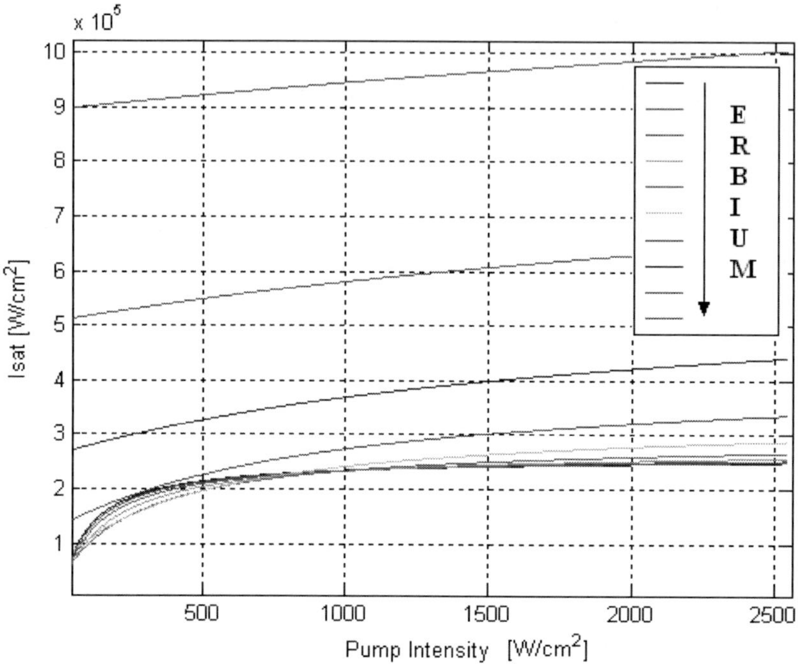

Figure 10. Saturation curves as a function of pump intensity evaluated for ten values of erbium concentration, in the range $10^{19} \div 10^{22}$ cm-3.

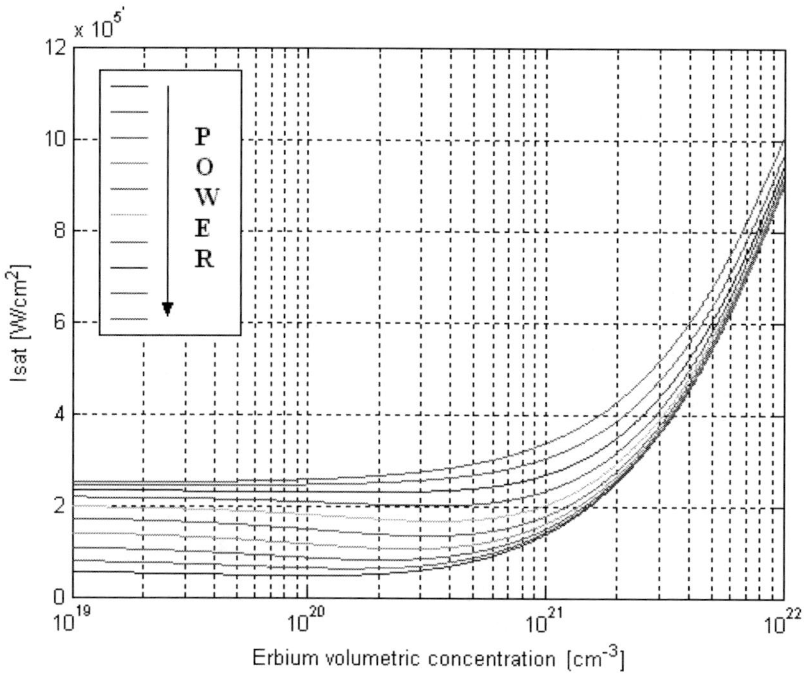

Figure 11. Saturation curves versus erbium volumetric concentration, evaluated for ten values of pump photon flux, in the range $10^{20} \div 10^{22}$ cm^{-2}/s.

3. MODELING OF ELECTRICAL PUMPING

Nowadays, an efficient electrical pumping of silicon-based light sources should be one of the biggest challenges in the Optoelectronics field. A silicon light source it is very important to allow the integration of both electrical and optical functions on the same chip in order to increase the elaboration speed of data. For example, if the data transmission is optical on a silicon chip, it should be possible to remove the parasitic capacitances of metallic wires, then increasing the speed.

In this section, the investigated material system is again Er:Si-nc:SiO_2. About the electric current mechanisms established in this material, various experimental results and related explanations are available. In general, the main observed conduction mechanisms are direct tunnelling (DT) and Fowler–Nordheim tunnelling (FN) [10], but other possible conduction mechanisms are also possible [11]. The dominant conduction mechanism depends on intensity of applied electric field and sample characteristics. Under an electric field of amplitude 8 MV/cm the tunneling regime is important, while above that FN tunneling dominates. These two physical mechanisms are characterized by different laws [12]:

$$J_d = \frac{AE^2}{\left[1-\sqrt{1+\frac{qV}{\phi_B}}\right]^2} \exp\left\{-\frac{B}{E}\left[1-\left(1-\frac{qV}{\phi_B}\right)^{3/2}\right]\right\} \tag{30}$$

$$J_{FN} = AE^2 \exp\left\{-\frac{B}{E}\right\} \tag{31}$$

where $A = \frac{q^3}{16\pi^2\hbar\phi_B}$ and $B = \frac{4}{3}\frac{\sqrt{2m^*}}{\hbar q}\phi_B^{3/2}$, ϕ_B is the effective height of potential barrier, m^* is the electron effective mass in SiO_2, h is the Planck's constant and E is the electric field. The excitation mechanism of erbium inside this material system is illustrated in Figure 12 [4].

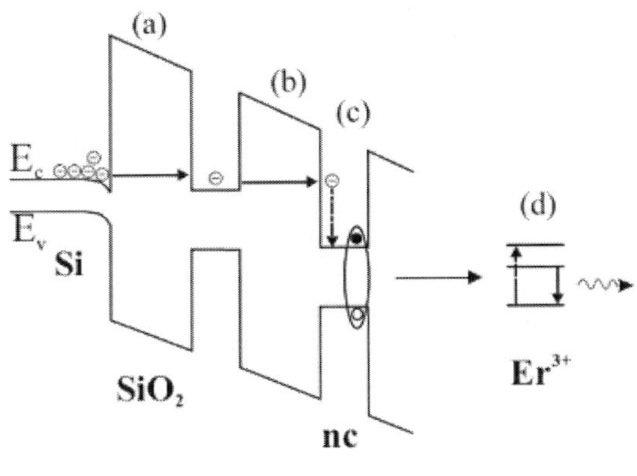

Figure 12. Energetic diagram with: (a) injection of an electron; (b) potential barrier tunnelling; (c) impact of an "hot" electron on a nanocluster with subsequent energy transfer that creates an electron-hole couple (exciton) inside the silicon cluster; (d) the "hot" electron is at the edge of conduction band and it is ready for a new process similar to (b) mechanism [4].

While the optical pumping is characterized by an increase of photoluminescence with increasing the nanoclusters population, in case of electrical pumping an increase of Si:nc volumetric density determines an electroluminescence decrement, as already proved by *Sun et al.* [10]. In fact, if the nanocluster population grows, the average distance between nanoclusters decreases and then the electrons will need lower energies to cross the potential barrier of silicon dioxide between them. Sometimes, this energy is not enough to create an exciton for impact ionization inside a nanocluster. This explanation is in good agreement with the experimental results found by *Castagna et al.* [13], which show that electroluminescence of erbium-doped silicon dioxide is larger than in erbium-doped silicon rich oxide (SRO). Anyway, the current flow inside the silicon dioxide causes damages to the dielectric structure, so these structures have very short lifetimes. On the contrary, structures with silicon percentages higher than the stoichiometric one can have larger lifetimes, due to the shorter average distance between nanoclusters that electrons have to cross (i.e. lower resistivity). Moreover, very high values of electric field are needed for the erbium-doped silicon dioxide structures, which in turn may cause the oxide breakdown.

In summary, the main physical points to be considered for a valid model of rate equations are the following:

- electroluminescence is due to radiative decay of erbium ions, directly excited by silicon nanoclusters;
- electroluminescence increases with electric field and decreases with volumetric concentration of nanoclusters;
- current flow quenches a part of luminescence because of a particular Auger mechanism, shown by *Priolo at al.* [11].

One could conclude that the model proposed for photoluminescence is still valid because the energy transfer mechanism between silicon nanoclusters and erbium ions still remains the same, but now another Auger de-excitation mechanism, related to the current flow, is also present. Moreover, in our model the direct impact excitation cross section is not considered. For the model derivation, the erbium and nanocluster concentrations are assumed as uniform in the sample volume, in order to consider only the temporal derivative of rate equations. The rate equation model for the electrical pumping of the material system Er:Si-nc:SiO_2 can be written as:

$$\frac{dn_b}{dt} = \sigma_{ab}\phi_e n_a - w_b n_b - \sum_{i=1}^{3} C_{bi} n_b N_i \tag{32}$$

$$\frac{dn_a}{dt} = -\sigma_{ab}\phi_e n_a + w_b n_b + \sum_{i=1}^{3} C_{bi} n_b N_i \tag{33}$$

$$\frac{dN_5}{dt} = \sum_{i=1}^{3} C_{bi} n_b N_i + C_3 N_3^2 - (w_{51} + w_{54}) N_5 \tag{34}$$

$$\frac{dN_4}{dt} = C_{b1}n_bN_1 + C_{up}N_2^2 + w_{54}N_5 - w_{43}N_4 \quad (35)$$

$$\frac{dN_3}{dt} = w_{43}N_4 - (w_{32} + w_{31})N_3 + C_{b3}n_bN_3 - 2C_3N_3^2 \quad (36)$$

$$\frac{dN_2}{dt} = w_{32}N_3 - (w_{21} + w_{Er})N_2 - \phi_s(\sigma_eN_2 - \sigma_aN_1) + \\ -w_AN_2 - 2C_{up}N_2^2 - C_{b2}n_bN_2 - C_An_bN_2 \quad (37)$$

$$\frac{dN_1}{dt} = (w_{21} + w_{Er})N_2 + \phi_s(\sigma_eN_2 - \sigma_aN_1) + w_AN_2 + \\ C_{up}N_2^2 + C_An_bN_2 + w_{31}N_3 + C_3N_3^2 + w_{51}N_5 - C_{b1}n_bN_1 \quad (38)$$

where ϕ_e and ϕ_s represent the electron flux and the signal flux generated at 1540 nm, respectively. The values of coefficients are summarized in the following Table 2. This model for predictions of electrical pumping, is rather conservative (as a worst case) because it considers all loss and recombination mechanisms, neglecting the direct excitation of erbium.

Table 2. Parameters used in the model of electrical pumping

Symbol	Parameter	Value (units)
N_c	Nanocrystals density	1×10^{19} cm^{-3}
σ_d	Direct absorption cross section	0
σ_{nc}	Nanocluster's impact cross section	1×10^{-14} cm^2
w_b	Nanocluster's inverse lifetime	4×10^4 s^{-1}
w_{21}	Erbium's inverse lifetime between 2nd and 1st energetic level	526.3 s^{-1}
w_{32}	Erbium's inverse lifetime between 3rd and 2nd energetic level	4.2×10^5 s^{-1}
w_{43}	Erbium's inverse lifetime between 4th and 3rd energetic level	1×10^7 s^{-1}
w_{54}	Erbium's inverse lifetime between 5th and 4th energetic level	1×10^7 s^{-1}
w_{31}	Auger non-radiative recombination rate	1.4×10^3 s^{-1}
w_A	Erbium's inverse lifetime between 3rd and 1st energetic level	7×10^4 s^{-1}
C_{b2}	Erbium-nanocluster coupling coefficient L2	3×10^{19} cm^3/s
C_{b3}	Erbium-nanocluster coupling coefficient L3	3×10^{19} cm^3/s
C_{up}	Cooperative up-conversion coefficient	5×10^{-17} cm^3/s
C_A	Auger recombination coefficient	3×10^{-19} cm^3/s
C_3	Cooperative up-conversion coefficient	1.8×10^{-17}cm^3/s
σ_{fca}	FCA effective cross section	1×10^{-18} cm^2
n	Refractive index	1.46

3.1. Simulations of Electrical Pumping

As for optical pumping, we have investigated the system gain and other parameters by considering the variation of erbium concentration inside the sample and the electron flux used to pump the material, taking fixed the volumetric concentration of silicon nanoclusters inside the sample. Er volumetric concentration has been varied in the range $1\times10^{19} \div 1\times10^{21}\,\text{cm}^{-3}$, while the variation range of electron flux has been changed in the range $1\times10^{18} \div 1\times10^{20}\,\text{cm}^{-2}/\text{s}$, corresponding to a current density in the range $1.6\times10^{-1} \div 16\,\text{A/cm}^2$. In Figure 13 the curves of net optical gain (red curves), total gain without considering FCA losses (blue curves) and FCA losses (green curves) are plotted. In next Figure 14, the number of erbium ions excited by a single nanocluster can be seen as a function of Er concentration, while Figure 15 shows the percentage of excited erbium ions. For each figure, the main physical considerations made for optical pumping are still valid.

To make a comparison with the optical pumping case, it is useful to investigate the effective excitation cross section (see Figure 16). The evaluation of the effective cross section is made using equation 2, but, at this time, considering the electron flux replacing the photon flux. As one can see from the curves in Figure 16, in case of electrical pumping the effective cross section is smaller that that for nanocluster excitation cross section (i.e. $1\times10^{-14}\,\text{cm}^2$). Moreover, the effective excitation cross section decreases with erbium volumetric concentration and its value is almost constant for lower values of erbium concentration and current density. The causes of this behavior are analogous to those already seen in case of optical pumping.

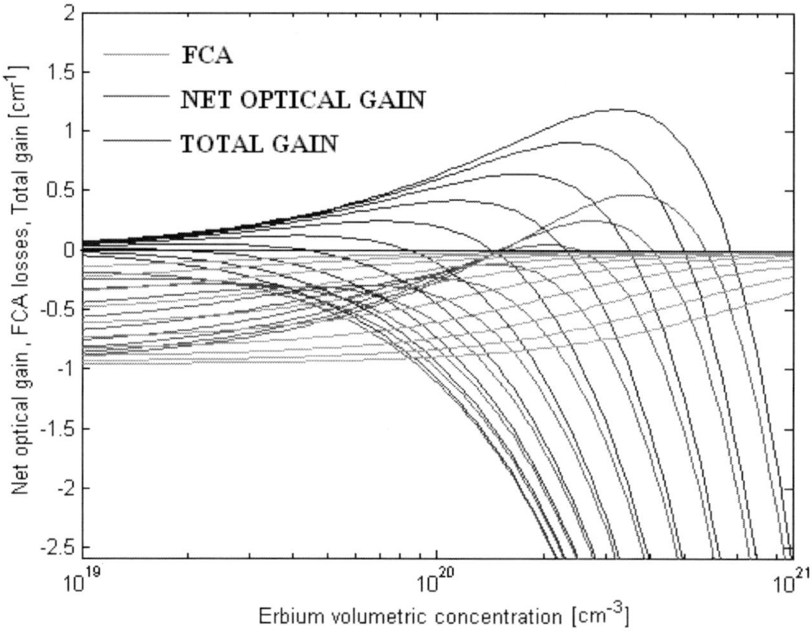

Figure 13. Comparison between total optical gain, net optical gain and FCA losses versus erbium volumetric concentration and electrical pumping power (ten values in the range $1\times10^{18} \div 1\times10^{20}\,\text{cm}^{-2}/\text{s}$).

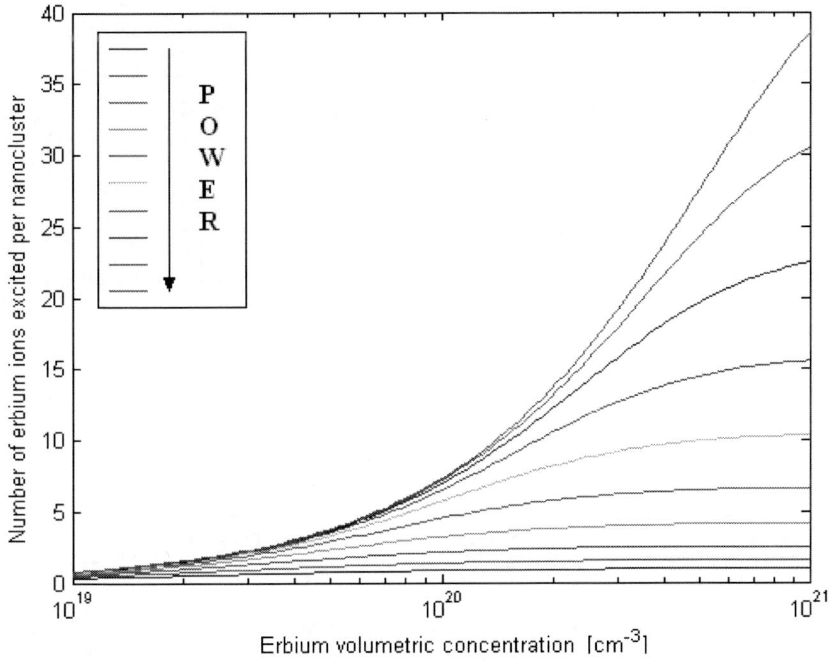

Figure 14. Number of erbium ions excited per nanocluster at level 2 versus Er concentration and electrical pumping power (ten values in the range $1\times10^{18} \div 1\times10^{20}$ cm^{-2}/s).

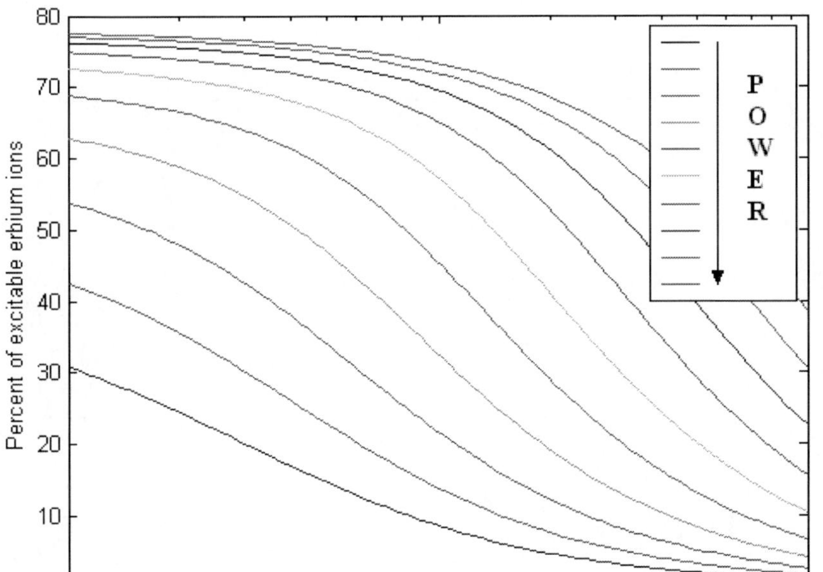

Figure 15. Percentage rates of erbium ions excitable versus Er concentration and electron flux.

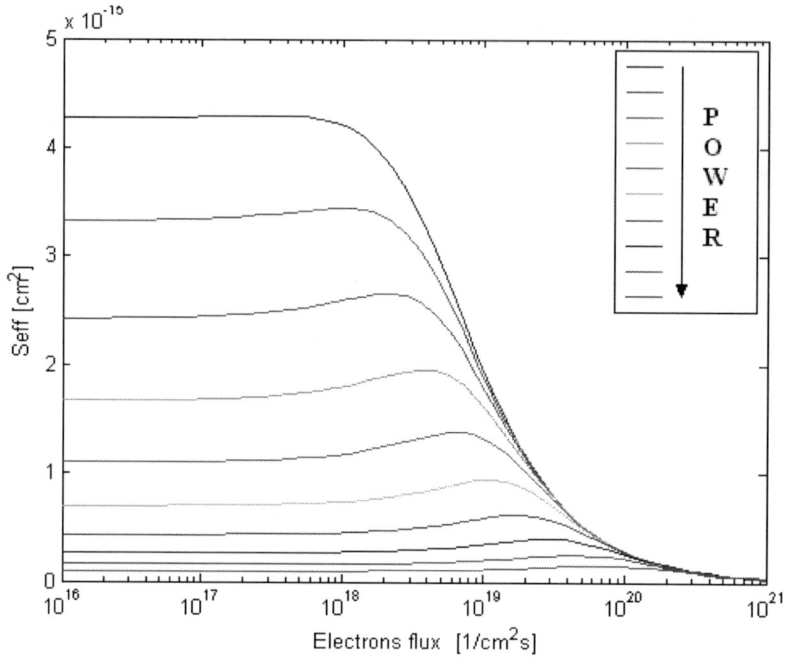

Figure 16. Effective excitation cross-section versus density of electron flux and erbium volumetric concentration. Electron flux is in the range $1\times10^{18}\div1\times10^{20}$ cm^{-2}/s (current density in the range $1.6\times10^{-1}\div16$ A/cm^2), while erbium concentration is in the range $1\times10^{19}\div1\times10^{21}$ cm^{-3}.

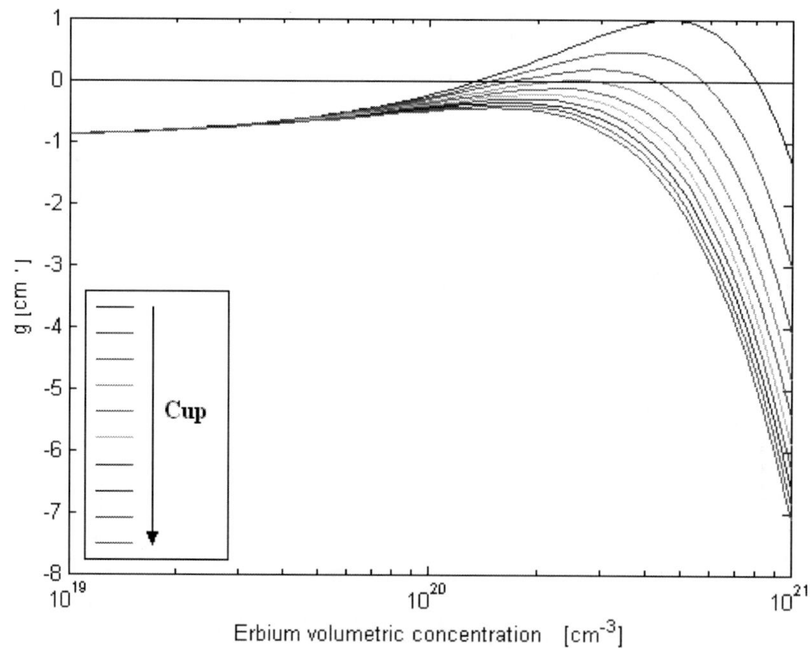

Figure 17. Net optical gain versus Er concentration, for different values of up-conversion coefficient in the range $1\times10^{-17}\div1\times10^{-16}$ cm^3/s.

Other simulations have been performed to investigate the parasitic effects, such as Auger non-radiative recombination and up-conversion. In Figure 17 the gain curves versus up-conversion coefficient in the range $1\times10^{-17} \div 1\times10^{-16}\,cm^3/s$ are sketched, keeping the electrons flux at $1\times10^{20}\,cm^{-2}/s$. The detrimental effect of up-conversion on gain clearly increases with erbium concentration, while FCA losses dominate for low values of erbium concentration. It is important to note that a positive gain is only possible for up-conversion coefficient values $> 4\times10^{-17}\,cm^3/s$.

In next Figure 18, the gain curves obtained for the same erbium and electrons flux concentrations are plotted as a function of Auger de-excitation rate in the range $1\times10^3 \div 1\times10^4 s^{-1}$, using parameters given in Table 2. As shown in this figure, it is always possible to obtain a positive gain for a certain range of erbium concentration. To avoid the negative effect of this Auger de-excitation mechanism, *Walters et al.* proposed a new mechanism of pumping based on sequential injection of electrons [14].

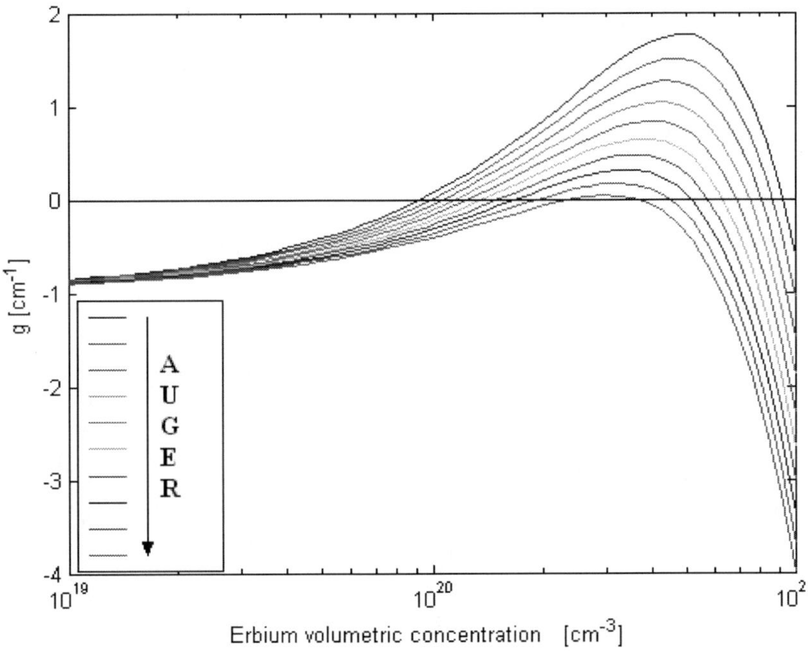

Figure 18. Gain curves as a function of Er concentration and evaluated for ten values of Auger de-excitation rate (ten values in the range $1\times10^3 \div 1\times10^4 s^{-1}$).

3.2. Simplified Model of Electrical Pumping

As in the case of optical pumping, a simplified model of rate equation for the electrical pumping is proposed too. The advantage and the drawbacks of this approach are similar to those illustrated above for the optical pumping. The only differences are the addition of an Auger de-excitation rate term and negligible direct excitation cross section.

Therefore, the equations of the modeling are written as follows:

$$N_2 = \frac{\left(w_{32} + w_{31}\right) N_3 + \phi_s \sigma_a N_1}{\left(w_{21} + w_{Er} + w_a\right) + 2C_{up} N_{t2} + C_{b2} n_b + C_A n_b + \phi_s \sigma_e} = aN_3 + bN_1 \quad (39)$$

$$N_1 = \frac{\left(w_{21} + w_{Er} + w_a + C_{up} N_{t2} + C_A n_b + \sigma_e \phi_s\right) N_2}{C_{b1} n_b + \sigma_a \phi_s} + \frac{C_3 N_{t3} N_3}{C_{b1} n_b + \sigma_a \phi_s} = cN_2 + dN_3 \quad (40)$$

$$N_3 = \frac{w_{43} N_4}{w_{32} + w_{31} + C_{b3} n_b + 2C_3 N_{t3}} = eN_4 \quad (41)$$

$$N_5 = \frac{C_{b2} n_b N_2 + (C_{b3} n_b + C_3 N_{t3}) N_3}{w_{54}} = gN_2 + hN_3 \quad (42)$$

The equation of the gain is the same as Eq. (24), with $\alpha, \beta, \gamma, \delta$ defined from equations (25)–(28).

3.3. Comparison between Simplified and Exact Models

In Figure 19 the gain curves calculated with the exact (blue) and simplified model (red) are compared. As one can see, the two models have minimum differences in the central region. This behavior is due to the approximation on the evaluation of an average transition rate, between levels 3 and 2 and between levels 2 and 1.

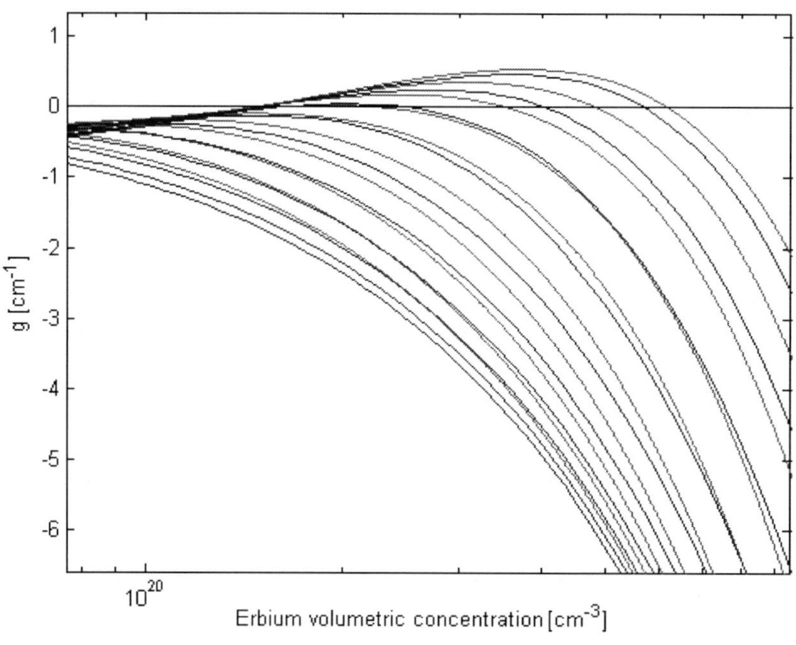

Figure 19. Comparison between exact (blue curves) and simplified (red curves) models.

The knowledge of the system saturation power is still important to characterize an optical amplifier because it gives its working limits. In Figure 20 the saturation curves calculated versus erbium volumetric concentration for ten values of electrons excitation flux are shown.

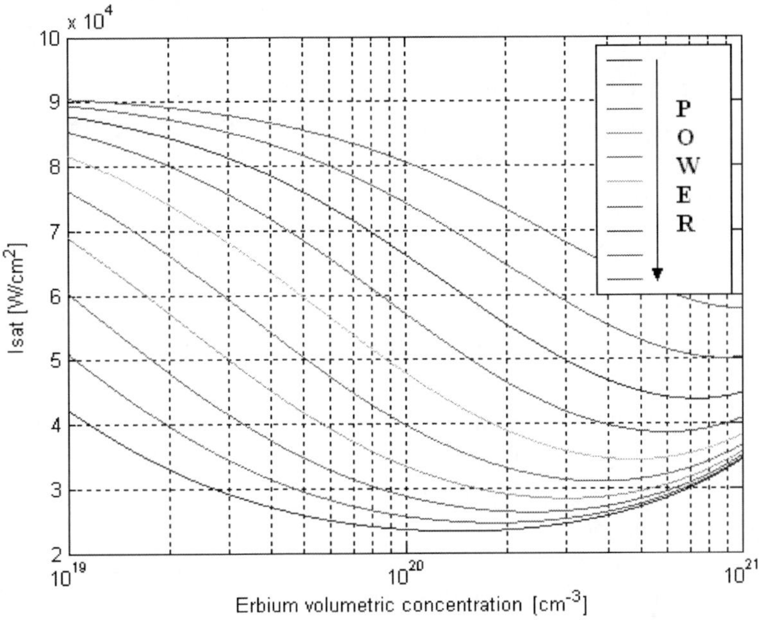

Figure 20. Saturation curves as a function of Er concentration for ten values of pump photon flux.

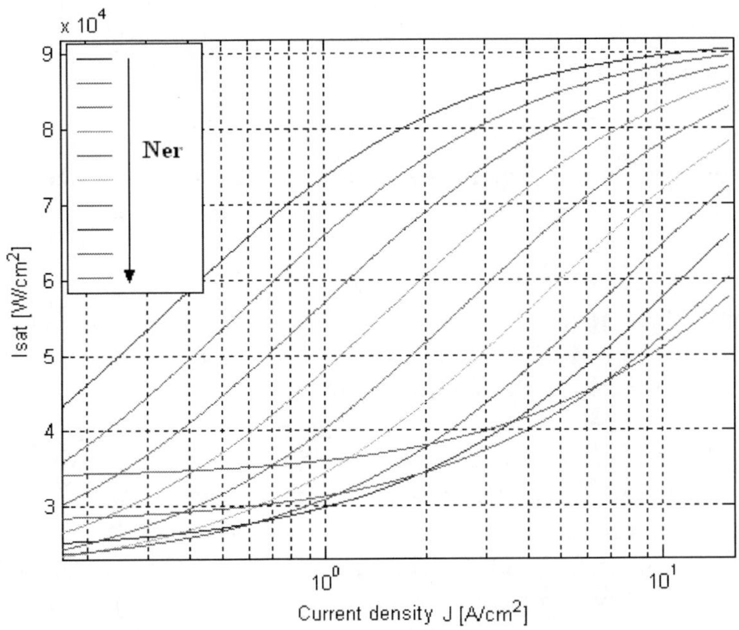

Figure 21. Saturation curves versus pump current density evaluated for ten values of erbium concentration.

It is worth noting that the intensity of the saturation signal increases with pump power, because the rate of erbium excitation increases too, but it decreases with Er concentration in the central zone of the plot, because of up-conversion and concentration quenching mechanisms, that empty the higher levels of excited erbium population. However, it tends to increase again with further increasing erbium concentration. This could be noted at the right edge of Figure 20.

Then, it can be derived an erbium concentration range where the probability to reach the saturation is inversely proportional to the erbium concentration itself. The same behavior is found in Figure 21 where the saturation curves calculated as a function of current density and evaluated for ten values of erbium population are given.

4. Electrically Pumped Er-doped Silicon Nanoclusters Laser

In this section a suitable laser cavity for electrical pumping of erbium doped silicon nanoclusters is proposed. As shown before, this material system has a positive gain only if some parasitic effects such as up-conversion, FCA and Auger de-excitation are lower than commonly are in this material.

Moreover, to obtain laser emission within this material system, cavity losses should be very low. To reduce material losses the light beam should propagate with a high confinement factor inside the active layer, but in this case the index contrast is moderate, having a refractive index close to 1.46.

Moreover, it is necessary the electrical injection too, then both p and n-doped silicon rails are simultaneously needed to contain the active layer. However, silicon has a refractive index of 3.48, then the beam tends to be confined inside the silicon rails, that are characterized by higher losses (mainly due to FCA).

The trade-off to obtain light confinement inside the active layer slotted within doped silicon rails could be met with a structure similar to that proposed by *Barrios et al.* [15]-[16], where the light confinement is enhanced by the larger index contrast between silicon rails and active layer. This structure has the advantage that could be fabricated exploiting a CMOS-compatible technology process.

In this kind of structure, light is confined in the low index layer, that in this case represents the active layer too, because of a big discontinuity of electric field components at interfaces between silicon and erbium-doped silicon dioxide. Reduced scattering losses at the silicon/active layer interfaces are also due to horizontal slot configuration.

The chosen laser structure is a distributed Bragg reflector (DBR), because of its highest mode selectivity. Its section, similar to that presented in [15-16], is shown in Figure 22 with its relevant sizes. Its 3D architectural scheme is sketched in Figure 23.

As one can see from Figure 23, the active region is large enough to increase the optical power but smaller to become bimodal. The confinement of optical field and 3D plot of fundamental mode are shown in Figure 24 and 25, respectively. The simulation have been carried out by finite element method (FEM).

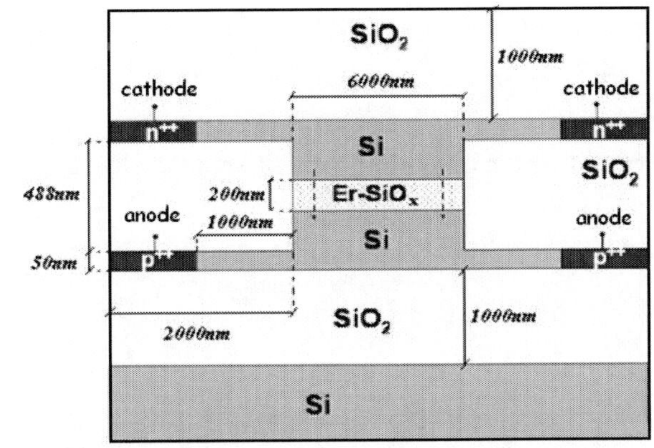

Figure 22. Cross section of proposed DBR laser with relevant sizes.

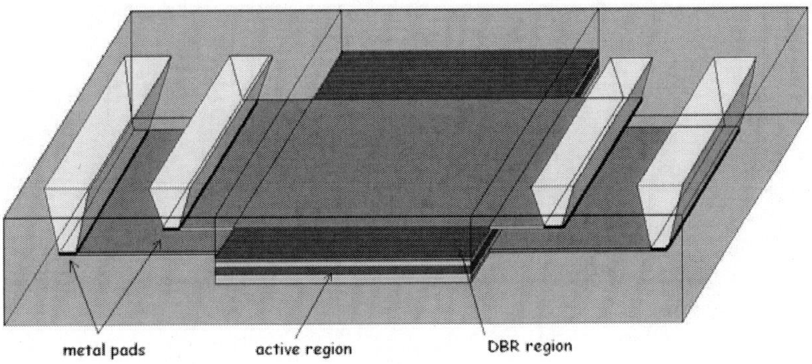

Figure 23. 3D scheme of DBR laser based on erbium-doped silicon nanoclusters.

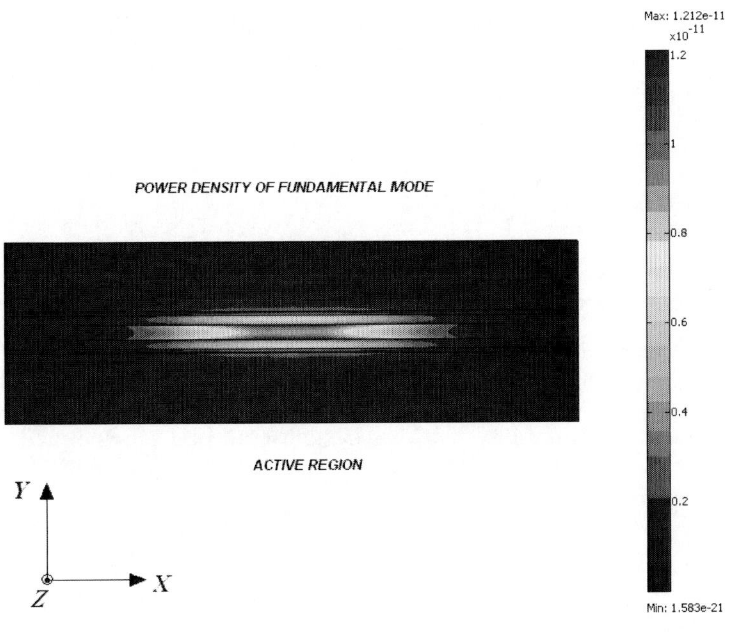

Figure 24. Optical field confined inside the silicon slot.

For an applied current density of $J = 16\,\text{Acm}^{-2}$ and an erbium concentration of $3.5938\times10^{20}\,\text{cm}^{-3}$, the calculated gain is 1.1916 cm^{-1}. However, the maximum current value estimated to avoid the breakdown of the junction is $43.1\,\text{A/cm}^2$, corresponding to a gain value of $g = 1.5977\,\text{cm}^{-1}$. Then, in order to obtain a positive gain, losses of laser cavity must be lower than this value.

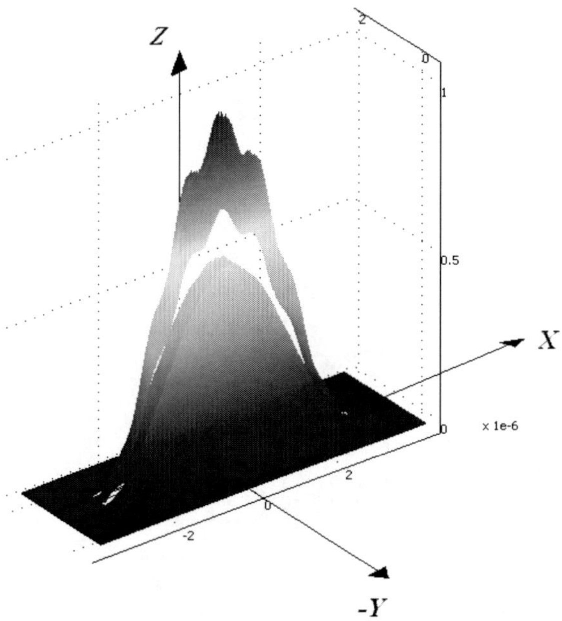

Figure 25. 3D plot of fundamental mode power density, propagating inside the slot.

The losses inside the active layer (FCA, scattering, output coupling) were estimated to be at least 0.7 cm^{-1}, then a positive gain is achievable. The optical power, shown in Figure 26, is calculated using the following relation:

$$P_{out} = \left(\frac{g_0}{g_{th}} - 1\right) A I_{SAT} \alpha_{out} \tag{43}$$

where g_0 is the non-saturated gain of the system, $g_{th} = \alpha_{FCA} + \alpha_{SCAT}/\Gamma + \alpha_{OUT}/\Gamma$ is the threshold gain, evaluated as the sum of FCA, scattering and output coupling losses ($\alpha_{OUT} = -\ln\sqrt{R_p R_t}/L_c$), A is the laser emission area and I_{SAT} is the saturation flux of signal photons. If we suppose for simplicity that this system could be schematized as a three-level system, we have:

$$I_{SAT} = \frac{h\nu}{\tau_s \sigma_e} \tag{44}$$

where τ_s is the effective saturation time of the population inversion, while σ_e is the effective emission cross section. For the three-level system, the effective saturation time is:

$$\tau_s = \tau_2 \frac{1+\sigma_a/\sigma_e}{1+W_p\tau_2} \quad (45)$$

where W_p is the pumping rate, expressed by the following relation:

$$W_p = \frac{1}{t_2}\frac{\sigma_a N_{Er} + g_0}{\sigma_a N_{Er} - g_0} \quad (46)$$

in which N_{Er} is the total erbium population. The lifetime of the second level τ_2 is equal to:

$$\tau_2 = 1/\left(w_{21} + w_{Er} + w_A + C_A n_b + 2C_{up}N_2 + C_{b2}n_b\right) \quad (47)$$

In the following Figure 26, the optical power generated by this laser is shown versus erbium concentration, for ten values of injected current density. The erbium concentration is changed in the range $1\times10^{19} \div 1\times10^{21}$ cm^{-3}, while the current density is varied from 4.3 A/cm^2 to 43 A/cm^2. As one can see from the figure, the optical power increases with erbium concentration until it reaches a peak, whose amplitude depends on the injected current, after that it decreases because of up-conversion and concentration quenching.

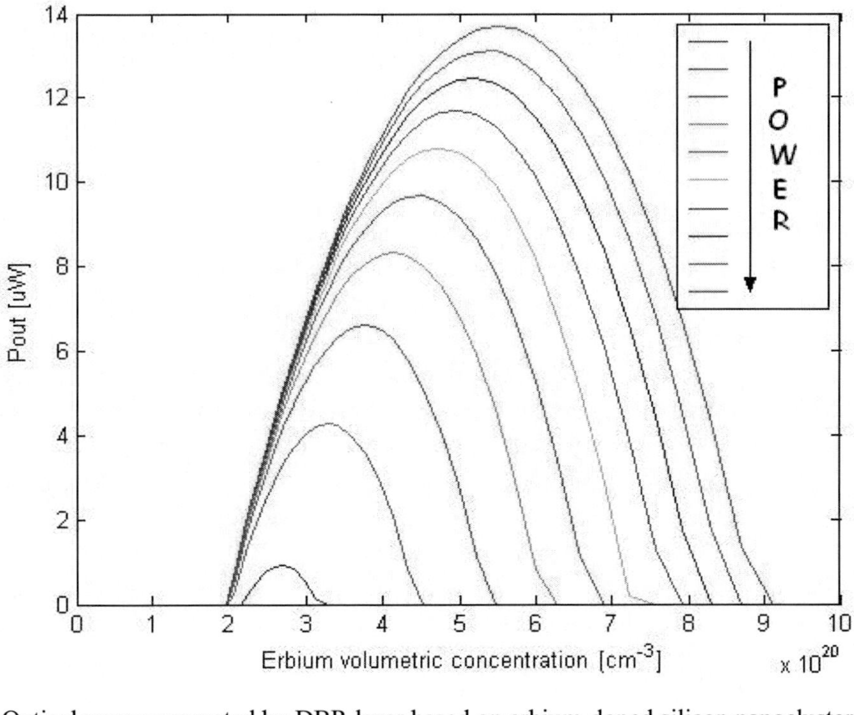

Figure 26. Optical power generated by DBR laser based on erbium-doped silicon nanoclusters.

The maximum power emission is theoretically about 14 μW, while the efficiency of the system is shown in Figure 27. As one can see in this figure, the wall plug efficiency should be around 0.1 ÷ 0.2%, well larger than that obtained experimentally by Kalkmann et al. [17], of the order of 0.0025%.

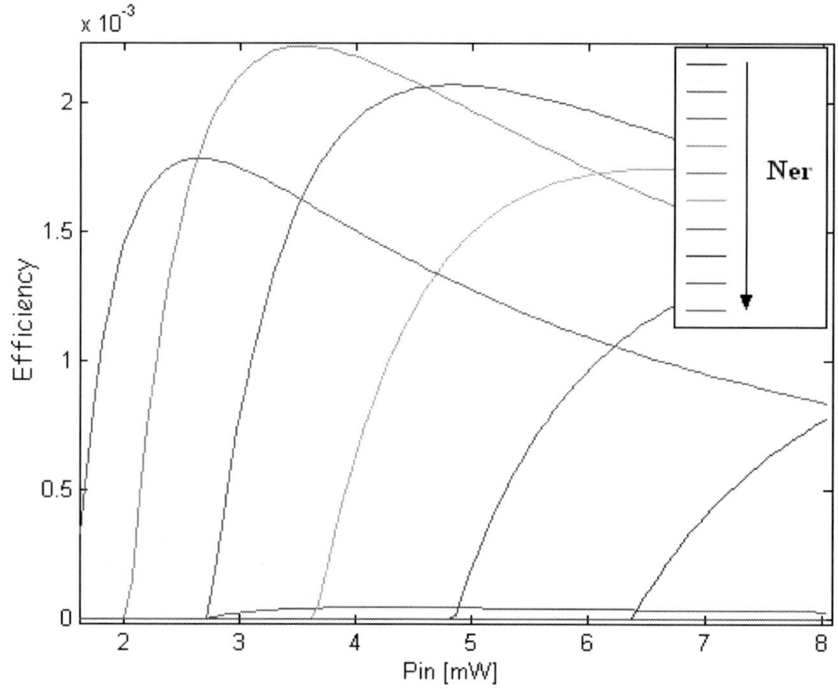

Figure 27. Wall plug efficiency of proposed DBR laser versus input power for various Er concentrations.

CONCLUSION

In this chapter, two accurate models of optical pumping and electrical pumping of AN erbium-doped silicon nanoclusters material system have been presented and discussed. The rate equation systems have been solved for both types of pumping mechanisms in order to find the suitable conditions for an optical net gain, with an investigation on the parasitic effects influencing the active medium performance. Moreover, an approximate model of rate equations has been introduced to simplify the calculation of system stationary conditions and to obtain a fast and accurate enough estimation of the optical saturation power as a function of both pump signal and erbium volumetric concentration. Finally, a DBR laser cavity design is proposed for the first time to the best of our knowledge, to achieve an electrical injection and obtain the light beam confinement inside the active medium with a refractive index lower than that of cladding regions. An efficiency much better than that experimentally presented in literature has been theoretically demonstrated.

REFERENCES

[1] Pacifici, D., Franzò, G., Priolo, F., Iacona, F., Dal Negro, L., *Physical Review B. Cond. Matter*, 2003, 67, 245301.

[2] Priolo, F., Franzò, G., Pacifici, D., Vinciguerra, V., Iacona, F., Irrera, A., *J. Applied Physics*, 2001, 89, 264-272.

[3] Shin, J. H., Lee, J., Han, H.-S., Jhe, J.-H., Chang, J. S., Seo, S.-Y., Lee, H., Park, N., *IEEE J. Selected Topics in Quantum Electron.*, 2006, 12, 783-796.

[4] Pacifici, D., *Erbium doped silicon nanoclusters for Microphotonics*, Ph.D. thesis, University of Catania, 2003.

[5] Han, H., Seo, S., Shin, J. H., Park, N., *Applied Physics Letters*, 2002, 81, 3720-3722.

[6] Lee, J., Shin J. H., Park, N., *J. Lightwave Technol.*, 2005, 23, 19-25.

[7] Daldosso, N., Navarro-Urrios, D., Melchiorri, M., Pavesi, L., Sada, C., Gourbilleau, F., Rizk, R., *Applied Physics Letters*, 2006, 88, 161901.

[8] Kik, P., Polman, A., Towards an Er-doped Si Nanocrystal Sensitized Waveguide Laser – The thin Line between Gain and Loss, *Proc. NATO Workshop OASIS*, 2002.

[9] Liu, Jia-Ming, *Photonic Devices*, Cambridge University Press, 2005.

[10] Sun, J. M., Skorupa, W., Dekorsy, T., Helm, M., Nazarov, A. N., *Optical Materials*, 2005, 27, 1050-1054.

[11] Priolo, F., Presti, C. D., Franzò, G., Irrera, A., Crupi, I., Iacona, F., Di Stefano, G., Piana, A., Sanfilippo, D., Fallica, P. G., *Physical Review*, 2006, B73, 113302.

[12] Depas, M., Vermeire, B., Mertens, P. W., Van Meirhaeghe, R. L., Heyns, M. M., *Solid-State Electronics*, 1995, 38, 1465-1471.

[13] Castagna, M. E., Coffa, S., Caristia, L., Messina, A., Bongiorno, C., Quantum Dot Materials and Devices for Light Emission in Silicon, ESSDERC, 2002, 439-442.

[14] Walters, R. J., Bourianoff, G. I., Atwater, H. A., *Nature Materials*, 2005, 4, 143-146.

[15] Barrios, C. A., Lipson, M., *Optics Express*, 2005, 13, 10092-10101.

[16] Almeida, V. R., Xu, Q., Barrios, C. A., Lipson, M., *Optics Letters*, 2004, 29, 1209-1211.

[17] Kalkman, J., Polman, A., Kippenberg, T. J., Vahala, K. J., Brongersma, L., *Nuclear Instrum. and Methods in Physics Research*, 2006, B 242, 182–186.

In: Modeling of Photonic Devices
Editor: Vittorio M. N. Passaro

ISBN 978-1-60456-980-3

Chapter 11

1D PHOTONIC BANDGAP AS DELAY LINE

Francesco De Leonardis*[*,1] and Vittorio M. N. Passaro[2]
[1]Photonics Research Group, Dipartimento di Ingegneria dell'Ambiente e per lo Sviluppo Sostenibile, Politecnico di Bari, viale del Turismo n. 8, 74100 Taranto, Italy
[2]Photonics Research Group, Dipartimento di Elettrotecnica ed Elettronica, Politecnico di Bari, via Edoardo Orabona n. 4, 70125 Bari, Italy

ABSTRACT

In this chapter, some aspects of the modeling of mono-dimensional (1D) photonic bandgap structures are considered. Floquet-Bloch theory is developed and applied to guided-wave multi quantum well structures. Particular applications for delay lines are presented, assuming an optimized multi quantum well waveguide.

Keywords: Photonic Bandgap Crystals, gratings, Multi Quantum Wells, Floquet-Bloch Theory, delay lines, modeling

1. INTRODUCTION

In the last years, a strong research effort has been devoted to the analysis and fabrication of photonic bandgap (PBG) crystals [1]. In fact, two-dimensional (2D) PBG structures offer the possibility to extend the functionality of integrated optics towards much more compact miniaturized devices, not influenced by bending radiation losses. In particular, the applications of PBG structures to Integrated Optics are of special interest for the design of a large number of photonic devices, including very narrow band filters, wavelength-division multiplexers, reduced-size guided wave devices and circuits, lasers with reduced factor of spontaneous emission [2], fiber lasers with Bragg gratings [3], microwave and millimeter wave devices [4], antennas [5], special optical fibers [6] and multi quantum well-based group

* E-mail address: passaro@deemail.poliba.it

delay lines [7]. Very recent applications of 1D and 2D PBGs also include THz designs, negative refraction structures [8-20] and sensors [21-23].

In general, the analysis of an integrated optical PBG configuration can follow a number of different approaches. The first of them involves the study of the grating as a mode coupler and a number of well-known methods have been derived such as the coupled-mode theory (CMT) [24], the impedance transfer combined with the Floquet's theorem [25], the guided Bloch mode approach [26] and the Green function approach combined with Floquet's theorem [27]. However, all these methods give only a very approximated evaluation of the losses as induced by the presence of the grating. The second approach looks for the spectral conditions under which the grating behaves as a perfect reflector, i.e. as a PBG component. Based on this approach, other methods have been proposed such as the mode-matching method [28], the bi-directional mode expansion and propagation (BEP) method [29-30] and the transfer matrix method (TMM) [31]. The methods derived by the perfect reflector conditions are either analytical or numerical, and make a number of assumptions in order to find the solutions of the electromagnetic problem. In particular, all the state-of-the-art numerical methods are based on a transfer matrix strategy which, using the local mode solutions, introduces an approximation on the evaluation of the spectral properties which increases with the grating length. Unfortunately, these methods cannot give a clear physical insight of the grating spectral behaviour in any operating regime. Finally, the third approach is essentially a simulation task, since it studies the propagation of the beam along the periodic structure either in the time or space domain. This approach includes some powerful simulation methods, such as the beam propagation method (BPM) [32], the scattering matrix method (SMM) [33-34], the Finite Difference Time Domain (FDTD) method [35-37] and the Finite Element Method (FEM) [38], which are based on the discretization of Maxwell's differential equations. In particular, FDTD can give very good results in any grating operating region, but is very heavy from a computational point of view and is not very useful for design purposes, since it loses the physical meaning of the involved phenomena. Some of these methods, including FDTD and plane waves and boundary-condition matching method [39], have been also used to numerically study both 1D- and 2D-photonic crystals including defects for the design of very narrow-band filters and other selective devices. Finally, the grating behaviour has also been described by using a number of Fourier modal techniques [40-41], ray optics [42], genetic algorithm [43], Dirichlet-to-Neumann maps [44], nonlinear Schroedinger equation [45]. Advances include also the grating investigation for TM polarization [46], finite incident beam and grating length in sub-wavelength resonant grating [47] and normal incidence [48]. In particular, the Floquet-Bloch theory (FBT) has been applied to the analysis of guided-wave output couplers [49], grating-assisted planar directional couplers [50], finite length fully and partially etched guided-wave 1D PBG [51], fiber Bragg gratings [52-53] and 2D PBG for filtering [54-55], showing large versatility, high accuracy and clear direct explanation of the grating physical effects for design purposes.

In this chapter, FBT formalism has been applied to model guided-wave planar photonic bandgap configurations including a multi quantum well (MQW) inside the periodic structure. A clear physical insight of the effects occurring in different operating regions can be so obtained, which is used to derive design criteria for a number of devices, in comparison with other methods presented in literature for analyzing periodic structures including MQWs, as the equivalent transmission line [56]. In particular, the attention is here focused on guided-wave MQW PBGs optimized as group delay lines. In section 2 the Floquet-Bloch theoretical analysis is applied to a finite length periodic structure. Section 3 summarizes the numerical results for a number of cases of MQW wave-guiding structures, including a brief comparison with other methods. In this chapter the MQW-based delay lines are particularly analyzed.

2. ANALYSIS OF 1D MQW PHOTONIC BANDGAP

The schematic diagram of the grating structure is shown in Figure 1. The periodic perturbation, having rectangular profile with width w, period Λ (duty cycle w/Λ), length L and arbitrary number of layers N_r, includes a multi quantum well (MQW) with an arbitrary number of wells (N_w) and barriers (N_b). Each layer is assumed isotropic, homogeneous and lossy, due to both propagation loss and exciton absorption. The structure is assumed infinite along the transverse direction y ($L_y >> \lambda_g$, with λ_g guided optical wavelength). Hereinafter, TE polarization is assumed, widely used in both active (lasers) and passive (delay lines) devices. However, it induces the exciton absorption peak with the heavy holes [56] and, then, its associated physical effects are more influenced by the optical losses.

A guided mode Ψ_i, incident on the grating region from the unperturbed planar waveguide, "sees" the interface at $z = -L/2$ (Figure 1), being n_{eff} its effective index. If the structure where the wave is travelling allows high confinement (as in III/V semiconductor MQW guiding structures), the scattering effect as induced by the grating at the first interface is moderate and the out-of-plane losses can be usually neglected. Therefore, the incident wave produces only two guided waves at the output of the grating region, one reflected and the other transmitted. Inside the grating region, the electromagnetic field is described as a superposition of two rigorous solutions (forward and backward modes) of lossy infinite structure, i.e. the guided leaky modes found according to the theory presented in literature [50-51].

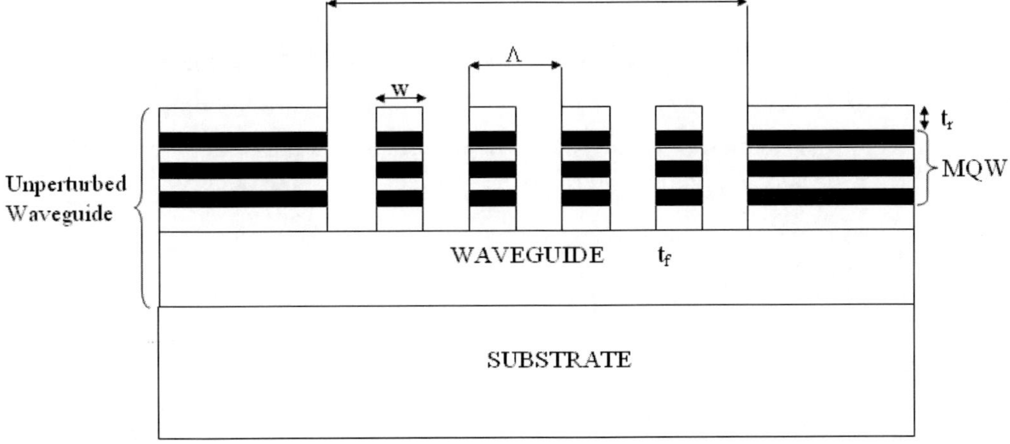

Figure 1. Schematic diagram of 1D MQW photonic bandgap.

These two solutions both satisfy the Maxwell's equations and are described as a superposition of space harmonics through the Floquet's theorem, each of them verifying the continuity conditions at each layer boundary. It is important to note that these solutions are obtained without any approximation for each guided-wave grating having high contrast ratio, both partially or fully etched. Moreover, the field description as a superposition of two leaky modes holds if $L = N_p\Lambda > \lambda_g$, where N_p is the number of periods. Therefore, since the matching condition at the first Brillouin zone requires $\lambda_g = 2\Lambda$, the minimum number of periods is very small ($N_p > 2$). Inside the grating, the modal solution travelling in positive z direction (forward mode) can be written as:

$$\Psi^{+}(x,z) = e^{-jk_{z0}z} \sum_{n=-N_{\max}}^{N_{\max}} f_n(x)e^{-jnKz} \tag{1}$$

where $f_n(x)$ represents the n-th Floquet space harmonic coefficient, $k_{z0} = \beta + j\alpha$ is the zero-order harmonic propagation constant, α is the leakage factor, $N_{\max}$ is the number of positive or negative space harmonics retained in the analysis, and K is the grating wave-vector. Since the dielectric constant satisfies the Hermitian property, i.e. $\varepsilon(x,z) = \varepsilon'(x,-z)$, the solution for the backward mode is simply given by:

$$\Psi^{-}(x,z) = e^{jk_{z0}z} \sum_{n=-N_{\max}}^{N_{\max}} f_n(x)e^{jnKz} \tag{2}$$

where the space harmonic coefficients are equal to those in Eq. (1). Therefore, the solution inside the grating can be written as:

$$\Psi_T = a^*\Psi^{+} + b^*\Psi^{-} \tag{3}$$

where the coefficients a^*, b^* depend on the boundary conditions at the input and output interfaces, $z = -L/2$ and $z = L/2$, respectively. If β_u and Ψ_u are the propagation constant and the transverse distribution of guided-wave unperturbed structure mode, respectively, then the incident (i), reflected (r) and transmitted (t) waves can be written as:

$$\Psi_i(x,z) = \Psi_u(x)e^{j\delta_u z}e^{-jKz} \tag{4}$$

$$\Psi_r(x,z) = \rho\Psi_u(x)e^{-j\delta_u z}e^{jKz} \tag{5}$$

$$\Psi_t(x,z) = \tau\Psi_u(x)e^{j\delta_u z}e^{-jKz} \tag{6}$$

where $\delta_u = K - \beta_u$, ρ is the amplitude reflection coefficient and τ is the amplitude transmission coefficient. By combining Eqs. (1), (2) and (3) with (4), (5) and (6), we obtain:

$$\Psi_T(x,z)=\left[a^*\cdot e^{(-jk_{z0}+jK)z}f_0(x)+a^*\cdot e^{-jk_{z0}z}\sum_{n=1}^{N_{\max}}f_n(x)e^{-j(n-1)Kz}+\right.$$
$$\left.+b^*\cdot e^{jk_{z0}z}\sum_{n=-N_{\max}}^{-1}f_n(x)e^{-j(|n|-1)Kz}\right]\cdot e^{-jKz}+$$
$$\left[b^*\cdot e^{(jk_{z0}-jK)z}f_0(x)+b^*\cdot e^{jk_{z0}z}\sum_{n=1}^{N_{\max}}f_n(x)e^{j(n-1)Kz}+\right.$$
$$\left.+a^*\cdot e^{-jk_{z0}z}\sum_{n=-N_{\max}}^{-1}f_n(x)e^{j(|n|-1)Kz}\right]\cdot e^{jKz} \tag{7}$$

Since optical power is not incident on the interface $z = L/2$ from the unperturbed waveguide, the coefficient of e^{jKz} operator must be zero, and then we derive:

$$\frac{b^*}{a^*}=-e^{-k_{z0}L}\zeta \tag{8}$$

where

$$\zeta=\frac{\sum_{n=-N_{\max}}^{-1}e^{j(|n|-1)K\frac{L}{2}}\left\langle\Psi_u(x),f_n(x)\right\rangle}{e^{-jK\frac{L}{2}}\left\langle\Psi_u(x),f_0(x)\right\rangle+\sum_{n=1}^{N_{\max}}e^{j(n-1)K\frac{L}{2}}\left\langle\Psi_u(x),f_n(x)\right\rangle} \tag{9}$$

In Eq. (9), the symbol $\langle\ \rangle$ designates the overlapping integral between the field distributions. In order to find the reflection coefficient, we impose the field continuity conditions at the interface $z = -L/2$. By combining Eqs. (4), (5) and (8) we have:

$$\rho=e^{-j\delta_u L}\frac{-\zeta\cdot e^{-2jk_{z0}L}\left[e^{jK\frac{L}{2}}\left\langle\Psi_u(x),f_0(x)\right\rangle+\sum_{n=1}^{N_{\max}}e^{-j(n-1)K\frac{L}{2}}\left\langle\Psi_u(x),f_n(x)\right\rangle\right]+}{\left[1-\zeta^2\cdot e^{-2jk_{z0}L}\right]\cdot\left[e^{-jK\frac{L}{2}}\left\langle\Psi_u(x),f_0(x)\right\rangle+\right.}$$
$$\frac{+\sum_{n=-N}^{-1}e^{-j(|n|-1)K\frac{L}{2}}\left\langle\Psi_u(x),f_n(x)\right\rangle}{\left.+\sum_{n=1}^{N_{\max}}e^{j(n-1)K\frac{L}{2}}\left\langle\Psi_u(x),f_n(x)\right\rangle\right]} \tag{10}$$

The transmission coefficient is found as:

$$\tau=\rho\frac{e^{-jk_{z0}\frac{L}{2}}\cdot e^{jK\frac{L}{2}}\langle\Psi_u(x),f_0(x)\rangle-\zeta\cdot e^{-jk_{z0}\frac{L}{2}}\cdot\sum_{n=-N_{\max}}^{-1}e^{-j(|n|-1)K\frac{L}{2}}\cdot\langle\Psi_u(x),f_n(x)\rangle+}{-\zeta\cdot e^{-3jk_{z0}\frac{L}{2}}\cdot e^{jK\frac{L}{2}}\langle\Psi_u(x),f_0(x)\rangle-\zeta\cdot e^{-3jk_{z0}\frac{L}{2}}\cdot\sum_{n=1}^{N_{\max}}e^{-j(n-1)K\frac{L}{2}}\cdot\langle\Psi_u(x),f_n(x)\rangle+}$$

$$\frac{+e^{-jk_{z0}\frac{L}{2}}\cdot\sum_{n=1}^{N_{\max}}e^{-j(n-1)K\frac{L}{2}}\cdot\langle\Psi_u(x),f_n(x)\rangle}{+e^{jk_{z0}\frac{L}{2}}\cdot\sum_{n=-N_{\max}}^{-1}e^{-j(|n|-1)K\frac{L}{2}}\cdot\langle\Psi_u(x),f_n(x)\rangle}\qquad(11)$$

Finally, the normalized modal loss of the whole structure has been calculated as $P_{loss}=1-|\rho|^2-|\tau|^2$.

3. Numerical Results

3.1. Comparison with Other Methods

First of all, we have applied our method to the structure described in Ref. [25] and results have been compared with predictions obtained by coupled mode theory (CMT) [24] and bidirectional eigenvalue propagation (BEP) method combined with the Floquet's theorem [25]. Then, the structure of Figure 1 has simply a single homogeneous guiding layer with n_f = 1.53 and t_f = 1.9 μm, substrate with n_s=1.52, air overlay $n_o=1$, grating layer number $N_r=1$, squared profile (w/Λ = 0.5), groove depth t_{r_1} = 0.5 μm and refractive index $n_{r_1}=1.53$ (grating index contrast Δn_g = 0.53), period Λ = 0.213 μm and variable length. The unperturbed waveguide is 2.4 μm thick with a refractive index of $n_f=1.53$. The spectrum calculated around 1.55 μm, follows the typical $sinc^2(\lambda)$-like law. Since CMT introduces the usual perturbation approximations and BEP an approximation given by the transfer matrix approach, which increases with the number of the grating periods, a significant discrepancy between our predictions and those obtained by CMT and BEP has been found. In particular, we have calculated a reflectivity value of 90% with 5000 periods at the resonance wavelength of 649.77 nm, while CMT estimates for the same reflectivity 1000 periods at 650.3 nm and BEP 10000 periods at 649.94 nm. High accuracy has been obtained by using only 5 space harmonics ($N_{\max}=2$).

3.2. MQW Photonic Bandgap Structures

We refer to an $Al_{0.3}Ga_{0.7}As/GaAs$ MQW nanostructure (named MQW1), with three wells and two barriers confined inside the multilayer grating ($N_r=7$) and surrounded by two cladding layers, both having refractive index 3.43 (Al title y = 0.3) and thickness $t_{r1} = t_{r7} = 25$ nm. The wells (w) have the same refractive index 3.63 (Al title x = 0) and same thickness $t_{r2} = t_{r4} = t_{r6} = 10$ nm, the barriers (b) have a refractive index 3.43 each and same thicknesses, $t_{r3} = t_{r5} = 10$ nm (overall groove depth $t_{r_{tot}} = 0.1$ μm, contrast ratio $\Delta n_g \approx 2.5$), the guiding layer index is 3.63 (GaAs, Al title $x_f = 0$) with thickness $t_f = 0.2$ μm, the Al_2O_3 substrate index is 1.6 and the grating period is 0.125 μm (Bragg wavelength at 0.85 μm).

First of all, the influence of exciton absorption in the guided-wave MQWs has been considered. According to Ref. [56], we have assumed that the absorption inside the MQW PBG is due to: 1) continuum of transitions between the free particle states, and 2) exciton transitions. The absorption spectrum has been calculated, considering only the first subband transition, assuming that the heavy (hh) and light holes (lh) are uncoupled [56] and applying then the superposition principle. Only the 1S exciton state has been considered, since it has a much higher oscillator strength than the other states. Lorentzian line shapes have been used to model the spectrum homogeneous broadening in both the continuum and the exciton absorption, as caused by phonon interaction and tunneling through the MQW barriers [56]. Therefore, the leakage factor α_{exc} due to the exciton absorption has been evaluated and included in lossy wells forming the MQW structure as the imaginary part of their dielectric constant, i.e. $imag(n) = \alpha_{exc}\lambda / 4\pi$. Moreover, the relevant changes of the refractive index real part have been considered by the Kramers-Kronig relationships [57]. Finally, the effect of chromatic dispersion in each layer of the photonic bandgap has been also taken into account in the model, its contribution being well larger than the Kramers-Kronig one (i.e. < 1.1 %). The Sellmeier relationships for the AlGasAs/GaAs refractive index dispersion have been taken into account [58].

The two peaks revealed in the absorption spectra for TE polarization are due to excitons formed between electrons and heavy holes (hh) and electrons and light holes (lh). Transition energies of hh and lh excitons are different due to different effective mass of lh's and hh's. In the absorption spectra, the first peak to higher wavelength corrisponds to the strongest hh exciton and the second to lh exciton. Comparisons of our calculations with those presented in Ref. [57] in case of zero electric field have shown a high agreement, the difference between the absorption peaks being lower than 4%.

3.3. MQW Leakage Factor and Influence of Duty Cycle

The method based on Floquet-Bloch formalism presents the important advantage, over other numerical methods, to allow a clear physical insight of the physical effects occurring in the photonic bandgap structures. Figure 2 shows the leakage factor spectrum of MQW1 structure for different duty cycles, 20%, 50%, and 80%, when the exciton absorption and chromatic dispersion are simultaneously taken into account. The curve for 50% duty cycle obtained without the effects of exciton absorption and chromatic dispersion is also shown for comparison. The PBG at Bragg wavelength (PBG1) is obtained around 0.85 μm for any duty cycle, shifting a little towards smaller (larger) wavelengths for smaller (larger) duty cycles (20% and 80%, respectively). The first consideration to be derived is that this structure, formed by high index III/V semiconductor layers, allows a strong field confinement within some wavelength regions, because the modal solutions have effective indices much larger than the substrate index.

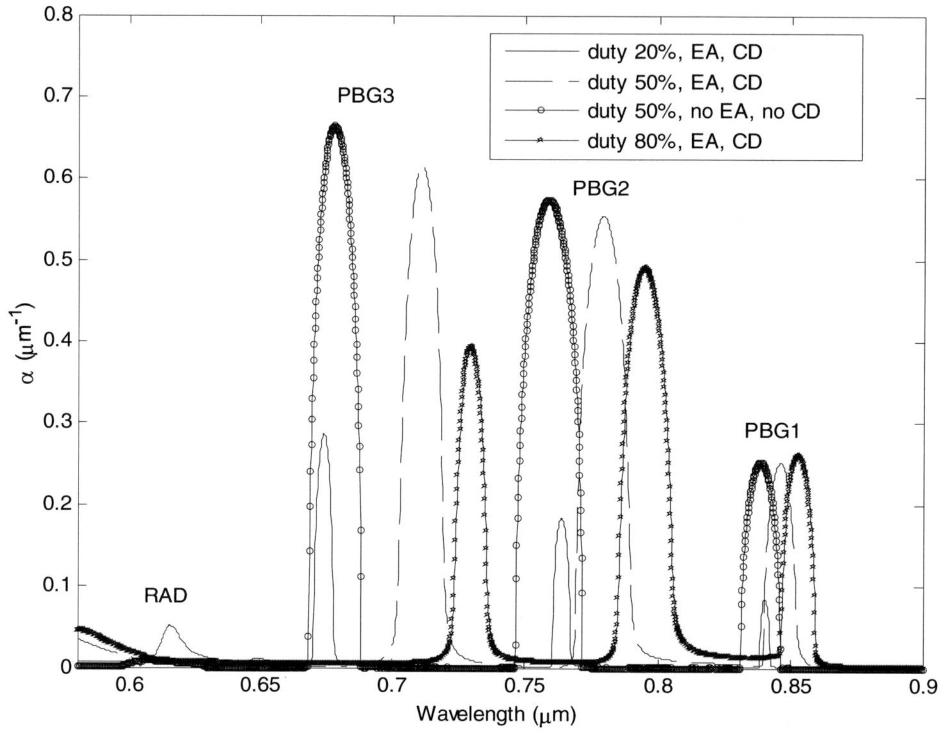

Figure 2. Leakage factor spectra for various duty cycles, with or without absorption and dispersion.

Under this circumstance, the electromagnetic field is located mainly inside the grating, generating higher-order photonic bandgaps (PBG2 and PBG3), whose solutions are orthogonal to each other (i.e. modes TE_0, TE_1, TE_2,...). The higher-order PBGs have been also revealed in bulk [59] and guided-wave structures [60]. In these cases, the space harmonics supporting the field can exist only in the grating region, since their propagation constants are much larger than that supported by the planar waveguide. When the optical wavelength falls in these regions, the field solutions are evanescent again ($\alpha > 0$), as in PBG at Bragg wavelength (PBG1), and are so forbidden (no extended states) in the infinite grating, as occurs for the surface modes in periodic layered media [61]. However, space harmonic radiation does not occur here.

Figure 2 also shows what happens when the duty cycle is changed, i.e. 20, 50 and 80%. In fact, where the duty cycle increases, the amount of high index material inside the grating increases, so inducing a stronger field confinement of spatial harmonics and shifting the PBGs regions towards larger wavelengths. For smaller wavelengths, a radiative (named RAD) region is also present, where some of the negative space harmonics radiate power towards air or substrate. The regions in Figure 2 are all well distinguished, although the MQW lossy structure induces a leakage factor $\alpha > 0$ even in the other neighbouring regions, due to exciton absorption loss. Seven space harmonics ($N_{max} = 3$) have been retained in these calculations and compared with results obtained by using 13 harmonics, the difference being less than 0.5%. High numerical stability of our method has been found for any wavelength.

3.4. Influence of Lower Waveguide Thickness and Index

Figure 3 shows the transmissivity ($T=|\tau|^2$) spectrum for different thicknesses and Al titles of the lower planar waveguide, $t_f = 0.2$ μm and $x_f = 0$ (solid line with squares), $t_f = 0.2$ μm and $x_f = 0.3$ (dotted line), $t_f = 0.055$ μm and $x_f = 0$ (solid line with crosses), $t_f = 0.055$ μm and $x_f = 0.3$ (dotted line), each for N_p=30. The normalized curve of exciton absorption is also shown (dashed line). The first two curves clearly show the presence of three PBGs, that at lowest energy (around the Bragg wavelength of 0.85 μm) and two higher-order ones. The PBG regions are shifted towards smaller wavelengths as large as the aluminium title, since the optical confinement in the lower waveguide (grating) is decreasing (increasing) and the stop-band effect of the grating is amplified. If the exciton peaks fall inside the PBG, as in Figure 3, the resultant effect is an additional modal loss of the waves which, in the infinite structure, should be forbidden. Therefore, the transmittivity curve shows reduced values but same shape.

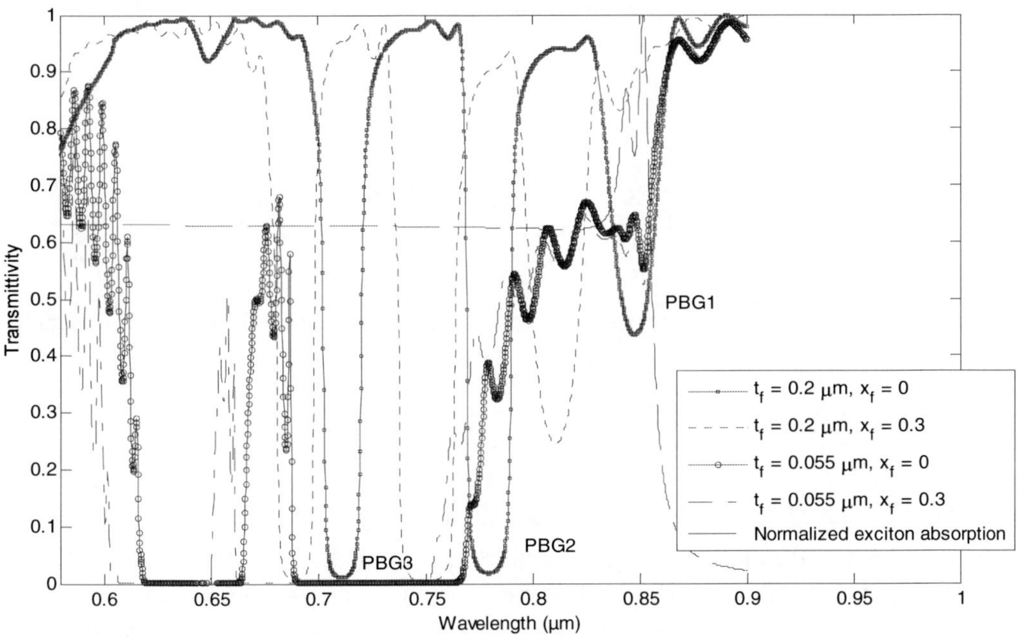

Figure 3. Transmittivity spectra for various waveguide parameters. Normalized exciton absorption is also shown.

In Figure 3 the strong enlargement of the PBG regions and their shift towards smaller wavelengths when the waveguide thickness decreases up to 0.055 μm is also shown. The comparison between the cases obtained for $x_f = 0$ and $x_f = 0.3$ can be observed, resulting only in an additional shift of the PBG region position. In these cases, the exciton peaks fall outside the PBG (on the right), changing significantly the shape of the transmittivity curves and reducing the transmission resonances. The curve relevant to $t_f = 0.055$ μm and $x_f = 0.3$ corresponds to a MQW with a quasi fully etched cladding layer and without any lower waveguide layer.

Figure 4 shows more clearly the influence of the exciton peaks over transmittivity and reflectivity curves when the absorption falls in the Bragg PBG, for $N_p = 30$, $t_f = 0.2$ μm and $x_f = 0$. A reduction and distortion of both curves can be noted, due to absorption.

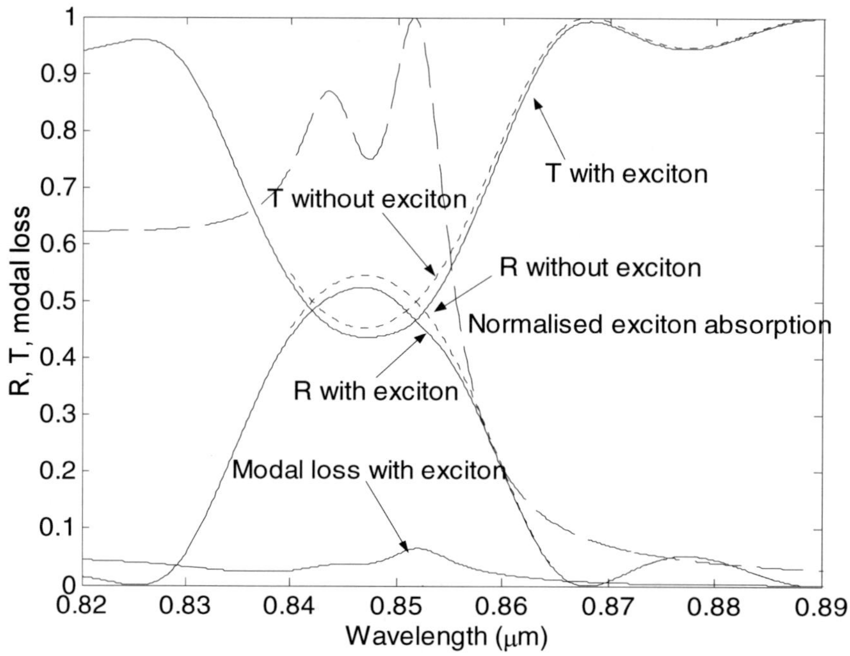

Figure 4. Reflectivity and transmittivity spectra around Bragg wavelength (PBG1), with and without excitons.

3.5. MQW Photonic Bandgap as Group Delay Line

We have evaluated the transmission phase, i.e. $\varphi = \arg(\tau)$. The results of calculations are sketched in Figure 5 for the same cases as before. It is clear that the MQW structure is more influenced by changes of the waveguide thickness than the material composition (Al title).

Moreover, it is to be noted the stronger slope of the curves inside the higher order PBG regions with respect to PBG at Bragg resonance. This is a very important property of these structures. In fact, we have calculated the group delay time suffered by an optical wave travelling along the MQW grating, as:

$$delay = \frac{L}{v_g} - \frac{L}{v_{gu}} = \frac{d}{d\omega}(\varphi + \beta_u L) - \frac{d}{d\omega}(\beta_u L) = \frac{d\varphi}{d\omega} \tag{12}$$

where v_g and v_{gu} are the group velocities inside and outside the grating, respectively. Results of group delay as a function of wavelength are shown in Figure 6 for $N_p = 30$ periods. It can be noted that the delay curves show a number of peaks relevant to the PBG region edges at the transmission resonances, where the phase curve slope is stronger. Moreover, the largest delays are obtained at the left side of PBG regions, in particular by reducing the lower waveguide thickness from 0.2 to 0.055 μm. Changes of material composition have less influence on the delay, as it occurs in the phase curves. Group delays approaching 600 fs can be then obtained, well larger than those presented in literature [62] using layered media. Of course, the results obtained by Eq. (12) represent the minimum values, the delay with respect to the free-space wave propagation being larger than about 50 fs. The physical effect of the negative time delay, which corresponds to wavelength regions with minimal transmission, is also shown in Figure 6. It has been well explained in Ref. [63], as due to the superluminal group velocity.

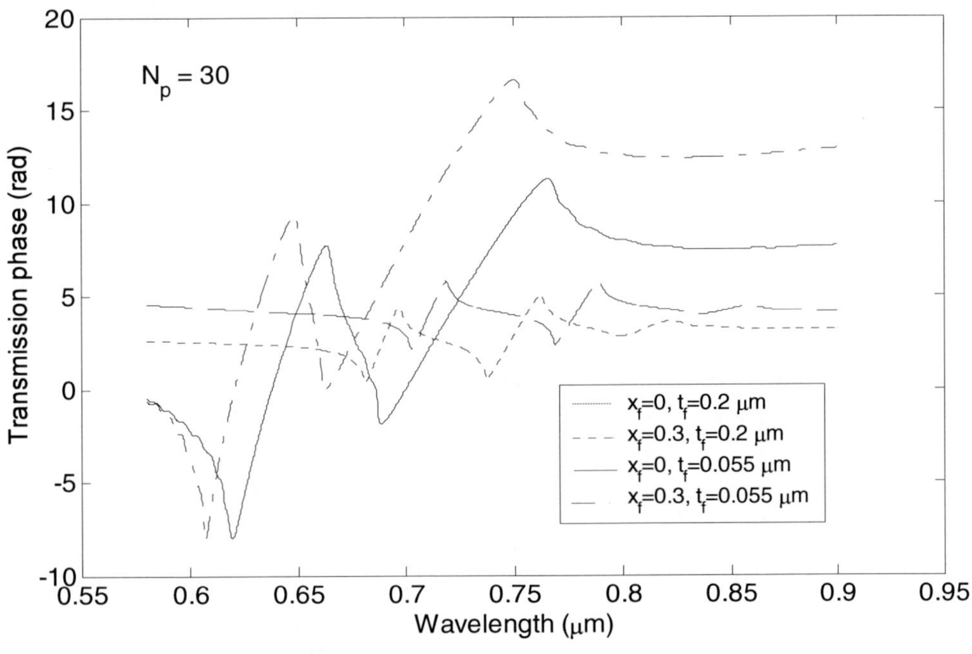

Figure 5. Transmission phase spectra for various waveguide parameters.

In Figure 7 the influence of hh exciton absorption peak is clearly shown when a greater number of periods is considered, $N_p = 150$, $t_f = 0.2$ μm and $x_f = 0$. In this case, the influence arises on the right edge of the first PBG region. This influence is given by a slope reduction in the transmission phase and, then, a significant decrease of the resultant first maximum of the delay time, from 726.7 to 333.7 fs. The other maxima of delay are also decreased, but the absorption effect decreases as large as the wavelength. Therefore, the delay reduction due to the absorption around PBG1 is one of the main reason of the larger delays, which can be revealed at higher-order PBGs [64].

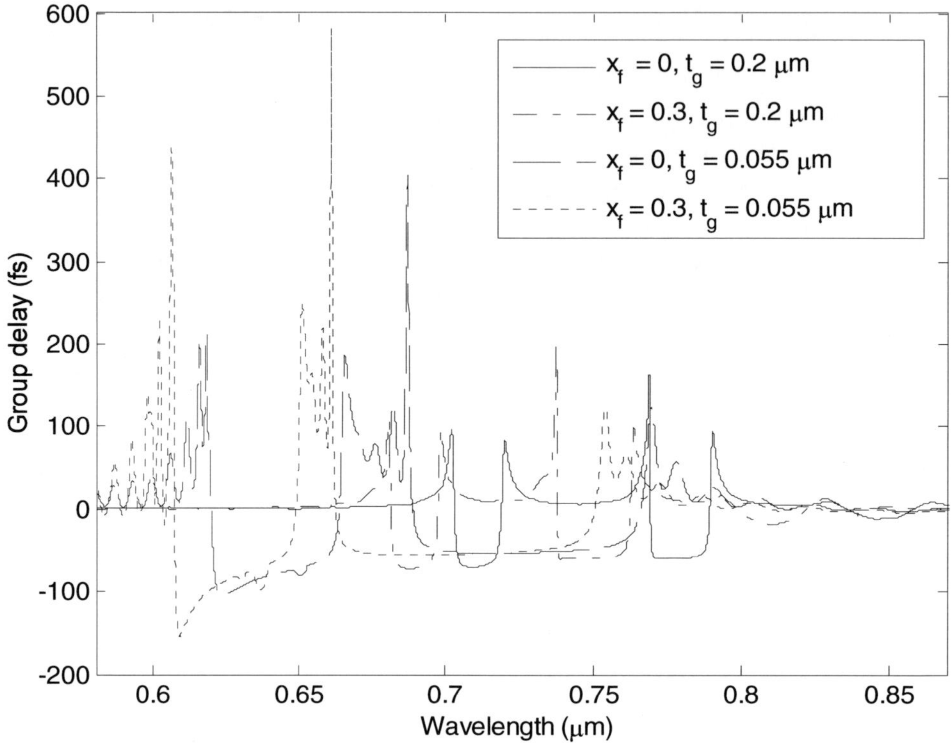

Figure 6. Group delay spectra for various waveguide parameters.

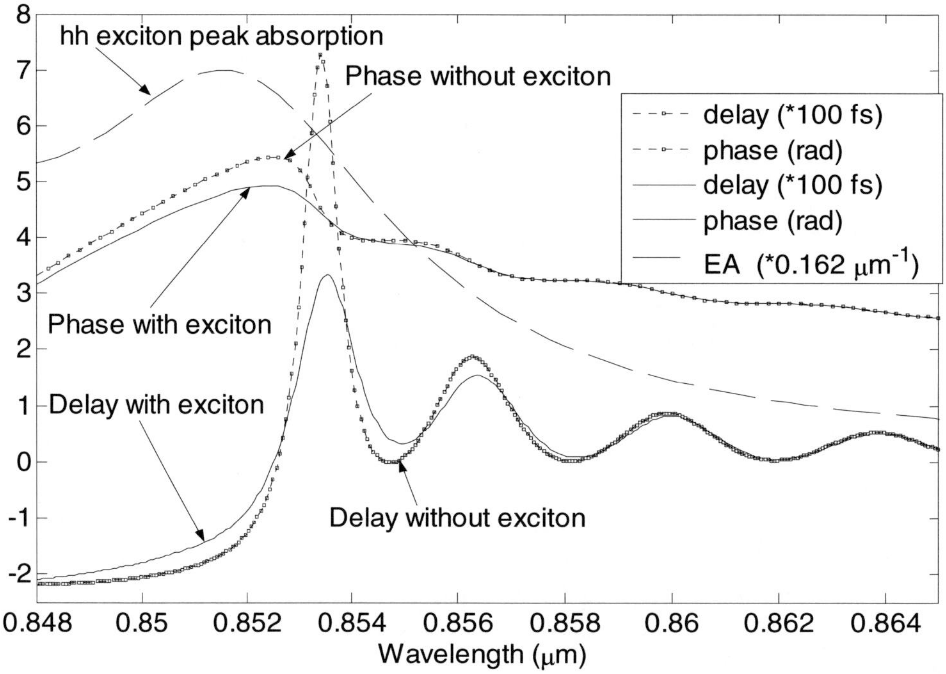

Figure 7. Delay and phase spectra, in presence or absence of exciton absorption

3.6. Influence of Grating Length

The peaks of group delay times have been evaluated as a function of the grating number of periods. Results taken at the PBG1 right side are shown in Figure 8 for different waveguide thicknesses and $x_f = 0.3$. The dependence of delay on the length follows a linear relation only after a minimal length, when the grating is long enough to behave as much as possible like a stop-band filter with constant bandwidth. Then, for larger lengths, the transmission phase grows linearly with length as $\varphi = \omega L / v^*$, where v^* has the dimension of a velocity.

From Figure 8 we can note that this minimal length is strongly reduced as small as the waveguide thickness, since the field confinement inside the grating strongly increases. Therefore, only the curves at 0.095 and 0.055 μm exhibit a linear behaviour after 100 and 60 periods, respectively. Since the higher-order PBGs allow a stronger field confinement than the Bragg PBG, their stopband effect is stronger and the minimal length is significantly reduced. This circumstance explains why the delays associated to higher-order PBGs are generally greater than those referred to zero-order one. For grating lengths lower than this minimal value, the delay monotonically grows with the number of periods without following a linear law or a constant ratio with the other delays, its behavior depending on the mutual influence of each PBG on the edges of the neighbouring PBG transmission curve (changing the positions and the values of its transmission resonances), and on the position of exciton absorption spectrum with respect to the same PBG.

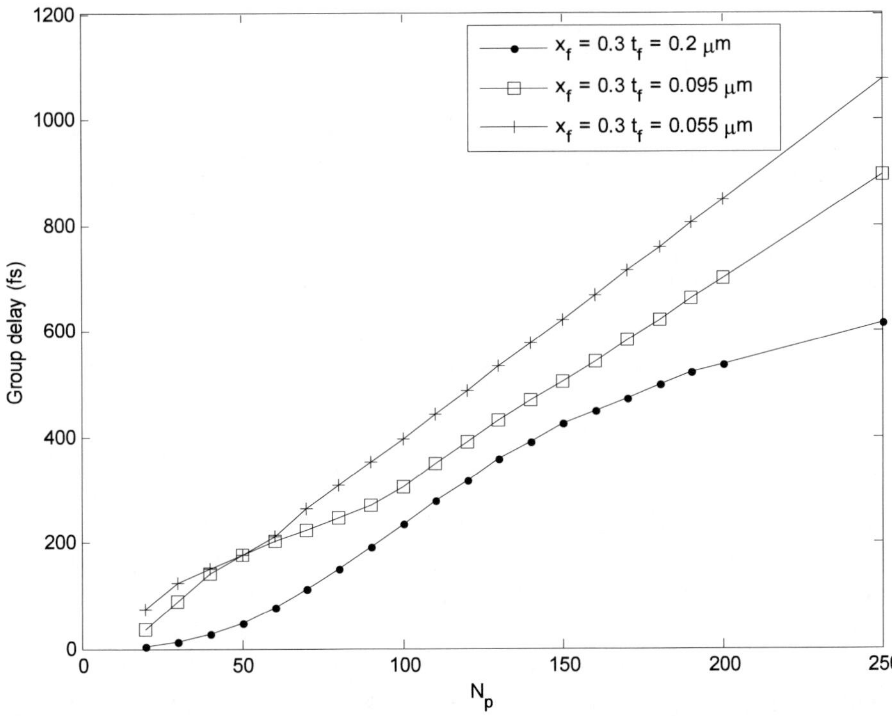

Figure 8. Group delay versus number of grating periods for various waveguide parameters.

The delays at the edges of each PBG have been calculated as a function of the grating length with changing the structure parameters, waveguide layer thickness and material. Figure 9 shows the delays associated to both PBGs of the structure, having $t_f = 0.055$ μm and $x_f = 0.3$. They represent the largest values which it is possible to find in these optimized structures. Results as large as 3.7 ps can be then obtained with $N_p = 200$. The wavelength position of these delay peaks can be controlled by changing the Bragg wavelength by tunable approaches, such as nonlinear effects or carrier injection.

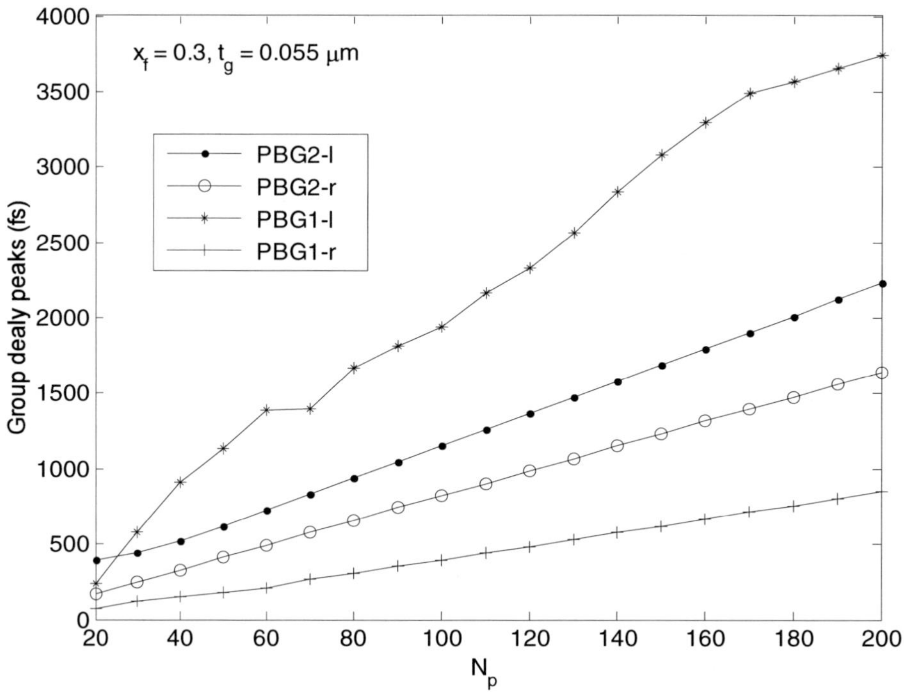

Figure 9. Group delay peaks versus number of periods at different PBG edges.

CONCLUSION

In this chapter, the design criteria of 1D guided-wave PBG structures incorporating a MQW have been presented. Most attention is focused on applications as group delay lines. The Floquet-Bloch formalism has been applied to evaluate both the leakage factor and the reflectivity and transmittivity curves of these lossy structures, finding the solution of the electromagnetic problem inside high contrast ratio gratings without any assumption. This formalism allows one to have a clear physical insight of the grating behaviour. All the basic physical effects generated by the interaction of the MQW periodic structure with the leaky modes have been so derived, including the higher-order PBG regions, the radiative region, the influence of the exciton absorption loss. The influence of lower waveguide thickness, material composition and grating length on the transmittivity characteristics, transmission phases and group delay times have been demonstrated, showing that the largest values of group delay can be obtained at the PBG left side for very small thicknesses. On the contrary, the influence of material appears to be very moderate. This approach is well suitable for the analysis of MQW photonic bandgap structures including defects, both for modeling active (e.g. lasers) and passive (e.g. filters, tuneable delay lines) devices, and it is also well able to investigate 2D guided-wave periodic structures.

REFERENCES

[1] see, for example, the Special Issue on "Electromagnetic Crystal Structures, Design, Synthesis and Applications", *J. Lightwave Technology*, 1999, 17, n. 11.

[2] Yablonovitch, E., *J. Optical Society of America B*, 1993, 10, 283-295.

[3] Sondergaard, T., *J. Lightwave Technology*, 2000, 18, 589-596.

[4] Yang, F.-R., Coccioli, R., Qian, Y., Itoh, T., *IEICE Trans. on Electronics*, 2000, E83-C, 687-696.

[5] Cheng, D., Biswas, R., Ozbay, E., McCalmont, S., Tuttle, G., Ho, K.-M., *Applied Physics Letters*, 1995, 67, 3399-3401.

[6] Knight, J. C., Birks, T. A., Russel, P. S. J., Atkin, D. M., *Optics Letters*, 1996, 21, 1547-1549.

[7] Nelson, T. R., Jr., Loehr, J. P., Xie, Q., Ehret, J. E., Van Nostrand, J. E., Gamble, L., Jones, D. K., Cole, S. T., Trimm, R. A., Diffey, W. M., Fork, R. L., Keys, A. S., *Proceedings SPIE*, 1999, 3714, 12-23.

[8] Lægsgaard, J., Mortensen, N. A., Riishede, J., Bjarklev, A., *J. Optical Society of America B*, 2003, 20, 2046-2051.

[9] Bahriz, M., Moreau, V., Colombelli, R., Crisafulli, O., Painter, O., *Optics Express*, 2007, 15, 5948-5965.

[10] Ao, X., He, S., *Optics Express*, 2004, 12, 978-983.

[11] Zhao, Y., Grischkowsky, D., *Optics Letters*, 2006, 31, 1534-1536.

[12] Awasthi, S. K., Malaviya, U., Ojha, S. P., *J. Optical Society of America B*, 2006, 23, 2566-2571.

[13] Sun, G., Kirk, A. G., *Optics Express*, 2008, 16, 4330-4336.

[14] Tandaechanurat, A., Iwamoto, S., Nomura, M., Kumagai, N., Arakawa, Y., *Optics Express*, 2008, 16, 448-455.

[15] [15] Makhan, M., Ramchurn, S. K., *J. Optical Society of America B*, 2007, 24, 3040-3047.

[16] Canning, J., Skivesen, N., Kristensen, M., Frandsen, L. H., Lavrinenko, A., Martelli, C., Tetu, A., *Optics Express*, 2007, 15, 15603-15614.

[17] Hu, X., Jiang, P., Yang, H., Gong, Q., *Optics Letters*, 2006, 31, 2777-2779.

[18] Wang, X., Young, J., Chen, Z., Weinstein, D., Yang, J., *Optics Express*, 2006, 14, 7362-7367.

[19] Rahachou, A. I., Zozoulenko, I. V., *J. Optical Society of America B*, 2006, 23, 1679-1683.

[20] Lousse, V., Fan, S., *Optics Express*, 2006, 14, 866-878.

[21] Gauvreau, B., Hassani, A., Fassi, F. M., Kabashin, A., Skorobogatiy, M. A., *Optics Express*, 2007, 15, 11413-11426.

[22] Kim, H. K., Digonnet, M. J F., Kino, G. S., *J. Lightwave Technology*, 2006, 24, 3169-3174.
[23] Cubillas, A. M., Hald, J., Petersen, J. C., *Optics Express*, 2008, 16, 3976-3985.
[24] Yariv, A., *IEEE J. Quantum Electronics*, 1973, QE-9, 919-933.
[25] Helfert, S. F., Pregla, R., *J. Lightwave Technology*, 1998, 16, 1694-1702.
[26] Atkin, D. M., Russell, P. St. J., Birks, T. A., Roberts, P. J., *J. Modern Optics*, 1996, 43, 1035-1053.
[27] Hadjicostas, G., Butler, J. K., Evans, G. A., Carlson, N. W., Amantea, R., *IEEE J. Quantum Electronics*, 1990, 26, 893-902.
[28] Hwang, R. B., Peng, S. T., *IEICE Trans. on Electronics*, 2000, E83-C, 705-712.
[29] Sztefka, G., Nolting, H.-P., *IEEE Photon. Technology Letters*, 1993, 5, 554-556.
[30] Ctyroky, J., *J. Optical Society of America A*, 2001, 18, 435-441.
[31] Ctyroky, J., Helfert, S., Pregla, R., *Optical and Quantum Electronics*, 1998, 30, 343-358.
[32] Scarmozzino, R., Gopinath, A., Pregla, R., Helfert, S., *IEEE J. Selected Topics in Quantum Electronics*, 2000, 6, 150-162.
[33] Yonekura, J., Ikeda, M., Baba, T., *J. Lightwave Technology*, 1999, 17, 1500-1508.
[34] Crocco, L., Cuomo, F., Isernia, T., *J. Optical Society of America A*, 2007, 24, A12-A22.
[35] Ziolkowski, R. W., Tanaka, M., *Optical and Quantum Electronics,* 1999, 31, 843-855.
[36] Kuang, W., Kim, W. J., O'Brien, J. D., *J. Lightwave Technology*, 2007, 25, 2612-2617.
[37] Pinto, D., Obayya, S. S. A., *J. Lightwave Technology*, 2007, 25, 440-447.
[38] Li, M.-J., West, J. A., Koch, K. W., *J. Lightwave Technology*, 2007, 25, 2463-2468.
[39] Smith, D. R., Dalichaouch, R., Kroll, N., Schultz, S., McCall, S. L., Platzman, P. M., *J. Optical Society of America B*, 1993, 10, 314-321.
[40] H. Shigesawa and M. Tsuji, *IEEE Trans. on Microwave Theory and Techniques*, 1989, 37, 3-14.
[41] P.-P. Borsboom and H. J. Frankena, *J. Optical Society of America A*, 1995, 12, 1134-1141.
[42] Li, J., Chiang, K. S., *J. Optical Society of America B*, 2007, 24, 1942-1950.
[43] Goh, J., Fushman, I., Englund, D., Vuckovic, J., *Optics Express*, 2007, 15, 8218-8230.
[44] Yuan, J., Lu, Y.Y., *J. Optical Society of America A*, 2006, 23, 3217-3222.
[45] Laegsgaard, J., *Optics Express*, 2007, 15, 16110-16123.
[46] Li, L., *J. Optical Society of America A*, 1996, 13, 1024-1035.
[47] Bendickson, J. M., Glytsis, E. N., Gaylord, T. K., Brundrett, D. L., *J. Optical Society of America A*, 2001, 18, 1912-1928.
[48] Jacob, D. K., Dunn, S. C., Moharam, M. G., *J. Optical Society of America A*, 2001, 18, 2109-2120.
[49] Passaro, V. M. N., Armenise, M. N., *IEEE J. Quantum Electronics*, 1995, 31, 1691-1697.
[50] Passaro, V. M. N., *J. Lightwave Technology*, 2000, 18, 973-984.
[51] Giorgio, A., Perri, A. G., Armenise, M. N., *J. Lightwave Technology*, 2001, 19, 1598-1613.
[52] Passaro, V. M. N., Diana, R., Armenise, M. N., *J. Optical Society America A*, 2002, 19, 1844-1854.
[53] Passaro, V. M. N., Diana, R., Armenise, M. N., *J. Optical Society America A*, 2002, 19, 1855-1866.

[54] Armenise, M. N., Ciminelli, C., De Leonardis, F., Diana, R., Passaro, V. M. N., Peluso, F., *VI Int. Conf. for Young Researchers on Wave Electronics and Its Applications in Information and Telecommunications Systems*, A2-12 ÷ A2-18, St. Petersburg, Russia, 7-11 Sept. 2003.
[55] Ciminelli, C., Peluso, F., Passaro, V. M. N., Armenise, M. N., *AIP Proc. School on "Microresonators as building blocks for VLSI photonics"*, 709, 439-440, Erice, 18-25 Oct. 2003.
[56] Yan, L., Jiang, M., Tamir, T., Koi, K.-K., *IEEE J. Quantum Electronics*, 1999, 35, 1870-1877.
[57] Stevens, P. J., Whitehead, M., Parry, G., Woodbridge, K., *IEEE J. Quantum Electronics*, 1988, 24, 2007-2016.
[58] Aspnes, D. E., in *Properties of Gallium Arsenide*, II Edition, EMIS Datareviews Series n.2, 157-160, INSPEC-IEE, London, 1990.
[59] Joannopoulos, J. D., Meade, R. D., Winn, J. N., *Photonic Crystals*, Princeton University Press, Princeton (NJ), 1995.
[60] Passaro, V. M. N., De Leonardis, F., Armenise, M. N., *ICO19 Int. Conf. "Optics for Quality of Life"*, Tech. Digest, SPIE Proc. n. 4829, 723-724, Firenze, 25-30 August 2002.
[61] Yariv, A., Yeh, P., *Optical Waves in Crystals*, John Wiley and Sons, New York, 1984.
[62] Gamble, L. J., Diffey, W. M., Cole, S. T., Fork, R. L., Jones, D. K., Nelson, Jr., T. R., Loehr, J. P., Ehret, J. E., *Optics Express*, 1999, 5, 267-272.
[63] Mojahedi, M., Schamiloglu, E., Agi, K., Malloy, K. J., *IEEE J. Quantum Electronics*, 2000, 36, 418-424.
[64] Passaro, V. M. N., De Leonardis, F., Ciminelli, C., Armenise, M. N., *11th European Conf. on Integrated Optics Proc.* (ECIO'03), 1, 33-36, Prague, 2-4 April 2003.

In: Modeling of Photonic Devices
Editor: Vittorio M. N. Passaro

ISBN 978-1-60456-980-3

Chapter 12

GRATING DEVICES FOR OPTICAL SENSING

Renzo Loiacono*[*,1] and Vittorio M. N. Passaro[2]
[1]Silicon Photonics Group, Advanced Technology Institute,
University of Surrey, Guildford, GU27XH, United Kingdom
[2] Photonics Research Group, Dipartimento di Elettrotecnica ed Elettronica,
Politecnico di Bari, via Edoardo Orabona n. 4, 70125 Bari, Italy

ABSTRACT

In this chapter, applications for a Bragg grating based photonic sensor are discussed. The main differences between fiber Bragg gratings and waveguide Bragg gratings are discussed and an overview on the current state of the technology is given for the latter category. Modeling of a refractometric sensor based on shallow Bragg gratings in silicon on insulator sub micrometric rib waveguides is introduced along with numerical results obtained from the model. The possibility of using third order instead of first order grating is discussed and performance compared, thus overcoming fabrication problems due to sub-micrometer scale features. A detection limit of approximately 10-4 RIU has been calculated for a 173 μm long grating. Strategies to further improve this value have been discussed, too. Finally, influence of fabrication tolerances on optimized gratings has been investigated.

Keywords: Integrated Optics, Bragg Gratings, Optical Sensors, Modeling, Silicon on Insulator

1. INTRODUCTION

The world of silicon photonics is witnessing an enormous development in the latest years, as the possibility of employing integrated photonic circuitry in the future would overcome some of the limitations of traditional silicon electronics The success in this field has been

---

[*] E-mail address: passaro@deemail.poliba.it

testified in the latest years by the increasing investments started by IT leading companies and the birth of many start up companies devoted to the commercialization of integrated silicon photonic devices. There are currently three broad areas [1] in which future silicon photonics is developing: the first one is that of the optical interconnects, which is crucial for managing the extremely fast data rates (up to 40GHz) [2]. The second one is the telecommunication market [3], where an integrated, low cost photonic technology will be required for applications such as fiber to the premises (FTTP – USA) and fiber to home (FTTH – Europe). The third major area of application for silicon photonics is the one represented by optical sensors. Currently there is no dominating technology for optical sensing, since there is a wide variety of devices based on different configuration and detection mechanisms. In order to secure a stable place in the global market the key attributes of these kinds of devices would be cost reduction, miniaturization, and mass production potential.

Photonic sensors based on Bragg gratings became increasingly popular in the latest years. A Bragg grating-based sensor can be fabricated in both fiber configuration and waveguide configuration. While the fiber configuration offers more versatility in terms of fabrication, the waveguide configuration ismore suited for integration, offering fast response and high sensitivities.

2. Bragg Grating-Based Sensors

Using a fiber Bragg grating as a sensory element has proved to be a very popular approach in the latest years [4] and a major part of fiber devices rely on this kind of configuration. Applications include measurement of temperature, strain, and pressure since these physical quantities directly affect the properties of the light propagating into the fiber (either by temporarily modifying the grating pitch or the effective index of the fiber mode) and can be quantified through the well-known equation for feedback gratings:

$$\Lambda = \frac{m}{2n_{eff}}\lambda_c \tag{1}$$

where Λ is the grating pitch, m is the grating order, n_{eff} is the effective index of the mode propagating into the fiber and λ_c is the radiation wavelength.

It has been demonstrated that these characteristics can be usefully employed to monitor a variety of physical effects in mechanical structures [5]-[6], specifically fiber sensors can be effectively employed to continuously check the condition of urban structures under stress. Moreover the measurement of acceleration and ultrasonic waves can be converted to strain, further broadening the range of applicability of these devices [7]-[8]. A different area of application for FBG includes biosensing. Detection of various organic solvents [9]-[10], degree of water salinity [11], or other specific chemical and biological reactions in aqueous environments that are associated to a known refractive index change have been investigated through the years. Bragg grating devices can also be used as simple refractometers since a change in the overlay refractive index alters the effective index of the propagating mode, thus inducing a central wavelength shift in the spectrum.

Since the detection mechanism in fiber Bragg gratings sensors depends solely on the wavelength, this prevents the measurement from being affected by power or signal amplitude fluctuations. Moreover, a system based on wavelength interrogation is suitable for quasi-distributed sensing approaches or for integration in WDM systems, since different Bragg gratings (with different resonant wavelengths) can be implemented in the same system [12].

In a WDM system, it is possible to introduce several sensors on the same fiber. Moreover the same device can be used several times by implementing a time division multiplexing scheme (TDM) so that each sensor is interrogated in pre-determined time intervals. This approach is shown in Figure 1(a), where several gratings are connected in a series on the same fiber. Although the devices are meant to operate separately, there is always the possibility that spectral components overlap in the same device resulting in output distortion. This effect is increased due to the fact that the light reflected from gratings constantly travels through the fiber. To overcome these limitations, a branching scheme can be adopted, as shown in Figure 1(b) and 1(c), to separate the different signals. The latter approaches reduce the overall efficiency in the system as a major complexity in implementation is required. Also, fiber grating sensors possess immunity to electromagnetic interferences, a light weight, performance stability over time, and low insertion losses (being embedded directly in the fiber).

Even if the number of integrated optical sensors based on Bragg gratings presented in literature is substantially less if compared to the corresponding fiber counterpart, the use of waveguide configurations offers the possibility of sub-micron integration for future integrated photonics circuits, that is not possible for fiber applications.

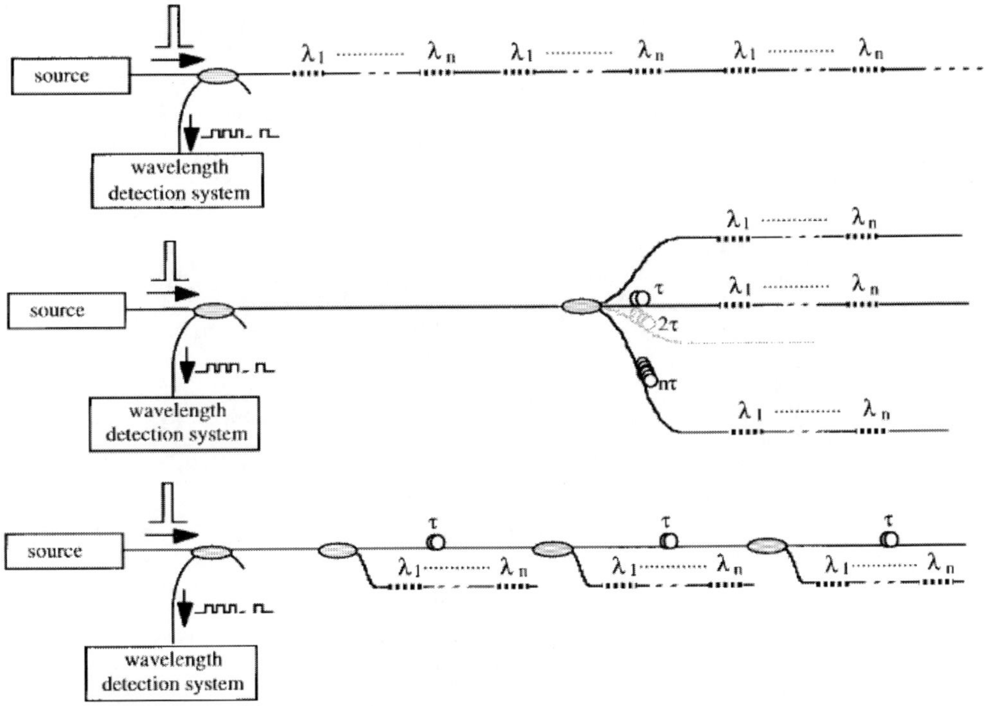

Figure 1. Different multiplexed approaches for fiber grating sensing schemes: (a) Serial configuration; (b) parallel configuration; (c) branching configuration.

Moreover, integrated grating sensors possess fast response and high sensitivities and are still compatible for WDM system integration, thus this chapter will mainly focus on the former category of devices. Some examples of compact hand held packages for grating based integrated sensors can be found in literature [13]-[14].

The idea of using waveguide Bragg gratings as sensing tools was introduced for the first time in 1983 by Tiefenthaler *et al.* [15]. Their first proposed sensor employs Bragg gratings on 120 nm thick SiO_2/TiO_2 waveguide on pyrex substrate (Figure 2). The device could be used to detect the humidity variation and the presence of organic solvent vapors such as ethanol or acetone.

The light is coupled in the device layer (*F*) through the transparent substrate (*S*), by a grating of length L_x. The optimal coupling condition depends on the effective index (*N*) of the mode travelling through the waveguide through the equation, which in turn can be influenced by variations in the grating overlay (*C*).

$$N = n\sin\alpha_1 + l\frac{\lambda}{\Lambda} \qquad (2)$$

where *n* is the air refractive index, *l* is the diffraction order of the grating, λ is the wavelength of the radiation and Λ is the grating pitch. In their first experiments, the authors noticed that the coupling condition could be disrupted by the presence of humidity on the grating. This was due to the change in the guided mode effective index (as the vapour induced a change in the optical thickness of the waveguide). This effect required re-adjusting the coupling angle (α_1) in order to keep the coupled power constant, thus a change in the coupling angle could be used to interpret surface modifications. This mechanism could actually be reversed by drying the device.

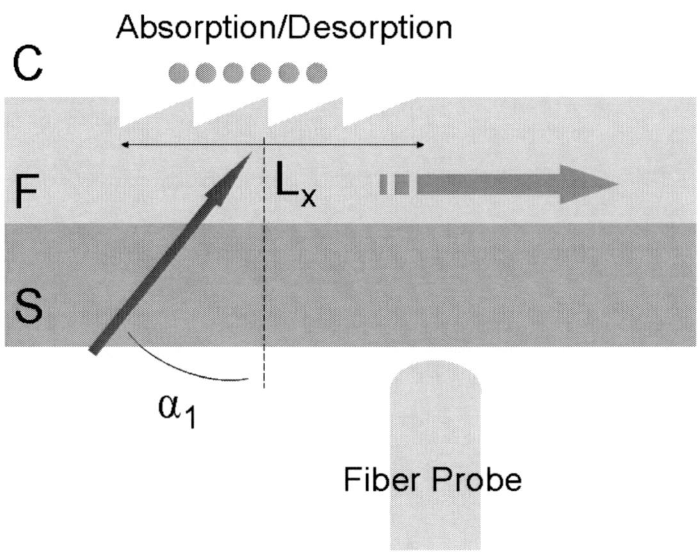

Figure 2. First grating sensing scheme proposed by Tiefenthaler et al [17].

A minimum refractive index change of 1×10^{-4} was reported as detectable by the device, provided that the grating had a minimum length of 5 mm and a grating pitch of 830 nm. The operating wavelength of the light was 514 nm. This kind of device did not have any kind of selectivity with respect to the composition of the sample. Starting from this idea, a number of grating devices for optical sensing have been presented in the subsequent years using hot embossing on polycarbonate thin film [16].

Moreover, from 1988 a number of authors extended the use of integrated Bragg gratings to a range of new applications related to biological and environmental sensing [18]. These techniques rely on treating the grating surface with an appropriate bio receptor [19], if the measured sample contains an analyte that can be "recognized" by the receptor it will bond with the grating surface, altering the effective optical thickness of the waveguide (Figure 3). This variation is related to a change in the effective index of the mode propagating in the waveguide, that in turn alters the coupling angle of the grating and its frequency response characteristics.

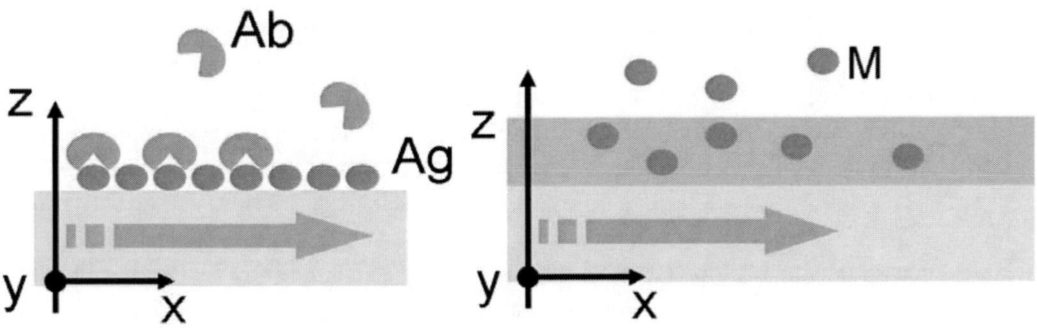

Figure 3. Schematic representation of detection mechanisms for bio-affinity sensors. The sensing surface of the device can be either treated with antigen sensible to the presence of a specific antibody (a) or with chemical layers responsive to specific molecules (b).

In 1990 Lukosz et al [20] proposed a SiO_2/TiO_2 grating sensor based on light outcoupling rather than incoupling, since this kind of device is reciprocal, Eq. (2) could also be used to predict the coupling of a guided mode from the waveguide to the cladding (Figure 4).

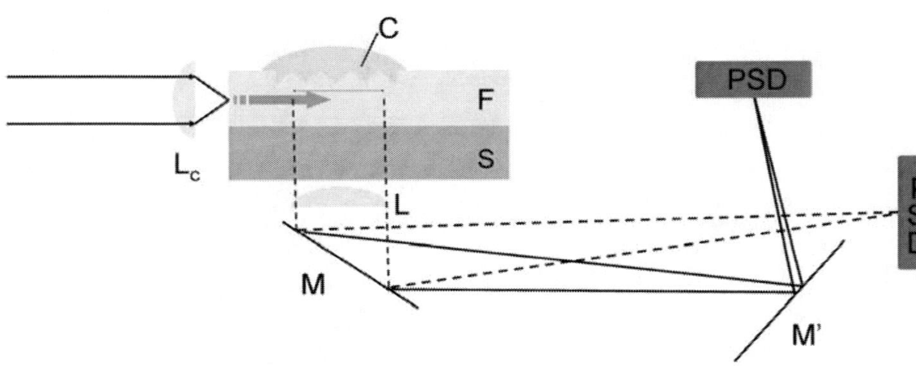

Figure 4. Outcoupling configuration proposed by Lukosz *et al* [20].

The device could be used as humidity sensor or as refractometer. Light was coupled in the guiding layer (*F*) through a lens near the device facet and out-coupled through the substrate (S) by the grating and directed towards a photodetector by a mirror (M-M'). A variation in the cover properties (C) would alter the outcoupling angle, which was used to track changes in the analyte. The grating was fabricated on the thin film by hot embossing, with grating pitch of 409 nm and total grating length of 2 mm. The coupling angle was chosen around 6°. In later work Lukosz *et al* [21] described how both input and output grating couplers could be effectively used as bio sensors by using immunochemistry methods (above all immobilizing an avidin layer on the grating surface for BSA detection). The authors reported a detection limit of a few nanomol/l for BSA (bovine serum albumin) by that time. In 1994 Clerc *et al* [22] reported a limit of 1.5×10^{-10} in detection concentration for BSA molecules with similar devices.

In 1996 Brandenburg *et al* [23] proposed a sensing scheme based on light reflected by a Bragg grating. The sensing scheme (Figure 5) included directing the light of an He-Ne laser on a grating and collecting the reflected light with a photodetector. The liquid sample could be placed in contact with the grating through a flow cell. The device was built on a Ta_2O_5 guiding film deposited on glass substrate. In this scheme a change in the sample on the surface of the grating induced a change in the optimal coupling angle for the grating, thus originating a minimum in the intensity of the reflected grating spectrum. By monitoring the position of the minimum it is possible to know if any bio affinity reaction is taking place in the sample. The source wavelength employed for this sensor is 633nm. A refractive index resolution of 3×10^{-6} is reported for this device.

A few years later the previous configuration was discussed again by Piehler et al [24] that proposed to use that scheme for pesticide sensing. A second order Ta2O5 waveguide Bragg grating was employed for the study, having a grating pitch of 750nm. A detection limit of 0.25 μg/l was reported by the authors.

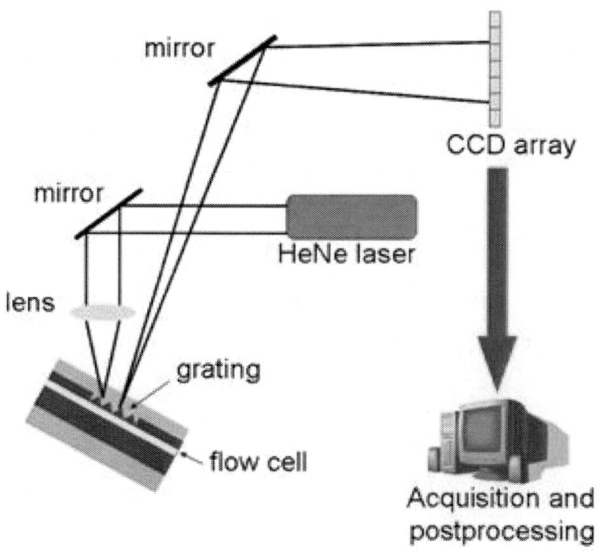

Figure 5. Setup configuration proposed by Brandenburg et al [23] for reflection measurements, and later used by Piehler [24].

In 1997 Veldhuis et al [25] proposed a grating sensor operating at 630 nm wavelength based on a Si3N4/SiO2 channel waveguide and a detection resolution of 2×10-5 in refractive index change. A 2.5 μm wide waveguide with a ridge height of 2 nm is defined by photolithography and wet etching of a layer of 120nm thick Si3N4 deposited by LPCVD on an oxidized silicon wafer (Figure 6). The grating is defined by holographic exposure on a 60 nm thick layer of photoresist, its pitch is 206 nm. After the resist has been developed the resist is removed through an O2 plasma process. The process oxidizes the zones that are not protected by the resist, thus creating a periodic refractive index change. Finally, the grating device is clad with an additional SiO2 layer which is locally removed in the sensing area.

When a liquid sample is deposited on the grating surface the spectral response of the grating shifts, this allows to measure possible variations in the composition of the fluid. The author reported a grating bandwidth of 0.2 nm at FWHM. Moreover the grating response displays a thermal dependence of 0.01 nm/°C, thus its operation as sensor needs to be stabilized in temperature.

In 1998 Dubendorfer *et al* [13] proposed high resolution PH sensor based on a polymer membrane (Figure 7a). The detection is based on a chirped grating coupler built by hot embossing on a on a polycarbonate sheet followed by TiO_2 132 nm thin film sputtering. The grating pitch of the input grating (Λ_i) varies between 420 nm and 423 nm whereas the output pitch (Λ_o) varies between 595 nm and 600 nm. The maximum grating length is about 950 μm and the chip width is 3mm. The device operates at 781 nm wavelength.

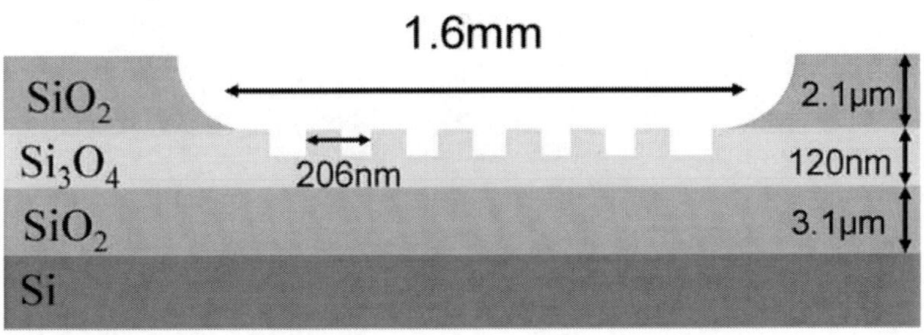

Figure 6. Schematic illustration of device proposed by Veldhuis *et al* [25].

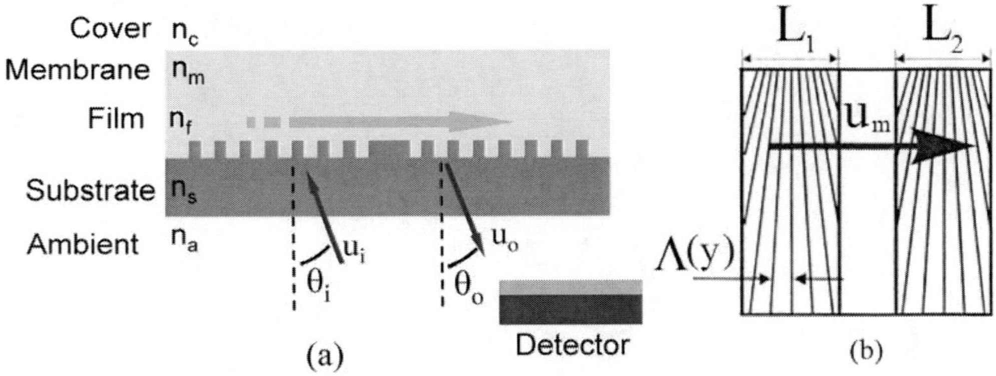

Figure 7. (a) PH sensor based on chirped Bragg gratings proposed by Dubendorfer et al [13] and (b) schematic of variable grating period.

Each sensing element make use of two chirped gratings, whose pitch $\Lambda(y)$ varies along the y direction (Figure 7b). The final device makes use of two sensing element. The first one is covered with a membrane whose properties are sensitive to changes in the PH, whereas the second sensing element is used as reference. The detection mechanism assumes that each input grating (on the left hand side) is illuminated from the substrate by one collimated light beam (u_i) at a fixed angle θ_i. Since there is a variation in the grating period, the resonant condition will be satisfied for a certain y position in the device (according to Eq. 2). This position is assumed to be the measuring variable for the device, since, as the input angle is kept constant, y varies with the properties of the gratings overlayer. The position of the coupled light beam can be measured by coupling the light from the output grating (on the right hand side) on a photodetector array at an angle θ_o. The shift in the y position can be normalized to the device where the membrane is absent. The authors reported a resolution of $\pm 1.1\times 10^{-4}$ in the measured value of PH.

In the same year Wiki *et al* [26] proposed a sensor based on multiple grating. The multiple grating structure was designed to improve the stability of the sensor towards mechanical stress (Figure 8) and is constituted of a central input grating (Λ_i) with two outcoupling grating on the sides (Λ_o). This structure is replicated on substrate and guiding layers. In order to perform biochemical measurement an analyte layer is deposited on the top of the device. Light is coupled in the centre of device from the substrate and excites an electromagnetic mode in the central high index layer (n_f). Depending on the overlay refractive index the outcoupling angle of the light is changed. Since there is a grating on each side of the device, the output light can be focused on two separate spots that possess a fixed distance D. This allows to directly determine if any reaction has take place in the liquid sample. Moreover the dual configuration makes the sensor more robust to mechanical misalignments or vibration of the measurement setup. The device was fabricated by hot embossing on a polycarbonate substrate (n = 1.581), the waveguide layer is constituted of a 139 nm thick TiO_2 deposited film (n = 3.279). The input grating Λ_i has a pitch of 350.4nm and a length L_i = 0.2mm, the side gratings Λ_o have a pitch of 433.4nm and a length L_o = 1.5mm.

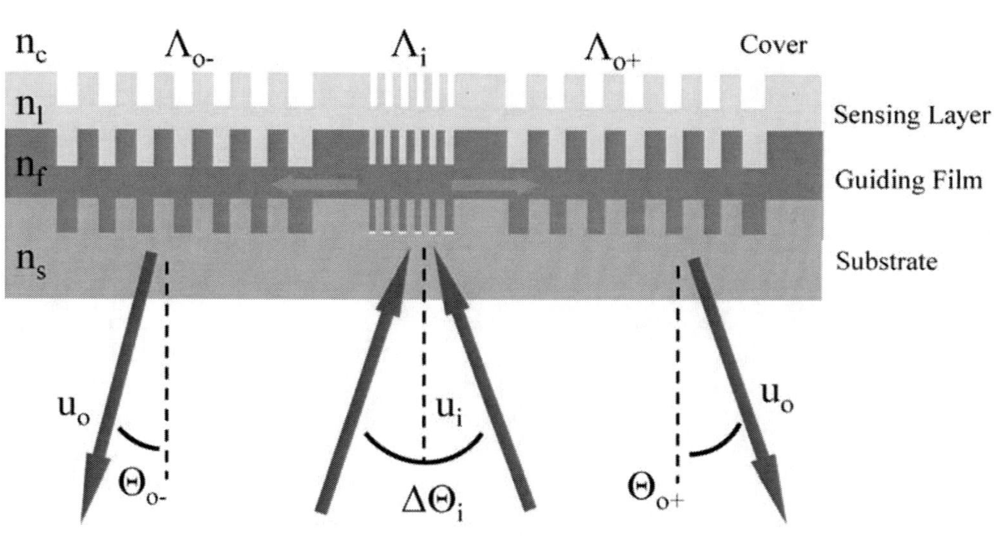

Figure 8. Triple grating sensor proposed by Wiki et al [26].

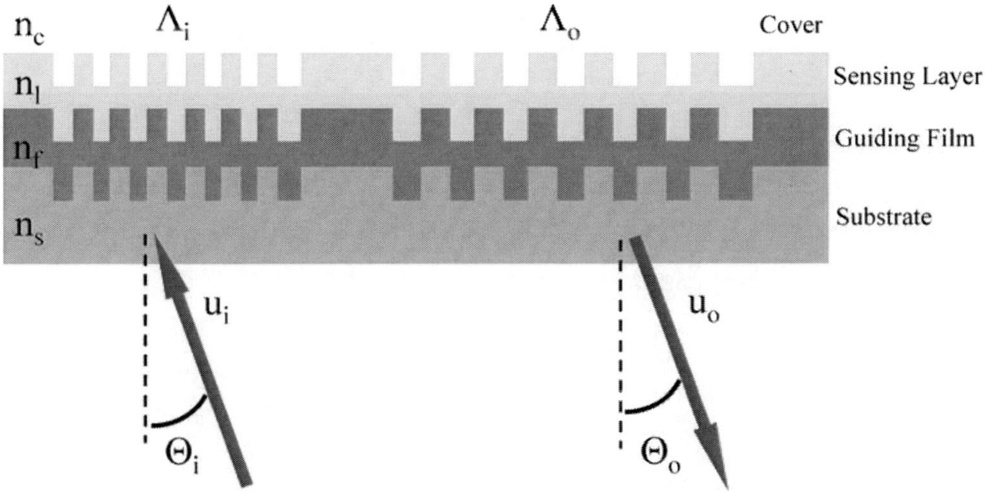

Figure. 9 Grating sensor based on separate input and output grating couplers with different periods, proposed by Wiki et al [14].

By altering the cover index between water (n = 1.33) and ethanol (n = 1.359) and measuring the variation in the out coupling angles the authors estimated a sensitivity of $\partial N_{eff} / \partial D = 1.17 \times 10^{-6}$ μm^{-1}. In 2000 the same author [14] proposed a similar device (Figure 9) based on the same concept. The latter device had two different period gratings on each layer, one serving as input coupler (Λ_i) and the other as output coupler (Λ_o). The device fabrication is very similar to that presented in [26], the input grating has a pitch of 420 nm whereas the output coupler has a pitch of 495nm. This device was reported to have a very high resolution (130 fg/mm^2 on the sensing surface) obtained through wavelength interrogation method. Moreover, the authors stated that it could be used both as a refractometer, or an affinity biosensor if provided with a suitable surface preparation. The input coupler (Λ_i) is illuminated through the substrate by a tuneable VCSEL laser with a fixed angle θ_i.

The operation wavelength for the device is around 764 nm. If the coupling condition is satisfied, the intensity of the guided mode in the middle layer will be maximum and could be coupled out of the device through the output grating (Λ_o). The sensing mechanism is based on detecting how much the output characteristics varies after an analyte has been placed on top of the grating.

In 2002 Voros *et al.* [27] presented a sensor based on an integrated grating coupler designed to investigate a range of bio affinity reactions such as DNA/RNA-protein binding and cell surface interaction. The technique used for fabrication is a one-step embossing process on a sol gel layer deposited on glass or plastic substrate. Grating depths for this kind of devices vary between 5 and 20 nm. The sensor consists of a grating built on a single mode planar waveguide (Figure 10) the light is introduced in the waveguide from the substrate through the coupling angle α and collected by photodiodes at each end of the waveguide.

When an analyte is introduced on the top of the grating surface, the effective index of the propagating mode changes according to the strength of the affinity reaction taking place in the sample. This means that the optimum incoupling angle for the grating is also modified, and the optical setup has to be readjusted continuously to keep the in-coupled light to a maximum. By monitoring the variation of α is possible to detect the molecular reaction in the analyte with a detection limit <1 ng/cm^2. Moreover, the author pointed out that fabrication of gratings through sol gel embossing technique would a viable way for mass production.

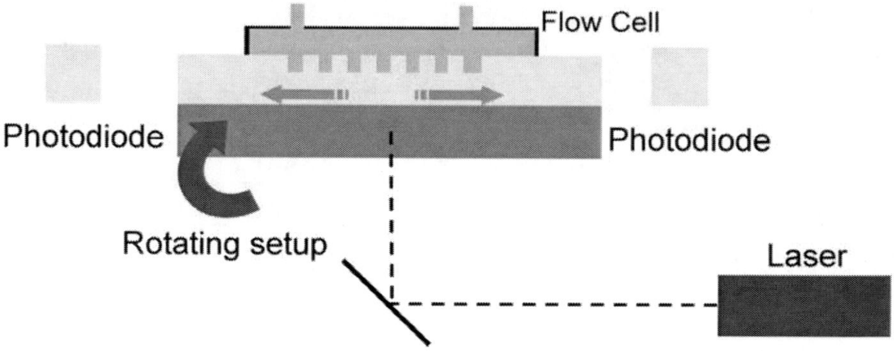

Figure 10. Schematic representation of the device proposed by Voros *et al* [27].

In 2004 Brioude *et al* [28] proposed a grating sensor fabricated on a Ta_2O_5 (n = 2.2) waveguide on a SiO_2 substrate (n = 1.44) by wet etching (Figure 11). The grating is optimized to work with the fundamental TM mode at a wavelength of 830 nm. The planar waveguide has a thickness of 194 nm. The grating device has a perturbation depth of 40 nm and a pitch Λ = 430 nm with a duty cycle of 34%.

After an analyte has been deposited on the grating, the light is coupled to the waveguide from the substrate. By varying the angle of incidence (θ) it is possible to achieve a destructive interference between the radiation propagating parallel to the grating and the in coupled wave. This will cause all the light to radiate out of the waveguide into the cover. If the amplitude of the 0th order transmitted wave and the radiated wave are the same the final result will be a zero transmission that takes name of abnormal reflection. In order to allow this phenomenon to take place, the condition $w\alpha \gg 2\pi$ must be satisfied, where w is the incoupling beam width and α is the radiation efficiency of the grating (leakage factor, as defined in [29]). The chemical changes in the analyte can be monitored by measuring the shift of the zero transmission peak. The authors reported a minimum detectable refractive index change of 2×10^{-6}.

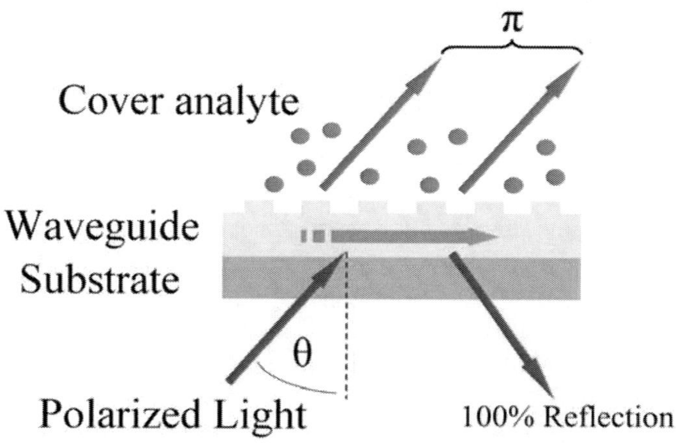

Figure 11. Abnormal reflection grating proposed by Brioude *et al* [28].

In 2005 Hopman *et al.* [30] proposed a Bragg grating configuration based on photonic wire waveguide realized on Si_3N_4 on SiO_2 substrate. The device is designed to work at 600nm wavelength and can be used as a thermally tuneable refractometer (Figure 12). The device is built on a ridge waveguide which is 2 μm wide, the guiding layer (n_g) is 212 nm thick and the grating has a depth of 22 nm and a pitch of 190 nm. The grating length is 76 μm. The authors reported a minimum detectable index variation of 10^{-4} and a spectral shift of 7 pm/K for a variation in temperature. Although this kind of devices ensure high sensitivities, they show very strong dependency on the signal polarization and have very strict fabrication requirements. In 2006 Dai *et al.* [31] (Figure 13) proposed a sensor configuration based on a SiO_2 Ge doped Bragg grating fabricated by flame hydrolysis deposition and reactive ion etching on a ridge waveguide. The detection mechanism in this device is based on measuring the spectral shift of the grating frequency response after a variation in the overlay refractive index n_t. The dimensions (a, b) of the waveguide varied from 5.7×5.6 μm^2 to 7.5×5.6 μm^2. The grating is etched through the full depth of the waveguide and is designed to operate around 1.55 μm wavelength.

Commercially available index matched liquids where used to characterize the wavelength shift induced by the overlay refractive index change. The device showed a sensitivity of 4×10^{-5} pm^{-1}. The authors also measured the variability of the grating frequency response to temperature, finding that the Bragg central wavelength shift is $d\lambda/dT = \pm 11\,\text{pm/°C}$. According to the authors, the device sensitivity could be improved by reducing its dimensions, although this would increase the modal birefringence in the waveguide and the optical losses.

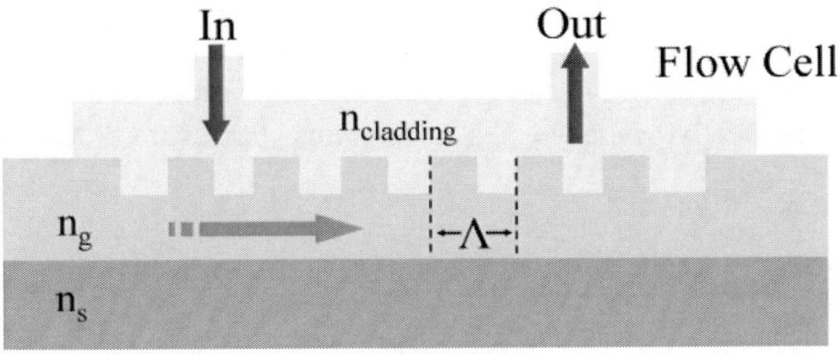

Figure 12. Thermally tuneable grating sensor on nanometer Si_3N_4 waveguide proposed by Hopman *et al* [30].

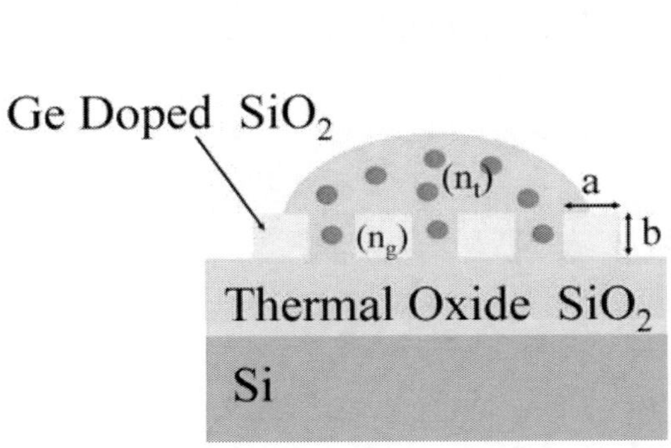

Figure 13. Bragg grating configuration proposed by Dai *et al*, based on Ge doped SiO2 ridge waveguide [31].

3. Theory

In the previous section, many integrated Bragg optical sensors have been presented, although none of them relies on a full SOI platform. We now present the design of an SOI optical sensor based on shallow etched Bragg gratings of 1st and 3rd order. To design the shallow SOI reflection gratings, a 2D mixed approach based on CMT and FEM has been developed, assuming a rectangular grating profile. Comparisons with rigorous Floquet-Bloch theory (FBT) [32] have demonstrated the good accuracy of this method in cases under investigation.

A Bragg grating can be generally described through a periodic perturbation of the waveguide refractive index along the propagating direction *z*:

$$n(x,y,z) = n_0(x,y) + \Delta n(x,y,z) \tag{3}$$

where $n_0(x,y)$ is the waveguide index distribution and $\Delta n(x,y,z)$ is the periodic refractive index change whose period Λ has to satisfy the following relation. It is known from general diffraction grating theory that a grating coupling between propagating optical modes happens when the wave vector of the input radiation (traveling at an angle θ_i) is phase matched with the wave vector of the diffracted radiation (diffracted at angle θ_d). Specifically the difference between k_i and k_d must equal the wave vector of the grating ($K = 2\pi/\Lambda$), meaning that:

$$n_d \frac{2\pi}{\lambda}\sin(\theta_d) - n_i \frac{2\pi}{\lambda}\sin(\theta_i) = m\frac{\lambda}{\Lambda} \tag{4}$$

where Λ is the grating period. In the case of feedback the optical radiation is coupled from a mode traveling with effective index n_{eff} to an identical backward traveling mode with the same effective index, thus

$$\Lambda = \frac{m}{2n_{eff}}\lambda_c \tag{5}$$

being m the grating order, λ_c the centre wavelength of the spectral response and n_{eff} the effective index of the propagating mode. Following the analysis developed in [33], the grating reflectivity can be expressed as:

$$R(\delta) = \frac{|\kappa|^2 \, tanh^2(\gamma L)}{\gamma^2 + \delta^2 tanh^2(\gamma L)} \tag{6}$$

where

$$\gamma = \sqrt{|\kappa|^2 - \delta^2} \tag{7}$$

and

$$\delta = \frac{2\pi}{\lambda} n_{eff} - \frac{\pi}{\Lambda} \tag{8}$$

where the quantity defined in Eq. (8) is also known as synchronicity factor and it is a measure of how efficiently the diffraction is close to the resonant condition. Moreover, λ is the generic optical wavelength, κ is the coupling coefficient between backward and forward propagating optical modes and L is the grating length. For $\lambda=\lambda_c$ maximum obtainable reflectivity R_{max} is given by:

$$R_{\max} = tanh^2 \left|\kappa L\right| \tag{9}$$

If the condition $\kappa L > 1$ holds, Bragg grating bandwidth can be approximated as the wavelength spacing between the first minima (nulls) of the spectrum [33]:

$$\Delta\lambda = \frac{|\kappa| \lambda_c^2}{\pi n_{eff}} \tag{10}$$

If we assume that each guided mode can be mainly described through its dominant electric field component (e_x for TE modes and e_y for TM modes) the grating coupling coefficient κ can be expressed as:

$$\kappa = \frac{\pi j}{2P\lambda\eta_o} \iint_{grating} \delta\varepsilon_m(x,y) \left|\mathbf{E}\right|^2 \, dxdy \tag{11}$$

where the rectangular grating perturbation is expressed through the Fourier series of a square wave having the same characteristics:

$$\delta\varepsilon_m(x,y) = \begin{cases} \left(n_{core}^2 - n_{clad}^2\right) \dfrac{\sin(m\pi\xi)}{m\pi} & \text{grating} \\ 0 & \text{otherwise} \end{cases} \tag{12}$$

ξ is grating duty-cycle, n_{core} is the guiding layer refractive index, n_{clad} is the overlay refractive index, P is the propagating mode optical power, η_0 is the free space impedance, j is the imaginary unit and **E** is the electric field vector associated to the optical mode propagating in the grating. The overlap integral in Eq. (11) has been directly calculated through FEM commercial tool [34]. In our case the built in asymmetric sparse matrices solver with real propagation constants has been used. The overlap integral is calculated over the cross section of an equivalent waveguide containing a grating region of depth d, as shown in the highlighted zone of Figure 14 inset.

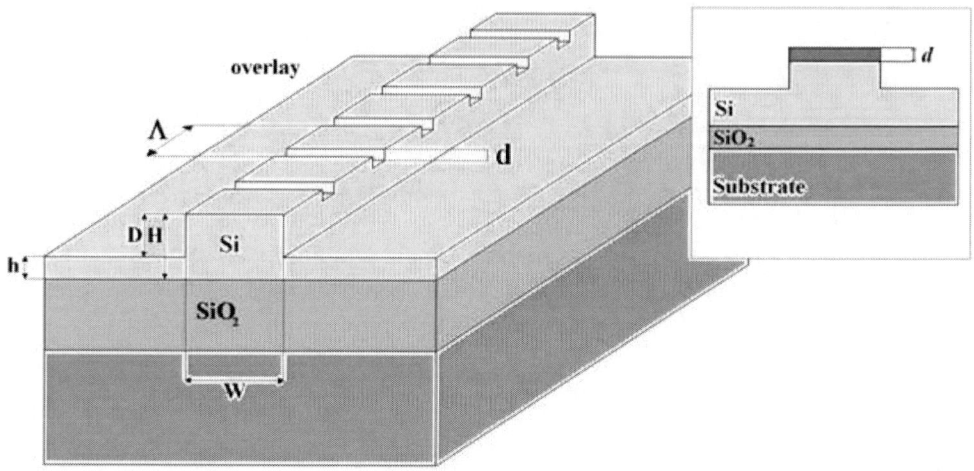

Figure 14. Schematic representation of the grating-based sensor device. The inset shows the reference structure used for coupling constant estimation.

The average refractive index of the perturbed region is defined as:

$$n_{eq} = \sqrt{\frac{n_{grating}^2 + n_{wg}^2}{2}} \tag{13}$$

for a 50% duty cycle grating. The propagating mode optical power (P) is evaluated by extending the integral domain to the whole waveguide cross section. Once these quantities are known, the coupling constants and grating spectrum can be analytically calculated from Eq. (11) and Eq. (6).

Grating devices are usually based on single first order gratings, or composite array structures [26], in order to maximize efficiency and employ short length chips. In general, performance is improved, but device fabrication is open to tolerance problems. By using the same waveguide design criteria stated above, it is possible to employ higher order gratings (m > 1) in order to reduce the influence of fabrication tolerances. From Eq. (11) it is possible to observe that the choice of a higher order grating implies a decrease in the grating efficiency. In order to compensate the loss of efficiency it is possible to increase the length of the device. Also from Eq. (4) is possible to notice that for $m > 1$ coupling to different radiation orders ($\theta_d \neq -90°$) becomes possible, thus new radiation angles are allowed. This further reduces the grating efficiency, since a percentage of the power is lost either in the substrate or in the overlay. The decrease in coupling coefficient (κ) for higher order grating devices is associated to a bandwidth reduction, as it can be noticed from Eq. (10). This effect can be seen as favorable for sensor application, especially in the perspective of employing a multi-wavelength system. Moreover, this kind of approximation is more consistent for shallow gratings (depth < 30nm). In order to describe the sensing efficiency of the device, we define the sensitivity as:

$$S = \partial\lambda_c / \partial n_{clad} \tag{14}$$

Using Eq. (14), sensitivity can be related to the spectral shift of the grating response as:

$$S = \frac{\partial \lambda_c}{\partial n_{clad}} = \frac{\partial \lambda_c}{\partial n_{eff}} \cdot \frac{\partial n_{eff}}{\partial n_{clad}} = \frac{\Lambda}{m} S_w = \frac{\lambda_c}{2n_{eff}} S_w \tag{15}$$

where S_w is the waveguide sensitivity defined as the first derivative of n_{eff} with respect to n_{clad} [35].

4. Numerical Results

Using the procedure described in the previous section, a first-order Bragg grating-based sensor (as in Figure 14) has been designed. We have assumed air as overlay and verified that the coupling coefficients, maximum reflectivity, and bandwidth of the designed gratings are practically not influenced (shift below 3%) by an overlay refractive index varying between 1 and 1.33 (case of aqueous solution). Grating duty cycle is set at 50% in order to maximize the coupling coefficient. This approach refers to the waveguide analysis reported by the authors in [36].

Three different rib waveguides, whose dimensions are reported in Table 1, have been adopted for the grating design. We have verified that these waveguide dimensions guarantee that only a quasi-TE and a quasi-TM mode are supported by the guiding structure even when n_{clad} =1.33 (i.e. case of aqueous solution adopted as overlay).

Moreover, for n_{clad} = 1.33, we have verified a good ZBR condition for these modes (birefringence equal to $8.54 \cdot 10^{-4}$, $3.73 \cdot 10^{-3}$, and $6.541 \cdot 10^{-3}$ for H = 1 μm, 0.75 μm and 0.5 μm, respectively). Although the difference of effective indices is relatively small, it still originates a slightly different spectral response for each polarization. Specifically, the difference in the central wavelength of the grating spectrum can be easily verified by referring to Eq. (5) and varies between 0.3 nm and 2.5 nm for quasi-TE or quasi-TM mode, respectively, for the aqueous overlay.

Table 1. Maximum waveguide sizes satisfying both ZBC and SMC

H [μm]	W [μm]	D [μm]	n_{eff}^{TE}	n_{eff}^{TM}	B
1	0.56	0.62	3.234771	3.234994	$-1.627 * 10^{-4}$
0.75	0.50	0.47	3.131629	3.130067	$1.562 * 10^{-3}$
0.5	0.42	0.36	2.851297	2.852205	$-7.895 * 10^{-4}$

First order grating period has been determined by using Eq. (5). The imposed central wavelength for the grating spectrum is 1550 nm. Table 2 reports the different grating pitches Λ for three considered values of H. If an aqueous solution (n_{clad} = 1.33) is deposited on top of the grating structure, the central wavelength of the spectrum will be shifted towards higher wavelength values as a consequence of the change of the guided mode effective index.

Table 2. First order grating parameters

H [µm]	d [nm]	L [µm]	Λ [nm]	κ [cm^{-1}]	$\Delta\lambda$ [nm]
1	30	607.67	240	49.25	1.7365
0.75	20	331.45	248	90.3	3.1836
0.5	10	172.57	272	173.44	6.1161

If the aqueous solution refractive index varies around 1.33 (we have assumed a variation range from 1.325 to 1.335) as a consequence of the presence of a specific liquid substance, the centre wavelength of the grating spectrum will experiment an additional shift. By measuring the new centre wavelength shift, a change in the composition of the solution can be detected.

For $H = 1\,\mu\text{m}$, grating spectral response dependence on overlay refractive index n_{clad} shift has been investigated by changing n_{clad} in the range from 1.325 to 1.335 and observing the relevant shift in λ_c (see Figure 15). The spectrum centre wavelength increases linearly with the overlay index value. The grating behaviour is qualitatively the same in other cases (for H = 0.75 µm and 0.5 µm).

Although guiding structures have been designed to exhibit zero-birefringence, coupling coefficients related to both polarizations are still different. This implies that the bandwidths of grating reflectivity spectra (but not their resonance center wavelengths λ_c) are influenced by incoming optical signal polarizations.

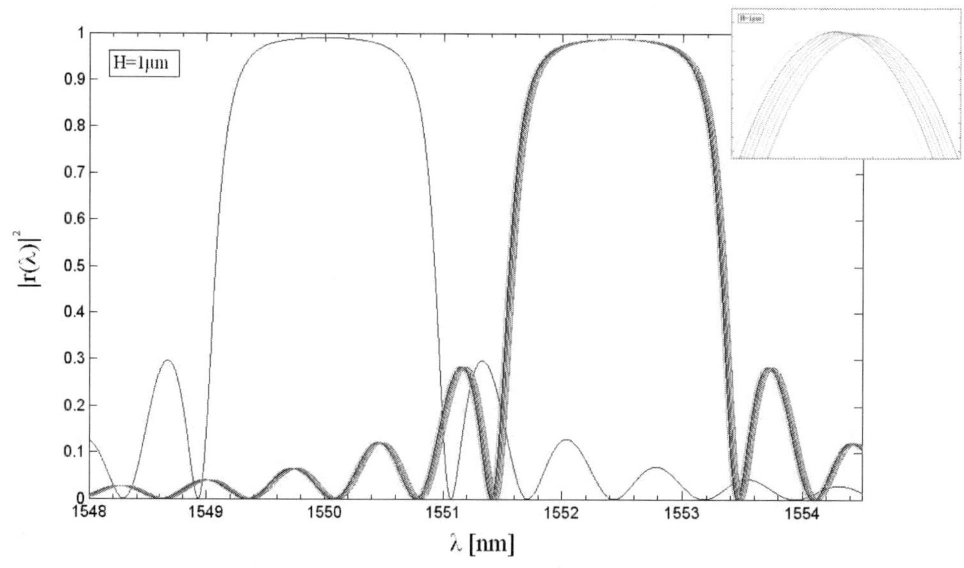

Figure 15. Spectral response of the first order grating for different values of the overlay refractive index. The blue curve is referred to air overlay ($n_{clad} = 1$) while the other spectra are for overlay indices ranging from 1.325 (rightmost response) to 1.335 (leftmost response). The inset shows a magnification of the reflectivity peaks.

Figure 16 shows the difference between coupling coefficients related to quasi-TM and quasi-TE modes versus grating depth *d*, for *H* = 1, 0.75 μm, 0.5 μm. The existence of this difference was first discussed in [37] and related to the different modal profiles of quasi-TE and quasi-TM modes. For the guiding structure having *H* = 0.5 μm, modal profiles are sketched in Figure 17 for both polarizations.

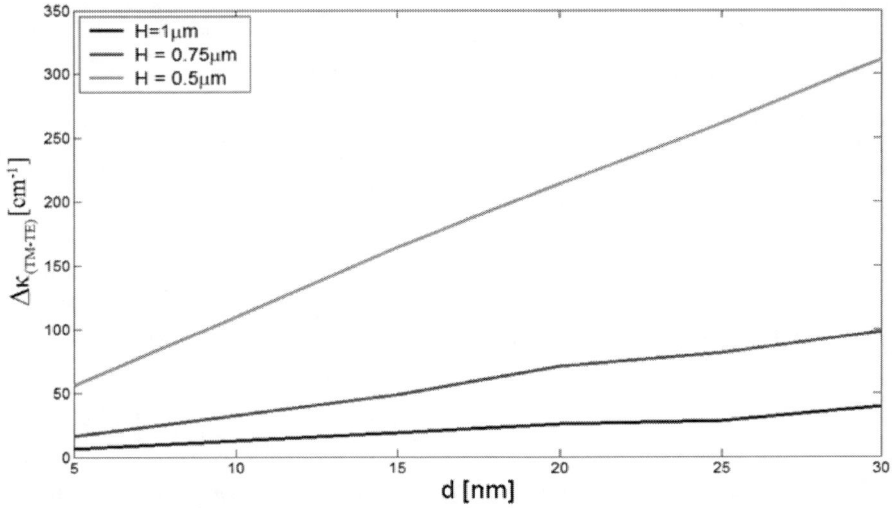

Figure 16. The curves show the difference between coupling coefficients related to quasi-TM and quasi-TE modes as the groove depth *d* varies, for three values of *H*.

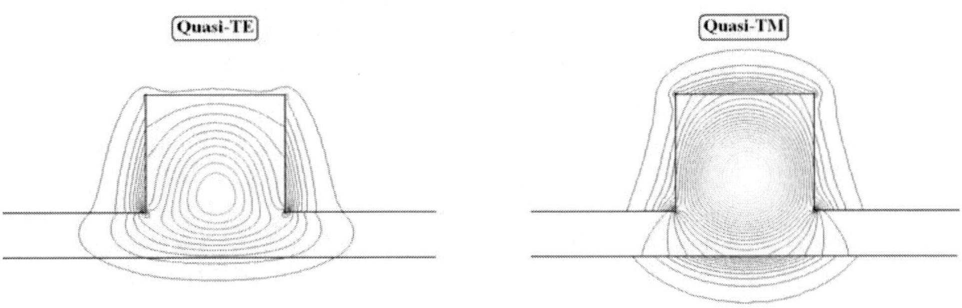

Figure 17. Electric field squared module distribution related to quasi-TE and quasi-TM modes for waveguide having *H* = 0.5 μm.

From this picture it is evident that quasi-TE and quasi-TM modes exhibit different electric field distributions. Since the grating coupling coefficient depends on the integral of the electric field squared module extended to the whole grating cross section (see Eq. (11)), the different modal profiles produce a difference between the coupling coefficients related to quasi-TM and quasi-TE modes. This difference becomes more relevant as the waveguide becomes smaller. Moreover, it also increases with the grating depth. Specifically, it is possible to observe that coupling coefficient κ is larger for quasi-TM rather than for quasi-TE modes. For this reason, higher reflectivity values are achieved in shorter gratings employing TM polarization.

Thus, d has been fixed as equal to 30 nm for $H = 1$ µm, 20 nm for $H = 0.75$ µm, and 10 nm for $H = 0.5$ µm, and all further results are referred to TM modes. In all cases it results $d << H$, so the grating grooves do not perturb the waveguide SMC, as it has been also verified by FBT approach [32]. For first order grating devices, the maximum reflectivity dependence on grating length has been also investigated for various H values, as shown in Figure 18.

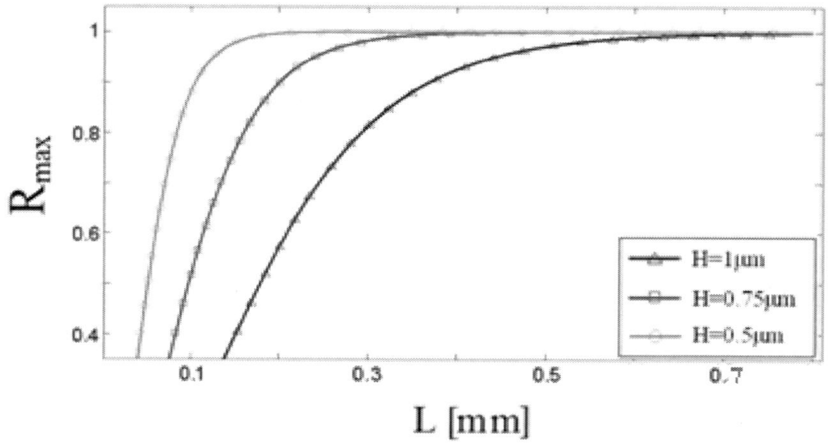

Figure 18. Maximum reflectivity versus grating length, for $H = 1$µm, 0.75µm and 0.5µm, in a first order grating ($d = 30$ nm for $H = 1$ µm, 20 nm for $H = 0.75$ µm and 10 nm for $H = 0.5$ µm, respectively).

Grating length has been chosen in order to obtain a maximum reflectivity of 99% for all cases, as reported in Eq. (9), with the aim to achieve a well definite grating response. All geometrical sizes and performance parameters of gratings are summarized in Table 2.

For the three considered sets of geometrical parameters, grating sensitivity has been calculated. Table 3 shows the estimated sensitivities for the three different cases, assuming an n_{clad} variation range from 1.325 to 1.335.

Maximum sensitivity can be obtained by employing the third waveguide configuration ($H = 0.5$ µm). Fabrication tolerances related to rib width and etch depth variations with respect to the nominal values have been investigated for this specific case. Etch depth D has been varied in the range from 0.3 µm to 0.36 µm (where SMC still holds) and grating maximum reflectivity and bandwidth have been calculated. The results are shown in Figure 19. We can observe that a decrease of D of 5% from nominal value induces a decrease of both maximum reflectivity (0.2%) and bandwidth (1.8%). Moreover, a 5% rib width W decrease from nominal value (keeping $D = 0.36$ µm) produces a 0.13 % coupling coefficient decrease and a consequent decrease of maximum reflectivity and bandwidth of 0.008 % and 0.13 %, respectively.

Table 3. First order grating sensitivities

H [µm]	Sensitivity [nm]
1	8
0.75	13.3
0.5	33.6

Thus, the bandwidth (coupling coefficient) is much more influenced by etch depth than rib width. If quasi-TE mode is excited instead of quasi-TM, maximum reflectivity decreases from 99% to 63.8 %, and bandwidth is reduced from 6.11 nm to 2.24 nm.

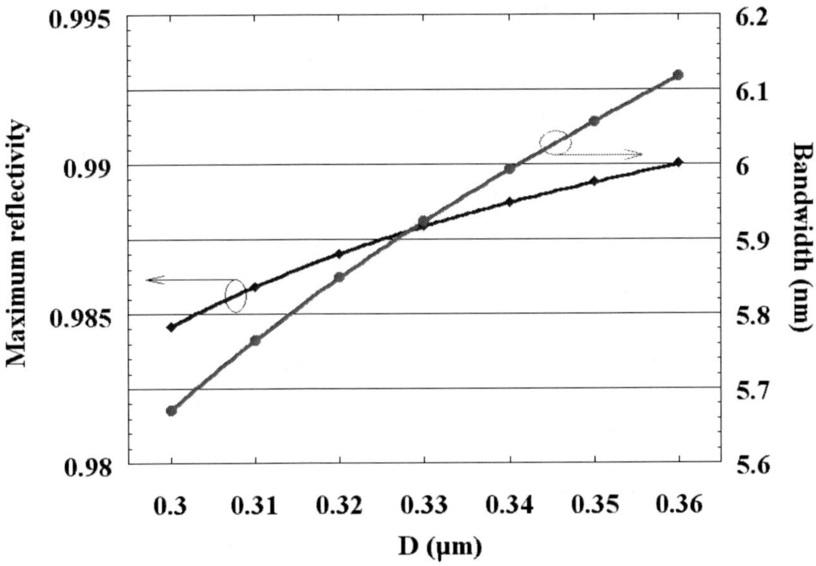

Figure 19. Maximum reflectivity and bandwidth versus etch depth. $H = 0.5$ μm has been considered.

Assuming to be able to detect a minimum λ_c shift of 5 pm, as in [38], the designed first order grating having $H = 0.5$ μm should allow to detect a minimum overlay refractive index change around 10^{-4}. However, detection scheme should be based on a wavelength interrogation by comparing the centre wavelength shift with the edge wavelength shift, rather than on power interrogation, because the shifting of a large grating spectrum could correspond to a power variation lower than the detector's resolution. This sensitivity is expected be less accurate of about one order of magnitude than that usually obtained by integrated sensors based on Mach-Zehnder interferometers (as for [39]). Moreover, our device is only 173 μm long, whereas Mach-Zehnder interferometer-based sensors are usually tens of millimetres long (superior by two orders of magnitude).

Moreover, proposed sensor detection limit could be significantly improved (up to one order of magnitude) by optimizing the guiding structure sensitivity. A detection limit similar with that reported for a micro-ring resonator SOI sensor described in [38] would be theoretically achievable. Since waveguide sensitivity depends on guiding structure geometrical dimensions, the sensitivity optimization (briefly discussed for SOI rib waveguides in [35]) becomes problematic if also ZBR condition has to be maintained. Also the adoption of a Si-wire waveguide (as proposed in [38] and [40]) may in principle allow an improvement of guiding structure sensitivity (and a consequent detection limit. decrease), effectively excluding the possibility of achieving ZBR. Birefringence of Si-wire waveguides adopted in [38] and [40] are equal to 0.753 and 0.534, respectively.

Other very compact device configurations, such as that reported in [41], allow a sensitivity of 465 nm in just 10 μm waveguide length, at the expenses of employing a planar guiding structure. However, this latter kind of devices does not allow lateral confinement of optical signal and thus is less suitable for integrated photonic circuit design. Moreover, optical losses reported in [41] are considerably high (> 10dB).

It is interesting to observe how the latter value of sensitivity could be still achieved by our architecture using a different guiding structure (such as Si-wire or slot waveguide, where S_w >1 [35]) which would allow sensitivity to exceed 490 nm, at the expenses of polarization dependence. Examples of grating-based slot structures have also been proposed [42], although they are not designed initially for optical sensing, it must be noted that hollow structure are potentially very suitable for sensing application.

Device sensitivity can also be significantly enhanced by increasing the operating wavelength (as shown in Eq. 15). Since silicon wave-guided components for long-wave infrared region are actually under development [28], nowadays this possibility appears to be realistic.

Grating devices are usually based on single first order gratings, or multiple grating structure, in order to maximize efficiency and employ short length chips. In general, performance is improved, but device fabrication is open to tolerance problems.

By using the same waveguide design criteria analyzed above, we discuss the design of a third order grating sensor. By using a larger grating period it is possible to reduce the influence of fabrication tolerances. Third order grating period is given by Eq. (6), by considering $m = 3$.

From Eq. (11) it is possible to observe that the choice of a third order grating implicates a decrease in the grating efficiency. Table 4 reports the relevant parameters corresponding to the third order device, where L has been fixed by imposing again a grating reflectivity of 99%.

Table 4. Third order grating parameters

H [μm]	d [nm]	L [μm]	Λ [nm]	κ [cm^{-1}]	$\Delta\lambda$ [nm]
1	30	1823	720	16.41	0.578
0.75	20	994.35	744	30.10	1.061
0.5	10	517.71	816	57.81	2.038

Figure 20 shows the maximum reflectivity calculated as the grating length increases. For a fixed length, the maximum reflectivity of a higher order grating will be lower when compared to the first order case. In order to obtain the same grating performance, it is necessary to increase the grating length. By choosing the length reported in Table 4, we obtain a spectral response (see Figure 21) similar to the first order cases, although the bandwidth reduction is apparent.

Table 5 shows the sensitivity calculated for each value of rib height H. Theoretically, sensitivity values are almost identical to those reported in the first order case (thus allowing similar detection limits for the spectral shift) as long as the grating length is increased appropriately. Therefore, it is possible to obtain a device performance very similar to the corresponding first order grating, gaining higher fabrication flexibility, at the cost of a longer device.

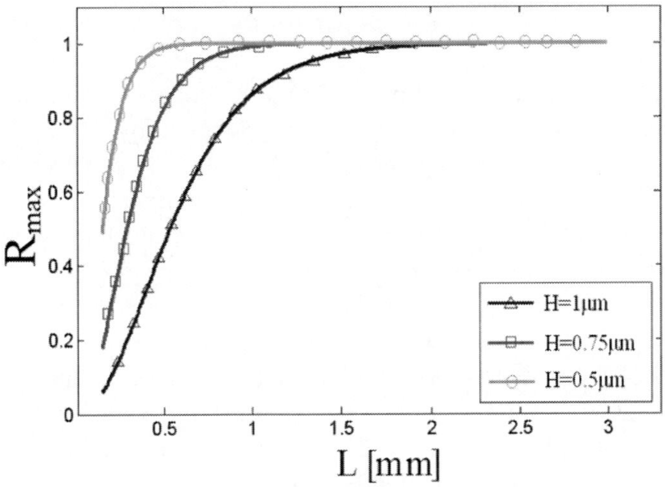

Figure 20. Maximum reflectivity versus grating length, for $H = 1\mu m$, $0.75\mu m$ and $0.5\mu m$, in a third order grating ($d = 30$ nm for $H = 1$ μm, 20 nm for $H = 0.75$ μm and 10 nm for $H = 0.5$ μm).

Table 5. Third order grating sensitivities

H [μm]	Sensitivity [nm]
1	7.08
0.75	11.955
0.5	32.22

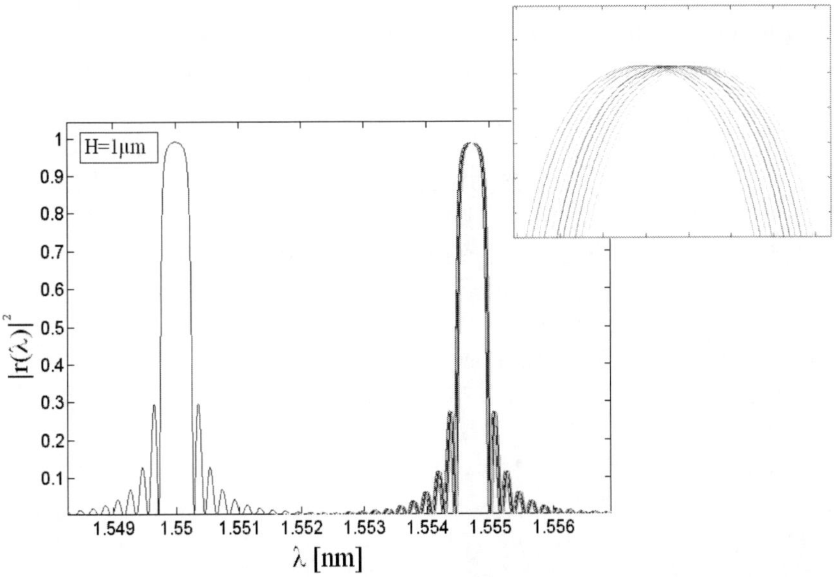

Figure 21. Spectral response of third order grating for overlay refractive index ranging from 1.325 (rightmost response) to 1.335 (leftmost response). The inset shows a magnification of reflectivity peaks. Considered waveguide has $H = 1$ μm.

CONCLUSION

In this chapter, a brief review of the use of Bragg gratings for sensor application has been given. The two mainstreams in this field are represented by fiber Bragg gratings and waveguide Bragg gratings. Fiber Bragg gratings offer many advantages and are suitable for a broad range of applications, although they are not the best choice in the perspective of on-chip future silicon photonics integration. On the other hand, waveguide gratings offer extreme compactness. Using a silicon on insulator system as a platform for Bragg gratings integrated sensors appears to be a relatively unexplored area, since most of the previous technology relies on other materials, preferably Ta_2O_5 on glass. We have investigated the possibility of using Bragg gratings fabricated on sub-micrometer SOI single mode rib waveguides to implement an integrated optical sensor. The approach of using a SOI sub-micrometer structure for sensing is relatively new. It permits one to improve the sensor performance dramatically, whereas it poses new challenges for fabrication issues. To analyze the device behavior, we employed a mixed method based on both FEM and CMT. This approximate model allows one to quickly estimate the efficiency of the Bragg grating, although it is limited to shallow perturbations.

As a complementary aspect of the sensor design, we have also considered the possibility to realize the sensor by using a third order grating. This strongly reduces the fabrication problems while increasing the device length. The grating coupling effect has proved to be more efficient for TM polarization. Specifically, a minimum detectable overlay refractive index change around 10^{-4} has been calculated for an optimized 173 μm long first order grating. By excluding fulfillment of the ZBR condition in the design, a significant improvement of this detection limit value could be achieved. This would allow optimizing the device sensitivity by scaling down the waveguide dimensions to those similar to silicon wire or slot, which are generally optimized for just one polarization, and also introduce different fabrication issues. Finally, the use of a labeled detection method could increase the device sensitivity, as well as allowing the investigation of a broader variety of biological reactions.

REFERENCES

[1] Jalali, B., Fathpour, S., *J. Lightwave Technology*, 2006, 24, 4600-4615.

[2] Pavesi, L., Guillot, G. (Eds.), *Optical Interconnects : The Silicon Approach*, Springer Series in Optical Sciences, Springer, New York, 2006.

[3] Paniccia, M., Koehl, S., *IEEE Spectrum*, 2005, 42, 30-35.

[4] Byoungho, L., *Optical Fiber Technology: Materials, Devices and Systems*, 2003, **9**, 57-79.

[5] Henderson, P. J., *et al.*. Kyongju, South Korea: SPIE-Int. Soc. Opt. Eng., 1999.

[6] Vohra, S. T., *et al. Fiber Bragg grating sensor system for civil structure monitoring: applications and field tests*, Kyongju, South Korea: SPIE-Int. Soc. Opt. Eng., 1999.

[7] Cusano, A., *et al.*, *IEEE Sensors Journal*, 2006, 6, 67-77.

[8] Fujihashi, K., *et al.*, Tokyo, Japan: IEEE, 2007.

[9] Falate, R., *et al.*, *Sensors and Actuators, B: Chemical*, 2005, 105, 430-436.

[10] Liang, W., *et al.*, *Applied Physics Letters*, 2005, 86, 151122.

[11] Pereira, D. A., Frazao, O., Santos, J. L., *Optical Engineering*, 2004, 43, 299-304.
[12] Kersey, A. D., *et al.*, *J. Lightwave Technology*, 1997, 15, 1442-1463.
[13] Dubendorfer, J., *et al.*, *Sensors and Actuators, B: Chemical*, 1998, B50, 210-219.
[14] Wiki, M., Kunz, R. E., *Optics Letters*, 2000, 25, 463-465.
[15] Lukosz, W., Tiefenthaler, K., *Directional switching in planar waveguides effected by adsorption-desorption processes*, Florence, Italy, IEE, 1983.
[16] Tiefenthaler, K., Lukosz, W., *Embossing technique for fabricating IO components in hard inorganic waveguiding materials*, Florence, Italy, IEE, 1983.
[17] Tiefenthaler, K., Lukosz, W., *Optics Letters*, 1984, 9, 137-139.
[18] Lukosz, W., Tiefenthaler, K., *Sensors and Actuators*, 1988, 15, 273-284.
[19] Vo-Dinh, T. (Ed.), *Biomedical Photonics Handbook*, CRC Press, 2003.
[20] Lukosz, W., *et al.*, *Output grating couplers on planar waveguides as integrated optical chemical sensors*, Montreux, Switzerland, 1990.
[21] Lukosz, W., Clerc, D., Nellen, P. H., *Input and output grating couplers as integrated optical biosensors*, Karlsruhe, West Germany, 1991.
[22] Clerc, D., Lukosz, W., *Sensors and Actuators, B: Chemical*, 1994, B19, 581-586.
[23] Brandenburg, A., *et al.*, *Sensors and Actuators, B: Chemical*, 1996, B30, 55-59.
[24] Piehler, J., *et al.*, *Applied Optics*, 1997, 36, 6554-6562.
[25] Veldhuis, G. J., *et al.*, *Pure and Applied Optics: Journal of the European Optical Society Part A*, 1998, **7**, 23-26.
[26] Wiki, M., *et al.*, *Biosensors and Bioelectronics*, 1998, 13, 1181-1185.
[27] Voros, J., *et al.*, *Biomaterials*, 2002, 23, 3699-3710.
[28] Brioude, V., *et al. Resonant grating biosensor platform design and fabrication*, St. Etienne, France: International Society for Optical Engineering, 2004.
[29] Tamir, T. (Ed.), *Integrated Optics*, Topics in Applied Physics, Springer-Verlag, 1979.
[30] Hopman, W., *et al.*, *Quasi 1-dimensional photonic crystals as building block for compact integrated optical sensors*, Wroclaw, Poland, IEEE, 2004.
[31] Dai, X., *et al.*, *Measurement Science and Technology*, 2006, 17, 1752-1756.
[32] Passaro, V. M. N., *J. Lightwave Technology*, 2000, 18, 973-984.
[33] Yariv, A., Yeh, P., *Optical Waves in Crystals*, John Wiley and Sons, 1983.
[34] Comsol Multiphysics® by COMSOL AB, *ver. 3.2, single license*, 2005.
[35] Dell'Olio, F., Passaro, V.M.N., *Optics Express*, 2007, 15, 4977-4993.
[36] Passaro, V. M. N. (Ed.), *Silicon Photonics*, Research Signpost, 2006.
[37] Wa Peng, W., Seng, C. K., *IEEE J. Quantum Electronics*, 2001, 37, 1138-1145.
[38] De Vos, K., *et al.*, *Optics Express*, 2007, 15, 7610-7615.
[39] Prieto, F., *et al.*, *Nanotechnology*, 2003, 14, 907-912.
[40] Densmore, A., *et al.*, *IEEE Photonics Technology Letters*, 2006, 18, 2520-2522.
[41] Debackere, P., *et al.*, *Surface plasmon interferometer in silicon-on-insulator: novel concept for an integrated biosensor*, Ottawa, Canada, IEEE, 2006.
[42] Riboli, F., Bettotti, P., Pavesi, L., *Optics Express*, 2007, 15, 11769-11775.

In: Modeling of Photonic Devices
Editor: Vittorio M. N. Passaro
ISBN 978-1-60456-980-3

Chapter 13

MULTI-QUANTUM-WELL SOLAR CELLS

***Francesco De Leonardis**[*1] **and Vittorio M. N. Passaro**[2]*

[1]Photonics Research Group, Ingegneria dell'Ambiente e per lo Sviluppo Sostenibile, Politecnico di Bari, viale del Turismo n. 8, 74100 Taranto, Italy

[2]Photonics Research Group, Dipartimento di Elettrotecnica ed Elettronica, Politecnico di Bari, via Edoardo Orabona n. 4, 70125 Bari, Italy

ABSTRACT

In this chapter, a generalized and self-consistent modeling for analyzing multi-quantum-well solar cells is presented. Continuity equations, Poisson, and Schrödinger equations are introduced in the model with the aim to consider the carriers dynamics inside the device. Theoretical investigations on behaviour and performance of multi-quantum-well solar cells are presented in terms of carrier concentrations and electric fields produced inside the device.

Keywords: Solar Cells, Multi Quantum Well, Photovoltaics, modeling

1. INTRODUCTION

Actually, photovoltaic sales are dominated by "first generation" silicon wafers, while various 'second generation' thin-films are now relatively mature including amorphous silicon, copper indium diselenide and cadmium telluride. All these technologies potentially offer substantially lower costs than "first generation" through greatly reduced material costs. Commercialisation has been slow, possibly due to low stabilised efficiency, manufacturability and concerns about toxicity, respectively. Historically, these technologies gave unexpectedly good results early in technological development, being difficulties apparent later. In contrast, the recently reported thin-film polycrystalline silicon on glass technology [1] was developed

---

* E-mail address: f.deleonardis@poliba.it

over a prolonged period using a "top-down" approach, stimulated by the desirable properties of the final product in terms of ultimate efficiency potential, durability, resource availability and non-toxicity, rather than immediate success [2]. In terms of energy conversion efficiency, this "second generation" technology could largely bridge the present gap between it and "first generation" product. As thin-film "second generation" technology matures, costs will become progressively dominated by those of the constituent materials, in this case, the top cover sheet and other encapsulants required for a 30-year operating life.

One surprising feature common to studies conducted over the last two decades on thin-film solar cells in large-scale manufacturing is the difficulty in differentiating between candidate technologies on the basis of costs [3-5]. This suggests that actual rather than assumed conversion efficiency will be the ultimate manufacturing cost differentiator between thin-films to reach large volume manufacture. This trend is already apparent with amorphous silicon cells with the transition from the single junction initial product releases to the double- and triple junctions, now offered by market-leaders.

Generalising this trend, it appears that, to reach its full potential after transition to “second generation” thin films, photovoltaics must evolve to a "third generation" with a number of features, including thin-film, high efficiency potential (greater than single junction), and abundant non-toxic materials. Advanced material technology and innovative device design concepts aim at bringing together scientists and engineers working in the field of artificially structured semiconductors and their exploitation in novel devices for third generation solar cells. In comparison with other optoelectronic devices, an idealized solar cell geometry can take on the simplest form of an active sheet to simultaneously have an area as large and a thickness as small as possible. Two possible technologies are candidates to realise third generation solar cells: silicon, and III-V semiconductors. Several investigations directed toward pushing theoretical limits of Si devices have been reported recently [6-9]. Progress seems possible by the superposition of different Si-based materials or different phases of the same Si material having complementary optoelectronic behaviours. A superposition of crystalline (c-Si) and amorphous (a-Si) phases, undertaken since the 1970s, has been complicated by an insurmountable electronic yield deterioration that accompanies the desired optical improvement.

Today an advantageous superposition of two Si phases is possible thanks to two methods: i) a-Si plasma enhanced chemical vapour deposition (PECVD) combined with hydrogenation [10-11] or ii) ion-beam implantation followed by adequate thermal treatment [12-13].

In case of III-V technologies, the multi quantum well (MQW) structures could guarantee an interesting improvement in the solar cell performance [14]. Generally speaking, the use of MQW structure into the intrinsic region of a p-i-n diode induces two conflicting effects: the short-circuit current is increased because of the additional absorption of low-energy photons in the lower band-gap quantum wells; and the open-circuit voltage is decreased because of the increased recombination of carriers trapped in the quantum wells [15]. However, several experimental results [16-17] have shown that the additional photocurrent resulting from the extension of absorption spectrum to lower energies can overcome the disadvantage of a reduction in the open-circuit voltage. More recently, results of photo-response calculations [18] have revealed that the insertion of quantum wells in a solar cell could lead to an improved photocurrent, without significant degradation in the open-circuit voltage. However, it seems evident that a mathematical model should include a large number of interdependent variables, as necessary to control the device behaviour.

A mathematical model based on the inclusion of quantum-well recombination and generation in an ideal diode model has been proposed [19]. The results proposed in [19] have induced people in the field to conclude that an improvement in efficiency is achieved only when the depth of the quantum well is less than about 200 meV. Similar results, using similar ideal diode equations, were reported [20]. Moreover, a work based on the solution of the Schrödinger equation [21], has indicated that while in InAs–InGaAs MQW solar cells the efficiency peaks could be obtained at a QW depth (or barrier height) of 450 meV, in GaAs–AlGaAs devices, the efficiency peaks are possible to achieve at zero QW depth, implying that GaAs single-gap cells are always more efficient than MQW GaAs–AlGaAs devices. Most recently, a model for MQW solar cell in which the dark current was characterized in terms of the quantum-well density of states, without considering the escape and capture rates of the wells, has been reported [22]. In this work, the photocurrent has been calculated by applying the diffusion equation to the photo-excited minority carriers in the well and bulk materials. However, the model proposed in [22], when compared to the experimental results, systematically overestimated the dark current. Finally, a self-consistent Poisson–Schrödinger-drift-diffusion mono-dimensional model for simulation of illuminated characteristics of a MQW solar cell has been presented [23].

In section 2 of this chapter, we introduce the mathematical modeling for the numerical investigations of solar cell performance and its behavior. In section 3 we present some numerical results and simulations applied to MQW p-i-n solar cells.

2. THEORY

The operation principle of MQW solar cells can be understood by observing Figure 1, where the energy band diagram is sketched. The structure is an GaAs/$Al_xGa_{1-x}As$ p-i-n photodiode with GaAs quantum wells in the undoped layer, and doped $Al_xGa_{1-x}As$ for the p, and n layers respectively. As schematically shown in Figure 1, the low-energy photons are absorbed in the GaAs quantum wells, resulting in generation of electron-hole pairs. The photo-excited carriers in the wells can either escape through thermo-ionic emission, or recombine predominantly through interface traps and direct recombination. The carriers that escape the quantum wells contribute to the current in the AlGaAs barrier region, or they may be captured back into a quantum well primarily through optical phonon scattering. Those carriers in the AlGaAs region that are not captured by any of the subsequent wells contribute to the current density. Thus, the goal of this section is to propose a mathematical model that self-consistently leads one to study the physical behaviour previously described. Generally speaking, the mathematical model has to include the drift and diffusion effects in the bulk AlGaAs layers, and the quantum confined effect into the MQW structure.

Thus, in AlGaAs regions of a device, where the electrons and holes behave as bulk carriers without any quantum confinement, the current continuity and current density equations are given by the traditional drift-diffusion equations:

$$\frac{dn_b}{dt} = G_b - U_b + \frac{n_{qw}}{\tau_{en}} - \frac{n_b}{\tau_{cn}} + \frac{1}{q}\frac{dJ_n}{dy} \tag{1}$$

$$\frac{dp_b}{dt} = G_b - U_b + \frac{p_{qw}}{\tau_{ep}} - \frac{p_b}{\tau_{cp}} - \frac{1}{q}\frac{dJ_p}{dy} \tag{2}$$

$$\bar{J}_n = -qn\mu_n \nabla\psi + qD_n \nabla n \tag{3}$$

$$\bar{J}_p = -qp\mu_p \nabla\psi - qD_p \nabla p \tag{4}$$

being y coordinate the direction perpendicular to the layer. In the previous equations, n_b and p_b are the bulk electron and hole densities, respectively, and n_{qw} and p_{qw} are the quantum-well electron and hole densities, respectively, The terms $\bar{J}_n$, and $\bar{J}_p$ are the electron and hole current densities, respectively. The terms n_{qw}/τ_{en} and p_{qw}/τ_{ep} are the rates by which the electrons and holes escape the quantum well and go into the bulk, whereas n_b/τ_{cn} and p_b/τ_{cp} are the rates by which electrons and holes are captured by the quantum wells from the bulk. The coefficients τ_{en} and τ_{ep} represents the electron and hole escape times, respectively, and τ_{cn} and τ_{cp} are the electron and hole capture times, respectively. Moreover, G_b and U_b represent the generation and recombination rates in the bulk, respectively. Finally, the term ψ indicates the electric potential, and can be determined by solving the Poisson equation:

$$\nabla \cdot (\varepsilon \nabla \psi) = q\left(N_A^- - N_D^+ + n_b - p_b\right) \tag{5}$$

being N_D^+, and N_A^- are the ionized donor and acceptor doping levels, and ε is the dielectric constant.

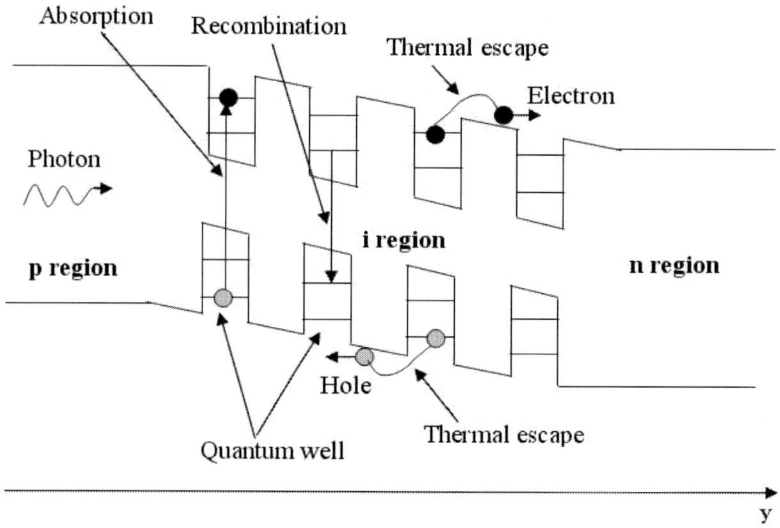

Figure 1. Band diagram of a MQW solar cell.

As outlined before, to accurately evaluate the performance of MQW solar cells it is necessary to consider both the process of transport and escape of carriers across and from the quantum wells. In this sense, to realise a self-consistent mathematical model, it is appropriate to include the Schrödinger equation [24] in the model:

$$\left[\frac{\hbar^2}{2}\frac{d}{dx}\frac{1}{m^*(x)}\frac{d}{dx}+V(x)\right]\Psi_i(x)=E_i\Psi_i(x) \tag{6}$$

being $V(x)$ is the potential profile in the device, and $m^*(x)$ represents a non-constant effective mass, used to account for different material systems throughout the bulk and quantum wells. The eigen-functions $\Psi_i(x)$ and eigen-energies $E_i(x)$ of carriers in the quantum wells are needed for calculation of quantum well absorption spectra, as well as the escape rate of carriers from the quantum wells.

Finally, the current continuity equations for the quantum-well system are written as the rate equations in which the net balance of four rates must vanish. These rate equations are given by [23]:

$$\frac{dn_{qw}}{dt}=G_{qw}-U_{qw}-\frac{n_{qw}}{\tau_{en}}+\frac{n_b}{\tau_{cn}}=0 \tag{7}$$

$$\frac{dp_{qw}}{dt}=G_{qw}-U_{qw}-\frac{p_{qw}}{\tau_{ep}}+\frac{p_b}{\tau_{cp}}=0 \tag{8}$$

The terms U_{qw} and G_{qw} are the modified Shockley–Hall-Read (SHR) recombination rate, and the photo-generation rate of the quantum wells, respectively. It is worth to note that the absence of current density terms in the earlier equations is due because the carriers in the quantum wells are assumed to be bound and, therefore, these carriers must first escape from the quantum wells to contribute to the current in the device. The considered equations represent a two-dimensional generalization of the mono-dimensional model proposed in literature [23]. This extension induces non trivial numerical complications that can be overcome using a software code based on Finite Element Method (FEM). Thus, the mathematical model here presented requires a link between FEM software used to solve Eqs. (1)-(5) and hand-made code needed to characterise the MQW structure (Eq. (6)). However, the increased numerical complexity of the model described in this section, with respect to the mathematical equations proposed in [23], is compensated by the advantage that the 2D approach give in terms of accuracy, in particular for predicting the electrode position influence.

3. Numerical Results

In this section, we will show some numerical results concerning the MQW solar cell with p-i-n architecture. Two assumptions about the model proposed in the previous section are very common in literature [23]. First, the escape and capture rates of the bulk are assumed to be equal to the capture and escape rates of the quantum wells. In other words, the carriers that escape the bulk are captured by the quantum wells, and those that escape the quantum wells are captured by the bulk. Second, the internal quantum efficiency is assumed unity, and the surface reflection and shading from metal contacts are considered negligible. In this hypothesis, the generation term (G_b or G_{qw}) is given by:

$$G(x)=\alpha(\lambda)N_{photon}(\lambda)e^{-\alpha(\lambda)x} \tag{9}$$

where $N_{photon}(\lambda)$ is the incident solar energy flux, x is the distance from the front surface, $\alpha(\lambda)$ is the absorption coefficient, and λ is the wavelength.

It is worth to note that to produce numerical simulations as induced by Eqs. (1)-(9), it is necessary to define the recombination terms both in quantum-wells and bulk regions. It is expected that non-radiative recombination processes in the quantum well will be dominant at room temperature [25]. Therefore, we neglect radiative recombination in the well, and model the recombination as occurring via non-radiative mechanisms. The recombination at the interfaces is usually modelled with a surface recombination velocity. However, such model requires the carriers to diffuse to the interface. Since confined carriers do not diffuse in the direction of the interface, the interface recombination in the quantum wells can be modelled as Shockley–Hall-Read recombination with appropriate choices for interface trap density and capture cross sections (σ_n , σ_p), as:

$$U_{qw}=U_{nrad}=\frac{\sigma_n\sigma_p v_T\left(n_{qw}p_{qw}-n_i^2\right)}{\sigma_n\left(n_{qw}+n_i\right)+\sigma_p\left(p_{qw}+n_i\right)}\frac{N_S}{d} \tag{10}$$

where v_T is the thermal velocity, and n_i is the intrinsic carrier concentration. The usual trap density in SHR recombination is replaced by the interface trap density N_s divided by the well width d to distribute the 2D interface traps throughout the well as a 3D trap density.

In the AlGaAs bulk, the recombination is modelled as the sum of a radiative term and a non-radiative term, as given by [23]:

$$U_b=U_{rad}+U_{nrad}=r\left(n_b p_b-n_i^2\right)+\frac{\sigma_n\sigma_p v_T\left(n_b p_b-n_i^2\right)}{\sigma_n\left(n_b+n_i\right)+\sigma_p\left(p_b+n_i\right)}N_T \tag{11}$$

where r is the radiative recombination coefficient of the barrier and N_T is the trap density. The trap density and capture cross sections in $Al_xGa_{1-x}As$ are assumed to be constant for all mole fractions of aluminum.

As it is evident from the theory section, the mathematical model involves partial differential equations that require appropriate boundary conditions to be used, even in stationary regime. In the device cross-section, we can distinguish two kinds of boundary, isolation and metallic (electrode). For the isolated boundaries, it is allowed to consider that both electric field and current density have normal components equal to zero (Neumann boundary condition):

$$n \cdot \nabla \psi = 0 \tag{12.1}$$

$$n \cdot J_n = 0 \tag{12.2}$$

$$n \cdot J_p = 0 \tag{12.3}$$

Differently, at the electrode boundaries we can assume that the electric potential is fixed and the recombination velocity at the electrode contact is infinite. In this hypothesis, we can express (at the contact boundaries) the electric potential, and the carrier concentrations as a function of applied voltage V_a and doping concentrations N, as:

$$\psi = V_a + \frac{kT}{q} ln \left(\frac{\frac{N}{2} + \sqrt{\left(\frac{N}{2}\right)^2 + n_i^2}}{n_i} \right) \tag{13.1}$$

$$n = \frac{N}{2} + \sqrt{\left(\frac{N}{2}\right)^2 + n_i^2} \tag{13.2}$$

$$p = -\frac{N}{2} + \sqrt{\left(\frac{N}{2}\right)^2 + n_i^2} \tag{13.3}$$

As mentioned before, the advantage of the generalization proposed in this chapter with respect to a previous work [23] is represented by the possibility to simulate 2D structures (see Figure 2). Figure 2 shows the cross section and relative sizes of device used in the following simulations. Since, as mentioned before, it is important to precisely set the boundary conditions, in the following Table 1 we list the main geometrical parameters of cross section shown in Figure 2.

Table 1. Geometrical parameters

Parameters	Value	Description
y_1	1.4 μm	Structure height
x_1	1.0 μm	Structure width
ju_1	0.2 μm	p-i junction depth
ju_2	0.8 μm	i-n junction depth
ac_1	0.4 μm	Anode left edge
ac_2	0.6 μm	Anode right edge

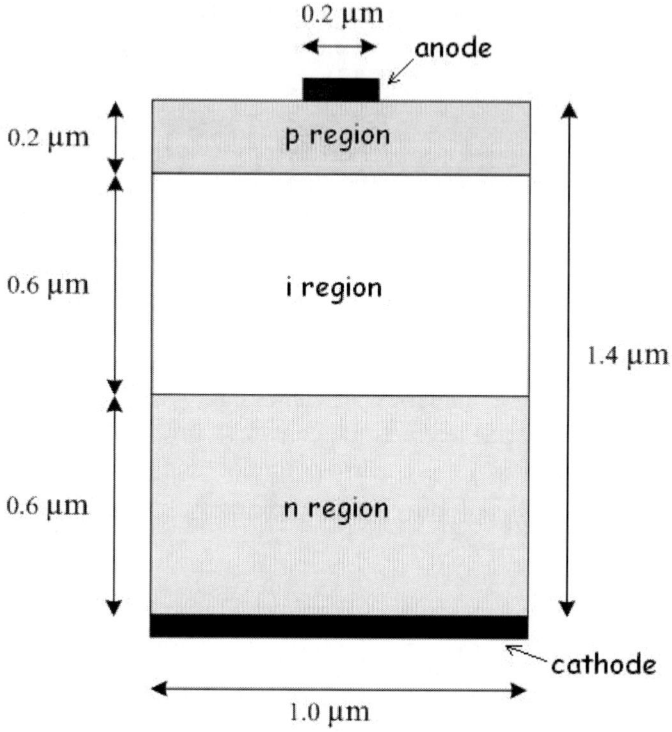

Figure 2. 2D p-i-n structure.

The previous geometrical coordinates are very useful to define the profile of the doping concentrations N in Eqs. (13.1)-(13.3). In fact, in our simulations we remove the simplification of step index profile made in [23], assuming a Gaussian profile. This hypothesis leads to better approximation of the doping distribution in real devices and, at the same time, leads to control the convergence problems in FEM approach [26]. Thus, the total expression of doping concentration N for the p-i-n structure is given by:

$$N = N_{Dn} + N_{Dnmax} G(x, -y, \infty, y_1) - N_{Ap} - N_{Apmax} G(x, y, x_A, 0) \tag{14}$$

where the values used in the simulation are listed in following Table 2 for the coefficients of Eq. (14).

Table 2. Doping parameters

Variable	Value
N_{Dn}	2.5×10^{21} m^{-3}
N_{Dnmax}	2.5×10^{23} m^{-3}
N_{Ap}	1.0×10^{18} m^{-3}
N_{Apmax}	9.0×10^{23} m^{-3}

Finally, the function G is given by:

$$G(x,y,ac_1,y_1)=\begin{cases} e^{-\frac{(x-x_A)^2}{ch_1^2}}e^{-\frac{(y-y_1)^2}{ch_1^2}} \forall x>x_A, y>y_1 \\ e^{-\frac{(x-x_A)^2}{ch_1^2}} \forall x>x_A, y\le y_1 \\ e^{-\frac{(y-y_1)^2}{ch_2^2}} \forall x\le x_A, y>y_1 \\ 1 \forall x\le x_A, y\le y_1 \end{cases} \tag{15}$$

being $x_A = ac_2 - ac_1$ the anode width, and terms ch_1 and ch_2 are given by:

$$ch_1 = \frac{ju_1}{\sqrt{ln\left(\frac{N_{Ap\,max}}{N_{Ap}}\right)}} \tag{16.1}$$

$$ch_2 = \frac{ju_2}{\sqrt{ln\left(\frac{N_{Ap\,max}\cdot N_{Ap}}{N_{Dn}}\right)}} \tag{16.2}$$

3.1. Comparison between MQW P-I-N and P-N Solar Cells

In this section, some numerical results showing the operation principle of MQW solar cells and their potential with respect to classical p-n cells, are presented. The physical parameters used in the following simulations are listed in Table 3. Figure 3 shows the hole concentration inside the structure as sketched in Figure 2, by assuming x_A = 0.2 μm, applied voltage $V_a = 0$, aluminium title $x = 0.1$, and number of quantum wells $N_{qw} = 20$.

Table 3. Physical parameters in simulations

Variable	Value
σ_b	5.0×10^{-19} m^{-2}
σ_{qw}	1.0×10^{-20} m^{-2}
$v_{T,AlGaAs}$	4.4×10^5-2.1×10^5 m/s
$v_{T,GaAs}$	4.4×10^5 m/s
r	1.0×10^4 m^3/s
N_T	5.0×10^{21} m^{-3}
N_S	1.0×10^{15} m^{-2}
d	200×10^{-10} m

Figure 4 shows the module of electric field inside the p-i-n device as a function of y-coordinate for different values of applied voltage. The plot indicates that the electric field module decreases with increasing the applied voltage. In addition, the electric field holds a constant value inside the intrinsic region (from 0.2 μm to 0.8 μm).

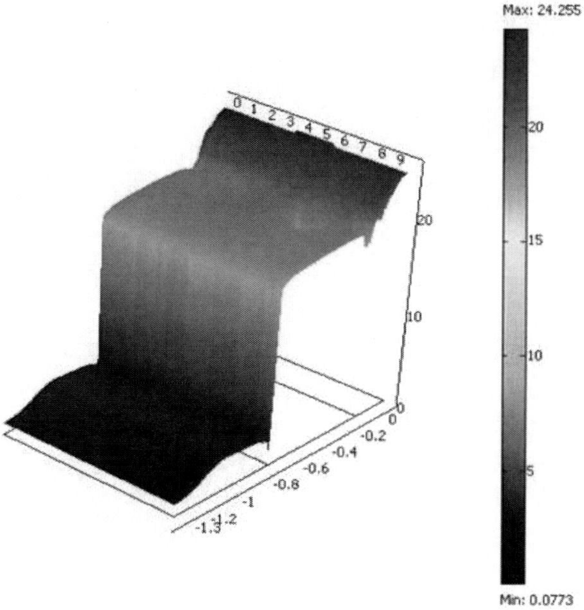

Figure 3. Hole concentration inside the p-i-n structure for $V_a = 0$ V.

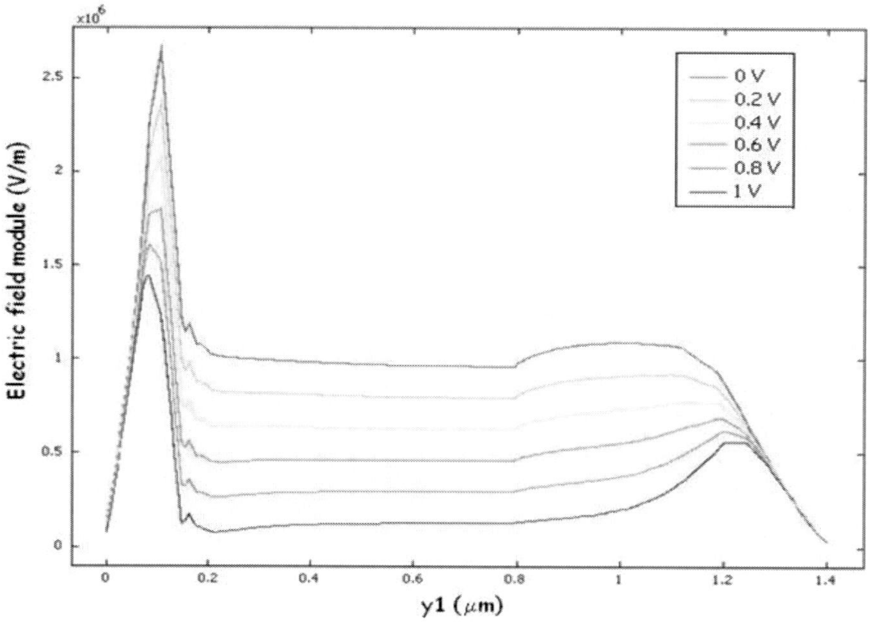

Figure 4. Electric field module versus y-vertical coordinate for different values of V_a.

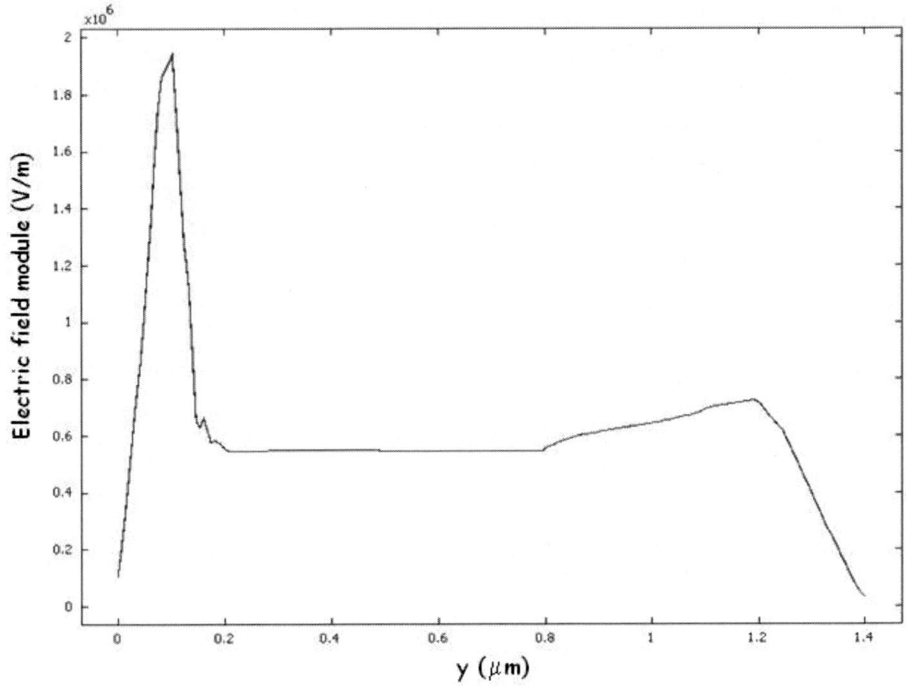

Figure 5(a). Electric field module versus y-vertical coordinate for p-i-n structure.

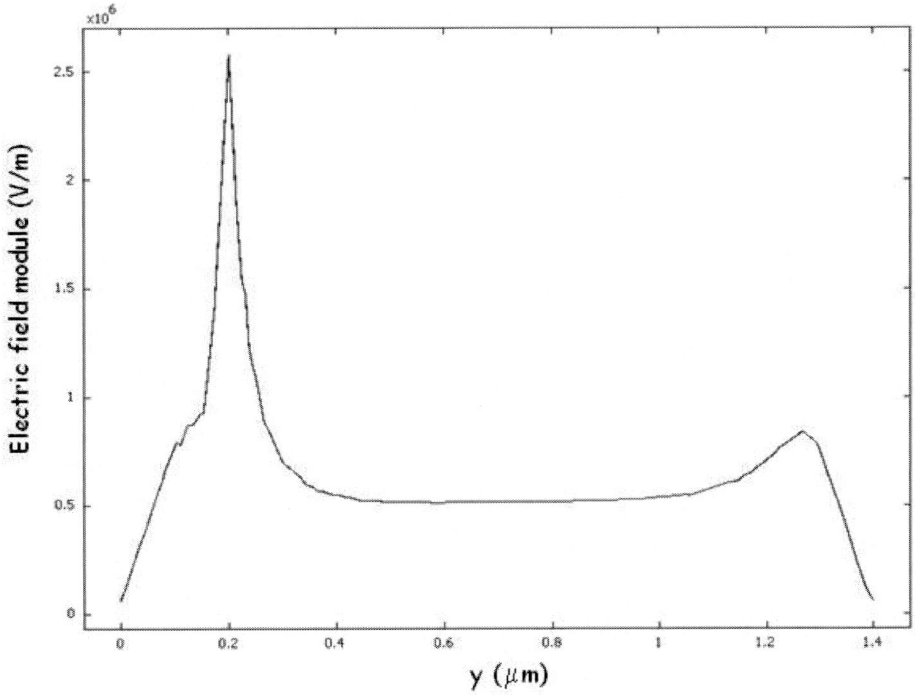

Figure 5(b). Electric field module versus y-vertical coordinate for p-n structure.

It is worth to note that the electric field module inside the solar cell can be used to evaluate the device performance in terms of dark current. In particular, low values of electric field induce low levels of dark current, and, then, a larger efficiency. In order to put into evidence the potential of the MQW solar cell with respect to p-n junction, Figure 5(a) and 5(b) show the electric field inside the device for p-i-n MQW diode and p-n diode, respectively. Assuming $V_a = 0.5$ V, Figures 5(a) and 5(b) indicate that the maximum value of electric field module is 2.6×10^6 V/m, and 1.9×10^6 V/m for p-n and p-i-n structures, respectively. This simulation induces to conclude that the insertion of MQW intrinsic region leads to lower the dark current with respect to the p-n solar cell, so improving the efficiency. In addition, as explained in [16-17], the presence of MQW in the intrinsic region allows a further advantage to be achieved, i.e. the improvement of absorption of low energy photons.

3.2. Effect of Anode Width

As mentioned before, the 2D model proposed in this chapter leads us to evaluate the influence of the anode width, too. Also in this case, we use the electric field as a figure of merit to evaluate the trend of performance change. Figures 6 and 7 show the electric field module versus y-vertical coordinate for $x_A = 0.2$ μm and $x_A = 0.4$ μm, respectively. In both simulations, the device total width has been assumed as equal to 1 μm.

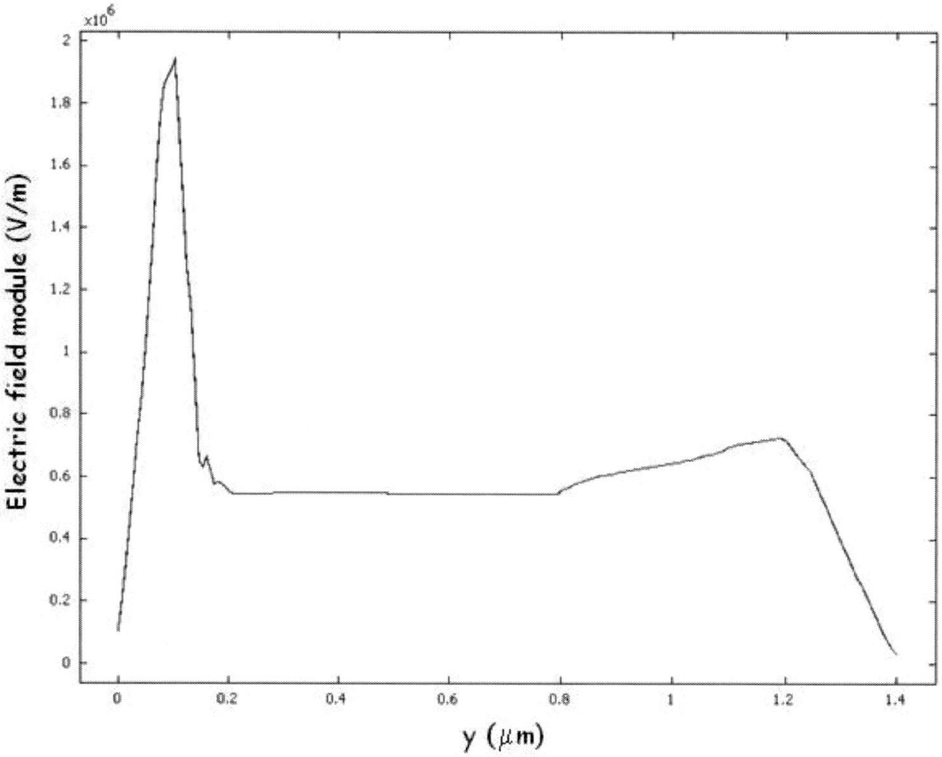

Figure 6. Electric field module versus y-vertical coordinate for p-i-n structure (xA = 0.2 μm).

Simulations indicate the maximum value of electric field module as 1.9×10^6 V/m for $x_A = 0.2$ μm, and 6.1×10^6 V/m for $x_A = 0.4$ μm. Thus, Figures 6-7 demonstrate that the dark current increases with increasing the anode width, inducing a falling of solar cell efficiency.

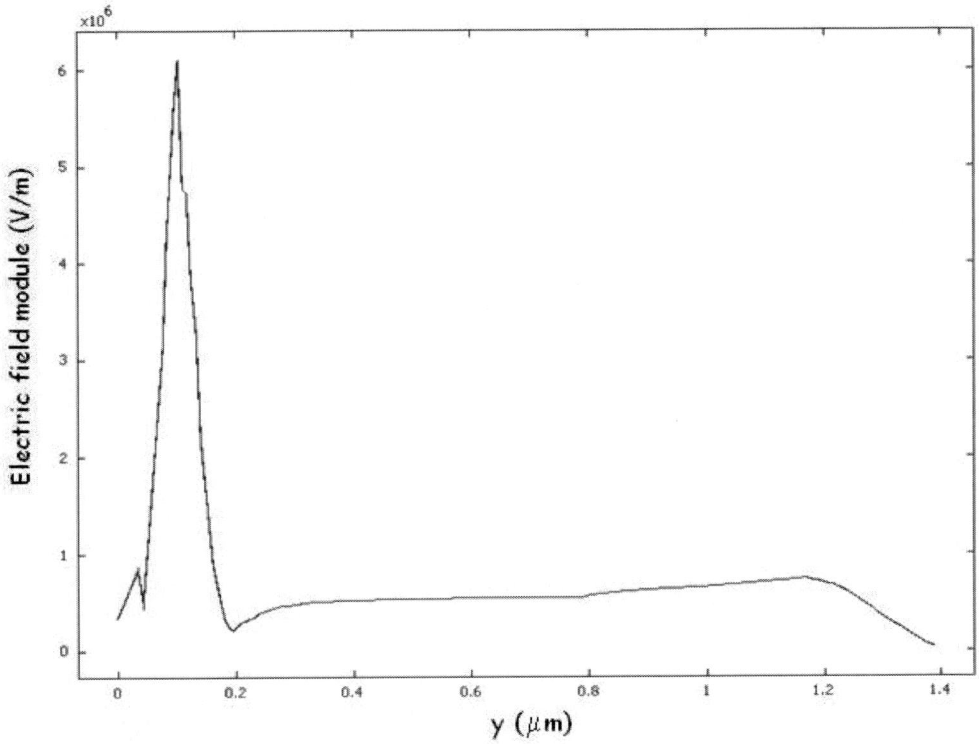

Figure 7. Electric field module versus y-vertical coordinate for p-i-n structure (xA = 0.4 μm).

3.3. Effect of Quantum Well Width

In this section, the influence of quantum well width is evaluated. In the following simulations, the number of quantum wells is hold as a constant ($N_{qw} = 20$). In addition, anode width, applied voltage, and aluminium title are assumed as 0.2 μm, 0.8 V and 0.1, respectively. The numerical results are shown in Figures 8-11, where the quantum well width has been set to values d = 200 Å, 150 Å, 100 Å, 50 Å, respectively. In this case, we have chosen as figure of merit the electric potential inside the structure. Thus, Figures 8-11 plot the electric potential versus the y-vertical coordinate.

The simulations indicate that the electric potential is weakly influenced by the quantum well width. In this case, it is possible to conclude that the solar cell efficiency can be also considered approximately independent from the quantum well width, as experimentally demonstrated in [23].

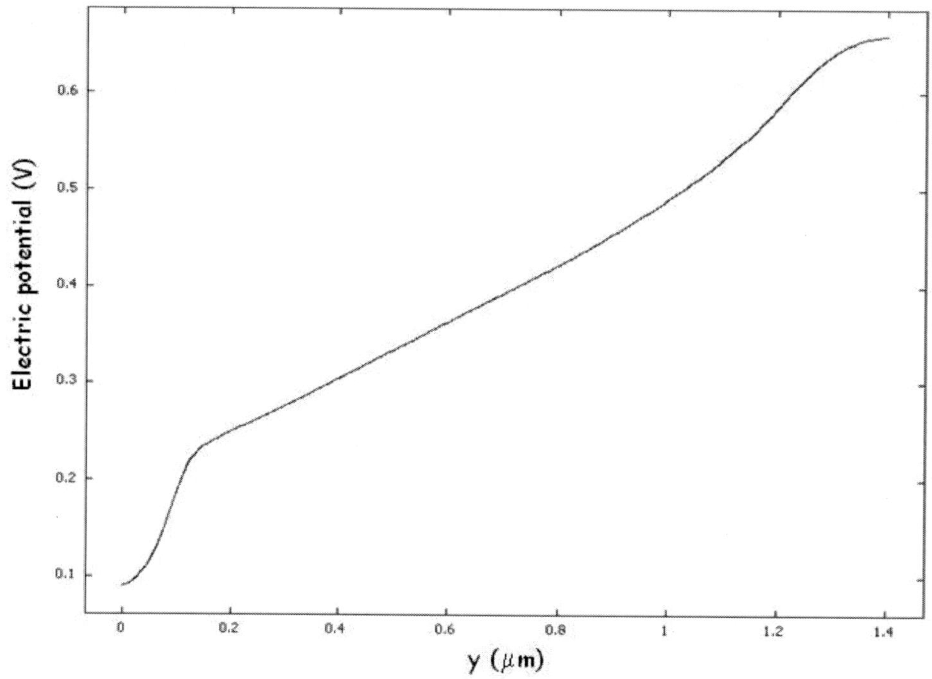

Figure 8. Electric potential inside p-i-n structure with d = 200 Å.

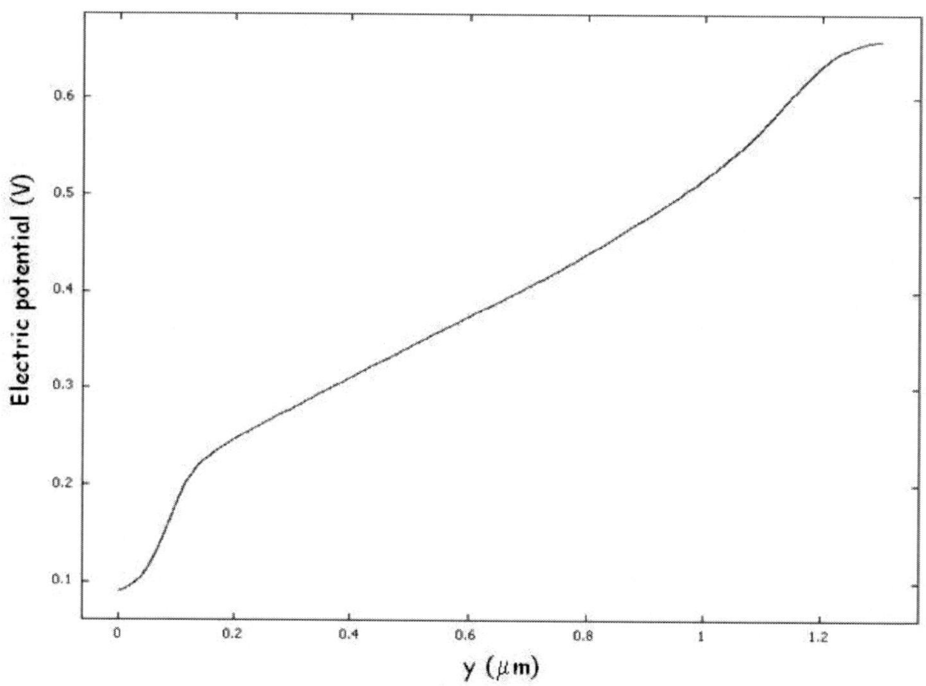

Figure 9. Electric potential inside p-i-n structure with d = 150 Å.

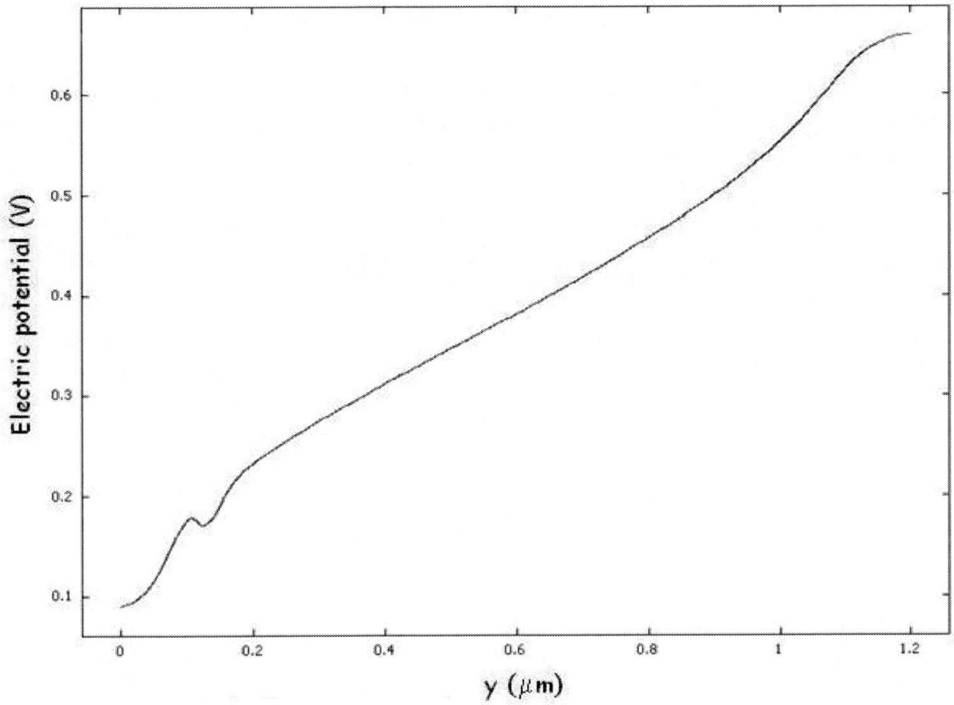

Figure 10. Electric potential inside p-i-n structure with $d = 100$ Å.

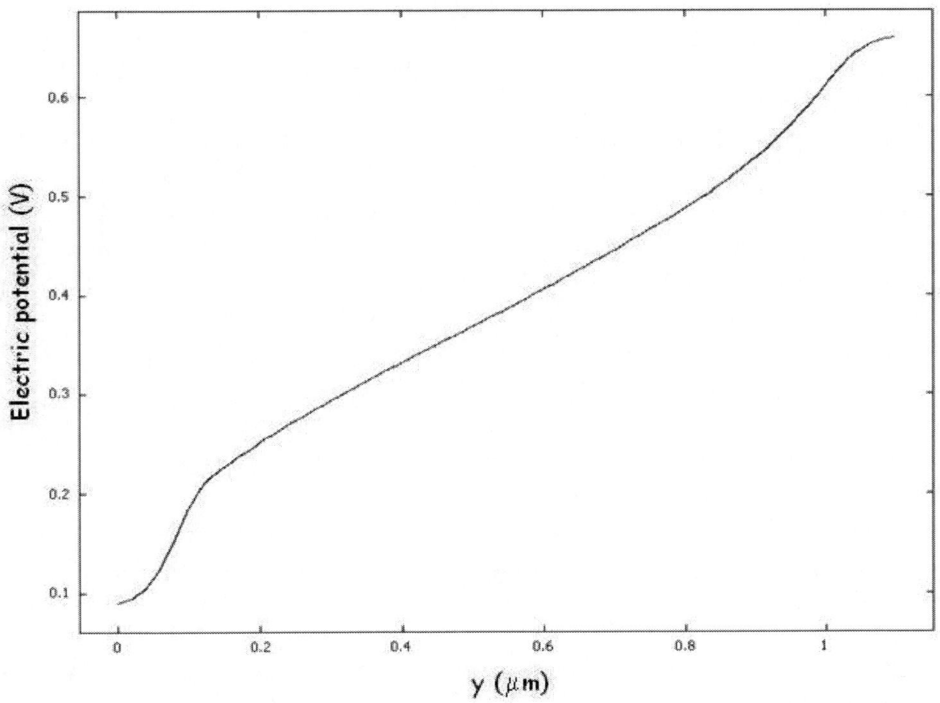

Figure 11. Electric potential inside p-i-n structure with $d = 50$ Å.

CONCLUSION

In this chapter, we have presented a generalization of 2D modeling to evaluate performance and behaviour of MQW solar cells for 3^d generation photovoltaics. The numerical results proposed have demonstrated that incorporating GaAs quantum wells in the intrinsic region of a p-i-n AlGaAs solar cell, can increase the efficiency of the single band-gap baseline cell. Thus, we have proposed 2D numerical simulations relevant to carrier concentrations and the electric field inside the MQW solar cell. In particular, our investigations demonstrate that the solar cell efficiency could be influenced by the anode width. Improvements in the efficiency can be strongly dependent on the number and width of quantum wells. However, for a aluminium title equal to 0.1, a substantial independence of solar cell efficiency from the quantum well width can be noted.

REFERENCES

[1] Basore, P. A., 29th *IEEE Photovoltaic Specialists Conference*, May 2002.

[2] Green, M., *Solar Cells*, 1989, 26, 1.

[3] Bamett, A., Rothwarf, A., *IEEE Trans. Electron Devices*, 1980, ED-27, 615.

[4] Woodstock, J., et al., 14th *European PV Solar En. Cod*, 1997, 857.

[5] Frantzis, L., et al., 16th *European Photovoltaic Solar Energy Conf.,* Glasgow, 2000, 2100.

[6] Green, M. A., Progress in Photovoltaics: *Research and Applications*, 2001, 9, 123-135.

[7] Luque, A., Marti, A., *Physical Review Letters*, 1997, 78, 5014-5017.

[8] Landsberg, P. T., Badescu, V., *Progress in Quantum Electronics*, 1998, 22**,** 211-230.

[9] Brendel, R., Wemer, J. H., Queisser, H. J., *Solar Energy Materials and Solar Cells*, 1996, 41/42, 419-425.

[10] Tanaka, M., Toguchi, M., Matsuyama, T., Sawada, T., Tsuda**,** S., Nakano**, S.,** Hanafusa H., Kuwano, Y., *J. Applied Physics*, 1992, 31, 3518-3522.

[11] Sakata, H., Nakai, T., Baba, T., Toguchi, M., Tsuge, **S.,** Uchihashi, K., Kiyama**, S.,** 28th IEEE Photovoltaic Spec. Conf., Anchorage, Alaska, USA, 16-23 September 2000, and *Progress in Photovoltaics*, 2000, 8, 503-513.

[12] Kuznicki, Z. T., Int. Seminar on Semiconductor Converters of Visible and Infrared Radiation, 25-26 April 1995, Rzeszow, Poland, *Optoelectronics Review*, 1995, 34. 79-85.

[13] Kuznicki, Z. T., Grob, J.-J., Provot, E., E-MRS, Spring Meeting, Strasbourg, France, 22-26 May 1995, *Nuclear Instruments and Methods in Physics Research* B, 1996, 112, 188-191.

[14] Barnham, K., Duggan, G., *J. Applied Physics*, 1990, 67, 3490–3493.

[15] Corkish, R., Green, M., Proc. 23rd *Photovoltaics Specialist Conf.*, 1993, 675-680.

[16] Barnham, K., Braun, B., Nelson, J., Paxman, M., *Applied Physics Letters*, 1991, 59, 135–137.

[17] Ragay, F., Wolter, J., Marti, A., Araujo, G., Proc. 12th *Eur. Community Photovoltaic Solar Energy Conf.,* 1994, 1429–1433.

[18] Renaud, P., Vilela, M., Freundlich, A., Bensaoula, A., Medelci, N., Proc. *IEEE Conf. Photovoltaic Energy Conversion*,. New York, 1995, 1787-1790.
[19] Anderson, N., *J. Applied Physics*, 1995, 78, 1850-1861.
[20] Rimada, J. C., Hernandez, L., *Microelectronics Journal,* 2001, 32, 719–723.
[21] Mohaidat, J. M., Shum, K., Wang, W. B., Alfano, R. R., *J. Applied Physics*, 1994, 79, 5533–5537.
[22] Connolly, J. P., Nelson, J., Barnham, K. W. J., Ballard, I., Roberts, C., Roberts, J. S., Foxon, C. T., Conf. Rec. 28th *IEEE Photovoltaic Specialists*, NJ, 2000, 1304–1307.
[23] Ramey, S. M., Khoie, R., *IEEE Transactions on Electron Devices*, 2003, 50, 1179-1188.
[24] Harrison, P., Quantum Wells, *Wires and Dots*, John Wiley and Sons, 2005.
[25] Gurioli, M., Vinattieri, A., Colocci, M., Deparis, C., Massies, J., Neu, G., Bosacchi, A., Franchi, S., Physical Review B, *Condensed Matter*, 1991, 44, 3115–3124.
[26] COMSOL Multiphysics by COMSOL AB, Stockholm, ver. 3.2, *single license*, 2005.

INDEX

A

Aβ, 205, 206, 207
ABC, 211
absorption, viii, 26, 31, 32, 40, 46, 63, 64, 71, 83, 92, 96, 103, 104, 105, 143, 145, 147, 155, 174, 175, 176, 181, 197, 207, 211, 231, 236, 251, 253, 254, 259, 311, 312, 314, 315, 316, 327, 343, 347, 348, 349, 350, 351, 352, 353, 354, 384, 387, 388, 394
absorption coefficient, 31, 46, 96, 175, 254, 388
absorption spectra, 347, 387
acceptor, 386
access, 143, 251
accidental, 108
accuracy, 22, 57, 64, 69, 71, 88, 91, 93, 96, 97, 100, 112, 198, 211, 212, 213, 237, 238, 239, 244, 259, 260, 262, 279, 288, 312, 342, 346, 370, 387
acetone, 362
achievement, 54
acoustic, 5, 12, 198
acoustic waves, 12
ADC, 8, 9, 10, 11
adsorption, 381
aerospace, 20
agent, 55
aging, 199
AIP, 357
air, 21, 54, 83, 90, 91, 93, 103, 104, 153, 167, 199, 215, 295, 296, 300, 346, 348, 362, 373, 374
aircraft, 60, 207
airplanes, 199
algorithm, 49, 63, 208, 238, 260, 342
alternative, 5, 22, 30, 32, 61, 84, 199, 212
alters, 360, 363
aluminium, 349, 391, 395, 398
aluminum, 294, 295, 296, 388
ammonia, 26, 52, 69
amorphous, 383, 384
amplitude, 21, 22, 23, 27, 35, 40, 41, 47, 81, 87, 147, 148, 149, 167, 195, 207, 212, 225, 226, 267, 269, 276, 280, 325, 337, 344, 361, 368
analog, 4, 8, 31
angular velocity, 26, 63, 64, 65, 71
anisotropic, 84, 201, 203, 207
annealing, 20, 24, 198, 312
anode, 172, 389, 391, 394, 395, 398
anomalous, 243, 245, 246, 247, 281, 285
antenna, 4, 10, 12, 21, 24, 26, 34, 35, 36, 40, 41, 207
antibody, 44, 45, 48, 49, 363
antigen, 44, 45, 48, 49, 363
anti-reflection coating, 194
application, 2, 3, 8, 30, 34, 40, 43, 60, 66, 84, 85, 87, 112, 143, 188, 190, 191, 192, 196, 199, 223, 265, 360, 372, 378, 380
aqueous solution, 47, 54, 55, 57, 93, 94, 373, 374
arsenide, 12, 22
artificial intelligence, 93
assumptions, 265, 277, 312, 321, 342, 387
astronomy, 8, 11
asymmetry, 122, 242, 244
atmosphere, 197
attention, viii, 3, 14, 26, 44, 54, 80, 105, 187, 192, 297, 342, 354
automotive, 3, 11, 43, 60
automotive applications, 11
autonomous, 61

B

backscattered, 149
backscattering, viii, 63, 64, 143, 145, 149, 150, 153, 160, 162, 163
bandgap, ix, 341, 342, 343, 347, 355
bandwidth, 2, 4, 7, 8, 12, 15, 17, 18, 20, 22, 24, 25, 26, 27, 29, 30, 31, 32, 36, 39, 41, 64, 95, 144,

173, 179, 195, 196, 198, 199, 251, 353, 365, 371, 372, 373, 376, 377, 378
barrier, 62, 325, 326, 385, 388
barriers, 343, 346, 347
batteries, 26
battery, 20, 26
beams, 12, 22, 34, 61, 62, 63, 64, 65, 160, 161, 162, 196, 197, 208, 253
behavior, viii, 24, 38, 41, 42, 43, 106, 143, 154, 163, 175, 248, 263, 279, 287, 298, 299, 300, 317, 323, 328, 332, 334, 353, 380, 385
behaviours, 384
bending, 61, 144, 146, 147, 154, 265, 272, 341
benefits, 2, 21
bias, 31, 32, 41, 63, 174, 175, 176
biochemical, vii, 1, 3, 25, 26, 44, 45, 47, 48, 51, 52, 55, 71, 85, 306, 366
biochemistry, 3, 112
biocompatible, 44
biological, 44, 45, 47, 207, 360, 363, 380
biologically, 44
biology, 112
biomedical, 44, 71
biomedical applications, 44, 71
biomolecules, 44
biosensors, 34, 44, 47, 49, 50, 51, 53, 54, 71, 381
birefringence, 62, 87, 88, 119, 121, 122, 124, 141, 173, 175, 182, 195, 196, 233, 235, 236, 238, 239, 242, 253, 258, 263, 282, 369, 373, 374
blocks, 51, 180, 181
Boltzmann constant, 168
bonding, 84, 105
bottlenecks, 21, 180
boundary conditions, 43, 168, 201, 208, 211, 223, 224, 226, 295, 296, 344, 388, 389
bovine, 51, 364
BPM, 3, 16, 40, 120, 166, 269, 342
Bragg grating, ix, 26, 30, 51, 52, 54, 55, 103, 145, 180, 193, 195, 196, 197, 198, 199, 341, 359, 360, 361, 362, 363, 364, 365, 369, 370, 371, 373, 380
branching, 361
breakdown, 174, 326, 336
broadband, 2, 7, 30, 41, 71, 198, 312
buffer, 23, 38, 62
building blocks, 18, 145, 252, 357
buildings, 199
bulk materials, 385
bypass, 12

C

cables, 198
CAD, 3, 12, 16, 166, 175, 227
cadmium, 383
calibration, 42
cancer, 29, 45
cancer treatment, 29
candidates, 143, 384
capacitance, 31, 39, 40, 41, 42, 173, 174
capacity, 2, 10, 18, 71
carbon, 47
carbon dioxide, 47
carcinogen, 45
carrier, viii, ix, 12, 14, 15, 18, 22, 37, 39, 41, 42, 43, 64, 85, 92, 143, 145, 147, 148, 154, 155, 158, 165, 167, 168, 169, 172, 173, 174, 175, 176, 181, 184, 231, 232, 236, 239, 251, 254, 259, 261, 262, 263, 270, 275, 288, 312, 315, 354, 383, 388, 389, 398
Cartesian coordinates, 208
cathode, 172
cavities, 17, 26, 54, 265, 270, 273
cell, 21, 22, 28, 29, 42, 208, 211, 364, 367, 384, 385, 394, 398
cell surface, 367
ceramics, 33
channels, 2, 20, 29, 105, 143, 144, 196, 199
charge density, 45
chemical, 3, 25, 26, 34, 44, 45, 55, 60, 66, 84, 105, 197, 198, 306, 360, 363, 368, 384
chemical composition, 197
chemical sensing, 55
chemical vapour, 84, 105, 384
chemical vapour deposition, 84, 105, 384
cladding, 32, 49, 62, 79, 81, 82, 83, 84, 85, 86, 88, 89, 90, 91, 93, 94, 95, 97, 99, 100, 103, 104, 105, 121, 146, 159, 160, 173, 195, 197, 238, 295, 298, 299, 300, 338, 347, 350, 363
cladding layer, 32, 49, 62, 84, 103, 160, 173, 295, 347, 350
classes, 5, 96
classical, 41, 42, 148, 153, 391
classification, 24
classified, 65, 190, 192, 253
clouds, 37, 38
CMOS, 3, 48, 49, 53, 54, 71, 83, 85, 86, 99, 293, 334
CO2, 26, 71, 197
coatings, 103
coherence, 30, 31
commercial, 3, 16, 21, 166, 175, 191, 199, 231, 252, 371
commercialization, 360
communication, 2, 8, 18
communication systems, 8
compatibility, 2, 3, 20, 25, 44, 71, 83, 207, 231, 252, 268
compensation, 87, 196, 242, 294

competition, 63, 65, 152, 160, 162
complementary, 44, 86, 380, 384
complexity, 21, 34, 85, 212, 237, 252, 361, 387
complications, 387
components, viii, 10, 14, 16, 35, 41, 60, 63, 64, 65, 72, 80, 81, 84, 87, 96, 104, 144, 145, 151, 167, 187, 191, 208, 211, 223, 225, 232, 236, 245, 252, 253, 334, 378, 381, 389
composite, 223, 372
composition, 350, 351, 354, 363, 365, 374
compounds, 12, 14, 27, 45, 48, 145, 191, 192
compression, viii, 21, 22, 23, 231, 242, 243, 246, 247, 248, 286, 287, 290
computation, 195, 211, 238
computational grid, 212
computer, 252
computers, 24, 34, 207, 252
computing, 207, 251
concave, 191
concentration, 15, 39, 42, 43, 44, 46, 47, 50, 52, 53, 55, 69, 71, 158, 166, 167, 169, 172, 173, 176, 313, 314, 315, 316, 317, 318, 319, 320, 323, 324, 326, 328, 329, 330, 331, 333, 334, 336, 337, 338, 364, 388, 390, 391, 392
conduction, 150, 294, 295, 325
conductivity, 149, 294, 295, 296, 297
conductor, 57
confidence, 39
configuration, 19, 22, 29, 35, 36, 52, 63, 65, 84, 85, 91, 93, 100, 103, 104, 144, 145, 190, 194, 334, 342, 360, 361, 363, 364, 366, 369, 376
confinement, 38, 40, 54, 55, 57, 80, 83, 84, 85, 87, 88, 89, 90, 91, 92, 93, 94, 95, 102, 103, 105, 106, 108, 111, 119, 154, 173, 214, 218, 238, 272, 295, 306, 334, 338, 343, 347, 348, 349, 353, 377
constituent materials, 384
constraints, viii, 143, 145, 163
consumption, viii, 2, 12, 15, 20, 22, 24, 25, 60, 61, 95, 166, 184, 293, 295, 305, 320
continuity, 81, 151, 170, 175, 344, 345, 385, 387
control, viii, 7, 10, 12, 22, 24, 61, 87, 143, 175, 188, 238, 252, 274, 285, 384, 390
controlled, 12, 17, 36, 144, 153, 197, 354
convergence, 97, 390
conversion, 2, 3, 4, 12, 18, 26, 65, 180, 232, 237, 248, 253, 260, 277, 290, 313, 314, 317, 318, 319, 323, 327, 330, 331, 334, 337, 384
cooling, 20, 193
copper, 3, 383
correlation, 22, 23, 153, 160, 161, 162
correlation function, 23, 153, 162
corrosion, 198
costs, 21, 383, 384
couples, 50
coupling constants, 372
covering, 12, 105, 175
CRC, 76, 228, 381
critical value, 105, 160, 280, 315
cross-phase modulation, viii, 231, 251
cross-talk, 144, 252
crystal, 22, 26, 27, 29, 30, 33, 34, 44, 87, 232
crystalline, 38, 39, 191, 384
crystals, 26, 34, 341
cycles, 98, 347, 348

D

damping, 194
data processing, 2, 20
decay, 45, 81, 82, 266, 274, 276, 315, 316, 317, 326
decomposition, 215
decoupling, 262, 299, 301
defects, 63, 105, 198, 342, 355
definition, 41, 92, 188, 267, 269, 274, 275
degenerate, 236
degradation, 384
degree, 91, 92, 198, 360
degrees of freedom, 25, 88
delays, 5, 6, 12, 30, 252, 351, 353, 354
delivery, 20, 21
delta, 255
density, 18, 39, 42, 45, 81, 148, 149, 150, 153, 167, 236, 252, 254, 261, 262, 275, 294, 296, 297, 314, 318, 319, 326, 327, 328, 330, 333, 334, 336, 337, 385, 387, 388, 389
deposition, 28, 84, 90, 108, 144, 192, 369
derivatives, 208
desorption, 381
detection, vii, ix, 1, 2, 7, 11, 14, 25, 28, 36, 40, 44, 45, 46, 48, 49, 50, 53, 54, 60, 61, 65, 359, 360, 361, 363, 364, 365, 366, 368, 369, 377, 378, 380
deviation, 196
diagnostic, 93
diamond, 191
dielectric, 26, 27, 45, 56, 80, 81, 93, 141, 153, 193, 195, 201, 211, 224, 326, 344, 347, 386
dielectric constant, 45, 81, 193, 224, 344, 347, 386
dielectric materials, 26
dielectric permittivity, 153
dielectrics, 207
differential equations, 153, 181, 195, 253, 320, 342
diffraction, viii, 21, 40, 187, 188, 191, 192, 193, 194, 212, 227, 362, 370, 371
diffusion, 24, 29, 154, 167, 239, 263, 385
diffusivities, 43
diffusivity, 294, 300
diode laser, 12, 193, 194

diodes, 192, 193, 194, 196
dipole, 28, 29, 35, 36, 41, 87, 148, 150
dipole moment, 87, 150
dipole moments, 87
Dirac delta function, 233, 254, 255
discontinuity, 50, 54, 81, 82, 170, 192, 213, 334
discretization, 212, 342
discrimination, 49
dispersion, viii, 14, 27, 37, 64, 87, 96, 103, 147, 155, 166, 167, 172, 181, 184, 191, 195, 196, 211, 231, 232, 233, 236, 237, 238, 239, 240, 243, 245, 246, 251, 253, 259, 260, 263, 265, 275, 281, 282, 283, 284, 285, 293, 347, 348
displacement, 167
distribution, 14, 51, 57, 58, 66, 82, 88, 95, 97, 100, 106, 150, 166, 167, 168, 169, 170, 177, 184, 198, 213, 215, 234, 241, 256, 274, 294, 295, 296, 297, 298, 299, 300, 307, 309, 344, 370, 375, 390
distribution function, 150
division, 143, 193, 198, 245, 341, 361
DNA, 26, 44, 45, 46, 367
DNA sequencing, 45
donor, 386
dopant, 166
doped, 38, 39, 69, 85, 172, 173, 196, 197, 199, 252, 311, 312, 315, 326, 334, 335, 337, 338, 339, 369, 385
doping, 25, 38, 39, 40, 42, 167, 172, 173, 174, 176, 184, 191, 192, 312, 323, 386, 389, 390
Doppler, 22, 24
drying, 362
durability, 198, 384
DUV, 84

E

earth, 20, 21, 24, 27, 71, 196, 311
eigenvalue, 212, 222, 296, 297, 346
eigenvalues, 169, 211
elaboration, 21, 325
electric current, 325
electric field, vii, ix, 1, 26, 27, 29, 30, 31, 32, 33, 35, 38, 40, 54, 56, 71, 75, 81, 82, 87, 96, 97, 99, 106, 137, 148, 149, 153, 169, 170, 171, 222, 223, 234, 241, 253, 254, 255, 256, 296, 307, 325, 326, 334, 347, 371, 375, 383, 389, 392, 394, 395, 398
electric potential, 386, 389, 395
electrical, viii, 2, 8, 10, 12, 14, 15, 16, 30, 31, 39, 41, 166, 176, 180, 198, 199, 251, 311, 312, 325, 326, 327, 328, 329, 331, 334, 338
electrical power, 15, 199
electrochemical, 26, 44, 71
electrodes, 10, 22, 24, 29, 30, 35, 37, 40, 42, 61, 86
electroluminescence, 326
electromagnetic, 2, 3, 4, 20, 25, 26, 27, 28, 29, 31, 44, 45, 71, 80, 96, 137, 146, 148, 169, 198, 207, 211, 225, 342, 343, 348, 354, 361, 366
electromagnetic fields, 26
electromagnetic wave, 45, 96, 211
electromagnetism, 207, 212
electron, 14, 15, 18, 42, 43, 84, 148, 150, 153, 166, 167, 169, 236, 325, 327, 328, 329, 330, 385, 386
electron beam, 84
electronic, iv, 2, 3, 4, 8, 9, 10, 12, 14, 18, 20, 24, 26, 27, 43, 64, 71, 198, 233, 254, 255, 384
electronic circuits, 9, 27, 71
electronics, 18, 34, 71, 293, 359
electrons, 39, 43, 175, 254, 326, 331, 333, 347, 385, 386
electrostatic, 42, 43, 87
embossing, 363, 364, 365, 366, 367, 368
emission, 45, 85, 92, 192, 193, 194, 196, 277, 283, 312, 313, 315, 316, 334, 336, 337, 338, 341, 385
EMP, 27, 28
employment, 3, 14, 49, 51, 104
encoding, 14, 180, 252
energy, 45, 144, 150, 160, 194, 233, 246, 263, 266, 267, 270, 272, 311, 315, 316, 325, 326, 349, 384, 385, 394
energy transfer, 246, 263, 325, 326
engineering, 3, 207, 222
enlargement, 349
environment, 16, 31, 166, 294
environmental, 3, 25, 41, 43, 44, 93, 199, 363
environmental conditions, 25, 199
environmental influences, 44
enzymatic, 44
enzyme, 44
epitaxy, 31, 84, 105
equilibrium, 39, 45
equipment, 2
erbium, viii, 85, 144, 191, 197, 199, 311, 312, 313, 314, 315, 316, 317, 318, 319, 320, 323, 324, 325, 326, 327, 328, 329, 330, 331, 333, 334, 335, 336, 337, 338
ESA, 24, 316
estimating, 297
etching, 83, 84, 89, 91, 108, 144, 162, 192, 219, 220, 365, 368, 369
ethanol, 55, 69, 362, 367
evidence, vii, 1, 65, 262, 283, 394
evolution, 147, 149, 153, 157, 158, 181, 182, 183, 232, 235, 242, 243, 244, 245, 247, 253, 258, 264, 281, 284, 286, 287
excitation, viii, 65, 143, 156, 158, 159, 162, 179, 182, 267, 268, 283, 285, 286, 312, 316, 317, 318, 325, 326, 327, 328, 330, 331, 333, 334

exciton, 315, 325, 326, 343, 347, 348, 349, 350, 351, 352, 353, 354
exploitation, 384
explosive, 198
exponential, 31, 272, 288, 298, 313
exposure, 33, 365
external environment, 47
external influences, 199
extinction, 16, 17, 32
extraction, 21
extrapolation, 97
extrinsic, 33

F

fabricate, 14, 17, 31, 44, 49, 55, 86, 91, 119, 145, 166, 192
fabrication, vii, ix, 7, 16, 24, 25, 26, 29, 48, 51, 60, 63, 71, 79, 83, 84, 85, 86, 88, 91, 93, 94, 99, 100, 102, 103, 105, 108, 112, 121, 141, 144, 145, 162, 188, 192, 197, 217, 236, 259, 306, 341, 359, 360, 367, 368, 369, 372, 378, 380, 381
FBG, 18, 193, 194, 195, 196, 197, 198, 360
feedback, 41, 65, 145, 188, 192, 193, 194, 360, 370
FEM, viii, 38, 39, 40, 42, 43, 55, 57, 58, 66, 69, 70, 71, 87, 96, 97, 100, 105, 112, 117, 118, 120, 126, 132, 135, 137, 140, 141, 154, 159, 160, 166, 167, 187, 188, 211, 212, 213, 214, 215, 232, 238, 239, 253, 260, 280, 281, 294, 303, 304, 306, 307, 334, 342, 370, 371, 380, 387, 390
Fermi, 150
Fermi-Dirac, 150
FFT, 49, 63
fiber, ix, 1, 18, 20, 29, 32, 33, 34, 44, 45, 46, 47, 60, 61, 65, 104, 192, 193, 194, 195, 196, 197, 198, 199, 224, 249, 252, 291, 341, 342, 359, 360, 361, 380
fiber Bragg grating, ix, 18, 193, 194, 195, 196, 197, 198, 224, 342, 359, 360, 361, 380
fiber optics, 20, 249, 291
fibers, 2, 20, 96, 117, 175, 190, 195, 197, 198, 232, 233, 255
fidelity, 27
film, 7, 28, 47, 48, 50, 69, 70, 84, 191, 364, 366, 383, 384
films, 96, 383, 384
filters, 4, 5, 6, 7, 8, 16, 19, 117, 143, 144, 146, 154, 158, 162, 191, 192, 194, 195, 196, 226, 293, 341, 342, 355
finite element method, vii, 38, 79, 117, 154, 166, 188, 211, 232, 238, 239, 253, 260, 280, 294, 309, 334
first generation, 383
flame, 369
flatness, 144
flexibility, 2, 3, 5, 222, 378
flow, 21, 326, 364
fluctuations, 361
fluid, 365
fluorescence, 45, 46
focusing, 3, 80, 105
Fourier, 49, 63, 204, 224, 225, 226, 237, 238, 244, 254, 260, 295, 342, 371
four-wave mixing, 236, 246
FTTH, 360
fulfillment, 380
fullerene, 96, 99
FWHM, 154, 158, 162, 181, 237, 243, 248, 259, 260, 261, 262, 263, 267, 284, 285, 365

G

GaAs, 11, 14, 20, 22, 27, 30, 36, 61, 63, 145, 146, 153, 160, 191, 347, 385, 391, 398
gallium, 12, 22
gas, 199
gases, 47
Gaussian, 86, 144, 153, 157, 158, 162, 199, 237, 242, 243, 244, 245, 246, 247, 248, 260, 284, 285, 286, 288, 390
gel, 367, 368
generalization, 141, 233, 266, 387, 389, 398
generation, viii, 2, 4, 7, 57, 87, 147, 180, 181, 191, 231, 232, 245, 252, 276, 286, 288, 296, 315, 383, 384, 385, 386, 387, 388, 398
generators, 4
genetic, 342
geometrical parameters, vii, 60, 79, 80, 87, 90, 93, 97, 98, 99, 100, 112, 141, 213, 376, 389
glass, 48, 50, 197, 364, 367, 380, 383
glasses, 48, 95, 191, 197
glucose, 26, 47, 52, 53, 55
government, iv
GPS, 61
graph, 90, 323
gratings, viii, ix, 25, 34, 35, 36, 187, 188, 189, 190, 191, 192, 193, 194, 195, 196, 197, 198, 199, 212, 217, 223, 227, 239, 294, 341, 354, 359, 360, 361, 362, 366, 367, 368, 370, 372, 373, 375, 376, 378, 380
grid resolution, 211
grids, 208
ground-based, 2
groups, 95
growth, 84, 192
guidance, 61, 232
guidelines, 39, 41, 251
gyroscope, 26, 61, 63, 65

H

H1, 200
H_2, 200
handling, 212
harmonics, 188, 224, 225, 226, 344, 346, 348
head, 26, 27, 34
heat, 294, 295, 296, 297, 304
heat capacity, 294, 296, 297
heat transfer, 294, 304
heating, 42, 297, 299, 303, 304, 305, 306, 307, 308
height, 6, 17, 38, 57, 66, 67, 80, 81, 82, 84, 88, 89, 91, 93, 94, 98, 100, 102, 104, 105, 118, 153, 154, 167, 182, 218, 233, 239, 240, 255, 261, 272, 278, 308, 309, 325, 365, 378, 385, 389
high resolution, 8, 84, 193, 199, 365, 367
high tech, 34
high-frequency, 4
high-speed, 14, 34, 166, 293
HIV, 46
homogeneous, 47, 48, 51, 53, 67, 93, 94, 95, 211, 222, 224, 225, 321, 343, 346, 347
human, 46, 51
human immunodeficiency virus, 46
humidity, 362, 364
hybrid, 96, 194
hybridization, 45
hydrogen, 197
hydrogenation, 384
hydrolysis, 369
hyperbolic, 208, 245
hyperthermia, 29
hypothesis, 181, 254, 275, 388, 389, 390

I

identification, 24
illumination, 22, 197
images, 21, 22, 24
imaging, 104
immune system, 44
immunity, 3, 4, 25, 44, 60, 61, 361
immunochemistry, 364
implementation, 5, 7, 8, 211, 212, 252, 265, 361
impulsive, 27
in vitro, 45
in vivo, 45
incidence, 211, 225, 342, 368
inclusion, 385
independence, 85, 398
independent variable, 87
indices, 38, 51, 57, 66, 67, 68, 71, 83, 88, 89, 92, 97, 121, 135, 137, 140, 169, 200, 215, 226, 239, 255, 280, 294, 297, 348, 373, 374
indium, 383
industrial, 20, 252
industry, 231, 252
inelastic, 45
inertial navigation system, 60
infinite, 81, 82, 121, 211, 224, 343, 348, 349, 389
infrared, 47, 80, 87, 92, 94, 104, 113, 197, 312, 378
infrared light, 197
infrastructure, 2, 14, 145
injection, viii, 15, 61, 64, 65, 85, 143, 145, 160, 162, 166, 325, 331, 334, 338, 354
innovation, 75
inorganic, 381
InP, 11, 12, 13, 14, 19, 32, 191, 213, 252
insertion, 7, 31, 361, 384, 394
insight, 201, 211, 323, 342, 347, 354
instability, 245
integrated circuits, 54, 231, 252
integrated optics, 14, 21, 112, 341
integration, 2, 3, 21, 24, 34, 43, 80, 84, 86, 105, 144, 162, 195, 252, 272, 293, 309, 325, 360, 361, 362, 380
Intel, 2, 39, 251, 290
intensity, 5, 8, 27, 32, 40, 57, 58, 82, 83, 87, 95, 96, 99, 160, 180, 184, 197, 244, 246, 253, 262, 269, 270, 287, 298, 300, 323, 324, 325, 334, 364, 367
interaction, viii, 12, 29, 30, 44, 45, 47, 50, 51, 52, 86, 148, 180, 181, 184, 189, 232, 246, 252, 253, 254, 261, 263, 266, 268, 269, 270, 271, 272, 279, 288, 289, 311, 347, 354, 367
interactions, 44, 96, 137, 201, 253, 312
interface, 39, 45, 47, 48, 49, 54, 79, 81, 82, 84, 154, 170, 192, 225, 239, 343, 345, 385, 388
interference, 3, 4, 11, 26, 27, 36, 44, 49, 71, 80, 85, 152, 153, 175, 191, 198, 263, 273, 368
interval, 199
intrinsic, 2, 20, 22, 24, 25, 36, 43, 62, 147, 167, 211, 265, 384, 388, 392, 394, 398
inversion, 39, 41, 337
ion implantation, 84
ionic, 385
ionization, 326
ions, viii, 172, 174, 311, 312, 316, 317, 318, 319, 320, 326, 328, 329
iron, 25
irradiation, 20, 197
isolation, 27, 295, 299, 300, 389
isotropic, 28, 29, 203, 208, 294, 343
iteration, 97, 98

K

kinetics, 45

L

L1, 180
L2, 180, 327
large-scale, 279, 384
laser, viii, 2, 3, 5, 6, 8, 9, 12, 13, 14, 19, 21, 22, 24, 26, 27, 28, 29, 40, 45, 46, 61, 62, 63, 64, 65, 96, 143, 145, 148, 149, 150, 151, 152, 153, 160, 161, 162, 163, 165, 188, 192, 193, 194, 196, 197, 198, 199, 212, 215, 217, 221, 223, 224, 252, 253, 275, 277, 278, 279, 280, 311, 312, 334, 335, 336, 337, 338, 364, 367
lasers, viii, 7, 19, 26, 64, 143, 144, 145, 146, 191, 192, 193, 194, 195, 196, 199, 226, 251, 252, 265, 277, 279, 312, 341, 343, 355
latency, 252
law, 46, 79, 167, 295, 312, 346
laws, 325
lead, 36, 38, 64, 83, 162, 196, 235, 258, 272, 384
leakage, 104, 105, 149, 153, 225, 226, 344, 347, 348, 354, 368
LED, 26, 85, 312
lens, 22, 194, 364
lifetime, 44, 51, 147, 148, 154, 155, 237, 239, 259, 261, 262, 263, 268, 270, 288, 314, 327, 337
light beam, 79, 87, 334, 338, 366
light scattering, 315
light transmission, 103, 195
limitation, 7, 30, 61, 102, 276, 278
limitations, 3, 8, 20, 22, 24, 45, 63, 103, 105, 199, 251, 252, 359, 361
linear, vii, 2, 7, 9, 12, 24, 29, 31, 34, 41, 42, 59, 79, 85, 87, 88, 89, 92, 96, 97, 98, 99, 100, 102, 106, 112, 147, 150, 151, 196, 200, 201, 226, 237, 244, 245, 246, 254, 260, 262, 266, 268, 269, 274, 277, 296, 297, 321, 353
linear dependence, 59, 297
linear function, 41
linear law, 353
links, 41
liquids, 103, 369
literature, vii, viii, 3, 16, 17, 20, 21, 22, 27, 61, 63, 65, 68, 71, 79, 80, 82, 99, 104, 112, 120, 125, 134, 141, 165, 166, 172, 176, 180, 182, 188, 223, 224, 232, 233, 243, 252, 253, 254, 261, 262, 268, 269, 277, 288, 290, 294, 297, 300, 305, 312, 313, 315, 318, 338, 342, 343, 351, 361, 362, 387, 388
lithium, 12, 20, 22, 24, 27, 29, 30, 31, 191, 252
lithography, 54, 84, 94
location, 10, 205, 206, 263
locus, 239
long-distance, 22
losses, 2, 3, 5, 7, 20, 24, 25, 38, 49, 63, 80, 83, 84, 95, 103, 105, 126, 144, 145, 147, 149, 154, 162, 166, 173, 176, 184, 197, 214, 218, 226, 244, 245, 265, 272, 277, 288, 293, 313, 315, 316, 318, 328, 331, 334, 336, 341, 342, 343, 361, 369, 378
low cost, 60, 61
low power, viii, 25, 60, 100, 166, 184, 232, 288, 293
luminescence, 326
luminosity, 85

M

M1, 65
machinery, 199
magnetic, 21, 25, 42, 45, 57, 82, 222, 255
magnetic field, 25, 42, 82, 222
maintenance, 18
management, 21
manufacturing, 112, 191, 231, 252, 384
market, 252, 360, 384
mask, 22, 94, 197
mathematical, vii, viii, 1, 145, 148, 165, 187, 212, 232, 236, 237, 248, 253, 255, 259, 265, 279, 294, 384, 385, 387, 388
matrices, 222
matrix, 3, 12, 22, 35, 40, 137, 138, 139, 146, 150, 151, 188, 201, 202, 205, 206, 222, 223, 238, 260, 294, 312, 342, 346
Mb, 13
measurement, 26, 27, 30, 44, 53, 61, 66, 71, 360, 361, 366
measures, 27, 198
mechanical, iv, viii, 1, 3, 43, 87, 143, 145, 148, 151, 191, 195, 198, 360, 366
mechanical stress, 366
mechanical structure, 360
mechanics, 41, 212
media, viii, 5, 45, 87, 96, 187, 191, 207, 211, 348, 351
medicine, 3, 43
MEH-PPV, 96, 99
memory, 18, 71, 211, 320
MEMS, 3, 43, 61, 231, 252
metal-oxide-semiconductor, viii, 15, 165, 166
microcavity, 34, 52, 53, 155, 265, 267, 268, 269, 270, 271, 272, 274, 276, 277, 279
microelectronics, 2, 14
micrometer, ix, 15, 51, 54, 66, 285, 290, 359, 380
microprocessors, 251
microsystem, 17
microwave, 3, 4, 5, 26, 71, 199, 207, 222, 341

millimeter-wave, 5
miniaturization, 61, 85, 360
minority, 39, 385
mirror, 34, 103, 192, 194, 199, 224, 364
missions, 24
mixing, 69
mobile users, 1
mobility, 39, 167
MOCVD, 32
modeling, vii, viii, ix, 1, 3, 64, 71, 80, 117, 141, 145, 165, 167, 184, 188, 231, 253, 278, 293, 311, 312, 331, 341, 355, 383, 385, 398
models, 166, 195, 211, 253, 332, 338
modern society, 34
modulation, viii, 2, 3, 9, 12, 14, 15, 29, 30, 31, 32, 51, 87, 165, 167, 172, 174, 175, 180, 181, 190, 192, 194, 195, 196, 225, 232, 253, 293, 301, 302, 303, 306
mole, 388
molecular structure, 44
molecules, 44, 45, 363, 364
monochromatic light, 191
monolithic, 3, 43, 144, 191, 194, 272
MOS, 15, 35, 37, 38, 39, 40, 41, 42, 166, 167, 170, 171, 172, 184
MRS, 398
multimedia services, 2
multiplexing, 2, 14, 193, 198, 245, 263, 361
multiplication, 5, 24
multiplicity, 45

N

nanoclusters, viii, 85, 311, 312, 314, 315, 316, 317, 318, 320, 321, 326, 328, 334, 335, 337, 338, 339
nanocrystals, 99, 252, 311, 314, 320, 327, 339
nanometer, 54, 57, 80, 87, 95, 146, 162, 369
nanometers, 49, 194, 195, 196, 197
nanoparticles, 311
NASA, 24
NATO, 339
natural, 61, 144
neglect, 154, 267, 275, 388
neoplastic cells, 45
network, 11, 12, 13, 14, 18, 21
Ni, 167
NIST, 28
nitride, 14, 17, 48, 49, 69, 84, 85
noise, 9, 12, 24, 25, 28, 41, 63, 65, 193, 194
nonlinear, vii, viii, 3, 9, 12, 18, 21, 43, 64, 65, 79, 80, 86, 87, 92, 95, 96, 97, 98, 99, 100, 102, 103, 112, 143, 144, 145, 146, 150, 152, 153, 154, 155, 157, 180, 181, 192, 201, 231, 232, 233, 234, 235, 236, 237, 238, 239, 242, 243, 244, 246, 247, 248, 252, 253, 254, 255, 257, 258, 260, 263, 264, 265, 267, 268, 276, 277, 279, 283, 284, 286, 290, 342, 354
nonlinearities, 99, 196
non-uniform, 211, 213
non-uniformity, 211
normal, 12, 63, 65, 81, 167, 200, 201, 202, 203, 204, 225, 246, 281, 295, 342, 389
normalization, 97, 157, 233, 234, 255, 256
normalization constant, 234, 256
normal-mode, 203
novelty, 80
n-type, 38, 39, 160, 167, 172, 173, 174
nuclei, 254
nucleic acid, 44
nucleotides, 45

O

oil, 199
omni-directional, 2
one dimension, 224
operator, 148, 150, 226, 345
optical absorption coefficient, 172
optical chemical sensors, 55, 381
optical communications, 191, 193
optical fiber, 2, 5, 12, 20, 27, 32, 33, 34, 45, 65, 105, 193, 194, 198, 227, 232, 244, 252, 294, 341
optical gain, 62, 313, 314, 315, 316, 318, 319, 328, 330
optical microscopy, 87
optical parameters, 39, 118, 278
optical properties, 87, 103, 198, 224
optical pulses, viii, 8, 143, 181, 232, 242, 245, 247, 253, 254
optical solitons, 242
Optical Time Division Multiplexing, 2
optical transmission, 18
optics, 21, 27, 194, 342
optimization, vii, 55, 66, 71, 79, 90, 95, 96, 100, 112, 377
optoelectronic, 34, 105, 144, 166, 187, 191, 384
optoelectronic devices, 191, 384
optoelectronics, 80, 113
ordinary differential equations, 260, 268
organic, 26, 45, 48, 95, 360, 362
organic compounds, 48
organic solvent, 26, 360, 362
organic solvents, 26, 360
orientation, 32, 194
orthogonal functions, 237, 260
oscillation, 19, 44, 45, 59, 145, 242
oscillator, 347

oxide, 16, 37, 38, 39, 40, 41, 42, 49, 54, 57, 69, 80, 83, 84, 88, 92, 98, 100, 118, 121, 154, 167, 170, 173, 239, 252, 295, 299, 307, 326
oxide thickness, 40, 167, 173, 295
oxygen, 47

P

packaging, 252
packet switching, 18
packets, 18
paper, 126
parameter, 17, 40, 89, 100, 121, 132, 160, 188, 199, 204, 207, 217, 218, 239, 240, 275, 279, 285
partial differential equations, 167, 181, 182, 211, 232, 238, 253, 260, 262, 268, 388
particle physics, 3, 43
passive, viii, 19, 27, 65, 117, 143, 145, 153, 165, 191, 194, 198, 252, 306, 343, 355
pathogens, 47
pattern recognition, 24
PDEs, 167, 212
performance, viii, ix, 1, 2, 6, 8, 12, 14, 17, 20, 21, 24, 25, 28, 30, 31, 32, 34, 41, 44, 54, 80, 89, 95, 96, 99, 108, 112, 143, 144, 145, 146, 154, 170, 175, 178, 184, 191, 193, 194, 217, 221, 265, 267, 285, 286, 293, 300, 338, 359, 361, 372, 376, 378, 380, 383, 384, 385, 387, 394, 398
periodic, viii, 3, 51, 63, 87, 144, 187, 188, 189, 190, 192, 194, 197, 202, 203, 204, 208, 212, 217, 223, 224, 225, 226, 245, 342, 343, 348, 354, 365, 370
periodicity, 187, 193, 224
permeability, 87, 254
permit, 2, 16, 49, 71
permittivity, 81, 87, 149, 153, 167, 225, 226, 254
perturbation, 26, 51, 144, 194, 195, 200, 201, 204, 212, 223, 224, 234, 245, 257, 343, 346, 368, 370, 371
perturbations, 30, 204, 380
pesticide, 364
pH, 44, 47
phase inversion, 7
phase shifts, 95, 184
phased array antennas, 4
phonon, 347, 385
phosphate, 27, 53, 197
photodetectors, 56, 64
photolithography, 29, 365
photoluminescence, 326
photon, viii, 45, 51, 64, 83, 96, 143, 145, 147, 154, 155, 181, 197, 231, 251, 253, 266, 276, 313, 314, 315, 316, 317, 318, 319, 323, 324, 328, 333
photonic, iv, vii, viii, ix, 1, 2, 3, 4, 5, 8, 13, 14, 18, 24, 25, 26, 27, 28, 32, 33, 34, 46, 54, 61, 64, 65, 66, 71, 80, 83, 84, 87, 93, 95, 117, 145, 159, 162, 165, 187, 188, 191, 197, 208, 211, 212, 224, 226, 227, 252, 294, 306, 341, 342, 343, 347, 348, 355, 359, 369, 378, 381
photonic crystal fiber, 197
photonic crystals, 80, 188, 342, 381
photonic devices, iv, vii, 2, 3, 14, 24, 64, 84, 117, 159, 188, 211, 224, 226, 227, 306, 341, 360
photonics, 2, 3, 4, 252, 357, 359, 361, 380
photons, 266, 317, 323, 336, 384, 385, 394
photosensitivity, 197
photovoltaics, 383, 384, 398
physical mechanisms, 197, 312, 325
physical properties, 207
physics, 175, 313
piezoelectric, 22, 33, 44, 191
piezoelectric properties, 191
pipelines, 199
pitch, 10, 360, 362, 363, 364, 365, 366, 367, 368, 369
planar, 21, 24, 25, 28, 117, 144, 192, 193, 212, 342, 343, 348, 349, 367, 368, 377, 381
plane waves, 208, 342
plasma, 14, 37, 45, 64, 84, 105, 147, 155, 166, 167, 172, 181, 184, 232, 236, 253, 263, 265, 275, 293, 365, 384
plastic, 367
platforms, 3, 11, 14, 20, 21, 22
play, 263
PMMA, 33, 69
Poisson, ix, 42, 175, 383, 385, 386
Poisson equation, 386
polarization, viii, 12, 20, 28, 29, 30, 38, 55, 57, 59, 62, 65, 68, 85, 87, 109, 117, 127, 129, 132, 145, 147, 148, 149, 150, 151, 152, 170, 182, 195, 196, 200, 201, 211, 225, 231, 232, 233, 236, 238, 239, 240, 243, 246, 247, 251, 253, 254, 255, 259, 263, 285, 286, 288, 301, 307, 309, 342, 343, 347, 369, 373, 375, 378, 380
polarized, 12, 21, 45, 81, 87, 106, 107, 108, 109, 170, 171, 172, 184, 239, 244, 256, 259
polarized light, 21, 170, 171, 172, 184
pollution, 26
polycarbonate, 363, 365, 366
polycrystalline, 383
polymer, 99, 197, 365
polymer optical fibers, 197
polymeric materials, 52, 53
polymers, 11, 14, 54, 96, 192, 294, 309
polynomial, 99, 102, 127
polyurethane, 96, 99
population, 317, 318, 320, 321, 322, 323, 326, 334, 337

porous, 252
ports, 16, 126, 137
powers, 56, 57, 95, 99, 127, 155, 193, 196, 247, 278, 316
prediction, 262, 274
preparation, iv, 367
pressure, 26, 44, 84, 105, 197, 198, 360
probability, 150, 315, 334
probe, 26, 145, 261, 263, 264, 266, 267, 268, 269, 270, 273, 275, 283, 284, 285, 286, 288
production, 47, 145, 165, 252, 295, 360, 368
production technology, 252
property, 5, 7, 33, 188, 297, 309, 344, 350
protein, 44, 45, 53, 367
protein binding, 367
proteins, 48
pseudo, 22, 24
p-type, 38, 39, 160, 172, 173, 174
pulse, 8, 22, 24, 64, 96, 144, 154, 155, 156, 157, 158, 180, 181, 182, 183, 194, 198, 233, 234, 236, 237, 238, 242, 243, 244, 245, 246, 247, 248, 253, 254, 255, 258, 259, 260, 261, 262, 263, 264, 265, 266, 267, 268, 269, 270, 271, 272, 273, 283, 284, 285, 286, 288, 290
pulsed laser, 8
pulses, viii, 8, 9, 10, 157, 158, 180, 181, 182, 184, 197, 199, 231, 232, 234, 235, 236, 239, 242, 243, 244, 245, 246, 247, 248, 251, 253, 254, 257, 258, 263, 264, 265, 267, 268, 275, 281, 285, 286, 287, 288, 290
pumping, viii, 311, 312, 313, 314, 316, 317, 325, 326, 327, 328, 329, 331, 334, 337, 338
pumps, 65
pyrene, 45

Q

QAM, 13
quantization, 9, 10, 11, 12
quantum, viii, ix, 31, 32, 41, 61, 63, 143, 145, 148, 150, 151, 153, 188, 212, 341, 342, 343, 383, 384, 385, 386, 387, 388, 391, 395, 398
quantum confinement, 385
quantum mechanics, 41
quantum phenomena, 148
quantum well, ix, 31, 61, 153, 341, 342, 343, 384, 385, 386, 387, 388, 391, 395, 398

R

race, 64, 75
radar, 2, 8, 10, 21, 22, 207
radial distance, 82
radiation, 25, 45, 144, 148, 149, 160, 162, 200, 207, 224, 265, 341, 348, 360, 362, 368, 370, 372
radio, 8, 11
radiofrequency, 2, 12
radius, 7, 16, 17, 19, 36, 40, 52, 53, 61, 63, 136, 143, 145, 149, 154, 155, 156, 160, 163, 269, 272, 273, 274, 278, 279, 280
Raman, v, viii, 45, 46, 64, 65, 76, 154, 155, 165, 166, 180, 181, 182, 184, 231, 232, 233, 235, 237, 247, 251, 252, 253, 254, 255, 258, 260, 261, 262, 263, 265, 269, 270, 271, 272, 273, 274, 275, 277, 278, 279, 281, 282, 283, 285, 286, 288, 290
Raman scattering, viii, 233, 235, 251, 255, 258, 275, 288
Raman spectroscopy, 45, 46
random, 22, 24
range, 7, 8, 9, 10, 11, 12, 14, 16, 21, 22, 23, 24, 25, 28, 29, 32, 33, 34, 40, 41, 42, 48, 56, 69, 71, 87, 102, 109, 119, 121, 122, 127, 144, 145, 157, 160, 162, 175, 193, 195, 196, 198, 199, 208, 232, 239, 247, 253, 272, 276, 296, 313, 314, 315, 316, 317, 318, 319, 323, 324, 328, 329, 330, 331, 334, 337, 360, 363, 367, 374, 376, 380
rare earth, 191
rare earths, 191
Rayleigh, 45, 212
reactive ion, 84, 369
real-time, 8, 20, 21
reciprocity, 200, 201, 202
recognition, 44
recombination, 18, 37, 41, 42, 62, 148, 153, 154, 167, 237, 239, 259, 261, 262, 263, 268, 270, 288, 314, 323, 327, 331, 384, 385, 386, 387, 388, 389
recombination processes, 37, 388
reconstruction, 21, 22
recovery, 18
rectification, 86
reduction, 1, 2, 9, 14, 20, 21, 22, 39, 51, 54, 65, 71, 88, 89, 93, 95, 100, 104, 154, 174, 190, 243, 252, 272, 295, 312, 350, 351, 360, 372, 378, 384
reflection, 12, 22, 35, 36, 49, 51, 80, 104, 189, 190, 192, 193, 194, 195, 196, 207, 212, 215, 221, 222, 225, 344, 345, 364, 368, 370, 388
reflectivity, 49, 103, 104, 149, 155, 157, 195, 196, 198, 207, 220, 224, 226, 346, 350, 354, 370, 371, 373, 374, 375, 376, 377, 378, 379
refraction index, 313
refractive indices, 49, 57, 66, 68, 81, 96, 97, 99, 103, 224, 295, 296
regeneration, 2
regular, 2, 10
rejection, 7, 17, 144

relationship, 14, 36, 40, 150, 160, 202, 237, 254, 259, 274, 288, 315
relationships, 235, 255, 258, 277, 296, 347
relative size, 389
relaxation, 45, 94, 153
relaxation process, 45
relaxation time, 153
reliability, 1, 2, 61
remote sensing, 20, 24
replication, 191
research, 2, 3, 17, 18, 34, 51, 144, 188, 191, 232, 238, 253, 311, 341
research and development, 191
researchers, 312
resistance, 1, 20, 42
resistive, 28
resistivity, 326
resolution, 8, 21, 22, 24, 26, 198, 321, 364, 365, 366, 377
resonator, viii, 7, 16, 17, 34, 35, 36, 37, 38, 40, 51, 52, 53, 54, 55, 60, 61, 63, 64, 65, 85, 86, 87, 135, 136, 137, 138, 140, 141, 143, 144, 145, 146, 147, 148, 149, 153, 154, 155, 160, 162, 193, 194, 196, 265, 266, 267, 268, 269, 270, 272, 273, 274, 275, 276, 277, 278, 279, 377
resource availability, 384
resources, 207
response time, 3, 24, 44, 297, 301, 305, 307
returns, 158
revolutionary, 95
RIE, 84
rings, 135
risk, 198
RNA, 367
robotics, 60, 61
robustness, 2, 3, 44, 222
room temperature, 388
rotations, 60
roughness, 40, 63, 84, 144, 145, 149, 153, 160, 161, 162, 265, 272
routing, 2, 14, 18

S

safety, 3, 43
Sagnac effect, 63, 64, 65
sales, 383
saline, 53
salinity, 26, 360
sample, 8, 44, 45, 46, 47, 85, 317, 318, 320, 325, 326, 328, 363, 364, 365, 366, 367
sampling, 5, 8, 9, 10, 11, 71
SAR, 21, 22, 23, 24, 26
satellite, 11, 20, 21, 24
saturation, ix, 7, 89, 152, 153, 160, 162, 262, 263, 269, 311, 320, 323, 333, 334, 336, 337, 338
scalar, 208, 209, 225, 296
scaling, 154, 380
scattering, 25, 35, 40, 45, 84, 103, 144, 147, 152, 207, 211, 222, 223, 226, 233, 266, 272, 334, 336, 342, 343, 385
Schrödinger equation, ix, 383, 385, 387
scientific, 212
scientists, 384
searches, 207
second generation, 383, 384
seed, 193
seismic, 199
SEL, 193
selecting, 19, 238
selectivity, 71, 194, 334, 363
self-consistency, 150
self-phase modulation, 242
semiconductor, viii, 7, 12, 14, 18, 21, 26, 27, 30, 31, 34, 39, 42, 48, 61, 103, 113, 143, 144, 145, 146, 150, 151, 153, 160, 175, 191, 192, 193, 194, 223, 232, 343, 347
semiconductor lasers, 150, 193, 194
semiconductors, 27, 31, 95, 191, 384
sensing, vii, 3, 25, 26, 27, 29, 30, 32, 35, 44, 47, 48, 49, 51, 52, 53, 55, 57, 58, 71, 85, 93, 103, 104, 112, 191, 196, 197, 198, 199, 227, 265, 306, 360, 361, 362, 363, 364, 365, 366, 367, 372, 378, 380
sensitivity, vii, 1, 20, 24, 25, 26, 27, 29, 30, 32, 33, 34, 36, 38, 39, 40, 41, 47, 48, 49, 51, 52, 53, 54, 56, 57, 59, 60, 61, 63, 65, 66, 67, 69, 70, 71, 80, 85, 86, 87, 93, 94, 95, 108, 109, 110, 111, 112, 145, 199, 265, 281, 306, 307, 308, 309, 367, 369, 372, 373, 376, 377, 378, 380
Sensitivity Analysis, 66
sensitization, 311
sensors, vii, 1, 2, 3, 20, 26, 27, 28, 29, 30, 32, 33, 34, 43, 44, 45, 47, 48, 49, 52, 54, 60, 61, 64, 65, 66, 71, 85, 87, 93, 192, 193, 195, 196, 198, 199, 294, 342, 360, 361, 362, 363, 364, 370, 377, 380, 381
separation, 63, 144, 234, 247, 256, 279, 299
sequencing, 46
series, 144, 198, 204, 215, 224, 225, 226, 237, 259, 361, 371
serum, 51, 364
serum albumin, 51, 364
services, iv, 2, 20
shape, 7, 22, 34, 36, 41, 84, 144, 155, 157, 162, 212, 233, 242, 243, 244, 245, 247, 255, 262, 263, 268, 270, 272, 273, 274, 280, 283, 284, 285, 349, 350
shaping, 11

Si3N4, 51, 52, 53, 57, 60, 69, 70, 84, 85, 104, 365, 369
SIGMA, 29
sign, 201, 202, 203, 204, 236, 242, 282, 295
signals, 2, 3, 4, 5, 6, 7, 8, 10, 12, 13, 14, 15, 17, 19, 23, 24, 25, 26, 30, 31, 45, 46, 47, 61, 63, 64, 71, 145, 261, 263, 361
signal-to-noise ratio, 22, 24
signs, 45
silica, 6, 7, 11, 14, 27, 32, 96, 191, 197, 198, 232, 233, 252, 255, 295, 296, 297, 298, 299, 300
silicate, 197
silicon dioxide, 180, 232, 252, 312, 326, 334
silicon nanocrystals, 54, 86, 96, 97, 98, 99
simulation, 3, 40, 57, 58, 97, 102, 112, 159, 160, 166, 167, 175, 176, 207, 210, 227, 245, 247, 262, 263, 264, 268, 273, 279, 283, 294, 303, 304, 309, 312, 317, 334, 342, 385, 390, 394
simulations, 39, 43, 99, 127, 146, 154, 155, 159, 160, 168, 211, 212, 240, 242, 248, 268, 270, 273, 279, 281, 285, 286, 287, 296, 313, 314, 318, 320, 322, 331, 385, 388, 389, 390, 391, 394, 395, 398
SiO2, 27, 51, 52, 53, 57, 60, 83, 84, 85, 86, 87, 88, 89, 90, 91, 97, 99, 104, 105, 121, 191, 192, 295, 312, 313, 325, 326, 362, 363, 365, 368, 369
software code, 387
solar, ix, 20, 383, 384, 385, 386, 387, 388, 391, 394, 395, 398
solar cell, ix, 383, 384, 385, 386, 387, 391, 394, 395, 398
solar cells, ix, 383, 384, 385, 387, 391, 398
solar energy, 388
solid state, 193
soliton, viii, 231, 232, 243, 244, 245, 248, 285, 286, 288, 290
solitons, 242, 243, 244, 245, 248
solutions, 21, 47, 61, 68, 105, 200, 201, 203, 205, 207, 211, 221, 222, 226, 232, 234, 246, 247, 253, 256, 260, 271, 272, 274, 275, 276, 280, 299, 320, 342, 343, 344, 348
space exploration, 2
space shuttle, 20
space-time, 180, 232, 237, 247, 253, 260, 281
spatial, 21, 24, 26, 190, 199, 200, 208, 210, 211, 241, 348
Specialists, 398, 399
species, 44, 48, 198
specificity, 44
spectra, 51, 140, 141, 190, 221, 232, 246, 281, 282, 283, 347, 348, 349, 350, 351, 352, 374
spectral component, 361
spectroscopy, 45, 193, 194
spectrum, 8, 26, 36, 37, 40, 71, 105, 154, 190, 195, 212, 222, 233, 240, 242, 255, 346, 347, 349, 353, 360, 364, 371, 372, 373, 374, 377, 384
speed, 4, 8, 15, 18, 30, 34, 37, 41, 45, 95, 145, 180, 210, 251, 325
spin, 45
SPR, 50, 54
sputtering, 365
square wave, 105, 371
SRD, 28
SRS, 64, 65, 180, 181, 232, 235, 246, 252, 254, 258, 260, 265, 267, 268, 269, 272, 275, 276, 280, 281, 284, 285
stability, 6, 41, 71, 210, 211, 212, 293, 349, 361, 366
stabilization, 61, 196
stable states, 19
standard deviation, 153, 160, 161, 162, 322
stimulus, 265
strain, 33, 105, 195, 198, 199, 360
strategic, 180
stratification, 224
streams, 2, 15, 22
strength, 32, 96, 144, 195, 198, 215, 347, 367
stress, 195, 360
structural changes, 197
subdomains, 212
substances, 1, 26, 44, 47
subtraction, 7, 24
suffering, 61
superconducting, 5
superposition, 82, 87, 88, 95, 188, 190, 197, 225, 343, 344, 347, 384
supply, 27
suppression, 145, 190, 196
surface modification, 362
surface roughness, 144
susceptibility, 25, 215, 237, 254, 255, 260
switching, viii, 2, 14, 17, 18, 34, 51, 144, 145, 173, 179, 180, 184, 293, 294, 297, 301, 381
symmetry, 242
synchronization, 24, 65
synchronous, 207
synthesis, 21
systems, vii, 1, 2, 3, 8, 10, 14, 18, 20, 21, 24, 28, 31, 51, 71, 83, 93, 104, 144, 148, 162, 180, 191, 192, 193, 194, 196, 226, 338, 361, 387

T

targets, 22
technological, viii, 2, 3, 14, 16, 20, 25, 71, 80, 88, 89, 102, 112, 120, 143, 145, 163, 165, 268, 293, 306, 383

technology, viii, ix, 2, 3, 6, 7, 11, 12, 14, 16, 19, 20, 22, 29, 34, 35, 41, 49, 53, 54, 60, 63, 65, 71, 80, 83, 85, 95, 105, 118, 145, 146, 165, 190, 199, 231, 236, 251, 252, 265, 272, 275, 286, 293, 294, 334, 359, 360, 380, 383, 384
Teflon, 55, 103
telecommunication, 2, 14, 88, 143, 360
telecommunication networks, 14
telecommunications, vii, 1, 2, 3, 8, 18, 20, 34, 51, 71, 144, 191, 193, 194, 226, 251, 293
teleconferencing, 2
telephony, 1
temperature, 17, 20, 26, 29, 40, 42, 44, 48, 103, 168, 193, 195, 196, 197, 198, 199, 294, 295, 296, 297, 298, 299, 301, 307, 309, 360, 365, 369
temperature dependence, 29, 103
temperature gradient, 295
temporal, 21, 22, 208, 210, 242, 243, 244, 245, 283, 284, 285, 326
terminals, 172
theoretical, vii, ix, 39, 60, 64, 65, 79, 144, 145, 162, 227, 232, 233, 252, 253, 262, 290, 294, 309, 311, 315, 342, 384
theory, ix, 3, 12, 40, 42, 141, 148, 188, 199, 211, 224, 225, 227, 253, 255, 312, 315, 341, 342, 343, 346, 370, 388
thermal, 20, 29, 33, 42, 50, 84, 167, 198, 293, 294, 295, 296, 297, 299, 300, 301, 307, 308, 309, 365, 384, 388
thermal expansion, 198
thermal properties, 307
thermal stability, 29
thermal treatment, 384
thermodynamics, 212
thin film, 28, 84, 363, 364, 365, 384
thin films, 384
third order, ix, 96, 102, 147, 254, 259, 260, 266, 288, 359, 378, 379, 380
third-order susceptibility, 255
three-dimensional, 195
threshold, 10, 25, 33, 61, 64, 65, 145, 160, 162, 221, 253, 265, 267, 268, 269, 272, 274, 275, 276, 277, 278, 279, 280, 281, 288, 336
threshold level, 276, 277, 279
thresholds, 279, 288
Ti, 12, 24, 25, 29, 30
time consuming, 71, 97
time increment, 208
time resolution, 253
TiO_2, 362, 363, 365, 366
TIR, 79, 81, 85, 104, 105
tissue, 44, 207
titanium, 24, 29
title, 347, 349, 350, 391, 395, 398
tolerance, 29, 94, 97, 98, 99, 100, 372, 378
top-down, 384
total internal reflection, 49, 51, 80, 103
toxic, 384
toxicity, 383
TPA, 64, 96, 145, 146, 147, 148, 153, 154, 155, 157, 158, 181, 232, 236, 239, 244, 245, 247, 253, 254, 259, 261, 272, 276, 280, 288
tracking, 11
trade, 29, 34, 41, 48, 60, 88, 93, 96, 196, 244, 262, 277, 334
trade-off, 29, 34, 41, 48, 60, 88, 93, 96, 196, 244, 262, 277, 334
traffic, 14
transducer, 22, 44
transduction, 47
transfer, 2, 3, 4, 5, 7, 9, 10, 27, 57, 127, 180, 181, 188, 206, 207, 222, 223, 256, 266, 303, 311, 315, 316, 342, 346
transformation, 38
transition, 45, 150, 317, 332, 347, 384
transition rate, 317, 332
transitions, 313, 347
translational, 234, 255
transmission, 2, 13, 18, 21, 22, 31, 32, 34, 35, 36, 51, 87, 141, 180, 190, 196, 223, 232, 246, 252, 293, 325, 342, 344, 345, 350, 351, 353, 354, 368
transparency, 18
transparent, 32, 103, 362
transport, 18, 42, 387
traps, 385, 388
trend, 34, 54, 88, 89, 100, 102, 252, 284, 384, 394
trial, 164
tunneling, 41, 325, 347
two-dimensional (2D), 21, 23, 117, 212, 224, 341, 387

U

ubiquitous, 34
ultrasonic waves, 360
ultraviolet (UV), 54, 113, 197
ultraviolet light, 197
uncertainty, 312
uniform, 97, 145, 172, 175, 176, 189, 223, 326
user data, 20
users, 2, 10, 20, 21

V

vacuum, 67, 81, 153, 167, 254
valence, 150
validation, 98

validity, 259
variability, 369
variable, 12, 82, 152, 199, 212, 234, 256, 312, 321, 346, 365, 366
variables, 42, 237, 260, 320, 321, 384
variation, 7, 41, 56, 69, 80, 85, 88, 91, 96, 106, 109, 110, 294, 313, 315, 328, 362, 363, 364, 366, 367, 369, 374, 376, 377
VCSEL, 193, 224, 226, 367
vector, 45, 56, 57, 104, 150, 151, 152, 225, 254, 344, 370, 371
vehicles, 20
velocity, viii, 26, 30, 63, 87, 147, 149, 154, 180, 181, 207, 211, 231, 235, 236, 237, 238, 239, 240, 254, 258, 260, 266, 267, 268, 275, 282, 283, 351, 353, 388, 389
versatility, 51, 342, 360
vertical-cavity surface-emitting lasers, 193
vibration, 366
vibrational, 233, 255
video, 2
visible, 197, 311

W

water, 26, 69, 104, 105, 360, 367
water vapour, 104
wave equations, 222
wave number, 81
wave power, 269
wave propagation, 210, 223, 294, 351
wave vector, 370
Wavelength Division Multiplexing, 2, 14
wavelengths, 9, 12, 20, 21, 40, 112, 145, 155, 191, 195, 196, 198, 211, 239, 241, 246, 247, 268, 270, 273, 279, 288, 293, 347, 348, 349, 361, 374
wells, 62, 160, 199, 343, 346, 347, 384, 385, 386, 387, 388, 395, 398
wet, 365, 368
wind, 199
wind turbines, 199
windows, 80
wireless, 1, 10, 34, 93
wireless networks, 1, 10
wires, 54, 80, 82, 83, 84, 85, 86, 87, 88, 91, 94, 97, 100, 102, 325
working conditions, 24
writing, 25, 196, 197, 268

Y

yield, 193, 201, 210, 384

Z

ZnO, 21